John David Jackson

A Course in Quantum Mechanics

Edited by Robert N. Cahn

John David Jackson

A Course in Quantum Mechanics

Edited by

Robert N. Cahn

Registered Office
John Wiley & Sons, Inc., 111 River Street, Hoboken, NJ 07030, USA

For details of our global editorial offices, customer services, and more information about Wiley products visit us at www.wiley.com.

Wiley also publishes its books in a variety of electronic formats and by print-on-demand. Some content that appears in standard print versions of this book may not be available in other formats.

A catalogue record for this book is available from the Library of Congress

Hardback ISBN: 9781119880387; ePub ISBN: 9781119880417; ePDF ISBN: 9781119880394

Cover Images: Courtesy of Dr. Robert N. Cahn, GeorgePeters/Getty Images
Cover Design by Wiley

Set in 9.5/12.5pt STIXTwoText by Integra Software Services Pvt. Ltd, Pondicherry, India

SKY10082413_082024

Contents

Preface

Prof. John David Jackson is known throughout the physics community for his text *Classical Electrodynamics,* which first appeared in 1962. More than 60 years later, in its third edition, it remains the standard text for graduate students and a frequent reference for researchers. As a faculty member at the University of California beginning in 1967, Jackson taught many other courses, from first-year physics to graduate courses in particle physics. It was this latter course that led me to become his graduate student. From time to time, he taught a year-long graduate course in quantum mechanics. His meticulous notes from these courses in 1978-1979 and 1988-89 were given to me by his daughter, Nan Jackson, after his death in 2016. Nearly the entirety of the text here is taken directly from his notes, supplemented occasionally with explanations he might have added in class, but which do not appear in the notes.

There is no lack of outstanding texts on quantum mechanics, starting with Dirac's book first published in 1930. Other classics include Schiff's book, based on the lectures of J. Robert Oppenheimer, and the Landau and Lifshitz book, part of their extraordinary series covering all of theoretical physics. Many other fine texts have appeared in the last 60 years. References to those by J. J. Sakurai, K. Gottfried, E. Merzbacher, and by H. Bethe and R. Jackiw and others appear in these lectures. Still more texts have appeared since Jackson delivered these lectures. Why, then, should we have another text? Relativity and quantum mechanics are the two transcendent achievements of twentieth-century physics. Quantum mechanics has more profoundly affected science and humanity generally. By the late 1920s, quantum mechanics provided a complete understanding of the immediate world around us, the world of atoms and molecules, an understanding beyond the reach of classical physics. The unique significance of quantum mechanics justifies the variety of treatments by exceptional physicists, each providing new insights, just as each new fine recording of the Beethoven string quartets provides truly new pleasures.

The problems in Jackson's *Classical Electrodynamics* are legendary. On the occasion of his 60th birthday, a group of Cornell physics graduate students composed a song to a tune from "The Sound of Music" with the line "How do you solve a problem out of Jackson in any less than geologic time?" Included here are many of the problems Jackson assigned in the quantum mechanics courses. In some instances, the problems are so essential to the development that I have included Jackson's solutions, as well. These are marked with an asterisk. In the other instances I have maintained Jackson's own policy of never making the solutions public. The students in the two courses were also assigned some problems from texts mentioned above. The diligent reader may also wish to seek additional problems there, though the ones included here will push the student to a firm understanding of the material.

Many students find *Classical Electrodynamics* particularly challenging because of the demands it places on mathematical preparation. Of course, part of the value of that text is that it provides the means to achieve a higher level of mathematical facility. Jackson's quantum mechanics courses also made mathematical demands, but rather fewer than in *Classical Electrodynamics*. Fourier transforms and integration by parts are the most frequent devices. As an aid, I have added a brief appendix giving the rudiments of complex analysis and contour integration. But just as in *Classical Electrodynamics,* the mathematical tools are applied for the purpose of addressing real and important physical problems. While Jackson's own research was primarily in particle physics, atomic physics plays a more central role in the text, with multiple references to two classic works: Condon and Shortley, *Theory of Atomic Spectra*, and Bethe and Salpeter, *Quantum Mechanics of One- and Two-electron Atoms.*

As the title suggests, this book is a course, not a comprehensive treatment of quantum mechanics. The goal is not so much to present physics, as it is to show how to do physics. For this reason, many derivations are given in greater detail than is conventional. A preference is given to concrete applications of quantum mechanics over formal development. References to complementary treatments are provided. It is assumed that the reader has some familiarity with quantum mechanics at the undergraduate level and with the associated mathematical techniques. Throughout, the intention is to emulate the informality of the actual lectures Jackson delivered to the Berkeley students. One manifestation of that informality is that in some instances $\hbar$ and c appear, while in many others they do not. Knowing how to restore $\hbar$ and c is just one aspect of the topic. No doubt some typographic errors or worse will have evaded scrutiny in my transcription of Jackson's notes. For these, I alone am responsible.

I hope this presentation of Jackson's notes will provide some of the joy and pleasure that he experienced himself and transmitted to his students in seeing the beauty of this subject, which is at once elegant, abstract, and essential to understanding the tangible world around us.

R. Cahn
March 2023

About the Companion Website

This book is accompanied by a companion website:

www.wiley.com/go/Jackson/QuantumMechanics

This website includes solutions to the problems, available to registered instructors

1

Basics

In 1900, Planck found that he could explain the spectrum of blackbody radiation by introducing a new physical constant with the dimensions of angular momentum. The Planck constant is the basis of quantum mechanics, determining when classical concepts no longer apply, specifying the inevitable uncertainties introduced by measurement, and fixing the size of atoms. The gradual development of wave mechanics led by Niels Bohr over the next twenty years was superseded by the extraordinarily rapid progress that followed de Broglie's proposal that particles were also waves. By 1927, the canon was largely in place and was codified by Dirac's magisterial text in 1930. Suddenly there was a comprehensive theory of atoms and molecules and their electromagnetic interactions: our material world.

We review some of the essential aspects of quantum mechanics, the wave function and Born's interpretation of its absolute square as a probability density. The Schrödinger equation prescribes the non-relativistic behavior of the wave function in space and time in terms of a Hamiltonian, which combines the kinetic and potential energies. A free particle already evidences the effects of quantum mechanics by gradually spreading from any initially confined configuration. The behavior of a particle in the field of a constant conservative force is closely related to what would be expected classically, with motion that is bounded or unbounded depending on the potential and the particle's energy. The Schrödinger equation brought with it the power of the well-developed mathematics of differential equations. A recurrent concept is that of an eigenvalue, a value of a parameter that allows the solution of a differential equation subject to boundary conditions. The Rayleigh-Ritz method provides a means of determining a good approximation to the smallest eigenvalue, generally the lowest energy, by varying parameters in a trial function.

1.1 Wave Mechanics of de Broglie and Schrödinger

Louis duc de Broglie, in his 1924 Ph.D. thesis, proposed that material particles of mass m, momentum $\mathbf{p}$ and energy $E = \sqrt{p^2c^2 + m^2c^4}$ have associated with them a wave vector and angular frequency through the Planck constant $\hbar = h/2\pi$

$$\mathbf{k} = \mathbf{p}/\hbar; \quad \omega = E/\hbar. \tag{1.1}$$

This leads us to associate with a particle of momentum $\mathbf{p}$ and energy E a *wave amplitude*

$$\psi_{\mathbf{p}}(\mathbf{x}, t) = e^{i\mathbf{k}\cdot\mathbf{x} - i\omega t} = e^{i\frac{\mathbf{p}\cdot\mathbf{x}}{\hbar} - i\frac{E}{\hbar}t}. \tag{1.2}$$

John David Jackson: A Course in Quantum Mechanics, First Edition. Robert N. Cahn.
© 2024 John Wiley & Sons, Inc. Published 2024 by John Wiley & Sons, Inc.
Companion website: www.wiley.com/go/Jackson/QuantumMechanics

de Broglie's insight led to the rapid development of quantum mechanics by Werner Heisenberg, Erwin Schrödinger, Paul Dirac, Max Born, and others. We review here the elements of quantum mechanics from the wave perspective. Quantum mechanics, like any physical theory, cannot be derived as if it were a mathematical theorem. Its validity can be judged only by its ability to predict correctly what is observed. Later, we will show how a quantum mechanical prescription can be derived from some reasonable assumptions. But such derivations are not proofs, but rather logical arguments in favor of a theory that must face the test of experiment.

The idea of localized particles ("at **x**, with momentum **p**") plus experience with superposition of waves, lead to consideration of a *general wave amplitude*

$$\psi(\mathbf{x}, t) = \int d^3 p \, A(\mathbf{p}) e^{i\frac{\mathbf{p}\cdot\mathbf{x}}{\hbar} - i\frac{E(\mathbf{p})}{\hbar}t} \tag{1.3}$$

in association with a free particle of mass m. For a one-dimensional case, as we show more completely below, if $A(p)$ is narrowly confined about $p = 0$ with a spread Δp, $\psi(x)$ will describe a particle with a spread $\Delta x \approx \hbar/\Delta p$. We shall return to the uncertainty principle later.

With classical ideas (sound intensity, Poynting vector, etc.) to guide us, we are led fairly directly to Born's probability interpretation of $|\psi|^2$, i.e. the probability of finding a particle "at **x**" at time t in a volume d^3x is proportional to $|\psi(\mathbf{x}, t)|^2 d^3x$. Thus we speak of $|\psi(\mathbf{x}, t)|^2$ as the *probability density at* $(\mathbf{x}, t)$.

1.2 Klein-Gordon Equation

With Einstein's relation $E = \sqrt{p^2c^2 + m^2c^4}$ we see that

$$\frac{\partial^2 \psi}{\partial t^2} = \int d^3 p \, A(\mathbf{p}) \left[-\frac{E^2}{\hbar^2} \right] e^{i\frac{\mathbf{p}\cdot\mathbf{x}}{\hbar} - i\frac{E(\mathbf{p})}{\hbar}t} = \int d^3 p \, A(\mathbf{p}) \left[-\frac{p^2c^2}{\hbar^2} - \frac{m^2c^4}{\hbar^2} \right] e^{i\frac{\mathbf{p}\cdot\mathbf{x}}{\hbar} - i\frac{E(\mathbf{p})}{\hbar}t}. \tag{1.4}$$

But we can write

$$\nabla^2 \psi = \int d^3 p \, A(\mathbf{p}) \left[-\frac{p^2}{\hbar^2} \right] e^{i\frac{\mathbf{p}\cdot\mathbf{x}}{\hbar} - i\frac{E(\mathbf{p})}{\hbar}t}. \tag{1.5}$$

Thus, independent of $A(\mathbf{p})$

$$\frac{1}{c^2} \frac{\partial^2 \psi}{\partial t^2} = \nabla^2 \psi - \left(\frac{mc}{\hbar} \right)^2 \psi. \tag{1.6}$$

This is the Klein-Gordon equation, a relativistic wave equation for free spinless particles, e.g. π mesons. We note that this equation allows solutions with $-E$ in place of E.

1.3 Non-Relativistic Approximation

If $\psi(\mathbf{x}, t)$ describes a state of motion of a particle, so also does $e^{i\alpha}\psi(\mathbf{x}, t)$, where α is independent of the motion of the particle. For example, we could take $\alpha = (10 \text{ eV}/\hbar)t$ or $\alpha = \pi/4$. This is permissible because only $|\psi|^2$ is relevant as a probability density.

Consider the non-relativistic limit

$$E(p) = \sqrt{p^2c^2 + m^2c^4} \approx mc^2 + \frac{p^2}{2m} \equiv mc^2 + E_{NR} \tag{1.7}$$

so that

$$\psi_{\mathbf{p}}(\mathbf{x}, t) = e^{i\frac{\mathbf{p}\cdot\mathbf{x}}{\hbar} - i\frac{E}{\hbar}t} \approx e^{-i\frac{mc^2}{\hbar}t} \cdot e^{i\frac{\mathbf{p}\cdot\mathbf{x}}{\hbar} - i\frac{E_{NR}}{\hbar}t}. \tag{1.8}$$

Thus, by superposition, in the non-relativistic limit

$$\psi(\mathbf{x}, t) = e^{-i\frac{mc^2}{\hbar}t} \int d^3p \, A(\mathbf{p}) e^{i\frac{\mathbf{p}\cdot\mathbf{x}}{\hbar} - i\frac{E_{NR}}{\hbar}t}. \tag{1.9}$$

Because the choice of phase is at our disposal, we define the non-relativistic wave amplitude

$$\psi_{NR}(\mathbf{x}, t) = e^{imc^2t/\hbar}\psi(\mathbf{x}, t). \tag{1.10}$$

Now dropping the subscript NR, we have

$$\psi(\mathbf{x}, t) = \int d^3p \, A(\mathbf{p}) e^{i\frac{\mathbf{p}\cdot\mathbf{x}}{\hbar} - i\frac{E_{NR}}{\hbar}t} \tag{1.11}$$

so that

$$\frac{\partial\psi}{\partial t} = \int d^3p \, A(\mathbf{p})\left[-\frac{ip^2}{2m\hbar}\right] e^{i\frac{\mathbf{p}\cdot\mathbf{x}}{\hbar} - i\frac{E_{NR}}{\hbar}t} \tag{1.12}$$

and as before

$$\nabla^2\psi = \int d^3p \, A(\mathbf{p})\left[-\frac{p^2}{\hbar^2}\right] e^{i\frac{\mathbf{p}\cdot\mathbf{x}}{\hbar} - i\frac{E_{NR}}{\hbar}t}. \tag{1.13}$$

Thus we obtain the general equation describing non-relativistic motion of free particles, the Schrödinger equation,

$$i\hbar\frac{\partial\psi}{\partial t} = -\frac{\hbar^2}{2m}\nabla^2\psi. \tag{1.14}$$

1.4   Free-Particle Probability Current

If $|\psi|^2$ is a density, there should be a probability current and a continuity equation to connect them because probability is conserved.

$$\int |\psi(\mathbf{x}, t)|^2 d^3x = \text{constant in time.} \tag{1.15}$$

Consider

$$\frac{\partial}{\partial t}|\psi|^2 = \psi^*\frac{\partial\psi}{\partial t} + \psi\frac{\partial\psi^*}{\partial t} \tag{1.16}$$

and use the Schrödinger equation to get

$$\frac{\partial}{\partial t}|\psi|^2 = -\frac{\hbar}{2mi}\left[\psi^*\nabla^2\psi - \psi\nabla^2\psi^*\right]$$

$$= -\frac{\hbar}{2mi}\nabla \cdot \left[\psi^*\nabla\psi - \psi\nabla\psi^*\right]. \tag{1.17}$$

So if we write

$$\mathbf{J} = \frac{\hbar}{2mi} (\psi^* \nabla \psi - \psi \nabla \psi^*). \tag{1.18}$$

we have a continuity equation,

$$\frac{\partial |\psi|^2}{\partial t} + \nabla \cdot \mathbf{J} = 0. \tag{1.19}$$

Note that $\mathbf{J}$ is arbitrary to the extent of adding a curl of some vector function, since the divergence of a curl is always zero. This is analogous to the arbitrariness of the Poynting vector in electromagnetism.

To find $\mathbf{J}$ for our general free-particle $\psi(\mathbf{x}, t)$, consider

$$\nabla \psi = \int d^3 p \, A(\mathbf{p}) \left(\frac{i\mathbf{p}}{\hbar} \right) e^{i \frac{\mathbf{p} \cdot \mathbf{x}}{\hbar} - i \frac{E(p)}{\hbar} t}. \tag{1.20}$$

Combining this with its complex conjugate, we find

$$\mathbf{J}(\mathbf{x}, t) = \frac{1}{2m} \int d^3 p \int d^3 p' \, A^*(\mathbf{p}') A(\mathbf{p}) [\mathbf{p} + \mathbf{p}'] \, e^{i \frac{(\mathbf{p} - \mathbf{p}') \cdot \mathbf{x}}{\hbar}} e^{-i(E - E')t/\hbar}. \tag{1.21}$$

The presence of the factor $(\mathbf{p} + \mathbf{p}')/(2m)$ shows that each component wave of definite momentum contributes a factor of $\mathbf{v} = \mathbf{p}/m$ to the flux, as expected. Note that there is a correspondence

$$\nabla \psi \leftrightarrow i \frac{\mathbf{p}}{\hbar} A(\mathbf{p}). \tag{1.22}$$

The identification by de Broglie of the momentum $\mathbf{p}$ with $\hbar \mathbf{k}$, where $\mathbf{k}$ occurs in a plane wave as $e^{i\mathbf{k} \cdot \mathbf{x}}$, leads through the Fourier transform with the identification of $\mathbf{p}$ with $\frac{\hbar}{i} \nabla$.

1.5 Expectation Values

If $|\psi|^2$ is the probability density in space, then it is natural to define averages of coordinates and functions of coordinates for the particular state of motion as integrals over all space, weighted with $|\psi|^2$. Suppose $F(\mathbf{x}, t)$ is a reasonable real function of $\mathbf{x}$ and t and assume $F|\psi|^2$ has a finite integral over all space. Then the average of (or *expectation value* of) F is

$$\langle F \rangle = \int d^3 x \, F(\mathbf{x}, t) |\psi(\mathbf{x}, t)|^2. \tag{1.23}$$

In general, $\langle F \rangle$ is a function of time through both the explicit time dependence of F and the time dependence of $|\psi|^2$. For example,

$$\langle \mathbf{x} \rangle = \int d^3 x \, \mathbf{x} |\psi(\mathbf{x}, t)|^2. \tag{1.24}$$

This all seems plausible, even obvious and necessary. If $|\psi(\mathbf{x}, t)|^2$ is big somewhere, the $\langle \mathbf{x} \rangle$ will reflect that. If the place where $|\psi|^2$ is big moves in time, then $\langle \mathbf{x} \rangle$ will move in time. This leads to the question: what is the expression for $\langle \mathbf{p} \rangle$?

We look at this in two ways: what is $m\langle\dot{\mathbf{x}}\rangle$ and what is $\langle\mathbf{p}\rangle$ in terms of $|A(\mathbf{p})|^2$? First, using the continuity equation, Eq. (1.19)

$$m\frac{d}{dt}\langle\mathbf{x}\rangle = m\int d^3x\,\mathbf{x}\frac{\partial}{\partial t}|\psi|^2$$

$$= -m\int d^3x\,\mathbf{x}(\nabla\cdot\mathbf{J}) = m\int d^3x\,\mathbf{J}(\mathbf{x},t). \tag{1.25}$$

The last step follows from integration by parts if the current $\mathbf{J}$ is confined to some finite portion of space:

$$\int d^3x\,x_i\nabla_k J_k = -\int d^3x\,J_k\nabla_k x_i = \int d^3x\,J_k\delta_{ik} = \int d^3x\,J_i. \tag{1.26}$$

The use of integration by parts is ubiquitous. This is inevitable because, when confronted with a formal integral expression, there is no way to proceed other than integration by parts.

Returning to the calculation at hand,

$$m\frac{d}{dt}\langle\mathbf{x}\rangle = m\int d^3x\,\mathbf{J}$$

$$= \frac{\hbar}{2i}\int d^3x\,[\psi^*\nabla\psi - \psi\nabla\psi^*]. \tag{1.27}$$

Once again invoking integration by parts,

$$\int d^3x\,\psi\nabla\psi^* = -\int d^3x\,\psi^*\nabla\psi. \tag{1.28}$$

We conclude that

$$m\frac{d}{dt}\langle\mathbf{x}\rangle = \int d^3x\,\left(\psi^*\frac{\hbar}{i}\nabla\psi\right). \tag{1.29}$$

On the other hand, if we think of $|A(\mathbf{p})|^2$ as the probability density in momentum space, then we would define

$$\langle\mathbf{p}\rangle = \int d^3p\,\mathbf{p}|A(\mathbf{p})|^2(2\pi)^3, \tag{1.30}$$

where our normalization has been

$$\int d^3p\,|A(\mathbf{p})|^2(2\pi)^3 = 1. \tag{1.31}$$

To check whether this is all consistent, we evaluate

$$m\int d^3x\,\mathbf{J} = \frac{1}{2}\int d^3p\int d^3p'\,A^*(\mathbf{p}')A(\mathbf{p})(\mathbf{p}+\mathbf{p}')(2\pi)^3\delta(\mathbf{p}-\mathbf{p}')$$

$$= \int d^3p\,(2\pi)^3|A(\mathbf{p})|^2\mathbf{p}. \tag{1.32}$$

This confirms our picture that

$$\langle\mathbf{p}\rangle = \int d^3x\,\psi^*\frac{\hbar}{i}\nabla\psi \tag{1.33}$$

and allows us to identify

$$\mathbf{p} = \frac{\hbar}{i}\nabla. \tag{1.34}$$

Using this identification, we can find the expectation value of a function of $\mathbf{p}$ as

$$\langle G \rangle = \int d^3x\, \psi^* G(\frac{\hbar}{i}\nabla)\psi. \tag{1.35}$$

To give this unambiguous meaning, $G(\mathbf{p})$ must be well-behaved, e.g. developable in a power series in $\mathbf{p}$.

1.6 Particle in a Static, Conservative Force Field

So far we have just considered the de Broglie relation and Born's probability interpretation of $|\psi|^2$ for free particles. Suppose a particle is moving in a potential $V(\mathbf{x})$. What equation governs the motion? Motivated by classical treatments of the index of refraction, Fermat's principle, etc. we modify our equations for a free particle:

$$i\hbar\frac{\partial\psi}{\partial t} = -\frac{\hbar^2}{2m}\nabla^2\psi = \frac{p^2}{2m}\psi, \tag{1.36}$$

where we used our identification $\mathbf{p}_{\text{operator}} = (\hbar/i)\nabla$, with the substitution

$$\frac{p^2}{2m} \rightarrow \frac{p^2}{2m} + V(\mathbf{x}). \tag{1.37}$$

In this way, we arrive at the time-dependent Schödinger equation

$$i\hbar\frac{\partial\psi}{\partial t} = \left[\frac{p^2}{2m} + V(\mathbf{x})\right]\psi. \tag{1.38}$$

This familiar equation will occupy us for some time. We note that the picture developed for the free particle requires no significant modification. The probability interpretation of Born is unaffected. The argument for the conservation of probability and the construction of the current survive as long as the potential V is real. The expression of expectation values and the representation of momentum as an operator are unchanged.

1.7 Ehrenfest Theorem

Consider the motion of a particle of mass m described by $\psi(\mathbf{x}, t)$, with ψ satisfying

$$i\hbar\frac{\partial\psi}{\partial t} = -\frac{\hbar^2}{2m}\nabla^2\psi + V(\mathbf{x})\psi. \tag{1.39}$$

We want to see to what extent we can make a connection between the Schrödinger equation and Newton's Laws, so let's calculate

$$
\begin{aligned}
\frac{d}{dt} \langle \mathbf{p} \rangle &= \frac{d}{dt} \int d^3x \left(\psi^* \frac{\hbar}{i} \nabla \psi \right) \\
&= \int d^3x \left[\frac{\partial \psi}{\partial t}^* (\frac{\hbar}{i} \nabla \psi) + \psi^* (\frac{\hbar}{i} \nabla \frac{\partial \psi}{\partial t}) \right] \\
&= \int d^3x \left[(\frac{\hbar}{i} \nabla \psi)(-\frac{i\hbar}{2m} \nabla^2 \psi^* + \frac{i}{\hbar} V \psi^*) - \psi^* \nabla (-\frac{\hbar^2}{2m} \nabla^2 \psi + V\psi) \right] \\
&= \int d^3x \left\{ \frac{\hbar^2}{2m} [\psi^* \nabla^2 \nabla \psi - (\nabla^2 \psi^*) \nabla \psi] + (\nabla \psi) V \psi^* - \psi^* \nabla(V\psi) \right\}.
\end{aligned}
\tag{1.40}
$$

Integration by parts, assuming the wave function vanishes fast enough at infinity, enables us to resolve this. Terms with any combination of ψ^* and ψ like

$$
\int d^3x \, \nabla f(\psi^*, \psi)
\tag{1.41}
$$

are zero as long as ψ vanishes fast enough at infinity. This is just the divergence theorem as is apparent if we introduce a constant vector $\mathbf{A}$

$$
\int d^3x \, \nabla \cdot (\mathbf{A} f(\psi^*, \psi)) = \int d\mathbf{S} \cdot \mathbf{A} f(\psi^*, \psi) = 0,
\tag{1.42}
$$

where the surface elements $d\mathbf{S}$ are at infinity. Thus

$$
\int d^3x \, \psi^* \nabla^2 \nabla \psi = - \int d^3x \, [\nabla \psi^*][\nabla^2 \psi] = \int d^3x \, [\nabla^2 \psi^*][\nabla \psi].
\tag{1.43}
$$

Thus, the term in square brackets in Eq. (1.40) vanishes with integration by parts. The last two pieces combine with an integration by parts to leave us with

$$
\frac{d}{dt} \langle \mathbf{p} \rangle = - \int d^3x \, \psi^*(\nabla V)\psi = - \langle \nabla V \rangle,
\tag{1.44}
$$

which looks a bit like Newton's equation of motion. A more direct connection would have $-\nabla V(\langle x \rangle)$ in place of $- \langle \nabla V(x) \rangle$. If $V(\mathbf{x})$ changes slowly enough in $\mathbf{x}$ and $\psi(\mathbf{x}, t)$ is reasonably localized, then we can show the correspondence with $m\ddot{\mathbf{x}}_{\text{classical}} = \mathbf{F}_{\text{classical}}$.

Consider the force at $\mathbf{x}$ evaluated by expanding about $\langle \mathbf{x} \rangle$.

$$
\mathbf{F}(\mathbf{x}) = \mathbf{F}(\langle \mathbf{x} \rangle) + (\mathbf{x} - \langle \mathbf{x} \rangle) \cdot \nabla \, \mathbf{F}(\langle \mathbf{x} \rangle)
$$
$$
+ \frac{1}{2} \sum_{i,j} (x_i - \langle x_i \rangle)(x_j - \langle x_j \rangle) \frac{\partial^2 \mathbf{F}}{\partial x_i \partial x_j} + \dots
\tag{1.45}
$$

Then

$$
\langle \mathbf{F}(\mathbf{x}) \rangle = \mathbf{F}(\langle \mathbf{x} \rangle) + \frac{1}{2} \sum_{i,j} \langle \Delta x_i \Delta x_j \rangle \frac{\partial^2 \mathbf{F}}{\partial x_i \partial x_j} + \dots
\tag{1.46}
$$

So if $\mathbf{F}$ doesn't vary too much over the spread of the wave amplitude, our correspondence holds.

1.8   Schrödinger Equation in Momentum Space

We can define a momentum-space wave amplitude by

$$\phi(\mathbf{p}, t) = \frac{1}{(2\pi\hbar)^{3/2}} \int d^3x \, \psi(\mathbf{x}, t) e^{-i\mathbf{p}\cdot\mathbf{x}/\hbar}. \tag{1.47}$$

What equation does ϕ satisfy? We compute

$$\begin{aligned}
i\hbar\frac{\partial\phi}{\partial t} &= \frac{1}{(2\pi\hbar)^{3/2}} \int d^3x \, i\hbar\frac{\partial\psi}{\partial t} e^{-i\mathbf{p}\cdot\mathbf{x}/\hbar} \\
&= \frac{1}{(2\pi\hbar)^{3/2}} \int d^3x \left[-\frac{\hbar^2}{2m}\nabla^2\psi + V\psi \right] e^{-i\mathbf{p}\cdot\mathbf{x}/\hbar} \\
&= \frac{p^2}{2m}\phi + \frac{1}{(2\pi\hbar)^{3/2}} \int d^3x \, V\psi e^{-i\mathbf{p}\cdot\mathbf{x}/\hbar},
\end{aligned} \tag{1.48}$$

where we integrated by parts to get the first term. Now suppose that V could be written in a series expansion

$$V = V_i x_i + V_{ij} x_i x_j + \dots, \tag{1.49}$$

where summation over repeated indices is assumed. Then

$$Ve^{-i\mathbf{p}\cdot\mathbf{x}/\hbar} = \left[V_i\left(i\hbar\frac{\partial}{\partial p_i}\right) + V_{ij}\left(i\hbar\frac{\partial}{\partial p_i}\right)\left(i\hbar\frac{\partial}{\partial p_j}\right) + \dots \right] e^{-i\mathbf{p}\cdot\mathbf{x}/\hbar}. \tag{1.50}$$

We see, then, that in momentum space the operators are $\mathbf{p}_{op} = \mathbf{p}, \mathbf{x}_{op} = i\hbar\nabla_{\mathbf{p}}$. So we can write the momentum-space Schrödinger equation as

$$i\hbar\frac{\partial\phi}{\partial t} = \frac{p^2}{2m}\phi + V(\mathbf{x}_{op})\phi. \tag{1.51}$$

1.9   Spread in Time of a Free-Particle Wave Packet

If we represent a particle by a wave packet, quantum mechanics, as represented by the Schrödinger equation, results in the spreading of that wave packet. If, in particular, we start with a normalized Gaussian wave packet,

$$\psi(x, t = 0) = \frac{e^{-x^2/4\Delta_x^2}}{(2\pi\Delta_x^2)^{1/4}} \tag{1.52}$$

so that

$$\langle x^2 \rangle_{t=0} = \Delta_x^2; \quad \langle p^2 \rangle_{t=0} = \left\langle -\hbar^2\frac{d^2}{dx^2} \right\rangle_{t=0} = \left(\frac{\hbar}{2\Delta_x}\right)^2, \tag{1.53}$$

we find (see Problem 1.6) that at time t

$$\int_{-\infty}^{\infty} dx \, \psi(x, t)^* x^2 \psi(x, t) = \Delta_x^2 + \frac{\hbar^2 t^2}{4m^2\Delta_x^2} = \Delta_x^2 + t^2\frac{\langle p^2 \rangle_{t=0}}{m^2}. \tag{1.54}$$

More generally, without assuming a Gaussian wave packet, for a given $(\Delta x)_0$, there is an uncertainty in the corresponding momentum $(\Delta p)_0 \sim \hbar/(\Delta x)_0$. This means that $(\Delta v)_0 \approx$

$\frac{\hbar}{m(\Delta x)_0} = \frac{(\Delta p)_0}{m}$. There will be built up in time t a smearing in x of an amount $t(\Delta v)_0$. Thus, the new spreading in the packet is $(\Delta x)_{\text{new}} = [t(\Delta p)_0]/m \sim \hbar t/[m(\Delta x)_0]$. If the original and new Δx's are combined quadratically (appropriate for uncorrelated effects), we get essentially the above result.

1.10 The Nature of Solutions to the Schrödinger Equation

A free particle can have any momentum and its corresponding energy. If the particle is otherwise free, but confined to a rectangular box by infinitely high potential walls, we constrain the function to vanish at the walls. The result is that only certain momenta are allowed. There are states of arbitrarily high energy, but the states can be counted one-by-one, from the lowest energy on up. The spectrum of states is discrete, whereas, for the completely free particle, any positive energy is possible and the spectrum of states is continuous. A simple inference can be drawn by considering classical analogs. If the motion is bounded, as for the particle in the box, the spectrum of states is discrete. If the motion is unbounded, the spectrum is continuous.

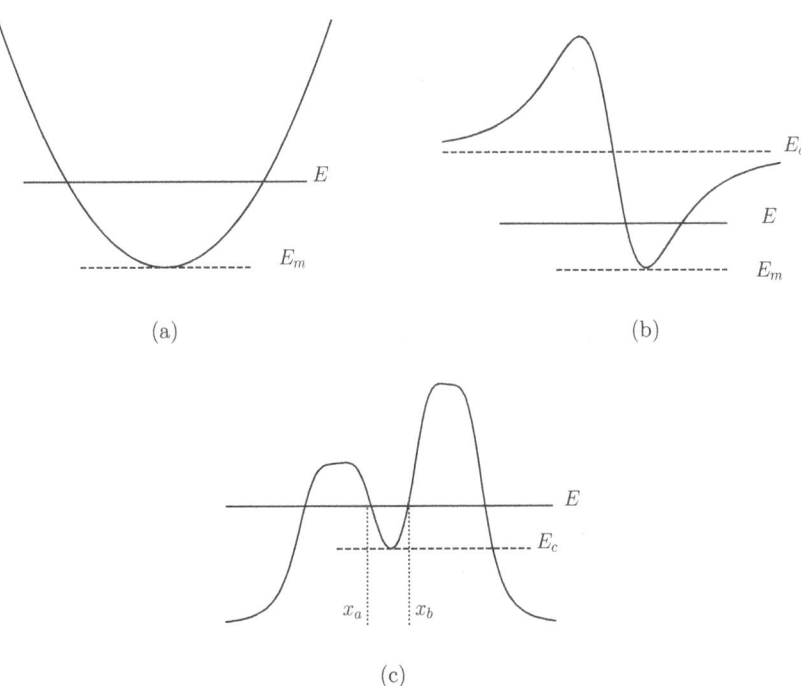

Figure 1.1 Connection between the classical motion of a particle in different potentials and its corresponding eigenvalue spectra. (a) Classical motion bounded in space, denumerable discrete energy eigenvalues with $E_0 > E_m$. (b) For $E_m < E < E_c$, classical motion bounded in space, so denumerable discrete eigenenergies. For $E > E_c$, unbounded classical motion, continuum E_j for $E_j \geq E_c$. (c) Classical motion is bounded between x_a and x_b, but same E has "orbits" that are unbounded. The energy spectrum is continuous. Figures adapted from Gottfried, *Quantum Mechanics: Fundamentals*, p. 50.

These simple observations can be extended to consider more complex potentials. This is illustrated in Figure 1.1.

1.11   A Bound-State Problem: Linear Potential

We consider a quite general Schrödinger equation defined by a Hamiltonian operator

$$H = \frac{p^2}{2m} + V(\mathbf{x}) = -\frac{\hbar^2}{2m}\nabla^2 + V(\mathbf{x}) \tag{1.55}$$

and its associated differential equation

$$H\psi_E(\mathbf{x}) = E\psi_E(\mathbf{x}). \tag{1.56}$$

A solution ψ_E that is physically reasonable in the sense described below determines an *eigenvalue*, E. For any eigenstate ψ_E, the expectation value of any operator not containing time dependence explicitly is independent of time. We therefore speak of a *stationary state*. There is no dispersion in energy since $\langle H^n \rangle = E^n$.

How should the problem of solving the time-independent Schrödinger equation be posed? We want ψ_E to be "reasonable." One criterion might be $\int |\psi_E|^2 d^3x < \infty$ because ψ_E is a probability density. This is often true, but it is too restrictive. For example, $e^{i\mathbf{p}\cdot\mathbf{x}/\hbar}$ is allowed in some circumstances. However, $\psi_E \propto e^{\alpha x}$ ($\alpha > 0, x \to \infty$) is not allowed because it says the particle is mostly at infinity.

Some experience can be found in determining numerically solutions to the Schrödinger equation for a linear potential: $V = V_0|x|/a$ for $x > 0$. This is a potential we shall encounter again.

$$-\frac{\hbar^2}{2m}\frac{d^2\psi}{dx^2} + V_0\frac{x}{a}\psi = E\psi. \tag{1.57}$$

With the substitutions

$$y = \left(\frac{2ma^2V_0}{\hbar^2}\right)^{1/3}\frac{x}{a}; \qquad \lambda = \left(\frac{2ma^2V_0}{\hbar^2}\right)^{1/3}\frac{E}{V_0}, \tag{1.58}$$

we have the dimensionless form

$$\frac{d^2\psi}{dy^2} - y\psi = -\lambda\psi. \tag{1.59}$$

There are two classes of solutions: ones that are even under $x \to -x$ ($y \to -y$) and ones that are odd under this symmetry. The ones that are odd must have $\psi(0) = 0$. To solve the differential equation numerically, one must in this case take some value for $\psi'(0)$ and then integrate in steps of Δx. If the solution is even, then $\psi'(0) = 0$. In either case, there is just one parameter to set at the origin and this is equivalent to setting the scale for ψ. Thus, the only open issue is the value of λ: For what values of λ do we get physically acceptable solutions? In particular, the solution must vanish in the limit $x \to \infty$.

Figure 1.2 Numerical integration of the linear potential, Eq. (1.59), shows that the first eigenvalue lies between 2.338 and 2.339 if the function is odd in x. The equation can also be solved analytically in terms of the Airy function $Ai(x)$, which will be encountered in a later chapter.

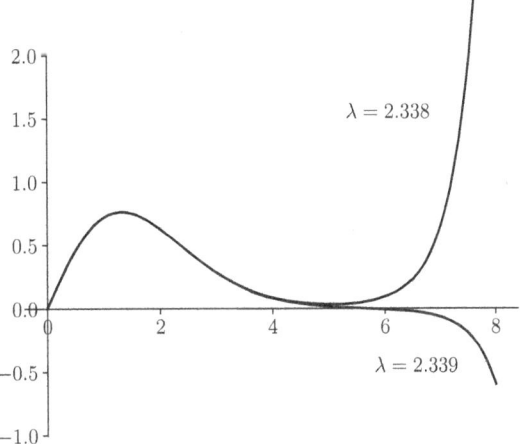

It is convenient to think of the second-order equation as a pair of first-order equations. Without loss of generality, one can take $a = 1$, which is to say that x is measured in units of a.

$$v = \frac{d\psi}{dx}$$

$$\frac{dv}{dx} = (x - \lambda)\psi. \tag{1.60}$$

Whatever the assumed value of λ, upon starting the integration at $x = 0$, at some point the integration reaches the value $x = \lambda$, where the motion is forbidden classically. In general, beyond this point the solution will be the sum of two pieces, one of which dies exponentially in x and the other of which grows exponentially in x. This latter is unacceptable physically. Only for certain values of λ will the solution be purely exponentially dying. These values are the eigenvalues. By varying λ and seeing whether the solution diverges to $+\infty$ or to $-\infty$, one can narrow down the range for λ until a good approximation for the eigenvalue is obtained. See Figure 1.2.

1.12 Sturm-Liouville Eigenvalue Problem

The time-independent Schrödinger equation is ubiquitous. It is therefore worthwhile to consider more generally the properties of the class of differential equations that arise from it.

Sturm-Liouville theory considers the ordinary differential equation in one dimension for $y(x)$:

$$\frac{d}{dx}\left(f(x)\frac{dy}{dx}\right) + g(x)y + \lambda h(x)y = 0 \tag{1.61}$$

for $a < x < b$ with f real and positive on this interval and λ a parameter. We impose boundary conditions $dy/dx = \alpha y$ at $y = a$ and $dy/dx = \beta y$ at $y = b$, with α, β independent of λ. These

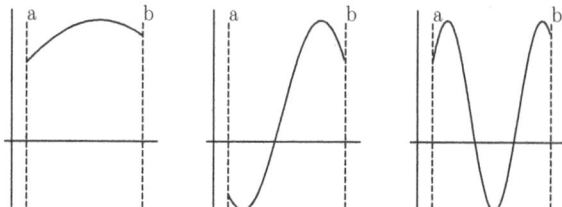

Figure 1.3 Examples showing the homogeneous boundary conditions at a and b: at a, $dy/dx = \alpha y(a)$ and at b, $dy/dx = \beta y(b)$. The chosen values have $\alpha > 0, \beta < 0$ and $\alpha = 2|\beta|$. With each increasing eigenvalue, the number of nodes of the eigenfunction increases by one.

are homogeneous boundary conditions in that if $y(x)$ is a solution, so is any constant times $y(x)$ (See Figure 1.3.). The Schrödinger equations we will encounter will often have $f(x) = 1$ and $\alpha = \beta = 0$ as special examples of the more general theory.

One can prove the following (see, e.g. Margenau and Murphy, *The Mathematics of Physics and Chemistry*, Vol.1, 2nd edition, pp. 270 ff., Morse and Feshbach, *Methods of Theoretical Physics*, pp. 736–739 or pp. 720–24, Courant and Hilbert, *Methods of Mathematical Physics*, Chap. V):

If $f(x)$ and $h(x)$ are positive on the interval, solutions $y_j(x)$ exist only for a countably infinite discrete set of eigenvalues λ_j with a smallest non-negative member λ_0. The eigenvalues have no upper bound as $j \to \infty$. The solutions are called eigenfunctions with eigenvalues λ_j. Eigenfunctions with increasing energy have an increasing number of nodes. We shall see this is a recurring feature of relevant solutions to the Schrödinger equation.

Moreover, if we order the solutions by $\lambda_{j+1} > \lambda_j$, then the number of nodes (zeros) of f_j in the interval (a, b) is j. Writing the equation twice and multiplying by y_i^* and y_j we find

$$\int_a^b dx \left\{ y_i^* \frac{d}{dx} \left(f(x) \frac{dy_j}{dx} \right) + [g(x) + \lambda_j h(x)] y_i^* y_j \right\} = 0$$

$$\int_a^b dx \left\{ y_j \frac{d}{dx} \left(f(x) \frac{dy_i^*}{dx} \right) + [g(x) + \lambda_i^* h(x)] y_i^* y_j \right\} = 0. \tag{1.62}$$

Now subtracting

$$\int_a^b dx \left\{ y_i^* \frac{d}{dx} \left(f(x) \frac{dy_j}{dx} \right) - y_j \frac{d}{dx} \left(f(x) \frac{dy_i^*}{dx} \right) + (\lambda_j - \lambda_i^*) h(x) y_i^* y_j \right\} = 0. \tag{1.63}$$

Integration by parts gives

$$\int_a^b dx \left\{ y_i^* \frac{d}{dx} \left(f(x) \frac{dy_j}{dx} \right) - y_j \frac{d}{dx} \left(f(x) \frac{dy_i^*}{dx} \right) \right\}$$

$$= f(x) \left[y_j \frac{dy_i^*}{dx} - y_i^* \frac{dy_j}{dx} \right]_a^b = 0$$

$$= -(\lambda_j - \lambda_i^*) \int_a^b dx \, h(x) y_i^* y_j. \tag{1.64}$$

The vanishing of the term at the end points is a consequence of the boundary conditions $dy/dx = \alpha y$ at $y = a$ and $dy/dx = \beta y$ at $y = b$, It follows that the eigenfunctions for different eigenvalues are orthogonal on the interval (a, b) with the weight $h(x)$. If instead $i = j$, we see that $\Im m \, \lambda_i = 0$, that is, the eigenvalues are real. It can be proved that the eigenfunctions in fact are complete: any "reasonable" function satisfying the same boundary conditions can be expanded as a linear combination of the eigenfunctions. For a proof, see, for example, Morse and Feshbach, *Methods of Theoretical Physics*, pp. 738–9.

1.13 Linear Operators on Functions

The Sturm-Liouville problem is an example of a more general eigenvalue problem with linear operators, i.e. an operator K on functions $f, g, \ldots$ such that for constants α and β

$$K(\alpha f(x) + \beta g(x)) = \alpha K(f(x)) + \beta K(g(x)). \tag{1.65}$$

Examples:

$$K(f(x)) = 5f(x); \quad K(f(x)) = \frac{d^3 f}{dx^3} + 7f(x). \tag{1.66}$$

Another particular example is the integral operator

$$K(f(x)) = \int_a^b dx' \, k(x, x') f(x'). \tag{1.67}$$

Suppose we define a scalar product of two functions by

$$(f, g) = \int_a^b dx \, g^*(x) f(x) = (g, f)^*. \tag{1.68}$$

Given a linear operator K, we define its adjoint by

$$(K^\dagger g, f) = (g, Kf). \tag{1.69}$$

In the case of our integral operator we see that we need to take

$$k^\dagger(x, x') = k(x', x)^*. \tag{1.70}$$

This is analogous to the usual meaning of adjoint for matrices: $A_{ij}^\dagger = A_{ji}^*$. An operator is self-adjoint or Hermitian if $K^\dagger = K$. In this case

$$(f, Kf) = (K^\dagger f, f) = (Kf, f) = (f, Kf)^*, \tag{1.71}$$

that is, the expectation value $\langle K \rangle$ of a Hermitian operator K is real. Conversely, if we want an operator to represent a physical quantity, it must be Hermitian. For an anti-Hermitian operator we have $K^\dagger = -K$. If K is Hermitian, then iK is anti-Hermitian.

1.14 Eigenvalue Problem for a Hermitian Operator

We suppose K is a Hermitian operator on functions on an interval (a, b) with some boundary conditions. If we indicate a function by ξ, then the eigenvalue problem is

$$K\xi_j = \lambda_j \xi_j. \tag{1.72}$$

We know the λ_j are real because

$$\lambda_j = \frac{(\xi_j, K\xi_j)}{(\xi_j, \xi_j)} \tag{1.73}$$

and both the numerator and denominator are real. If K is Hermitian then

$$(\xi_j, K\xi_i) = \lambda_i(\xi_j, \xi_i) = (K^\dagger \xi_j, \xi_i) = (K\xi_j, \xi_i) = \lambda_j^*(\xi_j, \xi_i). \tag{1.74}$$

Since the eigenvalues are real

$$(\lambda_i - \lambda_j)(\xi_j, \xi_i) = 0. \tag{1.75}$$

Thus if the eigenvalues are different, then the eigenfunctions are orthogonal. In general, it is possible that there are linearly independent eigenfunctions with the same eigenvalue, in which case orthogonality doesn't follow. This won't happen in the specific instance at hand, a one-dimensional differential equation with boundary conditions.

We now show how this formalism can incorporate the Sturm-Liouville problem. We take

$$K = \frac{d}{dx}\left(f(x)\frac{d}{dx}\right) + g(x) + \lambda \tag{1.76}$$

and ask under what conditions K is Hermitian.

$$(\eta, K\xi) = \int_a^b \eta^* \left[\frac{d}{dx}\left(f(x)\frac{d\xi}{dx}\right) + g(x)\xi + \lambda\xi\right] dx$$

$$= f\left(\eta^*\frac{d\xi}{dx} - \xi\frac{d\eta^*}{dx}\right)\Bigg|_a^b + \int_a^b \left[\frac{d}{dx}\left(f^*\frac{d\eta}{dx}\right) + g^*\eta + \lambda^*\eta\right]^* \xi dx. \tag{1.77}$$

We want the first term to vanish and the second term to be equal to $(K\eta, \xi)$. The latter is achieved if f, g, λ are all real. For the former, it suffices for $f = 0$ at a and b, or alternatively, for

$$\left(\frac{1}{\eta^*}\frac{d\eta^*}{dx}\right)_{a,b} = \left(\frac{1}{\xi}\frac{d\xi}{dx}\right)_{a,b}. \tag{1.78}$$

These are precisely the conditions we postulated in the treatment above of the Sturm-Liouville problem.

1.15 Variational Methods for Energy Eigenvalues

A powerful method of generating approximate energies of quantum mechanical systems is the so-called variational technique – Rayleigh-Ritz method is another term. Rayleigh used it in acoustics. We phrase it terms of the Schrödinger equation and energy eigenvalues, but it has very general applicability.

Consider a well-posed energy eigenvalue problem:

$$H\psi_n = E_n\psi_n \tag{1.79}$$

with a spectrum of eigenvalues $E_n = E_0, E_1 > E_0, E_2 > E_1, ...$, with corresponding eigenfunctions $\psi_0, \psi_1, \psi_2, ...$, but suppose we do not know the eigenvalues or eigenfunctions. Suppose that we are somehow given a function ψ that is not too far from the correct wave function ψ_0. We can form the quantity

$$[E] = \frac{\int dx\, \psi^* H\psi}{\int dx\, \psi^*\psi}, \tag{1.80}$$

where x might be one-dimensional or three-dimensional and ψ is some trial function, obeying the boundary conditions. What can we say about $[E]$ as an estimate of E_0? Two things:

(i) $[E] \geq E_0$
(ii) If ψ differs from the true solution by a small amount, characterized by ϵ, so $\psi = \psi_0 + \epsilon\phi$ then $[E] - E_0 = \mathcal{O}(\epsilon^2)$.

This means that $[E]$ is stationary with respect to small variations away from the true eigenfunction, ψ_0.

To prove this, expand ψ in the basis of true eigenfunctions

$$\psi = \sum_n a_n\psi_n, \tag{1.81}$$

where we assume orthonormality of the eigenfunctions

$$\int dx\, \psi_m^*\psi_n = \delta_{mn} \tag{1.82}$$

and that ψ is normalized:

$$\sum_n |a_n|^2 = 1. \tag{1.83}$$

Then immediately

$$[E] = \sum_n |a_n|^2 E_n = \sum_n |a_n|^2 E_0 + \sum_n |a_n|^2(E_n - E_0) \tag{1.84}$$

$$= E_0 + \sum_n |a_n|^2(E_n - E_0) \geq E_0. \tag{1.85}$$

Now consider a trial wave function that is near the true one

$$\psi = \psi_0 + \epsilon\phi \tag{1.86}$$

where ψ_0 and ϕ are normalized: $\int dx|\psi_0|^2 = 1; \int dx|\phi|^2 = 1$. It suffices to consider only a ϕ that is orthogonal to ψ_0 since any part of ϕ that is not can be absorbed into ψ, after which we can renormalize so that the coefficient of ψ_0 is unity. Now expand the expression for $[E]$, but impose the normalization by dividing by $\int dx\, |\psi|^2$.

$$[E] = E_0 + \delta E = \frac{\int dx\, \psi_0^* H \psi_0 + \epsilon^* \int dx\, \phi^* H \psi_0 + \epsilon \int dx\, \psi_0^* H \phi + |\epsilon|^2 \int dx\, \phi^* H \phi}{\int dx\, \psi_0^* \psi_0 + \epsilon^* \int dx\, \phi^* \psi_0 + \epsilon \int dx\, \psi_0^* \phi + |\epsilon|^2 \int dx\, \phi^* \phi}$$

$$= \frac{E_0 \int dx\, \psi_0^* \psi_0 + E_0 \epsilon^* \int dx\, \phi^* \psi_0 + E_0 \epsilon \int dx\, \psi_0^* \phi + |\epsilon|^2 \int dx\, \phi^* H \phi}{\int dx\, \psi_0^* \psi_0 + \epsilon^* \int dx\, \phi^* \psi_0 + \epsilon \int dx\, \psi_0^* \phi + |\epsilon|^2 \int dx\, \phi^* \phi}. \tag{1.87}$$

Then to order ϵ^2 we have

$$E_0 + \delta E \approx \left(E_0 + |\epsilon|^2 \int dx\, \phi^* H \phi \right)\left(1 - |\epsilon|^2 \right), \tag{1.88}$$

so that

$$\delta E = |\epsilon|^2 \left(\int dx\, \phi^* H \phi - E_0 \int dx\, \phi^* \phi \right) \geq 0, \tag{1.89}$$

where the inequality follows from the observation that if it were not true then ϕ would have lower energy than the ground state. If we can make a good guess at ψ_0 with ψ, then we will get a good estimate of E_0. This approach is systematized in the Rayleigh-Ritz method.

1.16 Rayleigh-Ritz Method

We construct a "trial" eigenfunction $\psi(\alpha_i, x)$, with α_i a set of n variational parameters. Then we form the functional

$$[E(\alpha_i)] = \frac{\int dx\, \psi^*(\alpha_i, x) H \psi(\alpha_i, x)}{\int dx\, \psi^*(\alpha_i, x) \psi(\alpha_i, x)} \tag{1.90}$$

of the parameters α_i. We know that $[E(\alpha_i)] \geq E_0$, with equality only for $\psi(\alpha_i, x) = \psi_0(x)$. We therefore vary the parameters to find a minimum, where we must have

$$\frac{\partial [E(\alpha_i)]}{\partial \alpha_j} = 0, \qquad j = 1, 2, \dots n. \tag{1.91}$$

The solution to these simultaneous equations yields the optimal value for E_0 possible within the limitations of the form of our trial function.

As a very simple example, let us seek the energy of the ground state in a one-dimensional box, $0 \leq x \leq 1$ and take $\psi(x) = [x(1-x)]^\alpha$. Then

$$[E] = \frac{\int dx\, \psi^* \frac{p^2}{2m} \psi}{\int dx\, \psi^* \psi} = \frac{\hbar^2}{2m} \frac{\int_0^1 dx\, |\psi'|^2}{\int_0^1 dx\, |\psi|^2}. \tag{1.92}$$

Since

$$\psi' = \alpha x^{\alpha-1}(1-x)^{\alpha-1}(1-2x) \tag{1.93}$$

we find

$$\frac{2m}{\hbar^2}[E] = \frac{\alpha^2 \int_0^1 dx\, x^{2\alpha-2}(1-x)^{2\alpha-2}(1-4x+4x^2)}{\int_0^1 dx\, x^{2\alpha}(1-x)^{2\alpha}}. \tag{1.94}$$

With the standard integral

$$\int_0^1 dx\, x^{m-1}(1-x)^{n-1} \equiv \beta(m,n) = \frac{\Gamma(m)\Gamma(n)}{\Gamma(m+n)}, \tag{1.95}$$

where the gamma function

$$\Gamma(z) = \int_0^\infty dt\, t^{z-1} e^{-t} \tag{1.96}$$

has the properties

$$z\Gamma(z) = \Gamma(z+1); \qquad \Gamma(n) = (n-1)!\ for\ n\ \text{a positive integer} \tag{1.97}$$

we find

$$\frac{2m}{\hbar^2}[E] = \frac{(2\alpha)(4\alpha+1)}{2\alpha-1}. \tag{1.98}$$

The requirement $\partial E/\partial\alpha = 0$ gives

$$\alpha^2 - \alpha - \frac{1}{8} = 0; \qquad a = \frac{1}{2}\left(1 + \sqrt{\frac{3}{2}}\right) = 1.11237 \tag{1.99}$$

and

$$\frac{2m[E]_{\min}}{\hbar^2} = 9.89898\ldots, \tag{1.100}$$

while the exact result is $\pi^2 = 9.86960\ldots$, a 0.3% error. The moral is that even simple choices for the trial wave function can give remarkably good results for the ground state energy.

References and Suggested Reading

There are numerous accounts of the history of quantum mechanics. Some particularly notable ones written by physicists are:

Gamow, G. *Thirty Years that Shook Physics*. New York: Doubleday.
Pais, A. *Inward Bound: Of Matter and Forces in the Physical World*. Oxford University Press.
Segrè, G. *Faust in Copenhagen: A Struggle for the Soul of Physics, Viking*. New York.

Every student of quantum mechanics should thrill to the remarkable story told in these books.

For classic presentations of mathematical physics:

Courant and Hilbert, *Methods of Mathematical Physics*.
Margenau and Murphy, *The Mathematics of Physics and Chemistry*, Vol.1, 2nd edition.
Morse and Feshbach, *Methods of Theoretical Physics*.

Problems

1.1 A one-dimensional wave amplitude $\psi(x, t)$ is given by

$$\psi(x, 0) = \frac{1}{\sqrt{2\pi\hbar}} \int_{-\infty}^{\infty} dp\, \phi(p) e^{ipx/\hbar} \qquad (1.101)$$

with

$$\phi(p) = \sqrt{\frac{2}{\pi\hbar}} \cdot \frac{\alpha^{3/2}}{(p - p_0)^2 + \alpha^2}. \qquad (1.102)$$

(a) Find $\psi(x, 0)$ and sketch $\phi(p)$ and $\Re\,\psi(x, 0)$. What is the significance of α and $1/\alpha$?
(b) Find $\langle p \rangle, \langle p^2 \rangle, \langle x \rangle$, and $\langle x^2 \rangle$.
(c) Show from the results of (b) that the r.m.s. deviations Δx and Δp satisfy $\Delta x \Delta p > \hbar/2$, independent of α.

1.2 A particle is in the ground state of a one-dimensional box of length L. with wave amplitude given by

$$\psi(x, t) = \sqrt{2/L} \sin(\pi x/L) e^{-i\pi^2\hbar t/(2mL^2)}. \qquad (1.103)$$

At $t = 0$ the walls are suddenly dissolved and the particle subsequently moves freely. What is the probability that, after $t = 0$, the particle's momentum is between p and $p + dp$? Sketch fairly accurately $|\phi(p)|^2$ versus p.

1.3 An anti-proton can be bound to a proton by electromagnetic interaction to form a hydrogenic atomic system. At short distances ($r \leq 10^{-13}$ cm) the interaction is augmented by strong forces, including annihilation.

(a) Neglecting annihilation and other strong-interaction effects, scale the $\bar{p}p$ system appropriately from the e^-p atom and compute the Bohr radius, and the spacing between the $1s$ and $2p$ levels, in keV.
(b) From the known $2p \rightarrow 1s$ transition probability in hydrogen ($\Gamma(2p \rightarrow 1s) = 6.2 \times 10^8$ sec^{-1}) and the scaling properties of the Larmor (electric dipole) radiation formula, find the corresponding transition probability for the $\bar{p}p$ atom.

1.4 A particle with mass m and charge e interacts with a real potential $V(\mathbf{x})$ and also an external electromagnetic field described by the real potentials $\mathbf{A}$ and Φ. The Schrödinger equation is

$$i\hbar \frac{\partial \psi}{\partial t} = \frac{1}{2m}\left(\mathbf{p}_{\mathrm{op}} - \frac{e}{c}\mathbf{A}\right)^2 \psi + V\psi + e\Phi\psi \qquad (1.104)$$

Show that the normal expression for the probability current $\mathbf{J}$ is augmented by the term

$$\Delta \mathbf{J} = -\frac{e}{mc}\psi^*\mathbf{A}\psi. \qquad (1.105)$$

Interpret.

1.5* **(a)** Show that a free-particle wave amplitude $\psi(\mathbf{x}, t)$ evolves in time from the amplitude $\psi(\mathbf{x}', t')$ according to

$$\psi(\mathbf{x}, t) = \int d^3x' \, \psi(\mathbf{x}', t') K(\mathbf{x} - \mathbf{x}', t - t'), \tag{1.106}$$

where the kernel (propagation function) is given by the Fourier integral

$$K(\mathbf{r}, \tau) = \frac{1}{(2\pi\hbar)^3} \int d^3p \, \exp\left(i\frac{\mathbf{p} \cdot \mathbf{r}}{\hbar} - i\frac{p^2\tau}{2m\hbar}\right). \tag{1.107}$$

(b) As a preliminary to evaluating this integral, prove that

$$\lim_{\epsilon \to 0} \int_{-\infty}^{\infty} d\xi e^{-i\xi^2} e^{-\epsilon\xi^2} = \sqrt{\pi} e^{-i\pi/4} = \sqrt{\pi/i}. \tag{1.108}$$

(c) Show that

$$K(\mathbf{r}, \tau) = \left(\frac{m}{2\pi i\hbar\tau}\right)^{3/2} \exp\left(-\frac{mr^2}{2i\hbar\tau}\right). \tag{1.109}$$

1.6 Suppose that at $t = 0$, a one-dimensional wave amplitude is

$$\psi(x, 0) = A e^{-x^2/4(\Delta x)_0^2}. \tag{1.110}$$

Using the one-dimensional analog of the solution to Problem 1.5, show that the amplitude at a later time t has the form

$$\psi(x, t) \propto \exp\left[-\frac{x^2}{4(\Delta x)_t^2}\left(1 - \frac{i\hbar t}{2m(\Delta x)_0^2}\right)\right]. \tag{1.111}$$

where

$$(\Delta x)_t^2 = (\Delta x)_0^2 + \frac{\hbar^2 t^2}{4m^2(\Delta x)_0^2} \tag{1.112}$$

is the r.m.s. spread in $|\psi(x, t)|^2$.

1.7 (a) By explicit construction with the Schrödinger equation, show that for any wave amplitude $\psi(\mathbf{x}, t)$

$$\frac{d}{dt}\langle x^2 \rangle = \frac{1}{m}\langle xp_x + p_x x \rangle, \tag{1.113}$$

where $p_x = (\hbar/i)\partial/\partial x$.

(b) Similarly, show that

$$\frac{d}{dt}\langle \mathbf{r} \cdot \mathbf{p} \rangle = 2\left\langle \frac{p^2}{2m} \right\rangle - \langle \mathbf{r} \cdot \nabla V \rangle. \tag{1.114}$$

This is the basis for the quantum mechanical virial theorem (for energy eigenstates only).

1.8 The power of the variational method can be illustrated by estimating the ground state energies of the hydrogen atom and the isotropic harmonic oscillator, using trial wave functions that are appropriate to the corresponding potential. For the ground state, the angular momentum is zero and the wave function depends only on r. We can write the Schrödinger equation as

$$\left(-\frac{\hbar^2}{2m}\frac{1}{r}\frac{d^2}{dr^2}r + V(r)\right)\psi = E\psi. \tag{1.115}$$

It is conventional to introduce a radial wave function $u = r\psi$ so that the equation is

$$-\frac{\hbar^2}{2m}\frac{d^2u}{dr^2} + V(r)u = Eu \tag{1.116}$$

and $u(r)$ must vanish at $r = 0$.

Use a variational principle for the radial wave-function $u(r)$ in an s-wave state.

(a) For the hydrogen atom, take a trial wave function of the form

$$u(r) = re^{-r^2/4b^2}, \tag{1.117}$$

where b is the parameter to be varied. Find the best value of b/a_0 and the corresponding binding energy. How accurate is the estimate?

(b) For the three-dimensional isotropic oscillator, take a trial wave function

$$u(r) = re^{-r/2d}, \tag{1.118}$$

where d is the variational parameter. Find the optimal value of d and the corresponding energy. How accurate is the estimate?

1.9 A hydrogen atom is confined in a spherical box of radius Ra_0, with the proton at the center. In atomic units, the radial equation for s-states is

$$-\frac{1}{2}\frac{d^2u}{dx^2} - \frac{u}{x} = Eu, \tag{1.119}$$

where $x = r/a_0$ and $u(0) = u(R) = 0$. Make a variational estimate of the ground state energy as a function of R using the trial wave function

$$u(x) = x(1 - x/R)^b, \tag{1.120}$$

where b is a variational parameter. Find the value of R when the atom has zero binding energy. Find the values of E for $R \gg 1$ and $R \ll 1$, and compare with known results. For positive binding energies, plot the energy as a function of R.

Hint: the variational energy takes the form, $E = R^{-2}f(b) - R^{-1}g(b)$. The stationary point is specified by a relation of the form $R = h(b)$. Rather than solving for b, choose a series of values of b, determine each corresponding R and $E(R(b), b)$.

2

Reformulation

In developing quantum mechanics, we can continue relying on wave functions and operators acting on them, but there are conceptual advantages to working in a more abstract
fashion. The basic tool is linear vector spaces. The general principle of quantum mechanics
is that a physical system can be represented by linear combinations of basis states, which
can be viewed as elements of a linear vector space. A given state of the system is represented by a ray in the vector space. By a ray, we mean a quantity defined up to an overall
constant phase, which is inessential because it always drops out in computing a physical
quantity.

Quantum mechanics requires complex numbers and thus we consider complex linear vector spaces. Observable quantities are represented by linear operators on the vector space
with the special property of being Hermitian (or self-adjoint), which guarantees that they
produce real, not complex, results. The rules for calculating the probabilities of physical
processes in quantum mechanics are clear. Interference phenomena are characteristic of
quantum mechanics and are quite different from those of classical mechanics, though they
are familiar from electrodynamics. Observables whose operators commute can be measured
simultaneously, or more precisely can be measured in either order with identical results.
Observables all of which commute are termed compatible. A complete set of compatible
observables is one that does not admit an additional independent operator. A complete set
of observables can label a set of linearly independent vectors that is a complete basis for the
states.

Classical mechanics provides a basis for postulating quantum mechanical relations.
Whether the resulting theory is correct is a matter to be tested experimentally. The results
over nearly one hundred years have been entirely positive. Symmetries of the Lagrangian in
classical mechanics lead to conservation laws in accordance with Noether's theorem. The
same is true in quantum mechanics, where symmetries play an even greater role, as they are
represented in linear vector spaces. Dirac developed a particularly powerful representation
of eigenstates of position $|q'\rangle$ and momentum $|p'\rangle$, which facilitates many analyses. The
Schrödinger and Heisenberg pictures provide complementary views of quantum mechanics. In the Schrödinger picture, the states evolve, while in the Heisenberg picture the states
are fixed but the operators evolve in time. Often a physical circumstance involves not a single state, but a superposition of states; then analysis can be facilitated with a density matrix
describing the mixture of states, either initial or final.

John David Jackson: A Course in Quantum Mechanics, First Edition. Robert N. Cahn.
© 2024 John Wiley & Sons, Inc. Published 2024 by John Wiley & Sons, Inc.
Companion website: www.wiley.com/go/Jackson/QuantumMechanics

2.1 Stern-Gerlach Experiment

The paradigm for measurement in quantum mechanics was established in 1922 when the "old quantum theory" still prevailed. In an experiment proposed by Otto Stern and executed by Walther Gerlach, a beam of neutral silver atoms was passed through a region of intense and inhomogeneous magnetic field, with the field gradient transverse to the line of flight. The force in such a region, classically, is given by $\mathbf{F} = \nabla(\mathbf{B} \cdot \boldsymbol{\mu})$, where $\boldsymbol{\mu}$ is the magnetic moment of the atom. The electronic configuration of the silver atom has a single s-wave electron outside a closed shell, so its magnetic moment is precisely that of the electron. Classically, if the magnetic moments of the atoms were randomly oriented, a band would be observed in the sensors stretching between the locations where atoms would end up if their moments were aligned with the field gradient to the point where they would end up if anti-aligned. What Stern and Gerlach observed in 1922 was two distinct, separated spots from their beam of silver.

The theory of linear vector spaces provides a particularly convenient description of such phenomena, and indeed all quantum mechanical phenomena.

2.2 Linear Vector Spaces

We begin by considering N-dimensional complex spaces whose members we can represent by an N-tuple of complex numbers $\phi = (\phi_1, \phi_2, ... \phi_N)$. Clearly multiplying ϕ by a complex number gives another such element. Similarly, the sum of two elements is itself an element.

We define an inner product of two elements ϕ and ψ by

$$(\phi, \psi) = \sum_{i=1}^{N} \phi_i^* \psi_i, \tag{2.1}$$

which is a complex number. The inner product has the properties

$$(\psi, \phi) = (\phi, \psi)^*; \quad (\phi, c\psi) = c(\phi, \psi); \quad (c\phi, \psi) = c^*(\phi, \psi). \tag{2.2}$$

We see that (ϕ, ϕ) is real and greater than zero unless $\phi = (0, 0, ... , 0)$. We define the norm of ϕ by

$$||\phi|| = \sqrt{(\phi, \phi)}. \tag{2.3}$$

If we have a set of N-tuples ϕ^α, $\alpha = 1, 2, ... N$ we say that the ϕ^α are linearly independent if the relation $\sum_\alpha c_\alpha \phi^\alpha = 0$ (here 0 means the N-tuple $(0, 0, ...0)$) for complex c_α requires that $c_\alpha = 0$ for every α. It is clear that if we take $\phi^1 = (1, 0, ... 0)$, $\phi^2 = (0, 1, 0, ... 0)$, etc., these N ϕ^α are indeed linearly independent, but not every set of N N-tuples will be linearly independent. A basis set for the linear space is a collection of N linearly independent vectors. Any vector can be expressed as a linear combination of the basis vectors.

Our reason for focusing on linear vector spaces is that they enable us to represent a fundamental aspect of quantum mechanics: particles and systems of particles show interference between distinct components that exist simultaneously and thus must be represented as linear combinations of basis states. We will adopt the notation introduced by Dirac, representing a vector by the "ket" $|\phi\rangle$. The reason for calling this a ket will become apparent shortly. Following the example of N-tuples, we assume that we can add two vectors and multiply them by complex numbers and the result is also a vector. Moreover, there is a zero vector $|0\rangle$ with the property

$$|\alpha\rangle + |0\rangle = |\alpha\rangle \tag{2.4}$$

for every $|\alpha\rangle$ and $0\,|\alpha\rangle = |0\rangle$. In our explicit example, of course, $|0\rangle$ is the N-tuple $(0, 0, ...0)$.

We proceed by analogy with the N-tuple example. We suppose we have basis states $|\xi_i\rangle$, the analogs of ϕ^1, ϕ^2 above. We continue the analogy by defining a scalar product of two vectors:

$$|\alpha\rangle = \sum_i \alpha_i \,|\xi_i\rangle; \qquad |\beta\rangle = \sum_j \beta_j \,|\xi_j\rangle$$

$$\langle\beta|\alpha\rangle = \sum_i \beta_i^* \alpha_i = \langle\alpha|\beta\rangle^*. \tag{2.5}$$

In the Dirac notation, we take this expression apart. Then we define the "bra" states $\langle\xi_j|$ as operators that act on "kets" to give complex numbers. Explicitly, we set

$$\langle\xi_j|\,(|\xi_i\rangle) = \langle\xi_j|\xi_i\rangle = \delta_{ij}. \tag{2.6}$$

If we take all linear combinations of the $\langle\xi_j|$, we get a linear vector space called the dual space. If we have a vector

$$|\beta\rangle = \sum_i \beta_i \,|\xi_i\rangle \tag{2.7}$$

we define its dual by

$$\langle\beta| = \sum_j \langle\xi_j|\, \beta_j^*. \tag{2.8}$$

This generates the required scalar product. Note that the meaning of $\langle\beta|$ is derived from the definition of the scalar product. If the basis states are to represent a physical situation, then the choice of basis states is not completely arbitrary as we shall see below.

The abstract vectors $|\alpha\rangle$ behave much like ordinary vectors in three-dimensional space. An example is the Cauchy-Schwarz inequality. Since for any $|\gamma\rangle$, $<\gamma|\gamma> \geq 0$ consider

$$[\langle\alpha| - \lambda^* \langle\beta|][|\alpha\rangle - \lambda\,|\beta\rangle] \geq 0 \tag{2.9}$$

and in particular take $\lambda = \langle\beta|\alpha\rangle\,/\,\langle\beta|\beta\rangle$. Then

$$\langle\alpha|\alpha\rangle - \lambda^*\,\langle\beta|\alpha\rangle - \lambda\,\langle\alpha|\beta\rangle + |\lambda|^2\,\langle\beta|\beta\rangle \geq 0$$

$$\langle\alpha|\alpha\rangle - \frac{|\langle\alpha|\beta\rangle|^2}{\langle\beta|\beta\rangle} - \frac{|\langle\alpha|\beta\rangle|^2}{\langle\beta|\beta\rangle} + \frac{|\langle\alpha|\beta\rangle|^2}{\langle\beta|\beta\rangle} \geq 0$$

$$\langle\alpha|\alpha\rangle\,\langle\beta|\beta\rangle \geq |\langle\alpha|\beta\rangle|^2. \tag{2.10}$$

The Cauchy-Schwarz relation allows us to think in geometrical terms. If $\langle\alpha|\alpha\rangle = \langle\beta|\beta\rangle = 1$ then we can think of $\langle\beta|\alpha\rangle$ as the cosine of the "angle" between α and β. If $\langle\beta|\alpha\rangle = 0$, then α and β are orthogonal.

Any vector $|\alpha\rangle$ can be expanded in our basis

$$|\alpha\rangle = \sum_i \alpha_i \, |\xi_i\rangle. \tag{2.11}$$

But then

$$\langle \xi_j | \alpha \rangle = \sum_i \langle \xi_j | \xi_i \rangle \, \alpha_i = \alpha_j \tag{2.12}$$

so that

$$|\alpha\rangle = \sum_i |\xi_i\rangle \langle \xi_i | \alpha \rangle. \tag{2.13}$$

Similarly,

$$\langle \beta | \alpha \rangle = \langle \beta | \left(\sum_i |\xi_i\rangle \langle \xi_i | \right) |\alpha\rangle. \tag{2.14}$$

Evidently, the sum of bras and kets (in the opposite to the usual order) can be identified as the unit operator or I, i.e.

$$\sum_i |\xi_i\rangle \langle \xi_i | \equiv I, \tag{2.15}$$

This is called the completeness relation and is dual to the orthogonality relation

$$\langle \xi_j | \xi_i \rangle = \delta_{ji}. \tag{2.16}$$

Suppose there are two orthonormal sets of basis vectors, $|\xi_i\rangle$ and $|\eta_j\rangle$. We can expand one set in terms of the other:

$$|\xi_i\rangle = \sum_{j'} |\eta_{j'}\rangle U_{j'i}; \qquad |\eta_j\rangle = \sum_{i'} |\xi_{i'}\rangle V_{i'j}. \tag{2.17}$$

The $U_{ij'}$ and $V_{ji'}$ are N × N matrices. Substituting one relation into the other, we have

$$|\xi_i\rangle = \sum_{i'} \sum_{j'} |\xi_{i'}\rangle V_{i'j'} U_{j'i} = \sum_{i'} |\xi_{i'}\rangle \sum_{j'} V_{i'j'} U_{j'i}. \tag{2.18}$$

The sum over j' is just matrix multiplication:

$$|\xi_i\rangle = \sum_{i'} |\xi_{i'}\rangle (VU)_{i'i} \tag{2.19}$$

so that

$$(VU)_{i'i} = \delta_{i'i}, \tag{2.20}$$

or as matrices

$$VU = I, \tag{2.21}$$

and similarly $UV = I$. Now if we compute

$$\langle \xi_{i'} | \xi_i \rangle = \delta_{i'i} = \sum_j \sum_{j'} \langle \eta_{j'} | \eta_j \rangle U_{j'i'}^* U_{ji} = \sum_j U_{ji'}^* U_{ji}, \tag{2.22}$$

we find that in matrix notation, if we define

$$U_{ji}^\dagger = U_{ij}^*, \tag{2.23}$$

then

$$(U^{\dagger}U)_{i'i} = \delta_{i'i}; \qquad U^{\dagger}U = 1. \tag{2.24}$$

Thus U is a unitary matrix – its adjoint is its inverse – as is $V = U^{-1} = U^{\dagger}$. We see that a change of basis is accomplished by a unitary transformation.

2.3 Linear Operators

Consider a relation that assigns to every vector $|\alpha\rangle$, another vector in the same linear space, and call it $A\,|\alpha\rangle$. This relation is a linear operator if it obeys the rules

$$A\big(c_1\,|\alpha_1\rangle + c_2\,|\alpha_2\rangle\big) = c_1 A\,|\alpha_1\rangle + c_2 A\,|\alpha_2\rangle. \tag{2.25}$$

We say that two operators A and A' are equal if $A\,|\alpha\rangle = A'\,|\alpha\rangle$ for every state $|\alpha\rangle$. We define a null operator X by $X\,|\alpha\rangle = |0\rangle$ for every $|\alpha\rangle$ and similarly an identity operator I by $I\,|\alpha\rangle = |\alpha\rangle$. We define the sum of linear operators in the obvious fashion: $(A+B)\,|\alpha\rangle = A\,|\alpha\rangle + B\,|\alpha\rangle$. The commutation and associative properties follow:

$$(A + B) = (B + A); \qquad (A + B) + C = A + (B + C). \tag{2.26}$$

It is clear that multiplication of operators is associative $A(BC) = (AB)C$, but there is no guarantee that it is commutative, $AB = BA$.

Just as there are operators that act on the space generated by the basis vectors $|\xi_i\rangle$, there are linear operators that act on the dual space, generated by the $\langle\xi_i|$. In fact, if we have an operator A acting on the original space, this will create an associated operator $A^{\dagger}$ acting on the dual space defined by the rule that if

$$A\,|\alpha\rangle = |\beta\rangle, \tag{2.27}$$

then we define $A^{\dagger}(\langle\alpha|) \equiv \langle\alpha|\,A^{\dagger}$ by

$$\langle\alpha|\,A^{\dagger} = \langle\beta|, \tag{2.28}$$

where $\langle\alpha|$ is the dual of $|\alpha\rangle$ and $\langle\beta|$ is the dual of $|\beta\rangle$. This abstract definition can be made concrete by considering

$$A\,|\xi_j\rangle = \sum_i A_{ij}\,|\xi_i\rangle; \qquad \langle\xi_i|A|\xi_j\rangle = A_{ij}, \tag{2.29}$$

where the A_{ij} are complex numbers, the matrix representation of A, and

$$\langle\xi_j|\,A^{\dagger} = \sum_i A^{\dagger}_{ji}\,\langle\xi_i|; \qquad \langle\xi_j|A^{\dagger}|\xi_i\rangle = A^{\dagger}_{ji}, \tag{2.30}$$

where similarly the $A^{\dagger}_{ij}$ are the representation of $A^{\dagger}$. But if $\langle\xi_j|\,A^{\dagger}$ is the dual of $A\,|\xi_j\rangle$, then

$$\sum_i A^{\dagger}_{ji}\,\langle\xi_i| = \sum_i A^{*}_{ij}\,\langle\xi_i|. \tag{2.31}$$

Thus, our abstract definition is equivalent to

$$A^{\dagger}_{ji} = A^{*}_{ij}. \tag{2.32}$$

We can write, quite generally,

$$\langle \beta |A^\dagger| \alpha \rangle = \langle \alpha |A| \beta \rangle^*. \tag{2.33}$$

which we might equally as well have taken as the definition of $A^\dagger$. From the rules of matrix multiplication it is apparent that $(AB)^\dagger = B^\dagger A^\dagger$.

If an operator A is equal to its adjoint $A^\dagger$, the operator is said to be self-adjoint or Hermitian, just as in the previous chapter, where we considered linear operators, K, on a space of functions. Of course, not every operator is Hermitian. If A is Hermitian, iA is anti-Hermitian: $(iA)^\dagger = -(iA)$. The abstract vector space here corresponds to the space of linear combinations of functions there, and the scalar product here was given there as an integral of the product of two functions. An advantage of the abstract presentation is that it will enable us to express conveniently entities like spin, which are not functions of position.

We recapitulate some results from the previous chapter on eigenvalues in the formalism of linear vector spaces. Suppose A is Hermitian and $|\alpha\rangle$ is an eigenvector with eigenvalue α. Then α is real. This is immediate from

$$\langle \alpha |A| \alpha \rangle = \alpha \langle \alpha|\alpha \rangle = \langle \alpha |A^\dagger| \alpha \rangle = \langle \alpha |A| \alpha \rangle^* = \alpha^* \langle \alpha|\alpha \rangle. \tag{2.34}$$

Suppose $|\alpha_1\rangle$ and $|\alpha_2\rangle$ are eigenvectors of a Hermitian operator A with eigenvalues $\alpha_1 \neq \alpha_2$. Then $|\alpha_1\rangle$ and $|\alpha_2\rangle$ are orthogonal: $\langle \alpha_1|\alpha_2 \rangle = 0$. This we can show by

$$\langle \alpha_2 |A| \alpha_1 \rangle = \alpha_1 \langle \alpha_2|\alpha_1 \rangle$$
$$\langle \alpha_2 |A^\dagger| \alpha_1 \rangle = \langle \alpha_1 |A| \alpha_2 \rangle^* = \alpha_2^* \langle \alpha_1|\alpha_2 \rangle^* = \alpha_2^* \langle \alpha_2|\alpha_1 \rangle. \tag{2.35}$$

But α_2 is real, so

$$\alpha_1 \langle \alpha_2|\alpha_1 \rangle = \alpha_2 \langle \alpha_2|\alpha_1 \rangle \tag{2.36}$$

and $\langle \alpha_2|\alpha_1 \rangle = 0$.

Eigenvectors with the same eigenvalues may or may not be orthogonal. For example, in the non-relativistic hydrogen atom the 2s and 2p states have the same energy eigenvalue, and while they are orthogonal, they are not orthogonal to linear combinations of the two.

We saw earlier that the completeness relation, involving outer products, $\sum_{\alpha'} |\alpha'\rangle\langle\alpha'| = I$, was an operator statement. Thus outer products appear to be operators. Consider the outer product $|\beta\rangle\langle\alpha|$ acting on the state $|\gamma\rangle$

$$\left(|\beta\rangle\langle\alpha| \right) |\gamma\rangle = |\beta\rangle\langle\alpha|\gamma\rangle. \tag{2.37}$$

Thus $|\beta\rangle\langle\alpha|$ maps $|\gamma\rangle$ along the direction of $|\beta\rangle$, with a length governed by the inner product $\langle\alpha|\gamma\rangle$. The adjoint of $|\beta\rangle\langle\alpha|$ is $|\alpha\rangle\langle\beta|$, as can easily be proved.

We have seen that the identity operator can be written as

$$\sum_{\xi'} |\xi'\rangle\langle\xi'|, \tag{2.38}$$

where the $|\xi'\rangle$ are an orthonormal basis. Suppose we sum over only a portion of the basis vectors, those in a set S.

$$P_S = \sum_{\xi' \in S} |\xi'\rangle\langle\xi'|. \tag{2.39}$$

We call P_S a projection operator, because it projects out just a portion of the whole vector space, the portion that can be made from the vectors in S. Note that since

$$\sum_{\xi' \in S} |\xi'\rangle\langle\xi'| \sum_{\xi'' \in S} |\xi''\rangle\langle\xi''| = \sum_{\xi' \in S} |\xi'\rangle \sum_{\xi'' \in S} \delta_{\xi''\xi'} \langle\xi''| = \sum_{\xi' \in S} |\xi'\rangle\langle\xi'|, \tag{2.40}$$

we see that

$$P_S^2 = P_S. \tag{2.41}$$

This is the signature of a projection operator. Now

$$(1 - P_S)^2 = 1 - 2P_S + P_S^2 = 1 - P_S, \tag{2.42}$$

so $1 - P_S$ is also a projection operator. It projects out all the vectors orthogonal to all the vectors in S.

2.4 Unitary Transformations of Operators

We have seen that a change of basis is effected by a unitary transformation:

$$|\xi_j\rangle = \sum_i |\eta_i\rangle\langle\eta_i|\xi_j\rangle = \sum_i |\eta_i\rangle U_{ij} \equiv U |\eta_i\rangle$$

$$|\eta_i\rangle = \sum_j |\xi_j\rangle\langle\xi_j|\eta_i\rangle = \sum_j |\xi_j\rangle U_{ji}^\dagger = U^\dagger |\xi_j\rangle. \tag{2.43}$$

We can think of unitary transformations in two different ways. The transformation can act on the basis vectors to change the basis. Alternatively, the transformation can actually change the states. Thus for ordinary three-dimensional vectors, a transformation (in this case, real and thus orthogonal, not simply unitary) might change the co-ordinate system or it might rotate the vectors themselves. In any event, the equations look the same:

$$|\alpha'\rangle = U |\alpha\rangle; \quad |\beta'\rangle = U |\beta\rangle. \tag{2.44}$$

Now the dual of $|\alpha'\rangle$ is $\langle\alpha| U^\dagger$, so

$$\langle\alpha'|\beta'\rangle = \langle\alpha |U^\dagger U| \beta\rangle = \langle\alpha|\beta\rangle. \tag{2.45}$$

If the states transform under U, how does an operator A transform? After A has acted on $|\alpha\rangle$ we have $UA |\alpha\rangle = UAU^\dagger U |\alpha\rangle$. So we conclude that the operator is transformed as $A \to UAU^\dagger$.

2.5 Generalized Uncertainty Relation for Self-Adjoint Operators

Let A and B be self-adjoint operators. Define the commutator $[A, B] = AB - BA$. Then since $[A, B]^\dagger = -[A, B]$ the operator $i[A, B]$ is self-adjoint. We can prove a theorem:
  If $\langle(\Delta A)^2\rangle = \langle\alpha |A^2| \alpha\rangle - \langle\alpha |A| \alpha\rangle^2$ and similarly for B, then

$$\langle(\Delta A)^2\rangle\langle(\Delta B)^2\rangle \geq \frac{1}{4} |\langle[A, B]\rangle|^2. \tag{2.46}$$

  It suffices to consider $\langle\alpha |A| \alpha\rangle = 0$ and $\langle\alpha |B| \alpha\rangle = 0$, because we can redefine A and B by subtracting $A \to A - \langle 0|A|0\rangle$, $B \to B - \langle 0|B|0\rangle$. Consider $(A + i\lambda B) |\alpha\rangle$ with λ real. Now

$(A + i\lambda B) \ket{\alpha}$ is a vector, and so taking the scalar product with itself must give a result greater than or equal to zero. Since A and B are self-adjoint, we have

$$\bra{\alpha} (A - i\lambda B)(A + i\lambda B) \ket{\alpha} = \bra{\alpha} A^2 + i\lambda[A, B] + \lambda^2 B^2 \ket{\alpha} \geq 0. \tag{2.47}$$

Because $i[A, B]$ is self-adjoint $\bra{\alpha} i[A, B] \ket{\alpha}$ is real, so this is a real quadratic equation in λ. Any real quadratic $ax^2 + bx + c$ has its minimum at $x = -b/(2a)$ and the value there is $c - b^2/(4a)$. Here we have

$$\bra{\alpha} A^2 \ket{\alpha} - \frac{|\bra{\alpha} i[A, B] \ket{\alpha}|^2}{4 \bra{\alpha} B^2 \ket{\alpha}} \geq 0$$

$$\bra{\alpha} A^2 \ket{\alpha} \bra{\alpha} B^2 \ket{\alpha} \geq \frac{1}{4} |\bra{\alpha} [A, B] \ket{\alpha}|^2. \tag{2.48}$$

Of course, this is most famous in the context in which $A = x$ and $B = p = p_x$. Then

$$\braket{(\Delta x)^2} \braket{(\Delta p)^2} \geq \frac{1}{4} \hbar^2. \tag{2.49}$$

Had this celebrated Heisenberg uncertainty principle instead been called the "principle of non-commuting operators," we would see it less frequently invoked to describe matters having nothing to do with quantum mechanics.

2.6   Infinite-Dimensional Vector Spaces - Hilbert Space

If our basis is thought of as being formed by the eigenkets of an operator A with discrete eigenvalues α_d' and a continuum of eigenvalues α_c', then the expansion of an arbitrary vector will involve the finite sum being replaced in general by

$$\sum_1^n \rightarrow \sum_{\alpha_d'} + \int d\alpha_c'. \tag{2.50}$$

The normalization conditions are

$$\braket{\alpha_d'' | \alpha_d'} = \delta_{\alpha_d'', \alpha_d'}; \qquad \braket{\alpha_c'' | \alpha_c'} = \delta(\alpha_c'' - \alpha_c'). \tag{2.51}$$

The completeness relation is

$$\sum_{\alpha_d'} \ket{\alpha_d'}\bra{\alpha_d'} + \int d\alpha_c' \ket{\alpha_c'}\bra{\alpha_c'}, \tag{2.52}$$

so the expansion of an arbitrary vector is

$$\ket{\xi} = \sum_{\alpha_d'} \ket{\alpha_d'}\braket{\alpha_d' | \xi} + \int d\alpha_c' \ket{\alpha_c'}\braket{\alpha_c' | \xi}. \tag{2.53}$$

Note that if the eigenkets forming the basis are simultaneously eigenkets for more than one (commuting) operator, then the index α' stands for more than one eigenvalue. For example, plane waves in three dimensions have $\alpha_c' = (k_x', k_y', k_z')$. The delta function is then the product of three delta functions. The index α may stand for some continuous and some discrete eigenvalues, e.g. plane waves with spherical wave expansion; $j_\ell(kr)Y_{\ell m}(\theta\phi)$ has $\alpha' = k$ (continuous), ℓ, m (discrete). In some situations, e.g. atoms, nuclei, etc. there are discrete and continuous eigenvalues - bound state and continuum states. Sometimes it is convenient to use the discrete approach for everything by putting the system in a large box with some sort of boundary conditions. Then, at some appropriate point, we convert to a continuum with a density of states.

The mathematics of Hilbert space required to deal with continuous spectra is more complicated than that required when the eigenvalues are discrete. The standard reference is von Neumann, *Mathematical Foundations of Quantum Mechanics*. However, as is typical, physicists generally evade the mathematical subtleties. We can argue that rather than considering all of three-dimensional space, we can just use a very large finite volume. Then the eigenvalues are necessarily discrete. This sort of pragmatism has served well. Had Dirac stopped to create the theory of distributions instead of proceeding with his delta function, the development of quantum mechanics could have been significantly delayed.

2.7 Assumptions of Quantum Mechanics

While physics is not a deductive enterprise, it is still worthwhile to state assumptions clearly for the purpose of checking whether they do lead to correct predictions.

1. A dynamical system with certain degrees of freedom (some with classical analogs, others without, e.g. isospin) is represented by a linear vector space (Hilbert space).
2. A state (actually, a pure state) of the dynamical system is in correspondence with a "ray" (a vector or set of vectors differing only by a multiplicative complex factor – if normalized, then just by a phase) in the Hilbert space. Such vectors are called state vectors.
3. Linear superposition holds, i.e. if $|\alpha\rangle$ and $|\beta\rangle$ each represent a possible state of the system, so does $|\alpha\rangle + |\beta\rangle$.
4. Physical observables or dynamical variables are represented by self-adjoint operators with associated (real) eigenvalues and eigenvectors that form a complete set of vectors. If a is the observable, its operator A has eigenkets $|a'\rangle$ such that $A |a'\rangle = a' |a'\rangle$, where $a' = a_1, a_2, ..$ are real eigenvalues. We assume orthonormality and completeness.

$$\langle a''|a'\rangle = \delta_{a'',a'}; \qquad or \qquad \delta(a'' - a')$$

$$\sum_{a'_d} |a'_d\rangle\langle a'_d| + \int da'_c\, |a'_c\rangle\langle a'_c| = I. \tag{2.54}$$

5. Measurements of the dynamical system involve the interaction of the system with a measuring apparatus. The result of the measuring process is some change in the apparatus that conveys information in the form of a record of the state of the system under study.
   (a) If an apparatus is measuring the dynamical variable described by operator A, the result must be one of the eigenvalues of A, namely a'.
   (b) In general, the measurement will disturb the system and the state vector $|\beta_i\rangle$ may be transformed into a different vector $|\beta_f\rangle$ by the measurement.
6. The average value of a large number of measurements of the observable A on an ensemble of systems that is described by the vector $|\beta\rangle$ (each sample of the system prepared identically) is $\langle A\rangle = \langle \beta |A| \beta\rangle / \langle \beta|\beta\rangle$.
   (a) If $|\beta\rangle = |a'\rangle$ is an eigenstate $A |a'\rangle = a' |a'\rangle$, then $\langle A\rangle = a'$. Note that $A^2 |a'\rangle = a'^2 |a'\rangle$ so that $\langle (\Delta A)^2\rangle = 0$.

(b) If $|\beta\rangle \neq |a'\rangle$, then we can expand it in the basis $|a'\rangle$:

$$|\beta\rangle = \sum_{a'} |a'\rangle \langle a'|\beta\rangle , \qquad (2.55)$$

and

$$\langle A\rangle = \frac{\langle \beta |A| \beta\rangle}{\langle \beta|\beta\rangle} = \frac{\sum_{a',a''} \langle \beta|a''\rangle \langle a''|A|a'\rangle \langle a'|\beta\rangle}{\sum_{a'} \langle \beta|a'\rangle \langle a'|\beta\rangle} = \frac{\sum_{a'} a'|\langle a'|\beta\rangle|^2}{\sum_{a'} |\langle a'|\beta\rangle|^2}. \qquad (2.56)$$

The quantities $\langle a'|\beta\rangle$ are called probability amplitudes. We have the probability of obtaining the result a' from a measurement of A on $|\beta\rangle$ as $p_{a'} = |\langle a'|\beta\rangle|^2$. Then

$$\langle A\rangle = \frac{\sum_{a'} a' p_{a'}}{\sum_{a'} p_{a'}}, \qquad (2.57)$$

as expected for an average.

7. "Collapse of the wave function" hypothesis. If a measurement is made on the dynamical system in the state $|\beta\rangle$ to measure the observable A with the result a', the act of measurement projects the state into the state $|a'\rangle$ with probability $p_{a'} = |\langle a'|\beta\rangle|^2$. The state of the system immediately after the measurement is given by the vector, $|\text{after}\rangle = |a'\rangle \langle a'|\beta\rangle$ up to an arbitrary phase.

8. Compatible and Incompatible Observables: A set of observables a_i are said to be compatible if there is a basis of states that are simultaneously eigenstates of all the operators corresponding to the observables. To be compatible, it is necessary that the set of operators commute, i.e.

$$[A_i, A_j] = 0 \qquad (2.58)$$

for all A_i and A_j in the set. If two operators, A and B do not commute, they represent incompatible observables. Note that for compatible observables A and B, a measurement of A yielding a' can be followed by a measurement of B yielding b'. Then a subsequent measurement of A will again yield a'. The measurement of B in between did not disturb A. In contrast, if $[A, B] \neq 0$, the B measurement will project the state onto some eigenstate of B, not an eigenstate of A. A second measurement of A will not necessarily yield a'.

2.8 Mixtures and the Density Matrix

In our discussion of measurements and average we have been considering always the same state of the system $|\beta\rangle$. The repetition (in our mind's eye) of the many measurements needed to define an average leads to the idea of an ensemble of dynamical systems. Our ensemble, defined by $|\beta\rangle$, is called a "pure ensemble" and the state $|\beta\rangle$ is called a pure state.

There are other kinds of ensembles, often closer to experimental reality. We need some mathematical apparatus to handle them. The tool is the density matrix, ρ, first introduced by von Neumann in 1927.

Before getting to ρ we need the concept of the trace of an operator. Given an orthonormal basis $|\xi_i\rangle$, we define the trace of the operator $\mathcal{O}$ as

$$\text{Tr}\,\mathcal{O} = \sum_i \langle \xi_i|\mathcal{O}|\xi_i\rangle = \sum_i \mathcal{O}_{ii}. \tag{2.59}$$

It is easy to see then that

$$\text{Tr}\,(AB) = \sum_i \sum_j A_{ij}B_{ji} = \sum_i \sum_j B_{ji}A_{ij} = \text{Tr}\,(BA). \tag{2.60}$$

We know that in a new basis related to the old one by a unitary transformation, $\mathcal{O}$ becomes $\mathcal{O}' = U\mathcal{O}U^\dagger$. But

$$\text{Tr}\,\mathcal{O}' = \text{Tr}\,U\mathcal{O}U^\dagger = \text{Tr}\,U^\dagger U\mathcal{O} = \text{Tr}\,\mathcal{O}. \tag{2.61}$$

Thus, the trace is independent of the choice of basis.

Now consider the average value of an operator B in the pure state $|a'\rangle$ and assume the state is normalized: $\langle a'|a'\rangle = 1$. Let the eigenstates of B be $|b'\rangle$. Then

$$\langle a'|B|a'\rangle = \sum_{b',b''} \langle a'|b''\rangle\langle b''|B|b'\rangle\langle b'|a'\rangle$$

$$= \sum_{b'} \langle a'|b'\rangle b'\langle b'|a'\rangle = \sum_{b'} b'\,|b'\rangle\langle b'|\,|a'\rangle\langle a'|$$

$$= \text{Tr}\,B\,|a'\rangle\langle a'|\,. \tag{2.62}$$

We thus get the expectation value or average value of B in the state $|a'\rangle$ by taking the trace of the product of B with the projection or measurement operator $|a'\rangle\langle a'|$.

Now suppose the kets $|a'\rangle$ are the eigenkets of A and that some filtration apparatuses (e.g. a Stern-Gerlach device) are selecting from several sources dynamical systems with different eigenvalues a' with probability $p_{a'}$ and that these selected systems are combined together to give the ensemble under consideration. We have

$$\sum_{a'} p_{a'} = 1. \tag{2.63}$$

The expectation value of B for this mixed ensemble is

$$\langle B\rangle = \sum_{a'} p_{a'}\langle a'|B|a'\rangle. \tag{2.64}$$

If we define the density matrix operator by

$$\rho = \sum_{a'} |a'\rangle p_{a'}\langle a'|, \tag{2.65}$$

we can write

$$\langle B\rangle = \text{Tr}\,(B\rho). \tag{2.66}$$

The density matrix ρ we have so far described has a number of properties:

1. $\rho^\dagger = \rho$.
2. ρ is diagonal in the $|a'\rangle$ basis by construction, but this will not necessarily be so in another basis.
3. The eigenvalues of ρ are between zero and unity and sum to 1, so $\text{Tr}\,\rho = 1$.

4. If all the eigenstates are equally probable, then $p_{a'} = 1/N$, where N is the number of eigenstates. In this case $\rho = (1/N)I$ and this is true in any orthonormal basis. This describes, for example, an unpolarized beam.

5.

$$\rho^2 = \sum_{a'} \sum_{a''} |a''\rangle \, p_{a''} \, \langle a''|a'\rangle \, p_{a'} \, \langle a'|$$

$$= \sum_{a'} |a'\rangle \, p_{a'}^2 \, \langle a'| \,. \tag{2.67}$$

(a) If one $p_{a'} = 1$, then all others vanish and $\rho^2 = \rho$.

(b) If there is no basis $|b'\rangle$ in which for some b', $p_{b'} = 1$, then $\rho^2 \neq \rho$ and $\operatorname{Tr} \rho^2 < 1$.

Thus we can say that only for a pure state is $\operatorname{Tr} \rho^2 = 1$. If $\operatorname{Tr} \rho^2 < 1$ we are dealing with a mixed state. This concept is independent of the basis.

2.9 Measurement

While there is no disagreement about the proper way to calculate the probabilities of various outcomes in quantum mechanics, there is an enduring discussion of how to understand the actual process of measurement. Here we only sketch the picture due to Kurt Gottfried. The interested reader is referred to his text and to a critique by John Bell, listed at the end of the chapter.

Consider a system under study with an observable A to be measured and let its eigenstates be indicated by $|a_i'\rangle$. The measuring apparatus itself has a line of cells at positions $z_{i=1,2,\dots}$, arranged so that an atom in the eigenstate denoted by a_i' will end up in the sensor at z_i. The archetype of such a setup is the Stern-Gerlach experiment. There an atom with a magnetic moment is passed through a region with a strong magnetic field gradient.

To describe this quantum mechanically, we need to represent both the atoms in the beam and the apparatus in terms of states. With each sensor, we associate a state $|n\rangle_i$, where n represents some excitation level. Initially for all the sensors we have $|0\rangle_i$. The atom has some eigenstates $|a_i'\rangle$, which for a Stern-Gerlach experiment would label the magnetic substate along the direction of the field gradient. The atom's wave function depends on time and position, as well at its internal state, which is some superposition of the states $|a_i\rangle$. For any single atom, the internal state can be written

$$\sum_i c_i \, |a_i'\rangle \,, \tag{2.68}$$

where the c_i are complex numbers normalized to $\sum_i |c_i|^2 = 1$. If, for example, the atom could be represented for this purpose as an angular momentum 1/2 state and the quantization axis z were along the gradient of the magnetic field, we would have, say, c_+ and c_-, and the expectation of J_z would be $\frac{1}{2}(|c_+|^2 - |c_-|^2)$, a continuous quantity.

The state of the atom would be a function of its location $\mathbf{R}$ and time t, but would have separate components for the magnetic substates.

$$|\beta(t)\rangle = c_+ \, |\mathbf{R}_+, t\rangle_{m=+1/2} + c_- \, |\mathbf{R}_-, t\rangle_{m=-1/2}. \tag{2.69}$$

Because of the magnetic field gradient, these two components would evolve differently: the amplitude as a function of $\mathbf{R}$ and t would differ for the two components. Before the atom reached the domain of the magnetic field though, the two components would have the same behavior as a function of $\mathbf{R}$ and t. However, after passing through the region of the inhomogeneous field, the wave functions associated with the two components would diverge. Nonetheless, $|\beta(t)\rangle$ would still be a pure state, the result of the time evolution of a single state. Its density matrix would be simply

$$\rho(t) = |\beta(t)\rangle\langle\beta(t)| = [c_+ \, |\mathbf{R}_+, t\rangle_{m=+1/2} + c_- \, |\mathbf{R}_-, t\rangle_{m=-1/2}]$$
$$\times [c_+^* \, \langle\mathbf{R}_+, t|_{m=+1/2} + c_-^* \, \langle\mathbf{R}_-, t|_{m=-1/2}].$$

We now need to specify the measuring apparatus in an appropriate fashion. In a very idealized way, we could imagine that each sensor at location i and with a volume ΔV_i would stop an atom "hitting it" and at the same time change the quantum state of the sensor, increasing its level by one unit. The measurement operator associated with this might be represented as

$$\mathcal{O}_i = \int_{\Delta V_i} d\mathbf{R}'' \, |\mathbf{R}''\rangle\langle\mathbf{R}''|. \tag{2.70}$$

The entire measurement apparatus might be represented as

$$\mathcal{O} = \prod_I \mathcal{O}_i. \tag{2.71}$$

The probability of a detection by sensor i is given by

$$\mathrm{Tr}\,\rho(t)\mathcal{O}_i. \tag{2.72}$$

Now because the experiment is arranged so that there is no overlap in the final positions of atoms with different magnetic quantum numbers, we can replace the full density matrix at the location of the sensors by the simpler one:

$$\hat{\rho}(t) = |c_+|^2 \, |\mathbf{R}_+, t\rangle_{m=+1/2} \, \langle\mathbf{R}_+, t|_{m=+1/2} + |c_-|^2 \, |\mathbf{R}_-, t\rangle_{m=-1/2} \, \langle\mathbf{R}_-, t|_{m=-1/2}. \tag{2.73}$$

The phases and interference have disappeared and the probabilities for the various outcomes conform to our expectations.

2.10 Classical vs. Quantum Probabilities

Consider a sequence of measurements of observables a, b, c. Classically, the conditional probability $P_{c'a'}$ that having found a' when a was measured, we will find c' when we measure c is given, if measurement b intervenes, by

$$P_{c'a'}^{classical} = \sum_{b'} P_{c'b'} P_{b'a'} \tag{2.74}$$

corresponding to the independence of the measurements and the observables.

Quantum mechanically,

$$P_{c'a'} = |\langle c'|a'\rangle|^2. \tag{2.75}$$

Let us set

$$\rho = |a'\rangle\langle a'| = \sum_{b'b''} |b''\rangle\langle b''|a'\rangle\langle a'|b'\rangle\langle b'|$$

$$= \sum_{b'b''} |b''\rangle |\langle b''|a'\rangle||\langle a'|b'\rangle| \langle b'| \times e^{i(\theta'-\theta'')}. \tag{2.76}$$

Then

$$P_{c'a'} = \text{Tr}\,(|c'\rangle\langle c'|\,\rho). \tag{2.77}$$

Now suppose we make an intermediate measurement of b with the result b'. The density matrix is now

$$\rho' = |b'\rangle\langle b'|\,P_{b'a'} \tag{2.78}$$

with

$$P_{b'a'} = \text{Tr}\,\rho\,|b'\rangle\langle b'|\,. \tag{2.79}$$

However, if we make a subsequent measurement of c, the probability that we find c' is

$$\text{Tr}\,|c'\rangle\langle c'|\,|b'\rangle\langle b'|\,P_{b'a'} = P_{c'b'}P_{b'a'}. \tag{2.80}$$

Thus,

$$\sum_{b'} P_{c'b'a'} = \sum_{b'} P_{c'b'}P_{b'a'} = P_{c'a'}^{classical}. \tag{2.81}$$

So if we measure b and sum over its possible values, we recover the classical result. However, in general,

$$P_{c'a'}^{qm} \neq P_{c'a'}^{classical}. \tag{2.82}$$

Without the measurement of b, we get the double sum over b' and b''. Only if there is a measurement do we eliminate the interference and get the classical composition law.

This is familiar from electromagnetism and optics because superposition applies to the electric and magnetic fields just as it does to wave functions. Let a, b, c refer to polarizers of light with a and c selecting orthogonal polarizations and b selecting a polarization between the two and placed between a and c. Then light that passes through a never passes through c with filter b removed, but light that passes from a to b will pass, at some level, through c. Interference is common to all wave phenomena.

2.11 Capsule Review of Classical Mechanics and Conservation Laws

With the methods of linear vector spaces and operators on them in hand, we now need to provide the dynamics in this framework. The only guidance we have is classical physics, so we review here the basics with an emphasis on conserved quantities.

Lagrangian mechanics for a system with n degrees of freedom uses as its dynamical variables positions q_i and velocities, $\dot{q}_i$, $i = 1, \ldots n$. The Lagrangian function, which is usually

$T - V$, the kinetic energy minus the potential energy, is written $L(q_i, \dot{q}_i, t)$. The principle of least action requires that, given the initial values and final values of the variables at t_1 and t_2 respectively, the integral of the Lagrangian is stationary under possible variations δq_i and $\delta \dot{q}_i$. An integration by parts shows that this yields

$$\int_{t_1}^{t_2} dt \left[\frac{\partial L}{\partial q_i} \delta q_i + \frac{\partial L}{\partial \dot{q}_i} \delta \dot{q}_i \right] = \int_{t_1}^{t_2} dt \left[\frac{\partial L}{\partial q_i} - \frac{d}{dt} \left(\frac{\partial L}{\partial \dot{q}_i} \right) \right] \delta q_i = 0, \tag{2.83}$$

where the endpoint contributions from the integration by parts vanish because the variables are fixed at t_1 and t_2. Since δq_i is otherwise arbitrary, we have the Euler-Lagrange equation

$$\frac{d}{dt} \left(\frac{\partial L}{\partial \dot{q}_i} \right) - \frac{\partial L}{\partial q_i} = 0. \tag{2.84}$$

It is useful to define the canonical momenta

$$p_i = \frac{\partial L}{\partial \dot{q}_i}. \tag{2.85}$$

Now consider some transformation of the coordinates

$$q_i \rightarrow q_i + \Delta q_i; \qquad \dot{q}_i \rightarrow \dot{q}_i + \Delta \dot{q}_i. \tag{2.86}$$

This will induce a change in the Lagrangian

$$\Delta L = \sum_i \left\{ \Delta q_i \frac{\partial L}{\partial q_i} + \Delta \dot{q}_i \frac{\partial L}{\partial \dot{q}_i} \right\} = \sum_i \{ \dot{p}_i \Delta q_i + p_i \Delta \dot{q}_i \}. \tag{2.87}$$

Now suppose that actually each q_i is, in fact, a three-vector position (so there are actually $3n$ degrees of freedom) and let the transformation be a translation of each position by the same fixed amount:

$$\Delta \mathbf{q}_i = \lambda \mathbf{a}; \qquad \Delta \dot{\mathbf{q}}_i = 0. \tag{2.88}$$

The change in the Lagrangian is then

$$\Delta L = \lambda \mathbf{a} \cdot \sum_i \dot{\mathbf{p}}_i \equiv \lambda \mathbf{a} \cdot \frac{d}{dt} \mathbf{P}_{\text{tot}}. \tag{2.89}$$

If the Lagrangian is invariant under translations, as is true, for example, if the dynamical system is isolated and involves forces depending only upon the relative co-ordinates and velocities, then $\Delta L = 0$ and $\mathbf{P}_{\text{tot}}$ is constant. Conservation of momentum follows from translational invariance. Next consider a change of all coordinates specified by rotation though a small angle $\delta \omega$ about an axis $\hat{\omega}$.

$$\mathbf{q}_i \rightarrow \mathbf{q}_i + \delta \omega \times \mathbf{q}_i$$
$$\dot{\mathbf{q}}_i \rightarrow \dot{\mathbf{q}}_i + \delta \omega \times \dot{\mathbf{q}}_i. \tag{2.90}$$

Then the change in the Lagrangian is

$$\Delta L = \sum_i \{ (\delta \omega \times \mathbf{q}_i) \cdot \dot{\mathbf{p}}_i + (\delta \omega \times \dot{\mathbf{q}}_i) \cdot \mathbf{p}_i$$
$$= \delta \omega \cdot \frac{d}{dt} \left(\sum_i \mathbf{q}_i \times \mathbf{p}_i \right) \equiv \delta \omega \cdot \frac{d}{dt} \mathbf{L}_{\text{tot}}. \tag{2.91}$$

Again, if the Lagrangian is invariant under rotations, then the total angular momentum is conserved.

The Hamiltonian formulation is traditionally chosen for quantum mechanics. The Hamilton classical equations are thus of importance. The essence is a change of variables from $(q_i, \dot{q}_i)$ to (q_i, p_i). This is accomplished with a Legendre transformation

$$H(q_i, p_i, t) = \sum_i \dot{q}_i p_i - L(q_i, \dot{q}_i, t), \tag{2.92}$$

where the right-hand side must be eventually expressed in terms of q_i and p_i.

To find the Hamilton equations of motion, consider the variation

$$\Delta H = \sum_i \left(\frac{\partial H}{\partial q_i} \Delta q_i + \frac{\partial H}{\partial p_i} \Delta p_i \right) + \frac{\partial H}{\partial t} \Delta t. \tag{2.93}$$

Substituting the Legendre transformation definition of H,

$$\Delta H = \sum_i (\dot{q}_i \Delta p_i + p_i \Delta \dot{q}_i) - \sum_i \left(\frac{\partial L}{\partial q_i} \Delta q_i + \frac{\partial L}{\partial \dot{q}_i} \Delta \dot{q}_i \right) - \frac{\partial L}{\partial t} \Delta t$$

$$= \sum_i (\dot{q}_i \Delta p_i + p_i \Delta \dot{q}_i) - \sum_i \left(\dot{p}_i \Delta q_i + p_i \Delta \dot{q}_i \right) - \frac{\partial L}{\partial t} \Delta t$$

$$= \sum_i \left(\dot{q}_i \Delta p_i - \dot{p}_i \Delta q_i \right) - \frac{\partial L}{\partial t} \Delta t. \tag{2.94}$$

Comparing Eqs. (2.93) and (2.94) gives the Hamilton equations

$$\dot{q}_i = \frac{\partial H}{\partial p_i}; \quad \dot{p}_i = -\frac{\partial H}{\partial q_i}; \quad \frac{\partial H}{\partial t} = -\frac{\partial L}{\partial t}. \tag{2.95}$$

The total time derivative of the Hamiltonian is obtained as

$$\frac{dH}{dt} = \sum_i \left[\frac{\partial H}{\partial p_i} \dot{p}_i + \frac{\partial H}{\partial q_i} \dot{q}_i \right] + \frac{\partial H}{\partial t} = \sum_i \left[\dot{q}_i \dot{p}_i - \dot{p}_i \dot{q}_i \right] + \frac{\partial H}{\partial t}$$

$$= \frac{\partial H}{\partial t}. \tag{2.96}$$

Thus, unless the Hamiltonian has explicit time dependence, the Hamiltonian is conserved. This is simply energy conservation.

A key step in the development of quantum mechanics was the recognition that the Poisson bracket of classical mechanics

$$[u, v]_{\text{PB}} \equiv \sum_i \left(\frac{\partial u}{\partial q_i} \frac{\partial v}{\partial p_i} - \frac{\partial u}{\partial p_i} \frac{\partial v}{\partial q_i} \right) \tag{2.97}$$

provides the key to inferring quantum rules from their classical analogs.

Suppose F is a function of q_i, p_i and t. Then

$$
\begin{aligned}
\frac{dF(q_i, p_i, t)}{dt} &= \sum_i \left(\frac{\partial F}{\partial q_i} \dot{q}_i + \frac{\partial F}{\partial p_i} \dot{p}_i \right) + \frac{\partial F}{\partial t} \\
&= \sum_i \left(\frac{\partial F}{\partial q_i} \frac{\partial H}{\partial p_i} - \frac{\partial F}{\partial p_i} \frac{\partial H}{\partial q_i} \right) + \frac{\partial F}{\partial t} \\
&= [F, H]_{\mathrm{PB}} + \frac{\partial F}{\partial t}.
\end{aligned}
\tag{2.98}
$$

This is just a hint of the role of Poisson brackets in classical mechanics. They are invariant under canonical transformation in which (q_i, p_i) become (Q_i, P_i). See Goldstein, *Classical Mechanics*, 2nd edition, pp. 297 ff.; Dirac, *Quantum Mechanics*, 4th edition, pp. 84 ff. Some examples of classical Poisson brackets are

$$
[q_i, p_j]_{\mathrm{PB}} = \delta_{ij}
\tag{2.99}
$$

$$
[L_i, x_j]_{\mathrm{PB}} = \epsilon_{ijk} x_k; \qquad L_i = \epsilon_{ijk} x_j p_k
\tag{2.100}
$$

$$
[L_i, L_j]_{\mathrm{PB}} = \epsilon_{ijk} L_k.
\tag{2.101}
$$

These parallels with quantum mechanical commutation relations are the basis of Dirac's general quantization rules:

$$
[A, B]_{\mathrm{PB}} \rightarrow \frac{1}{i\hbar}[A, B],
\tag{2.102}
$$

where the latter [,] is a commutator of operators. Thus the most fundamental of all quantum mechanical equations emerges:

$$
[q_i, p_j] = i\hbar \delta_{ij}.
\tag{2.103}
$$

2.12 Translation Invariance and Momentum Conservation

We choose to make the connection of translations and momentum that is manifest in Lagrangian mechanics with its quantum mechanical counterpart. Consider a quantum-mechanical system that is isolated, so there are no external forces. The dynamics of the system are invariant under translations and the classical discussion above leads us to expect momentum conservation as a consequence.

Suppose we have a state $|\alpha\rangle$ and we translate its position with the resultant state being $|\alpha_T\rangle$. This should not change the normalization of the state, so we expect this transformation to be unitary:

$$
|\alpha_T\rangle = U |\alpha\rangle.
\tag{2.104}
$$

We can write $U = e^{-iK}$, where K is a self-adjoint operator. This form is desirable because there is a class of state vectors that are equivalent eigenkets of U

$$
U |\text{special class}\rangle = e^{i\phi} |\text{special class}\rangle,
\tag{2.105}
$$

where ϕ is a real phase angle (eigenvalues of U are numbers of modulus unity).

Let's be more specific. Consider a translation in the x direction by an amount a. Since we can build up amount a in a set of small steps, each of a/n, $U(a) = [U(a/n)]^n$, the phase must be proportional to a. Specifically, we write

$$U(a) = e^{-iKa}. \tag{2.106}$$

Then if $|k'\rangle$ is eigenket of K with eigenvalue k'

$$U(a) |k'\rangle = e^{-ik'a} |k'\rangle. \tag{2.107}$$

The minus sign in the exponent of U corresponds to the picture where the state is actually moved, rather than moving the axes. It turns out that for large systems (or "large" eigenvalue k') that $\hbar k'$ is the x-component of the total momentum. The ket $|k'\rangle$ describes uniform motion of the system as a whole with a conserved momentum (in the x direction) of $\hbar k'$.

This is the quantum analog of the classical invariance of the Lagrangian under translations leading to conservation of momentum. It is natural to identify the self-adjoint operator $\hbar K = p_x$ with the observable, the x component of $\sum_i \mathbf{p}_i = \mathbf{p}_{total}$. We speak of the momentum operator as the generator of translations.

Consider the coordinate operator for the m-th particle in the system $\mathbf{r}_m$, with x-component x_m. If the system is translated unchanged by an amount a along the x-axis, we expect

$$\langle \beta_T |x_m| \alpha_T \rangle = \langle \beta |x_m| \alpha \rangle + a. \tag{2.108}$$

Thus,

$$U^\dagger x_m U = x_m + a. \tag{2.109}$$

Make a infinitesimal and write $U = e^{-iKa} \approx 1 - iK_x a$.

Then we have

$$(1 + iK_x a)x_m(1 - iK_x a) = x_m + a$$
$$x_m + ia[K_x, x_m] = x_m + a \tag{2.110}$$

or

$$[x_m, K_x] = i. \tag{2.111}$$

For the y and z coordinates we have $U^\dagger y_m U = y_m$ so $[y_m, K_x] = 0$ and similarly for z.

We now imagine that we have a K_x separately for each particle, which we indicate by k_{xm} so $K_x = \sum_m k_{m_x}$. If we make the reasonable assumption that the coordinate and momentum observables associated with different particles commute, we have the basic commutation relations for $\mathbf{x}_m$ and $\mathbf{p}_m = \hbar \mathbf{k}_m$

$$[(\mathbf{x}_m)_i, (\mathbf{p}_n)_j] = i\hbar \delta_{mn} \delta_{ij}. \tag{2.112}$$

2.13 Dirac's p's and q's

While we have used the bra-ket notation so far to indicate normalizable states, Dirac introduced a powerful notation that utilizes non-normalizable states. We define a position operator q and indicate an eigenstate with eigenvalue q' by $|q'\rangle$. Thus,

$$q\,|q'\rangle = q'\,|q'\rangle. \tag{2.113}$$

These states are normalized, so

$$\langle q''|q'\rangle = \delta(q'' - q'). \tag{2.114}$$

We assume, quite reasonably, that the $|q'\rangle$ form a complete basis and any state $|\xi\rangle$ can be written

$$|\xi\rangle = \int dq'\,|q'\rangle\langle q'|\xi\rangle. \tag{2.115}$$

This is really nothing other than a means of describing the wave function associated with the state $|\xi\rangle$:

$$\langle q'|\xi\rangle = \psi_\xi(q'). \tag{2.116}$$

Proceeding further, we consider an inner product

$$\langle \eta|\xi\rangle = \int dq' \int dq''\,\langle \eta|q''\rangle\langle q''|q'\rangle\langle q'|\xi\rangle$$

$$= \int dq'\,\psi_\eta^*(q')\psi_\xi(q'). \tag{2.117}$$

Similarly, we can represent operators

$$\langle \eta\,|A|\,\xi\rangle = \int dq' \int dq''\,\langle \eta|q''\rangle\langle q''\,|A|\,q'\rangle\langle q'|\xi\rangle$$

$$= \int dq' \int dq''\psi_\eta^*(q'')a(q'',q')\psi_\xi(q'). \tag{2.118}$$

The function $a(q'',q')$ is a q-basis representation of A. It can be thought of as a matrix, albeit with continuous rows and columns. As explicit examples, consider

$$\langle q''\,|q|\,q'\rangle = q'\delta(q' - q'')$$
$$\langle q''\,|F(q)|\,q'\rangle = F(q')\delta(q' - q''), \tag{2.119}$$

so

$$\langle \eta\,|F(q)|\,\xi\rangle = \int dq'\psi_\eta^*(q')F(q')\psi_\xi(q'). \tag{2.120}$$

Similarly, we can introduce state $|p'\rangle$ such that

$$p\,|p'\rangle = p'\,|p'\rangle$$
$$\langle p''|p'\rangle = \delta(p' - p'')$$
$$I = \int dp'\,|p'\rangle\langle p'|. \tag{2.121}$$

The question arises: what is $\langle q'|p'\rangle$? This must be determined by the fundamental connection $[q,p] = i\hbar$. See Problem 2.1. A heuristic derivation follows from the recollection that

$e^{i\mathbf{k}\cdot\mathbf{x}}$ is a plane wave representing momentum $\mathbf{p} = \hbar\mathbf{k}$. But $\langle q'|p'\rangle$ is the wave function at q' for momentum p', so we expect

$$\langle q'|p'\rangle \propto e^{iq'p'/\hbar}. \tag{2.122}$$

Since

$$\langle p'|p''\rangle = \delta(p' - p'') = \int dq'\, \langle p'|q'\rangle\langle q'|p''\rangle \tag{2.123}$$

and

$$\int dx e^{-ik'x}e^{ik''x} = 2\pi\delta(k' - k''), \tag{2.124}$$

we conclude that the proper normalization is

$$\langle q'|p'\rangle = \frac{1}{\sqrt{2\pi\hbar}}e^{iq'p'/\hbar}. \tag{2.125}$$

Consider the eigenvalue problem

$$A\,|a'\rangle = a'\,|a'\rangle \tag{2.126}$$

for the operator A. Then

$$\langle q'\,|A|\,a'\rangle = \int dq''\, \langle q'\,|A|\,q''\rangle\langle q''|a'\rangle = a'\,\langle q'|a'\rangle. \tag{2.127}$$

Defining $\langle q'|a'\rangle = \psi_{a'}(q')$ as the q-basis wave function and $\langle q'\,|A|\,q'''\rangle = a(q', q'')$ as the representative of A in the q-basis, we have

$$a'\psi_{a'}(q') = \int dq''\, a(q', q'')\psi_{a'}(q''). \tag{2.128}$$

This is analogous to an operator A in a finite-dimensional space, where we diagonalize A by a unitary transformation:

$$U^{\dagger}AU = A', \tag{2.129}$$

where A' is diagonal. Then $AU = UA'$ has the explicit realization in the set of algebraic equations

$$\sum_{\xi'} A_{\xi\xi'}U_{\xi'\eta} = \lambda_{\eta}U_{\xi\eta}, \tag{2.130}$$

where $A'_{\xi\xi'} = \lambda_{\xi}\delta_{\xi\xi'}$. The column vector $U_{\xi'\eta}$, for each fixed η value, is the eigenvector of A' associated with the eigenvalue λ_{η}.

We see that Eq. (2.127) has the same form, except that one index (q'') is continuous and the other may be discrete or continuous. Thus

$$
a' \longrightarrow
$$

$$
U = \begin{array}{c} q' \\ \downarrow \end{array} \left(\langle q'|a' \rangle = \psi_{a'}(q') \right). \tag{2.131}
$$

Solving an eigenvalue problem is the same as finding the unitary transformation that diagonalizes the operator. The wave functions $\psi_{a'}(q')$ for all a' and q' are the coordinate representations of the unitary operator U.

2.14 Time Development of the State Vector

The time evolution of a dynamical system is a central concern in quantum mechanics, as in classical mechanics. Suppose the system at $t = t_0$ is in the state described by $|t_0\rangle$. We wish to know the state $|t\rangle$ for $t > t_0$.

(i) Assume the existence of an operator that maps all the states at time t_0 into the states at time $t \geq t_0$:

$$
|t\rangle = U(t, t_0) \, |t_0\rangle. \tag{2.132}
$$

(ii) What properties should U have?
  (a) Linear operator (linear superposition)
  (b) Unitary operator (conservation of probability)
  (c) $U(t, t_0) = U(t - t_0)$ for an isolated system
  (d) Group property: $U(t - t_0) = U(t - t')U(t' - t_0)$
  (e) $U(0) = I$

(iii) Infinitesimal transformation: $U(\delta t) = I + \delta U$
  Consider now

$$
U^\dagger U = (I + \delta U^\dagger)(I + \delta U) = I + \delta U^\dagger + \delta U + \mathcal{O}(\delta U^2). \tag{2.133}
$$

We must have $\delta U^\dagger + \delta U = 0$, i.e. δU is anti-Hermitian. So

$$
\delta U = i \times \text{(Hermitian operator)}.
$$

For small time intervals, we must have $\delta U \propto \delta t$ so we write

$$
\delta U = -i \frac{H}{\hbar} \delta t, \tag{2.134}
$$

where H is a self-adjoint operator with dimensions of energy. We call H the generator of infinitesimal time displacements. Experience tells us that H is the Hamiltonian of the system:
(a) The Bohr frequency condition : $\omega_{12} = (E_2 - E_1)/\hbar$
(b) Schrödinger energy eigenstates have $\psi_{E_n} = \psi_{E_n}(\mathbf{x})e^{-iE_n t/\hbar}$

(iv) Differential equation for $U(t - t_0)$:

$$\frac{\partial U}{\partial t}\delta t = U(t + \delta t) - U(t) = (U(\delta t) - 1))U(t) = -i\frac{H}{\hbar}\delta t\, U. \qquad (2.135)$$

Thus U satisfies the differential equation

$$i\hbar\frac{\partial U}{\partial t} = HU. \qquad (2.136)$$

If H is independent of t, this can be integrated immediately to give

$$U(t, t_0) = e^{-iH(t-t_0)/\hbar}. \qquad (2.137)$$

(v) We can find the equation for the time dependence of the ket $|t\rangle$. Consider

$$i\hbar\frac{\partial}{\partial t}\,|t\rangle = i\hbar\frac{\partial U}{\partial t}U(t, t_0)\,|t_0\rangle$$

$$= HU(t, t_0)\,|t_0\rangle$$

$$= H\,|t\rangle. \qquad (2.138)$$

Thus we have a generalization of the Schrödinger equation

$$\left(i\hbar\frac{\partial}{\partial t} - H\right)|t\rangle = 0. \qquad (2.139)$$

(vi) If there are external fields or for some other reason H is time-dependent, we can still construct a solution for U in terms of H as follows: we write an integral equation equivalent to our differential equation, namely

$$U(t, t_0) = 1 - \frac{i}{\hbar}\int_{t_0}^{t} dt' H(t') U(t', t_0). \qquad (2.140)$$

By inspection, this is formally a solution to the differential equation. We can then substitute iteratively to find

$$U(t, t_0) = 1 - \frac{i}{\hbar}\int_{t_0}^{t} dt' H(t') + \left(\frac{-i}{\hbar}\right)^2 \int_{t_0}^{t} dt'' \int_{t_0}^{t'} dt'' H(t') H(t'') + \dots \qquad (2.141)$$

This is called the Dyson series expansion. If $[H(t''), H(t')] = 0$, this reduces to our previous result. This is often the case in non-relativistic quantum mechanics, but not in quantum electrodynamics and similar domains. One writes formally

$$U(t, t_0) = P\exp\left(-\frac{i}{\hbar}\int_{t_0}^{t} dt' H(t')\right), \qquad (2.142)$$

where P signifies "time-ordered product," i.e. the factors $H(t'), H(t''), \dots$ are ordered so that the earlier times always occur to the right of later times. The $1/n!$ that we normally see in an exponential is absent because of the possible $n!$ orders of n times, only one is present.

2.15 Schrödinger and Heisenberg Pictures

Alternative descriptions of quantum mechanics arise because the time dependence can be associated with the states or the operators. In the Schrödinger picture, the operators are fixed unless they have explicit time dependence, while the states evolve. In the Heisenberg picture the states are fixed and the time dependence is attributed to the operators.

(a) Let $a = a_1, a_2, ..., a_n$ be a complete set of compatible observables and let $|a'\rangle$ be the simultaneous eigenkets of the corresponding commuting operators. We assume that $|a'\rangle$ are independent of time. Now consider a general $|\xi, t\rangle$ and form the a-representative $\langle a'|\xi, t\rangle$. Then we have the Schrödinger equation for

$$i\hbar\frac{\partial}{\partial t}\langle a'|\xi, t\rangle = \langle a'|H|\xi, t\rangle = \sum_{a''}\langle a'|H|a''\rangle\langle a''|\xi, t\rangle. \tag{2.143}$$

Here we have the a-representative of H independent of time (if $\partial H/\partial t = 0$) and the "wave function" $\langle a'|\xi, t\rangle$ changing in time. This is the essence of the Schrödinger picture and is independent of the basis chosen for the representation.

If we suppose that $a = q_\alpha$, a set of generalized coordinates, with corresponding canonical momenta p_α, then

$$i\hbar\frac{\partial}{\partial t}\langle q'_\alpha|\xi, t\rangle = \prod_\alpha\int dq''_\alpha\langle q'_\alpha|H(q_\beta, p_\beta, t)|q''_\alpha\rangle\langle q''_\alpha|\xi, t\rangle. \tag{2.144}$$

With the canonical commutation relations giving the representatives

$$\langle q'_\alpha|q_\alpha|q''_\alpha\rangle = q'_\alpha\prod_\beta\delta(q'_\beta - q''_\beta)$$

$$\langle q'_\alpha|p_\alpha|q''_\alpha\rangle = \frac{\hbar}{i}\frac{\partial}{\partial q'_\alpha}\prod_\beta\delta(q'_\beta - q''_\beta) \tag{2.145}$$

the equation becomes

$$i\hbar\frac{\partial}{\partial t}\langle q'_\alpha|\xi, t\rangle = H(q'_\alpha, \frac{\hbar}{i}\frac{\partial}{\partial q'_\alpha}, t)\langle q'_\alpha|\xi, t\rangle. \tag{2.146}$$

This is the Schrödinger equation for n degrees of freedom, familiar from basic quantum mechanics.

The time dependence of matrix elements in the Schrödinger picture is given by $\langle \eta, t|A|\xi, t\rangle$, where A is a "time-independent" operator. Since $|\xi, t\rangle = U(t, t_0)|\xi, t_0\rangle$, we can write

$$\langle \eta, t|A|\xi, t\rangle = \langle \eta, t_0|U^\dagger AU|\xi, t_0\rangle. \tag{2.147}$$

Now the time-dependence has been transferred to the operators. This leads to the Heisenberg picture.

(b) Let the operator corresponding to the observable a be denoted by A_S in the Schrödinger picture and by A_H in the Heisenberg picture. Then we have

$$A_H = U^\dagger A_S U, \tag{2.148}$$

where $U(t, t_0)$ satisfies

$$i\hbar\frac{\partial U}{\partial t} = HU. \tag{2.149}$$

Now the state vectors are independent of time: $|\xi, t_0\rangle = |\xi\rangle$. We determine the equation of motion for A_H

$$\frac{dA_H}{dt} = \frac{\partial U^\dagger}{\partial t} A_S U + U^\dagger \frac{\partial A_S}{\partial t} U + U^\dagger A_S \frac{\partial U}{\partial t}$$

$$= -\frac{1}{i\hbar} U^\dagger H A_S U + U^\dagger A_S \frac{1}{i\hbar} H U + U^\dagger \frac{\partial A_S}{\partial t} U. \tag{2.150}$$

We define the Hamiltonian in the Heisenberg representation by $H_H = U^\dagger H U$ and set

$$\frac{\partial A_H}{\partial t} \equiv U^\dagger \frac{\partial A_S}{\partial t} U. \tag{2.151}$$

Then we have

$$\frac{dA_H}{dt} = \frac{1}{i\hbar}[A_H, H_H] + \frac{\partial A_H}{\partial t}. \tag{2.152}$$

We note the similarity to the classical Poisson bracket expression

$$\frac{dA}{dt} = [A, H]_{P.B.} + \frac{\partial A}{\partial t}. \tag{2.153}$$

In the event that the Hamiltonian has no explicit time dependence, then $U = e^{-iHt/\hbar}$ and $H_H = H_S$.

The Heisenberg equation of motion has a close correspondence to the classical equations of motion. Our methods of solution to the classical equations are often applicable directly to the equations of motion of the quantum mechanical operators. As an example, we consider an electron of charge $e = -|e|$ in a magnetic field $B\hat{z}$. The electron's spin is traditionally represented by the Pauli matrices

$$\sigma_x = \begin{pmatrix} 0 & 1 \\ 1 & 0 \end{pmatrix}; \quad \sigma_y = \begin{pmatrix} 0 & -i \\ i & 0 \end{pmatrix}; \quad \sigma_z = \begin{pmatrix} 1 & 0 \\ 0 & -1 \end{pmatrix} \tag{2.154}$$

with $\mathbf{s} = \frac{1}{2}\boldsymbol{\sigma}$. The interaction Hamiltonian is

$$H = -\frac{e\hbar B}{m_e c} s_z = \hbar\omega s_z, \tag{2.155}$$

where

$$\omega = \frac{|e|B}{m_e c}. \tag{2.156}$$

The Heisenberg equations of motion are easily determined from the commutation relations

$$[s_x, s_y] = i s_z; \quad [s_y, s_z] = i s_x; \quad [s_z, s_x] = i s_y. \tag{2.157}$$

Thus,

$$\frac{ds_z}{dt} = \frac{1}{i\hbar}[s_z, H] = 0; \qquad s_z(t) = s_z(0)$$

$$\frac{ds_x}{dt} = \frac{1}{i\hbar}[s_x, H] = -i\omega[s_x, s_z] = -\omega s_y$$

$$\frac{ds_y}{dt} = \frac{1}{i\hbar}[s_y, H] = -i\omega[s_y, s_z] = \omega s_x.$$

$$\tag{2.158}$$

Then

$$\frac{d^2 s_x}{dt^2} + \omega^2 s_x = 0. \tag{2.159}$$

If the initial conditions for the operators $\mathbf{s}(t)$ are $\mathbf{s}(t = 0) = \frac{1}{2}\boldsymbol{\sigma}$, we see that the solution is

$$s_x(t) = \frac{1}{2}\sigma_x \cos \omega t - \frac{1}{2}\sigma_y \sin \omega t$$

$$s_y(t) = \frac{1}{2}\sigma_x \sin \omega t + \frac{1}{2}\sigma_y \cos \omega t$$

$$s_z(t) = \frac{1}{2}\sigma_z. \tag{2.160}$$

Suppose at $t = 0$ the spin is aligned with the x direction. With states quantized along the z direction,

$$|\xi, t = 0\rangle = \frac{1}{\sqrt{2}} |+1/2\rangle + \frac{1}{\sqrt{2}} |-1/2\rangle \tag{2.161}$$

then the time behavior is

$$\langle s_x(t) \rangle = \frac{1}{2} \cos \omega t$$

$$\langle s_y(t) \rangle = \frac{1}{2} \sin \omega t$$

$$\langle s_z(t) \rangle = 0. \tag{2.162}$$

The spin vector precesses in the $x - y$ plane.

In the Schrödinger picture we focus on

$$H = \hbar \omega s_z = \frac{\hbar \omega}{2}\sigma_z. \tag{2.163}$$

If at the outset

$$|\xi, 0\rangle = \frac{1}{\sqrt{2}} |+\rangle + \frac{1}{\sqrt{2}} |-\rangle, \tag{2.164}$$

then later

$$|\xi, t\rangle = \frac{e^{-i\omega t/2}}{\sqrt{2}} |+\rangle + \frac{e^{i\omega t/2}}{\sqrt{2}} |-\rangle. \tag{2.165}$$

To find the probability that the spin is found in the positive x direction at time t, we calculate

$$p(s_x = 1/2, t) = \left| \frac{1}{\sqrt{2}} (\langle +| + \langle -|) |\xi, t\rangle \right|^2$$

$$= \left| \frac{1}{2} (e^{-i\omega t/2} + e^{i\omega t/2}) \right|^2 = \cos^2 \frac{\omega t}{2}. \tag{2.166}$$

So the probability of finding $s_x = -1/2$

$$p(s_x = -1/2, t) = \sin^2 \frac{\omega t}{2}. \tag{2.167}$$

The expectation value of s_x is simply

$$\langle s_x(t) \rangle = \frac{1}{2}(p(s_x = 1/2) - p(s_x = -1/2)) = \frac{1}{2}(\cos^2 \frac{\omega t}{2} - \sin^2 \frac{\omega t}{2}) = \frac{1}{2}\cos \omega t. \quad (2.168)$$

2.16 Simple Harmonic Oscillator

The simple harmonic oscillation is ubiquitous. Because of the power of operator methods, it is traditional to elucidate the problem with that technique. Before presenting that approach, we pause to address the harmonic oscillator directly with differential equations. The Hamiltonian is

$$H = \frac{p^2}{2m} + \frac{1}{2}m\omega_0^2 q^2, \quad (2.169)$$

with the corresponding Schrödinger equation

$$\left(-\frac{\hbar^2}{2m}\frac{d^2}{dq^2} + \frac{1}{2}m\omega_0^2 q^2\right)\psi(q) = E\psi(q). \quad (2.170)$$

With a change of variables

$$z = \sqrt{\frac{m\omega_0}{\hbar}}q; \qquad \epsilon = \frac{E}{\hbar\omega_0}, \quad (2.171)$$

we have

$$\left(-\frac{d^2}{dz^2} + z^2\right)\psi(z) = 2\epsilon\psi(z). \quad (2.172)$$

At large $|z|$, the dominant behavior must be $e^{\pm z^2/2}$, so try $\psi = e^{-z^2/2}F(z)$. The equation for F is

$$F'' - 2zF' + (2\epsilon - 1)F = 0. \quad (2.173)$$

For large z it seems that $F \propto e^{z^2}$, which would ruin the combined behavior of ψ. This is avoided if F is a finite polynomial. Trying $F \propto z^n$, we see that at large z we must have $2n = 2\epsilon - 1$, where $n = 0, 1, 2...$ Thus $\epsilon = n + 1/2$ or $E = (n + 1/2)\hbar\omega_0$.

The Hermite polynomials can be defined by

$$H_n(z) = (-1)^n e^{z^2}\frac{d^n}{dz^n}e^{-z^2}. \quad (2.174)$$

To see that this is indeed a solution to Eq. (2.173), write $D = d/dz$ and note that $[D^n, z] = nD^{n-1}$. Then writing

$$G = e^{z^2}D^n e^{-z^2}, \quad (2.175)$$

we find

$$G' = -2n\, e^{z^2}D^{n-1}e^{-z^2}$$

$$G'' = -4nz\, e^{z^2}D^{n-1}e^{-z^2} - 2n\, e^{z^2}D^n e^{-z^2}, \quad (2.176)$$

so, in fact, $H_m(z) = (-1)^m G(z)$ is a solution of Eq. (2.173). Now consider

$$g(s, z) = e^{-s^2+2sz} = \sum_m A_m s^m. \tag{2.177}$$

So a Taylor expansion shows that

$$A_m = \frac{1}{m!}\frac{d^m}{ds^m}e^{-(s-z)^2+z^2}\bigg|_{s=0} = \frac{(-1)^m}{m!}e^{z^2}\frac{d^m}{dz^m}e^{-(s-z)^2}\bigg|_{s=0} = \frac{1}{m!}H_m(z). \tag{2.178}$$

Thus $g(s, z)$ is a generating function for the $H_m(z)$:

$$g(s, z) = e^{-s^2+2sz} = \sum_m \frac{s^m}{m!}H_m(z). \tag{2.179}$$

This expression enables us to find still another representation of the Hermite polynomials. We can extract the coefficient of s^m by contour integration. Quite generally, if $f(s) = \sum_m a_m s^m$ is analytic for $|s| < R$, except for a pole at $s = 0$, then for any contour encircling $s = 0$ counterclockwise once with $|s| < R$ (See Appendix B.1)

$$\oint ds \frac{f(s)}{s^n} = 2\pi i a_{n-1}, \tag{2.180}$$

so

$$\oint ds\, e^{-s^2+2sz} s^{-m-1} = \frac{2\pi i}{m!}H_m(z) \tag{2.181}$$

or

$$H_m(z) = \frac{2^m m!}{2\pi i}\oint dt\frac{e^{-t^2/4+zt}}{t^{m+1}}. \tag{2.182}$$

The first few Hermite polynomials are

$$H_0(z) = 1$$
$$H_1(z) = 2z$$
$$H_2(z) = 4z^2 - 2$$
$$H_3(z) = 8z^3 - 12z$$
$$H_4(z) = 16z^4 - 48z^2 + 12. \tag{2.183}$$

As is apparent, we have the symmetry relation

$$H_m(-z) = (-1)^m H_m(z). \tag{2.184}$$

The orthogonality relation for Hermite polynomials arises from considering

$$\int_{-\infty}^{\infty} dz e^{-z^2} H_m(z)H_n(z) = (-1)^{m+n}\int_{-\infty}^{\infty} dz \left(\frac{d^m}{dz^m}e^{-z^2}\right)\left(e^{z^2}\frac{d^n}{dz^n}e^{-z^2}\right). \tag{2.185}$$

Now if $m \geq n$, integrate by parts m times. The contributions at the end points all vanish because there are more factors of e^{-z^2} than of e^{z^2}. In the end we will have

$$(-1)^n\int_{-\infty}^{\infty} dz\, e^{-z^2}\frac{d^m}{dz^m}\left(e^{z^2}\frac{d^n}{dz^n}e^{-z^2}\right) = (-1)^{m+n}\int_{-\infty}^{\infty} dz\, e^{-z^2}\frac{d^m}{dz^m}H_n(z). \tag{2.186}$$

Now $H_n(z)$ is a polynomial of order n so if $m > n$, this vanishes. If $m = n$ and if the highest order term of $H_n(z)$ is a_n, the integral is just $\sqrt{\pi} n!$. It is clear from Eq. (2.173) that, in fact, $a_n = 2^n$. Altogether

$$\int_{-\infty}^{\infty} dz e^{-z^2} H_m(z) H_n(z) = \delta_{mn} 2^n n! \sqrt{\pi}. \tag{2.187}$$

Combining all these results we determine the wave function for the nth state

$$\psi_n(q) = \frac{1}{\sqrt{2^n n!}} \left(\frac{m\omega_0}{\hbar \pi}\right)^{1/4} H_n\left(\sqrt{\frac{m\omega_0}{\hbar}} q\right) \exp\left(-\frac{m\omega_0}{2\hbar} q^2\right), \tag{2.188}$$

Having solved the problem as a straight-forward exercise with the Schrödinger equation, we address it with an operator technique. First recall the classical analysis of Eq. (2.169). From the Hamilton equations

$$\dot{q} = p/m; \qquad \dot{p} = -m\omega_0^2 q \tag{2.189}$$

from which

$$q = A\cos(\omega_0 t + \alpha); \qquad p = -m\omega_0 A \sin(\omega_0 t + \alpha). \tag{2.190}$$

If we now define

$$a = m\omega_0 q + ip = m\omega_0 A(\cos(\omega_0 t + \alpha) - i\sin(\omega_0 t + \alpha)) = m\omega_0 A e^{-i(\omega_0 t + \alpha)}, \tag{2.191}$$

we see that $|a|$ is constant and

$$|a|^2 = m^2 \omega_0^2 q^2 + p^2 = 2mE. \tag{2.192}$$

Quantum mechanically, we have

$$H = \frac{p^2}{2m} + \frac{1}{2} m\omega_0^2 q^2, \qquad \text{with } [q, p] = i\hbar. \tag{2.193}$$

Just as we did classically, we change from p, q to $a, a^\dagger$, where

$$a = \frac{1}{\sqrt{2m\hbar\omega_0}} (m\omega_0 q + ip)$$

$$a^\dagger = \frac{1}{\sqrt{2m\hbar\omega_0}} (m\omega_0 q - ip). \tag{2.194}$$

Using the canonical commutation relation, we find $[a, a^\dagger] = 1$. Solving for p and q,

$$p = -i\sqrt{\frac{m\hbar\omega_0}{2}} (a - a^\dagger)$$

$$q = \sqrt{\frac{\hbar}{2m\omega_0}} (a + a^\dagger) \tag{2.195}$$

and the Hamiltonian becomes

$$H = \left(a^\dagger a + \frac{1}{2}\right) \hbar\omega_0 = \left(aa^\dagger - \frac{1}{2}\right) \hbar\omega_0 = \frac{1}{2} \left(aa^\dagger + a^\dagger a\right) \hbar\omega_0. \tag{2.196}$$

From this we compute

$$[a, H] = [a, a^\dagger a \hbar \omega_0] = u \hbar \omega_0$$
$$[a^\dagger, H] = [a^\dagger, a^\dagger a \hbar \omega_0] = -a^\dagger \hbar \omega_0. \tag{2.197}$$

We seek the energy eigenvalues E' of H:

$$H \left| E' \right\rangle = E' \left| E' \right\rangle. \tag{2.198}$$

Assume such a state exists and examine the effects of operating with a on both sides:

$$aH \left| E' \right\rangle = aE' \left| E' \right\rangle = E'a \left| E' \right\rangle$$
$$= ([a, H] + Ha) \left| E' \right\rangle = (Ha + a\hbar \omega_0) \left| E' \right\rangle$$
$$H \left(a \left| E' \right\rangle \right) = (E' - \hbar \omega_0) \left(a \left| E' \right\rangle \right). \tag{2.199}$$

Thus, we can evidently find another eigenstate with lower energy. Analogously, we will find that $a^\dagger$ creates a state with energy greater by $\hbar \omega_0$. For this reason we call a and $a^\dagger$ lowering and raising operators. We cannot go on endlessly reducing the energy since the Hamiltonian is positive, as we demonstrate:

$$\left\langle \alpha \left| H \right| \alpha \right\rangle = \hbar \omega_0 [\left\langle \alpha \left| a^\dagger a \right| \alpha \right\rangle + 1/2] = \hbar \omega_0 [\left\langle \alpha' | \alpha' \right\rangle + 1/2] \geq \hbar \omega_0 / 2. \tag{2.200}$$

It follows that there is a state $\left| 0 \right\rangle$ such that $a \left| 0 \right\rangle = 0$, that is, it is annihilated by a. But then

$$H \left| 0 \right\rangle = \left(a^\dagger a + \frac{1}{2} \right) \hbar \omega_0 \left| 0 \right\rangle = \frac{1}{2} \hbar \omega_0 \left| 0 \right\rangle. \tag{2.201}$$

By applying $a^\dagger$ n times to $\left| 0 \right\rangle$ we can create a state with energy $(n + 1/2)\hbar \omega_0$, which state we label $\left| n \right\rangle$. Thus,

$$(a^\dagger)^n \left| 0 \right\rangle = A_n \left| n \right\rangle. \tag{2.202}$$

To normalize the states so that $\left\langle n' | n' \right\rangle = 1$, we compute A_n recursively. Now first note that

$$[a^\dagger a, a^\dagger] = a^\dagger; \qquad [a^\dagger a, (a^\dagger)^n] = n(a^\dagger)^n \tag{2.203}$$

and thus

$$a^\dagger a \left| n \right\rangle = n \left| n \right\rangle. \tag{2.204}$$

That is, $a^\dagger a$ 'counts', while

$$[aa^\dagger, (a^\dagger)^n] = (n + 1)(a^\dagger)^n; \qquad aa^\dagger \left| n \right\rangle = (n + 1) \left| n \right\rangle. \tag{2.205}$$

These relations allow us to fix the normalization.

$$\left| n + 1 \right\rangle = \lambda a^\dagger \left| n \right\rangle$$
$$\left\langle n + 1 | n + 1 \right\rangle = |\lambda|^2 \left\langle n \left| aa^\dagger \right| n \right\rangle = |\lambda|^2 (n + 1). \tag{2.206}$$

So $|\lambda|^2 = 1/(n + 1)$, $\lambda = 1/\sqrt{n + 1}$. By iteration,

$$A_n = \frac{1}{\sqrt{n!}}; \qquad \left| n \right\rangle = \frac{1}{\sqrt{n!}} (a^\dagger)^n \left| 0 \right\rangle. \tag{2.207}$$

Let us examine the uncertainty relation on the state $|n\rangle$. From Eq. (2.195),

$$p^2 = -\frac{m\hbar\omega_0}{2}(a - a^\dagger)(a - a^\dagger)$$

$$q^2 = \frac{\hbar}{2m\omega_0}(a + a^\dagger)(a + a^\dagger). \tag{2.208}$$

Because p and q either raise or lower a state, they have no diagonal components; $\langle n|p|n\rangle = 0$, $\langle n|q|n\rangle = 0$. Thus

$$(\Delta p)_n^2 (\Delta q)_n^2 = \langle n|p^2|n\rangle\langle n|q^2|n\rangle. \tag{2.209}$$

We can compute these explicitly

$$\langle n|p^2|n\rangle = -\frac{m\hbar\omega_0}{2}\langle n|(a - a^\dagger)(a - a^\dagger)|n\rangle$$

$$= \frac{m\hbar\omega_0}{2}\langle n|aa^\dagger + a^\dagger a|n\rangle = (n + 1/2)m\hbar\omega_0. \tag{2.210}$$

The analogous calculation shows

$$\langle n|q^2|n\rangle = \frac{(n + 1/2)\hbar\omega_0}{m\omega_0^2}. \tag{2.211}$$

Altogether,

$$(\Delta p)_n^2 (\Delta q)_n^2 = (n + 1/2)^2 \hbar^2 \tag{2.212}$$

or, in conformity with the uncertainty principle,

$$(\Delta p)_n (\Delta q)_n = (n + 1/2)\hbar \geq \frac{1}{2}\hbar. \tag{2.213}$$

We can connect the operator method to the Schrödinger equation by determining the coordinate representation of the eigenstate $|n\rangle$.

$$\psi_n(q') = \langle q'|n\rangle = \frac{1}{\sqrt{n!}}\langle q'|(a^\dagger)^n|0\rangle$$

$$= \frac{1}{\sqrt{n!}}\left(\frac{1}{2m\hbar\omega_0}\right)^{n/2}\langle q'|(m\omega_0 q - ip)^n|0\rangle$$

$$= \frac{1}{\sqrt{n!}}\left(\frac{1}{2m\hbar\omega_0}\right)^{n/2}(m\omega_0 q' - \hbar\frac{\partial}{\partial q'})^n\psi_0(q'), \tag{2.214}$$

where $\psi_0(q')$ is defined by the condition that $a|0\rangle = 0$

$$(m\omega_0 q' + \hbar\frac{\partial}{\partial q'})\psi_0(q') = 0. \tag{2.215}$$

Thus, properly normalized,

$$\psi_0(q') = \left(\frac{m\omega_0}{\pi\hbar}\right)^{1/4}\exp\left(-\frac{m\omega_0}{2\hbar}q'^2\right). \tag{2.216}$$

In terms of the variable $z = \sqrt{\frac{m\omega_0}{\hbar}}q'$, we have

$$\psi_n(z) = \frac{\pi^{-1/4}}{\sqrt{2^n n!}}\left(z - \frac{d}{dz}\right)^n e^{-z^2/2}. \tag{2.217}$$

Previously, we found

$$\psi_n(z) = \frac{\pi^{-1/4}}{\sqrt{2^n n!}}e^{-z^2/2}(-1)^n e^{z^2}\frac{d^n}{dz^n}e^{-z^2}, \tag{2.218}$$

so consider successively commuting

$$e^{z^2/2}\frac{d^n}{dz^n}e^{-z^2} = (\frac{d}{dz} - z)e^{z^2/2}\frac{d^n}{dz^{n-1}}e^{-z^2}$$
$$= (\frac{d}{dz} - z)^n e^{z^2/2}e^{-z^2}, \tag{2.219}$$

recovering the result of the operator treatment.

References and Suggested Reading

For foundations of quantum mechanics the classics are:

Dirac, *Quantum Mechanics*, 4th edition, Chapters 1, 2, 3.
von Neumann, *Mathematical Foundations of Quantum Mechanics*.

For discussion of measurement in quantum mechanics:

Gottfried, *Quantum Mechanics: Fundamentals* pp. 165-190.
Bell, *Speakable and Unspeakable in Quantum Mechanics*, pp. 213–231.

Problems

2.1 The operators q and p corresponding to a generalized coordinate and its canonical momentum satisfy the commutation relation $[q, p] = i\hbar I$.

(a) Show that the coordinate representation of the operator p satisfies

$$(q' - q'')\langle q' | p | q''\rangle = i\hbar\delta(q' - q''). \tag{2.220}$$

(b) By considering the coordinate representatives

$$\langle q' | p | \xi\rangle = \int dq'' \langle q' | p | q''\rangle\langle q'' | \xi\rangle \tag{2.221}$$

and the corresponding expression of $\langle q' | [q, p] | \xi\rangle$, show that

$$\langle q' | p | q''\rangle = i\hbar\frac{\partial}{\partial q'}\delta(q' - q'') \tag{2.222}$$

aside from a possible additive term $f(q')\delta(q' - q'')$.

(c) Show that

$$\langle q'|p'\rangle = Ce^{i\Phi/\hbar}, \tag{2.223}$$

where

$$\Phi = p'q' - \int^{q'} dq'' \, f(q''). \tag{2.224}$$

Show that $|q'\rangle$ can be redefined to remove the extra phase.

2.2 Show that the density matrix operator for a spin-1/2 system can be written

$$\rho = \frac{1}{2}(I + \mathbf{P} \cdot \boldsymbol{\sigma}), \tag{2.225}$$

where $\mathbf{P}$ is a real vector (called the polarization vector) and $\boldsymbol{\sigma}$ has as components the Pauli matrices

$$\sigma_x = \begin{pmatrix} 0 & 1 \\ 1 & 0 \end{pmatrix}; \quad \sigma_y = \begin{pmatrix} 0 & -i \\ i & 0 \end{pmatrix}; \quad \sigma_z = \begin{pmatrix} 1 & 0 \\ 0 & -1 \end{pmatrix}. \tag{2.226}$$

Show that the components of $\mathbf{P}$ are related to the matrix elements of ρ as follows:

$$P_x = 2\Re\langle 1|\rho|2\rangle; \quad P_y = -2\Im\langle 1|\rho|2\rangle; \quad P_z = \langle 1|\rho|1\rangle - \langle 2|\rho|2\rangle. \tag{2.227}$$

2.3 A beam of spin-1/2 particles is a mixed ensemble with 50% of the particles having spin up in the positive z-direction and 50% spin up along the positive x-direction.
(a) Find the density matrix, ρ and the polarization vector $\mathbf{P}$.
(b) Given the unit vector $\mathbf{n} = \sin\beta\cos\alpha\,\hat{\mathbf{x}} + \sin\beta\sin\alpha\,\hat{\mathbf{y}} + \cos\beta\,\hat{\mathbf{z}}$, use the relation $\langle A\rangle = \mathrm{Tr}\,(A\rho)$ to find the average value of $\boldsymbol{\sigma} \cdot \mathbf{n}$. Sketch your result versus β for $\alpha = 0$.

2.4 Two operators $A(q, p)$ and $B(q, p)$ can be expressed as a power series in q and p, with q and p obeying the usual commutation relation $[q, p] = i\hbar I$. Show by purely operator methods that

$$\lim_{\hbar \to 0} \frac{1}{i\hbar}[A, B] = [A, B]_{P.B.}, \tag{2.228}$$

where the right side is calculated as a classical Poisson bracket.

2.5 Show that the operators $U_1 = e^{iK}$ and $U_2 = (1 + iK)(1 - iK)^{-1}$ are unitary if K is self-adjoint. Also show that $U_2' = (1 - iK)^{-1}(1 + iK)$ is equal to U_2.

2.6* (a) Let A and B be two operators that commute with their commutator, $[A, B]$. Prove that

$$[A, B^n] = nB^{n-1}[A, B]$$

$$[A^n, B] = nA^{n-1}[A, B]. \tag{2.229}$$

(b) For any operator A, the operator e^A is defined by the power series

$$e^A = 1 + A + \frac{A}{1!} + \frac{A^2}{2!} + \frac{A^3}{3!} + \cdots \tag{2.230}$$

Prove the identity

$$e^A B e^{-A} = B + [A, B] + \frac{1}{2!}[A,[A,B]] + \frac{1}{3!}[A[A,[A,B]]] + \cdots \tag{2.231}$$

Hint: Consider $F(s) = e^{sA}B e^{-sA}$.

(c) For two operators A and B that commute with their commutator $[A, B]$, prove the identity

$$e^A e^B = e^{A+B+1/2[A,B]}. \tag{2.232}$$

Hint: Consider $F(s) = e^{sA}e^{sB}$.

These operator identities involving exponential are valuable in calculations, as in the next problem, for example.

2.7 Let q and p be canonically conjugate variables, with commutator $[q, p] = i\hbar I$. Let two vectors describing physical states $|\alpha\rangle$ and $|\xi\rangle$ be related by

$$|\xi\rangle = e^{-i\lambda p/\hbar} |\alpha\rangle \tag{2.233}$$

with λ real.

(a) Compare the expectation values $\langle \alpha |p| \alpha \rangle$ and $\langle \xi |p| \xi \rangle$.

(b) Compare the expectation values $\langle \alpha |q| \alpha \rangle$ and $\langle \xi |q| \xi \rangle$.

(c) Discuss the meaning of the operator $e^{-i\lambda p/\hbar}$ in light of your results for parts (a) and (b).

2.8 For a simple harmonic oscillator in one dimension, the Heisenberg operators for $q(t)$ and $p(t)$ are

$$q(t) = q(0) \cos \omega t + \frac{p(0)}{m\omega} \sin \omega t$$
$$p(t) = p(0) \cos \omega t - m\omega q(0) \sin \omega t. \tag{2.234}$$

Using the translation operator $\exp(-i\ell\, p(0)/\hbar)$ and results from Problem 2.6 calculate the expectation values of $q(t)$ and $p(t)$ in the nth oscillator state, displaced initially by a distance ℓ in the positive direction. Interpret your results.

2.9* The ground state of a one-dimensional harmonic oscillator is given an initial impulse at $t = 0$. Discuss the time development of the state by abstract operator methods (and later describe the coordinate-representation wave functions). Formally, the state vector at any time t is

$$|t\rangle = e^{-iHt/\hbar}e^{ikq(0)} |0\rangle, \tag{2.235}$$

where $\hbar k$ is the initial impulse.

(a) Show that

$$|t\rangle = e^{-i\omega t/2}e^{ikq(-t)} |0\rangle. \tag{2.236}$$

(b) Using the explicit form of $q(-t)$ from Problem 2.8 and a result from Problem 2.6, show that the coordinate representative of $|t\rangle$ can be written

$$\langle q'|t\rangle = e^{i\phi(q',t)} \langle q' |\exp[-i(k/(m\omega)) \sin \omega t\, p(0)]| 0\rangle \tag{2.237}$$

where

$$\phi(q',t) = -\omega t/2 + kq'\cos\omega t - \frac{\hbar k^2}{4m\omega}\sin 2\omega t. \tag{2.238}$$

(c) Complete the calculation indicated in part (b) and show that

$$|\langle q'|t\rangle|^2 = \sqrt{\frac{m\omega}{\pi\hbar}}\exp\left[-\frac{m\omega}{\hbar}\left(q' - \frac{\hbar k}{m\omega}\sin\omega t\right)^2\right]. \tag{2.239}$$

Interpret. What modification occurs if the state is the nth excited state instead of the ground state?

2.10 For the initially struck oscillator ground state of Problem 2.9, evaluate the expectation value of the energy operator H and its square H^2. Determine the root-mean-square uncertainty of the energy. Are these quantities time-dependent? Is there a classical limit in which $\Delta E \ll E$? What are the restrictions on the impulse $\hbar k$ for such a limit?

3

Wentzel-Kramers-Brillouin (WKB) Method

Rather few potentials admit closed-form solutions to the Schrödinger equation, most notably the simple harmonic oscillator and the Coulomb problem. The WKB (Wentzel, Kramers, Brillouin) technique enables us to address with a very good approximation a wide class of circumstances. Used in a physics context first by Lord Rayleigh, but much earlier by Liouville, this approximation is also called JWKB, adding the name of Jeffreys. It is useful whenever the index of refraction or the density of the medium (and so the sound speed) or the potential energy changes slowly over distances of order one wavelength. For the Schrödinger equation, this means that locally a particle of definite energy has a relatively well-defined momentum given by the classical relation. With the WKB method, we can explore the nature of bound states and of tunneling without being forced to pick a specific potential. For those potentials where complete solutions are available, WKB often provides a remarkably accurate approximation.

3.1 Semi-classical Approximation

Because the momentum is given classically by

$$p(\mathbf{x}) = \sqrt{2m(E - V(\mathbf{x}))}, \tag{3.1}$$

it is plausible that the wave amplitude should be something like $e^{i\mathbf{p}_{\text{local}} \cdot \mathbf{x}/\hbar}$.

These considerations suggest the following approach: for the time-independent Schrödinger equation,

$$-\frac{\hbar^2}{2m}\nabla^2\psi_E + V(\mathbf{x})\psi_E = E\psi_E \tag{3.2}$$

write

$$\psi_E(\mathbf{x}, t) = e^{i\overline{S}(\mathbf{x},t)/\hbar}, \tag{3.3}$$

where $\overline{S}$ is a complex function that is the quantum analog of Hamilton's principal function. (See Goldstein, *Classical Mechanics*, 2nd edition, Chapter 10).

We substitute into the time-dependent Schrödinger equation,

$$i\hbar\frac{\partial\psi}{\partial t} = \left(-\frac{\hbar^2}{2m}\nabla^2 + V\right)\psi \tag{3.4}$$

and find

John David Jackson: A Course in Quantum Mechanics, First Edition. Robert N. Cahn.
© 2024 John Wiley & Sons, Inc. Published 2024 by John Wiley & Sons, Inc.
Companion website: www.wiley.com/go/Jackson/QuantumMechanics

$$-\frac{\partial \overline{S}}{\partial t} = \frac{1}{2m}\left[(\nabla\overline{S})^2 + \frac{\hbar}{i}\nabla^2\overline{S}\right] + V. \tag{3.5}$$

In the limit of $\hbar \to 0$, or said differently, in the limit that $\hbar|\nabla^2\overline{S}| \ll |\nabla\overline{S}|^2$, we get the lowest order approximation for $\overline{S}$, called $\overline{S}_0$, satisfying

$$\frac{\partial \overline{S}_0}{\partial t} + \frac{1}{2m}(\nabla\overline{S}_0)^2 + V = 0. \tag{3.6}$$

This equation is, in fact, the Hamilton-Jacobi equation of classical mechanics and $\overline{S}_0(\mathbf{x}, t)$ is Hamilton's principal function. With $\overline{S}_0 = S_0 - Et$, we have

$$E = \frac{1}{2m}(\nabla S_0)^2 + V \tag{3.7}$$

and we can identify $\mathbf{p}(\mathbf{x}, t) = \nabla S_0/m$. Hamilton's work in 1834 showed how optics and classical mechanics were related. Schrödinger recalled Hamilton's 1834 paper in his discovery of wave mechanics. Let us now consider $S(\mathbf{x}) = \overline{S}(\mathbf{x}, t) + Et$ and so study the equation

$$(\nabla S)^2 - i\hbar\nabla^2 S = 2m(E - V(\mathbf{x})) \equiv p^2(\mathbf{x}). \tag{3.8}$$

This is an exact equation still. We are looking for an approximation that holds when the change in the potential is small in one wavelength, i.e.

$$\lambda|\nabla V| \ll |E - V|. \tag{3.9}$$

Since $\lambda = \hbar/p$, we can imagine that taking $\hbar \to 0$ achieves this. We thus try expanding $S(\mathbf{x})$ in powers of $\hbar$:

$$S(\mathbf{x}) = S_0(\mathbf{x}) + \hbar S_1(\mathbf{x}) + \hbar^2 S_2(\mathbf{x}) + \cdots \tag{3.10}$$

Then we obtain a set of equations by equating equal powers of $\hbar$ in the expression

$$i\hbar(\nabla^2 S_0 + \hbar\nabla^2 S_1 + \cdots)$$
$$- (\nabla S_0 + \hbar\nabla S_1 + \cdots) \cdot (\nabla S_0 + \hbar\nabla S_1 + \cdots) + p^2(\mathbf{x}) = 0, \tag{3.11}$$

which yields

$$\hbar^0 : \quad -(\nabla S_0)^2 + p^2(x) = 0$$
$$\hbar^1 : \quad i\nabla^2 S_0 - 2\nabla S_0 \cdot \nabla S_1 = 0$$
$$\hbar^2 : \quad i\nabla^2 S_1 - 2\nabla S_0 \cdot \nabla S_2 - (\nabla S_1)^2 = 0. \tag{3.12}$$

3.2 Solution in One Dimension

While in principle the set of equations can be solved in three dimensions, in practice the WKB approximation is employed in one dimension. Then the $\hbar^0$ equation reads

$$\left(\frac{dS_0(x)}{dx}\right)^2 = p^2(x) \tag{3.13}$$

with solution

$$S_0(x) = \pm\int^x p(x')dx'. \tag{3.14}$$

Note that $p(x) = \sqrt{2m(E - V)}$ is real in classically allowed regions and imaginary in classically forbidden regions. Thus $e^{iS_0/\hbar}$ oscillates in classically allowed regions and decays or grows exponentially in forbidden regions.

The $\hbar^1$ equation is now

$$i\frac{d^2 S_0}{dx^2} = 2\frac{dS_0}{dx}\frac{dS_1}{dx} \tag{3.15}$$

or

$$i\frac{dS_1}{dx} = -\frac{1}{2}\frac{d^2 S_0/dx^2}{dS_0/dx} = -\frac{1}{2}\frac{dp/dx}{p(x)} = -\frac{d}{dx}\left(\ln\sqrt{p(x)}\right). \tag{3.16}$$

Thus,

$$iS_1(x) = \ln\left(\frac{1}{\sqrt{p(x)}}\right) + constant. \tag{3.17}$$

Correct to first order, we have

$$iS(x)/\hbar = \pm\frac{i}{\hbar}\int^x p(x')dx' + \ln\frac{1}{\sqrt{p(x)}} \tag{3.18}$$

or

$$\psi(x) = \frac{A}{\sqrt{k(x)}}e^{\pm i\int^x k(x')dx'}, \tag{3.19}$$

where

$$k(x) = \sqrt{\frac{2m}{\hbar^2}(E - V(x))} \tag{3.20}$$

is the local wave number. The oscillatory wave function is just a straightforward manifestation of de Broglie waves with a potential. The factor out front is understandable, too. Since $|\psi|^2$ is proportional to probability density $P(x)$, we have $P(x)dx \propto dx/k(x) \propto dx/v(x) \propto dt$. Thus, the probability in any interval is proportional to the time spent in the interval, just as with the classical probability. This is the solution in the classically allowed (oscillatory) region. In the classically forbidden region, we write

$$\kappa(x) = \sqrt{\frac{2m}{\hbar^2}(V(x) - E)}. \tag{3.21}$$

Then the solutions are

$$\psi(x) = \frac{A}{\sqrt{\kappa(x)}}e^{\pm\int^x \kappa(x')dx'} + \mathcal{O}(\hbar). \tag{3.22}$$

The solution is valid provided the local wavelength is small compared to the distance over which V varies. If the local wavelength diverges, i.e. $k(x) \to 0$ or $\kappa \to 0$, the solution fails, as is apparent from the leading coefficient in the formula. The problem occurs at the classical turning point, that is, at the end of the classically allowed region. The WKB solution fails there because the expansion in $\hbar$ no longer is valid.

In one sense, these are innocuous problems because the square of the wave function is still integrable. However, because the solutions diverge at the turning point, we cannot use

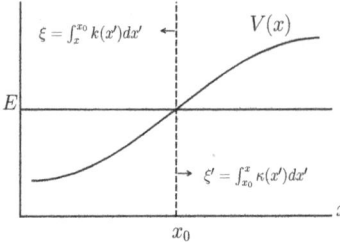
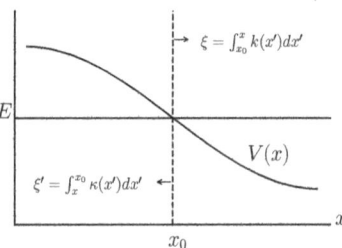

Figure 3.1 The quantities ξ and ξ', which define the functions needed for the WKB connection formulas. Both functions are positive and increasing as the argument moves away from the turning point as shown.

the usual technique of matching the value and first derivatives of the solutions on the two sides of the boundary. The problem is intrinsic to the WKB approximation. Nothing untoward happens in the full differential equation.

The crux of the problem is that we need be able to connect oscillatory functions, crudely $\sin kx, \cos kx$ with exponential functions, again crudely $e^{\pm \kappa x}$. We cannot do this directly because all the WKB solutions diverge at the turning point. While the power of the WKB approximation derives from its ability to make predictions that do not depend on the details of the potential, but rather more general features, over any short interval the potential can be approximated by a linear function. The resolution of the problem then follows from comparing the WKB solutions to exact solutions for a linear potential.

To address the solution to the Schrödinger equation in the neighborhood of a classical turning point, we study the behavior of $\int k(x)dx$ and $\int \kappa(x)dx$ near the turning point. To that end, if increasing x takes us from the classically forbidden region $x < x_0$ to the allowed region we define

$$\xi = \int_{x_0}^{x} k(x')dx'; \qquad \xi' = \int_{x}^{x_0} \kappa(x')dx'. \tag{3.23}$$

If increasing x takes us from the classically allowed region $x < x_0$ to the forbidden region we define

$$\xi = \int_{x}^{x_0} k(x')dx'; \qquad \xi' = \int_{x_0}^{x} \kappa(x')dx'. \tag{3.24}$$

With these definitions, both $\xi(x)$ and $\xi'(x)$ increase as x moves away from the turning point. See Figure 3.1

3.3 Schrödinger Equation for the Linear Potential

We consider the linear potential because near the turning point where $p(x) = 0$, we have $p^2(x) = 2m(E - V(x)) \approx 2mCx$, where we redefined x so the turning point is at $x = 0$. If the potential is falling at the turning point, so $dV/dx < 0$, then $C > 0$ and the classically allowed region is $x > 0$. See Figure 3.2. Now the Schrödinger equation is locally

$$\left(\frac{p^2}{2m} - Cx \right) \psi = 0. \tag{3.25}$$

Figure 3.2 The linear potential, $V(x) = E - Cx$ leading to a Schrödinger equation $(p^2/2m - Cx)\psi(x) = 0$, where $C > 0$. The classically allowed region, where the wave function is oscillatory is $x > 0$.

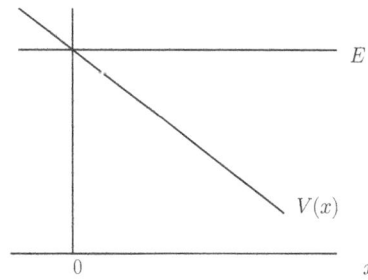

Since the potential is linear, it is better to solve this in momentum space using the representation

$$\phi(x) \propto \int dp\, \tilde{\phi}(p) e^{ipx} \tag{3.26}$$

and

$$x = i\frac{d}{dp}, \tag{3.27}$$

where we have set $\hbar = 1$. Then the momentum-space Schrödinger equation is

$$\left(\frac{p^2}{2m} - iC\frac{d}{dp}\right)\tilde{\phi}(p) = 0, \tag{3.28}$$

whose solution is

$$\tilde{\phi}(p) \propto e^{-ip^3/(3(2mC))}, \tag{3.29}$$

Thus the solution to the Schrödinger equation is

$$\phi(x) \propto \int dp\, e^{-ip^3/(3(2mC))} e^{ipx} \propto \int dq\, e^{-q^3/3 + iqx(2mC)^{1/3}}. \tag{3.30}$$

If we define

$$\psi(y) = \frac{1}{2\pi}\int dq\, e^{i(qy - q^3/3)}, \tag{3.31}$$

then our solution is

$$\phi(x) \propto \psi(x(2mC)^{1/3}). \tag{3.32}$$

We have not yet specified the limits of integration for $\psi(y)$. Because we started with a second-order differential equation, we anticipate that there will be two linearly-independent solutions. In fact, since the original equation is real, we can choose the solutions to be real. We are free to interpret the variable q in the integral Eq. (3.31) as a complex variable and to consider different integration paths in the complex plane. Different paths will lead to different outcomes.

We choose as our base two contours C_R and C_L. Both begin at $+i\infty$ and extend down the positive imaginary axis to the origin, where C_R proceeds to ∞ along the positive real axis,

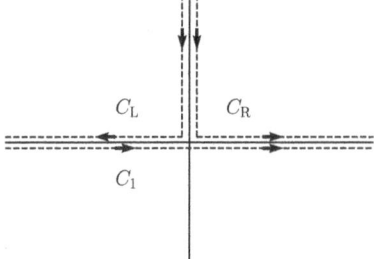

Figure 3.3 Contours used find solutions to the Schrödinger equation for a linear potential. The contour C_1 is equivalent to $C_R - C_L$. The contour used to define ψ_2 is equivalent to $C_R + C_L$.

while C_L moves along the negative real axis to $-\infty$. See Figure 3.3. Using these two, we define two functions:

$$\psi_1(y) = \frac{1}{2\pi} \left[\int_{C_R} - \int_{C_L} \right] dq \, e^{i(qy - q^3/3)} = \frac{1}{2\pi} \int_{-\infty}^{\infty} dq \, e^{i(qy - q^3/3)}$$

$$= \frac{1}{\pi} \int_0^{\infty} dq \, \cos(qy - q^3/3). \tag{3.33}$$

This is equal to the Airy function $Ai(-y)$. See, for example, Abramowitz and Stegun, *Handbook of Mathematical Functions with Formulas, Graphs, and Mathematical Tables*, p. 447. For the second solution, we take the sum of the two contours, multiplied by i, to give a result that is real:

$$\psi_2(y) = \frac{i}{2\pi} \left[\int_{C_R} + \int_{C_L} \right] dq \, e^{i(qy - q^3/3)}. \tag{3.34}$$

On the positive imaginary axis we have $q = i\tau, dq = id\tau$.

$$i(qy - q^3/3) = -(\tau y + \tau^3/3) \tag{3.35}$$

Now combining the various pieces:

$$\psi_2(y) = \frac{1}{\pi} \int_0^{\infty} d\tau e^{-(\tau y + \tau^3/3)}$$

$$+ \frac{i}{2\pi} \int_0^{-\infty} dq \, e^{i(qy - q^3/3)} + \frac{i}{2\pi} \int_0^{\infty} dq \, e^{i(qy - q^3/3)}$$

$$= \frac{1}{\pi} \int_0^{\infty} dq \left[e^{-(qy + q^3/3)} + \sin(q^3/3 - qy) \right]. \tag{3.36}$$

This function is equal to $Bi(-y)$, the second Airy function. See Figure 3.4.

We will use these solutions to the problem of the linear potential to join our WKB solutions across the turning points. To do this, we need to find the asymptotic forms of ψ_1 and ψ_2 for large positive and negative values of their argument y. This is accomplished with the saddle point and stationary phase techniques, which we sketch below.

Steepest descents uses a path in the complex plane where the dominant contribution to the integral is confined to a small region and analyzes that region carefully. Finding that path is analogous to finding the easiest way over a pass, a saddle point, in a mountain hike. The saddle point method is often useful when we have a large parameter in the problem. Consider

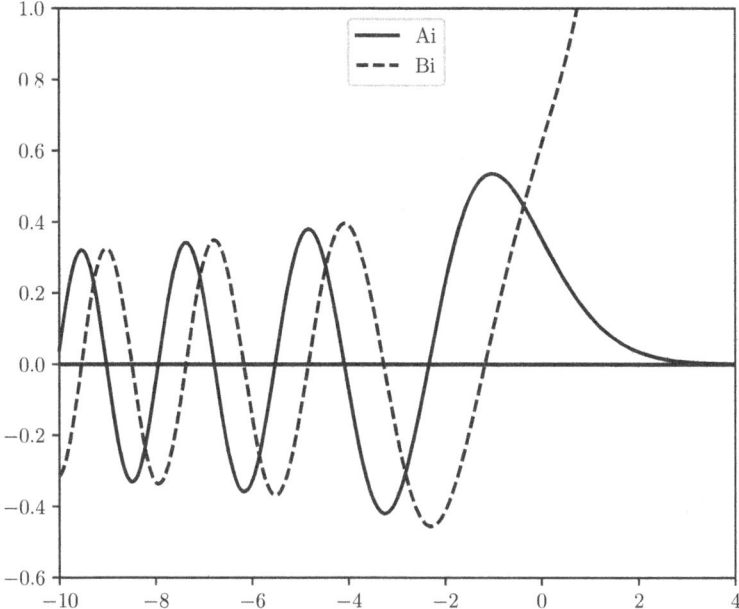

Figure 3.4 The Airy functions, *Ai(x)* and *Bi(x)*, which are used to determine the connection formulae for WKB solutions at the turning points. Our $\psi_1(x)$ is *Ai(−x)* and $\psi_2(x)$ is *Bi(−x)*.

quite generally,

$$I = \int dz \, e^{f(z)}, \tag{3.37}$$

where $f(z)$ contains some large parameter. We look for the point z_0 where f is maximal and expand

$$f(z) = f(z_0) + \frac{1}{2}f''(z_0)(z - z_0)^2 + \dots \tag{3.38}$$

Now we take our integration path near z_0 to be $z - z_0 = \rho e^{i\phi}$ and then choose ϕ so that $f''(z_0)e^{2i\phi}$ is real and negative. In this case,

$$f(z) \approx f(z_0) - \frac{1}{2}|f''(z_0)|\rho^2. \tag{3.39}$$

Now if the large parameter causes $f''(z_0)$ to be itself large, the integral is dominated by the region near $\rho = 0$ and we find

$$I \cong \int_{-\infty}^{\infty} d\rho \, e^{f(z_0) - (1/2)|f''(z_0)|\rho^2} = \frac{\sqrt{2\pi}e^{f(z_0)}}{|f''(z_0)|^{1/2}}. \tag{3.40}$$

Of course, if there is more than one saddle point along the path, each makes a contribution.

First consider the case $y < 0$ so that, expressing ψ_1 in exponential form

$$\psi_1(-|y|) = \frac{1}{2\pi} \int_{-\infty}^{\infty} dq \, e^{i(-q|y| - q^3/3)}. \tag{3.41}$$

That is,

$$f = -i(|y|q + q^3/3). \tag{3.42}$$

We look for a place q_0 where the derivative of f vanishes:

$$|y| + q_0^2 = 0; \qquad q_0^2 = -|y|; \qquad q_0 = \pm i\sqrt{|y|}. \tag{3.43}$$

At q_0, we find

$$f''(q_0) = -2iq_0 = \pm 2|y|^{1/2}$$

$$f(q_0) = \pm\frac{2}{3}|y|^{3/2}. \tag{3.44}$$

For the contour $C_1 = C_R - C_L$, we use the saddle point at $q_0 = -i|y|^{1/2}$. There $f''(q_0) = -2|y|^{1/2}$. This is already negative, so we want to take a path through this point with $q - q_0 = \rho$, introducing no additional phase. To do this, we distort the contour to pass through the point $-i|y|^{1/2}$ so that at that point the path is parallel to the real axis. See Figure 3.5. In this way, we find

$$\psi_1(y) \cong \frac{e^{-(2/3)|y|^{3/2}}}{2\sqrt{\pi}|y|^{1/4}}; \qquad |y| \gg 1, \quad y < 0. \tag{3.45}$$

For ψ_2 we use the saddle point at $+i|y|^{1/2}$. See Figure 3.5. We have $f''(q_0) = +(2/3)|y|^{3/2}$ and so that $(q - q_0)f''(q_0)$ can be negative we must take $q - q_0 = i\rho$, so our paths both run down the positive imaginary axis. Combining the two paths,

$$\psi_2(y) \cong \frac{e^{(2/3)|y|^{3/2}}}{\sqrt{\pi}|y|^{1/4}}. \tag{3.46}$$

These give the behavior of the solutions for $y < 0$ and $|y| \gg 1$. The results are in agreement with our expectations: one exponentially increasing solution, one exponentially decreasing solution. Note that the pre-factors differ by a factor 2 because of the two pieces of the vertical contour for ψ_2.

For the case of $y > 0$, where the behavior is oscillatory, we use the closely related method of stationary phase. Here the idea is that if the phase of the integrand is changing rapidly, the contribution to the integral vanishes so the dominant contribution comes from a place where the phase isn't rapidly varying.

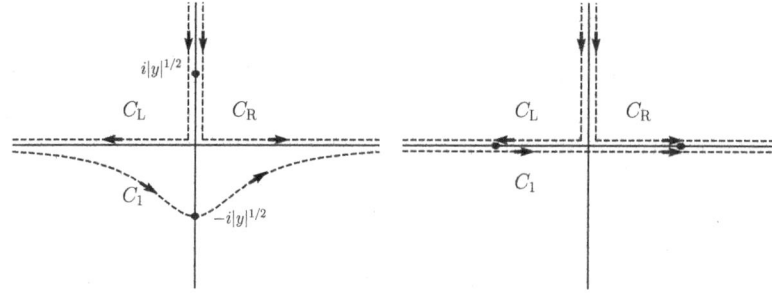

Figure 3.5 Contours used find the asymptotic values of $\psi_1(y)$ and $\psi_2(y)$. Left, for $y \ll 0$, Right, for $y \gg 0$. The critical points are indicated by dark circles. The straight portions of the contours are actually directly on the axes. For ψ_1, the contour C_1 is given by $C_R - C_L$, while for ψ_2, $C_R + C_L$ is used.

Now we have

$$\psi_1(y) = \frac{1}{2\pi} \int_{-\infty}^{\infty} dq \, e^{i(qy - q^3/3)}, \tag{3.47}$$

so the condition $f'(q_0) = 0$ gives $q_0 = \pm\sqrt{y}$, $f(q_0) = \pm i(2/3)y^{3/2}$, and $f''(q_0) = \mp 2i\sqrt{y}$. We see that along the real axis of q, if $|y| \gg 1$ the phase varies rapidly except in the region near q_0. Writing $\rho = q - q_0$ and expanding around q_0, we have contributions from both points on the real axis. See Figure 3.5.

$$\psi_1(y) \cong \frac{1}{2\pi} \int_{-\infty}^{\infty} d\rho \, e^{i(2/3)y^{3/2}} e^{-iy^{1/2}\rho^2} + \text{complex conjugate}. \tag{3.48}$$

Now we will need

$$\int_{-\infty}^{\infty} dx \, e^{-ix^2}, \tag{3.49}$$

which we find by inserting a convergence factor, which we ultimately remove:

$$\int_{-\infty}^{\infty} dx e^{-(\epsilon+i)x^2} = \frac{\sqrt{\pi}}{\sqrt{\epsilon + i}} \to e^{-i\pi/4}\sqrt{\pi}. \tag{3.50}$$

Armed with this result, we find

$$\psi_1(y) \cong \frac{1}{2\sqrt{\pi}} \frac{e^{i(2/3)y^{3/2}-i\pi/4}}{y^{1/4}} + \text{complex conjugate}$$

$$\cong \frac{1}{\sqrt{\pi}y^{1/4}} \cos\left(\frac{2}{3}y^{3/2} - \pi/4\right); \qquad y \gg 1. \tag{3.51}$$

To find $\psi_2(y)$ in the region $y > 0$, we must use the same contours, C_R and C_L, as before. Now the paths from $i\infty$ to the origin do not contribute, since the critical points are now on the real axis. The path C_R proceeds from the origin to $+\infty$, while C_L goes from the origin to $-\infty$. Thus we find

$$\psi_2(y) \cong \frac{i}{2\sqrt{\pi}} \left[\frac{e^{i(2/3)y^{3/2}-i\pi/4}}{y^{1/4}} - \text{complex conjugate} \right]$$

$$\cong -\frac{1}{\sqrt{\pi}y^{1/4}} \sin\left(\frac{2}{3}y^{3/2} - \pi/4\right); \qquad y \gg 1. \tag{3.52}$$

3.4 Connection Formulae for the WKB Method

With the solutions to the Schrödinger equation for the linear potential in hand, we can address the problem of the WKB solution at the turning point, where our approximation failed. In the classically forbidden region,

$$\psi_{WKB} \sim \frac{A}{\sqrt{\kappa(x)}} e^{\pm \int_x^0 \kappa(x')dx'} \tag{3.53}$$

with $\kappa(x) = \sqrt{2m(V(x) - E)}$. For the linear potential in the classically forbidden region $x' < 0$,

$$\kappa(x') = \sqrt{-p^2(x')} = \sqrt{2mC(-x')}. \tag{3.54}$$

Thus

$$\xi' = \int_x^0 \kappa(x')dx' = \sqrt{2mC} \int_0^{|x|} (x')^{1/2}dx' = \frac{2}{3}\sqrt{2mC}|x|^{3/2}$$

$$= \frac{2}{3}|y|^{3/2}, \tag{3.55}$$

where we make the identification from Eqs. (3.31,3.32) of $y = x(2mC)^{1/3}$ as the argument of the Airy function. Similarly,

$$\frac{1}{\sqrt{\kappa(x)}} \propto \frac{1}{|y|^{1/4}}. \tag{3.56}$$

We see that the WKB ξ' factors give exactly the same form as the asymptotic form of the exact solutions to the linear potential.

We found that, in the classically allowed region,

$$\psi_{WKB} \sim \frac{A}{\sqrt{k(x)}} e^{\pm i \int^x k(x')dx'} \tag{3.57}$$

with $k(x) = \sqrt{2m(E - V(x)/\hbar^2}$. We define

$$\xi = \int_0^x k(x')dx' = \int_0^x dx \sqrt{2Cmx} = \frac{2}{3}\sqrt{2mC}x^{3/2}$$

$$\xi = \frac{2}{3}y^{3/2}. \tag{3.58}$$

In the classically allowed region,

$$\psi_1(y) \cong \frac{1}{\sqrt{\pi}y^{1/4}} \cos\left(\frac{2}{3}y^{3/2} - \frac{\pi}{4}\right). \tag{3.59}$$

Using these results, we can now exhibit the WKB connection formulas for a linear turning point:

$$\frac{1}{2}\frac{1}{\sqrt{\kappa(x)}} e^{-\xi'} \rightarrow \frac{1}{\sqrt{k(x)}} \cos(\xi - \frac{\pi}{4}) \tag{3.60}$$

$$\frac{1}{\sqrt{\kappa(x)}} e^{+\xi'} \leftarrow -\frac{1}{\sqrt{k(x)}} \sin(\xi - \frac{\pi}{4}). \tag{3.61}$$

The first relation comes from following ψ_1 from the classically-forbidden to the classically-allowed region. The second comes from following ψ_2 from the classically-allowed to the classically forbidden region. Note that the connection formulas are unidirectional. From exponentially growing behavior (Eq. (3.61)), one cannot exclude the possibility of a $\cos(\xi - \frac{\pi}{4})$ piece, which would generate an exponentially dying behavior. Similarly, from a $\cos(\xi - \frac{\pi}{4})$ behavior, one cannot conclude that the corresponding classically forbidden behavior would be exponentially decreasing, because a very small component of $\sin(\xi - \frac{\pi}{4})$ could generate an exponentially growing component.

Our study of the solutions to the linear potential have enabled us to connect oscillatory and exponential behavior at the turning point when this would not have been possible, simply matching function values and derivatives. We shall see how these connection formulae enable us to sew together a complete solution.

3.5 WKB Formula for Bound States

Now imagine we are trying to solve a one-dimensional bound-state problem where the classically allowed region is from x_1 to x_2, as shown in Figure 3.6. In the classically allowed region, we define

$$\xi_1 = \int_{x_1}^{x} k(x')dx'; \quad \xi_2 = \int_{x}^{x_2} k(x')dx' \tag{3.62}$$

while in the classically forbidden region, we define

$$\xi_1' = \int_{x}^{x_1} \kappa(x')dx'; \quad \xi_2' = \int_{x_2}^{x} \kappa(x')dx', \tag{3.63}$$

so that in the classically allowed region both ξ_1 and ξ_2 are positive and similarly in the classically forbidden regions ξ_1' and ξ_2' are positive.

In the region $x < x_1$, we must have an exponentially decreasing function as x decreases:

$$\psi = \frac{A}{2\sqrt{\kappa(x)}} e^{-\xi_1'}. \tag{3.64}$$

This corresponds to the ψ_1 function.

Now we know that ψ_1 in the allowed region is

$$\psi(x) = \frac{A}{\sqrt{k(x)}} \cos(\xi_1 - \pi/4). \tag{3.65}$$

Figure 3.6 Potential with turning points for classical motion indicated as used in WKB determination of bound states.

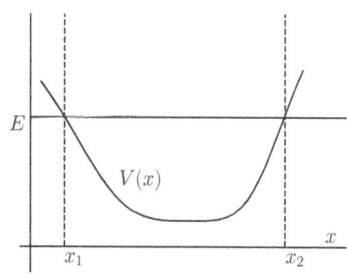

Now consider the point x_2. Again, we need the exponentially decreasing function this time as x increases.

$$\psi(x) = \frac{A'}{2\sqrt{\kappa(x)}}e^{-\xi_2'}. \tag{3.66}$$

But we know this connects with

$$\psi(x) = \frac{A'}{\sqrt{k(x)}}\cos(\xi_2 - \pi/4). \tag{3.67}$$

Now we have two separate solutions. When do they join smoothly? We must have

$$|A| = |A'|$$
$$\cos(\xi_1 - \pi/4) = \pm\cos(\xi_2 - \pi/4) = \pm\cos(\pi/4 - \xi_2 - \xi_1 + \xi_1)$$
$$= \pm\cos[\xi_1 - \pi/4 - (-\pi/2 + \xi_1 + \xi_2)]. \tag{3.68}$$

See Figure 3.7.

Thus we conclude that

$$\xi_1 + \xi_2 - \pi/2 = n\pi, \quad n = 0, 1, 2, ... \tag{3.69}$$

But

$$\xi_1 + \xi_2 = \int_{x_1}^{x_2} k(x')dx'. \tag{3.70}$$

So,

$$\int_{x_1}^{x_2} k(x')dx' = (n + 1/2)\pi. \tag{3.71}$$

This is the WKB requirement for a bound state, replacing the eigenvalue condition on the Schrödinger equation.

If we multiply by $2\hbar$, we can write it as

$$\oint pdq = (n + 1/2)(2\pi\hbar) = (n + 1/2)h. \tag{3.72}$$

The $\oint$ indicates an orbit, here from one turning point to the other and back. This differs from the Wilson-Sommerfeld quantization rule of the old quantum theory only by having $n + 1/2$ in place of n. The added $1/2$ is on account of the wave nature of quantum particles and the penetration into the classically forbidden region. There, the penetration is equivalent

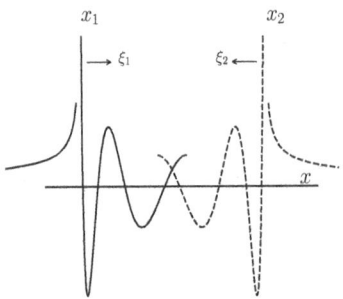

Figure 3.7 The WKB quantization condition follows from requiring that the two oscillatory functions merge smoothly. This determines $\xi_1 + \xi_2 - \pi/2 = n\pi, n = 0,1,2,...$

to $\lambda/8$ (i.e. $\pi/4$) at each end of the classically allowed region. This phenomenon occurs in classical wave problems, e.g. the lowering of the resonant frequency of a cavity because of finite conductivity and penetration by a skin depth into the metal.

3.6 Example of WKB with a Power Law Potential

Assume $V(x) = V_0|x/a|^\beta$. The quantization rule is

$$I = 2\int_0^{x_1} \sqrt{\frac{2mE}{\hbar^2}\left(1 - (V_0/E)|x/a|^\beta\right)}dx = (n + 1/2)\pi. \tag{3.73}$$

The substitution $t = (V_0/E)|x/a|^\beta$ gives

$$x = (E/V_0)^{1/\beta}at^{1/\beta}; \quad dx = (E/V_0)^{1/\beta}(a/\beta)t^{1/\beta - 1}dt. \tag{3.74}$$

This gives us

$$I = \frac{2a}{\beta}(E/V_0)^{1/\beta}\sqrt{\frac{2mE}{\hbar^2}}\int_0^1 t^{-1+1/\beta}\sqrt{1-t}\,dt. \tag{3.75}$$

We can do this integral, but we don't care about anything except the dependence on E.

$$(n + 1/2)\pi \propto E^{(1/2)+1/\beta}. \tag{3.76}$$

This is enough to tell us

$$E_n \propto (n + 1/2)^{2\beta/(2+\beta)}. \tag{3.77}$$

It is satisfying that we find for the simple harmonic oscillator, $E_n \propto (n + 1/2)$. For the Coulomb we have $\beta = -1$ and $E_n \propto (n+1/2)^{-2}$, which is correct in the large n limit, though we note that this is a stretch since the integral in I does not converge for $\beta = -1$.

In the case of the harmonic oscillator, we can do the integrals explicitly. Writing the potential as

$$V(x) = \frac{1}{2}m\omega_0^2x^2, \tag{3.78}$$

the quantization condition becomes

$$(n + \frac{1}{2})\pi = I = \int_{-x_0}^{x_0} k(x)dx = \int_{-x_0}^{x_0}\sqrt{\frac{2m}{\hbar^2}\left(E - \frac{1}{2}m\omega_0^2x^2\right)}dx, \tag{3.79}$$

where x_0 is the maximal value of x classically: $E = V(x_0)$. The substitution $\cos\theta = \sqrt{m\omega_0^2/(2E)}x$ yields the result

$$(n + \frac{1}{2})\pi = \pi\frac{E}{\hbar\omega_0}, \tag{3.80}$$

reproducing the well-known result $E = (n + \frac{1}{2})\hbar\omega_0$.

3.7 Normalization of WKB Bound State Wave Functions

For large n, the wave function undergoes many oscillations in the classically allowed region. The average of its absolute square is very nearly 1/2. If we write the WKB wave function as

$$\psi_{WKB}(x) = \frac{A_n}{\sqrt{k(x)}} \cos(\xi_1 - \pi/4), \tag{3.81}$$

then the normalization condition is, to a good approximation,

$$1 = A_n^2 \int_{x_1}^{x_2} \frac{dx}{2k(x)}. \tag{3.82}$$

For a potential with a single minimum, Furry proved that this is the proper formula, even when one keeps the turning point regions, etc. (Phys. Rev. **71**, 360 (1947)). This has an obvious interpretation:

$$\int_{x_1}^{x_2} \frac{dx}{k(x)} = \frac{\hbar}{m} \int_{x_1}^{x_2} \frac{dx}{v(x)} = \frac{\hbar \tau_{\text{classical}}}{2m} \tag{3.83}$$

where $\tau_{\text{classical}}$ is the classical period of the motion. Thus,

$$|A_n|^2 = \frac{4m}{\hbar \tau_{\text{classical}}} = \frac{2m\omega_{\text{classical}}}{\pi \hbar} \tag{3.84}$$

In the classically allowed region, the average probability density (over one wavelength) is

$$\rho_{WKB}(\tau)dx = \frac{|A_n|^2 dx}{2k(x)} = \frac{2m}{\hbar k(x)} \frac{dx}{\tau} = \frac{2}{\tau}dt. \tag{3.85}$$

This is just the normalized classical probability.

3.8 Bohr's Correspondence Principle and Classical Motion

Bohr's correspondence principle states that, in some sense, the limit of large quantum numbers corresponds to classical physics and, conversely, characteristics that emerge as black and white at the classical level, e.g. a symmetry property or special aspect of the motion, will manifest itself as a selection rule at the quantum level. The principle is a powerful tool in understanding quantum properties.

3.8.1 Energy Level Spacings and the Classical Frequency of Motion

A key aspect is the behavior of energy levels for large n and the relation to classical motion. The WKB quantization condition is an ideal vehicle for exploring the connection. We have

$$\int_{x_1(E)}^{x_2(E)} \sqrt{\frac{2m}{\hbar^2}(E - V(x))} \, dx = (n + \frac{1}{2})\pi. \tag{3.86}$$

We can think of n as a continuous variable. Then this equation defines $E = E(n)$. Evidently, the energy level spacing is given by $E_{n+1} - E_n \simeq \frac{dE}{dn}(E_n)$.

We can compute dE/dn by differentiating the left- and right-hand sides of Eq. (3.86). Since the left-hand side is a function E, we take

$$\frac{d}{dn}(LHS) = \frac{dE}{dn}\frac{d}{dE}\int_{x_1(E)}^{x_2(E)}\sqrt{\frac{2m}{\hbar^2}(E-V(x))}dx$$

$$= \frac{dE}{dn}\left\{\frac{2m}{\hbar^2}\times\frac{1}{2}\int_{x_1(E)}^{x_2(E)}\frac{dx}{k(x)} + \left[k(x)\Big|_{x_2}\frac{dx_2}{dE} - k(x)\Big|_{x_1}\frac{dx_1}{dE}\right]\right\}. \tag{3.87}$$

The piece in square brackets vanishes because $k(x)$ vanishes at the endpoints. But we showed previously that

$$\int_{x_1(E)}^{x_2(E)}\frac{dx}{k(x)} = \frac{\hbar\tau_{cl}}{2m}. \tag{3.88}$$

Thus,

$$\frac{d}{dn}(LHS) = \frac{\tau_{cl}}{2\hbar}\cdot\frac{dE}{dn}$$

$$\frac{d}{dn}(RHS) = \pi, \tag{3.89}$$

from which we learn

$$\frac{dE}{dn} = \frac{2\pi}{\tau_{cl}}\hbar = \hbar\omega_{cl}. \tag{3.90}$$

This establishes a theorem: in the limit of large quantum numbers, the spacing between adjacent levels is equal to $\hbar\omega_{cl}$, where $\omega_{cl} = 2\pi/\tau_{cl}$ is the classical frequency of the motion at that (average) energy.

3.8.2 Classical Particle Motion Versus the Schrödinger Wave Function de-Scription

When we speak of energy eigenstates, we focus on each state separately and consider

$$\psi_n(x,t) = \psi_n(x)e^{-iE_n t/\hbar}. \tag{3.91}$$

It is then not easy to see a connection with a particle oscillating back and forth in a classical picture. We choose to make up a wave packet to describe as well as we can the classical particle:

$$\psi(x,t) = \sum_n A_n\psi_n(x)e^{-iE_n t/\hbar}. \tag{3.92}$$

We center the expansion on $N = n_0$ and write $n = n_0 + m$. We assume that n_0 is large enough that $1 \ll |m_{max}| \ll n_0$, e.g. $n_0 = 10^6$, $m_{max} = 10^3$. (It will work approximately for $n_0 = 10, m_{max} = 3$.) Then we have

$$\psi(x,t) = \sum_m A_n\psi_n(x)e^{-iE_{n_0+m}t/\hbar}. \tag{3.93}$$

Multiply both sides by $e^{iE_{n_0}t/\hbar}$. Then we have

$$e^{iE_{n_0}t/\hbar}\psi(x,t) = \sum_m A_m \psi_{n_0+m} e^{-i(E_{n_0+m}-E_{n_0})t/\hbar}. \tag{3.94}$$

Now we can expand

$$E_{n_0+m} - E_{n_0} = \frac{dE}{dn}(E_{n_0})m + \frac{1}{2}\frac{d^2E}{dn^2}m^2 + \cdots \tag{3.95}$$

Assuming we can drop the non-leading pieces,

$$e^{iE_{n_0}t/\hbar}\psi(x,t) = \sum_m A_m \psi_{n_0+m} e^{-im\omega_{\mathrm{cl}}t/\hbar}. \tag{3.96}$$

We pick the A_m to form a localized wave packet at, say, $x = x_1$, at $t = 0$. The structure of the wave function shows that wave function is periodic with the classical period τ_{cl}. It will thus represent the classical motion of a particle oscillating back and forth.

3.8.3 Selection Rules for Radiative Transitions

Classically, an accelerating charged particle radiates. In the simplest circumstance, when a charged particle executes sinusoidal periodic motion, it radiates only at the basic frequency of the motion, i.e. $\omega = \omega_{\mathrm{cl}}$. The situation is different if the motion is relativistic or periodic but not sinusoidal. For example, an electron in a classical elliptical orbit has coordinates that can be expanded in a Fourier series

$$x(t) = \sum_m X_m(\epsilon)e^{im\omega_{\mathrm{cl}}t}$$

$$y(t) = \sum_m Y_m(\epsilon)e^{im\omega_{\mathrm{cl}}t}. \tag{3.97}$$

where the coefficients depend on the eccentricity ϵ of the orbit and the energy.

In general, dipole radiation will be emitted at the fundamental frequency, ω_{cl} and multiples of ω_{cl}. (See Jackson, *Classical Electrodynamics*, 3rd edition, Problem 14.22.) Only for a circular orbit will there be just the fundamental (because then there is only one $|m|$ value in each series).

Bohr's correspondence principle relates the classical occurrence of radiation to radiative transitions in the quantum system. Recall (a) Bohr's frequency condition $\hbar\omega_{\mathrm{rad}} = E_n - E_{n'}$ and (b) at high excitation, level spacing is $\Delta E = \hbar\omega_{\mathrm{cl}}$. This means that transitions with $\Delta E = m \times \hbar\omega_{\mathrm{cl}}$ will emit radiation with frequency $m \times \omega_{\mathrm{cl}}$. The correspondence principle says that the intensity of radiation at $\omega = m\times\omega_{\mathrm{cl}}$ in the classical motion will, at high quantum numbers, be equal to the quantum intensity for a transition $E_n \to E_{n'}$, where $E_n - E_{n'} = m \times \hbar\omega_{\mathrm{cl}}$.

The $\Delta\ell = \pm1$ selection rule for dipole radiation emerges from an action-angle description. Circular orbits, where ℓ is maximal for a given energy level, can radiate down only to the next lower circular orbit. However, elliptical orbits, where ℓ is not maximal, can radiate $2\hbar\omega_{\mathrm{cl}}, 3\hbar\omega_{\mathrm{cl}}$, while still obeying the restriction $\Delta\ell = \pm1$. See Figure 3.8.

3.8.4 WKB Approximation to the Virial Theorem

In Problem 1.7, we showed that

$$\left\langle \frac{p^2}{2m} \right\rangle = \frac{1}{2}\langle \mathbf{x} \cdot \nabla V \rangle \tag{3.98}$$

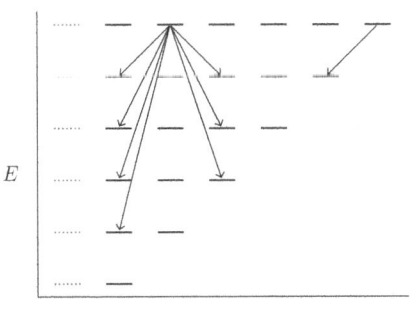

Figure 3.8 The semi-classical and quantum pictures of dipole radiation agree at large quantum numbers, in accord with Bohr's Correspondence Principle. Circular orbits decay only to circular orbits at the classical frequency, but elliptical orbits - ones with less than maximal angular momentum - can decay to multiple energy levels, all in keeping with the restriction $\Delta \ell = \pm 1$.

for energy eigenstates that are localized. There is a WKB version of this relation that follows directly by considering the kinetic energy. Now the square of the WKB bound-state wave function averages to

$$|\psi_{WKB}(x)|^2 \simeq \frac{1}{2} \frac{|A_n|^2}{k(x)} \tag{3.99}$$

within the classically allowed region, so

$$\langle T \rangle_n = \frac{\hbar^2}{2m} \langle k^2 \rangle_n \simeq \frac{\hbar^2}{2m} \frac{|A_n|^2}{2} \int_{x_1}^{x_2} \frac{k^2(x)dx}{k(x)}$$

$$= \frac{\hbar^2}{4m} |A_n|^2 \times (n + \frac{1}{2})\pi, \tag{3.100}$$

where we used the bound-state condition. But since

$$|A_n|^2 = \frac{2m}{\pi \hbar^2} \cdot \frac{dE}{dn} = \frac{2m}{\pi \hbar^2} \cdot \hbar \omega_{cl}, \tag{3.101}$$

we see that

$$\langle T \rangle_n \simeq \frac{1}{2}(n + \frac{1}{2})\hbar\omega_{cl}. \tag{3.102}$$

Similarly, we use

$$\hbar^2 k^2(x) = 2m(E - V(x)) \tag{3.103}$$

to show

$$\frac{dV}{dx} = -\frac{\hbar^2}{2m} \frac{d}{dx} k^2(x) = -\frac{\hbar^2 k(x)k'(x)}{m}. \tag{3.104}$$

So

$$\left\langle x \frac{dV}{dx} \right\rangle = -\frac{\hbar^2 |A_n|^2}{2} \int_{x_1}^{x_2} dx x \frac{k'(x)}{m} = \frac{\hbar^2 |A_n|^2}{2} \int_{x_1}^{x_2} dx \frac{k(x)}{m}$$

$$= 2 \langle T \rangle_n, \tag{3.105}$$

where we used the vanishing of $k(x)$ at the endpoints.

3.9 Power of WKB

The power of the WKB of approximation can be seen by comparing the WKB wave function with the exact solution at $x = 0$ for some particular cases. For the simple harmonic oscillator, the wave function is at a local maximum at $x = 0$ for symmetric solutions, that is, for n even. The exact solutions are

$$\psi_{2n}(q) = \frac{1}{2^n \sqrt{(2n)!}} \left(\frac{m\omega_0}{\hbar\pi} \right)^{1/4} H_{2n}\left(\sqrt{\frac{m\omega_0}{\hbar}} q \right) \exp\left(-\frac{m\omega_0}{2\hbar} q^2 \right). \tag{3.106}$$

The values of the even Hermite polynomials at $x = 0$ are given by Eq. (2.179).

$$H_{2n}(0) = \frac{(2n)!(-1)^n}{n!}. \tag{3.107}$$

Thus for the harmonic oscillator, the wave function at the origin is

$$\psi_{2n}(0) = \frac{(-1)^n \sqrt{2n!}}{\pi^{1/4} 2^n n!} \left(\frac{m\omega_0}{\hbar} \right)^{1/4}. \tag{3.108}$$

To compare with the WKB solution, we compute explicitly the WKB wave function. In the classically allowed region we find with $x_0^2 = \hbar/(m\omega_0)$

$$x_0^2 k^2(x) = (2n + 1 - (x/x_0)^2) = (x_t^2 - x^2)/x_0^2 \tag{3.109}$$

with $x_0 = \hbar/(m\omega_0)$. Then the turning points are at $x = \pm\sqrt{2n+1}\, x_0 = \pm x_t$.

$$\xi_2(x) = \int_x^{x_t} dx'\, k(x') = \frac{x_t^2}{x_0^2} \left\{ \frac{\pi}{4} - \frac{1}{2} \left[\sin^{-1}(x/x_t) + \frac{x}{x_t}(1 - (x/x_t)^2)^{1/2} \right] \right\}. \tag{3.110}$$

In particular, for $x = 0$, $\xi_2 = (2n + 1)\pi/4$, just half of the bound state condition, as expected since this is half the integral that occurs in the bound state relation. Moreover, since $\psi_{WKB} \propto \cos(\xi_2 - \pi/4)$, we see that the WKB wave function vanishes at the origin for odd n, as it should. The normalization is given by Eq. (3.84): $A_n = \sqrt{2/\pi}/x_0$, while $\sqrt{k(x)} = (x_t^2 - x^2)^{1/4}/x_0$. Altogether, in the classically allowed region

$$\psi_{WKB}(x) = \sqrt{\frac{2}{\pi}} (x_t^2 - x^2)^{-1/4} \cos(\xi_2(x) - \pi/4), \tag{3.111}$$

and in particular,

$$\psi_{WKB,2n}(0) = \sqrt{\frac{2}{\pi}} \frac{(-1)^n}{(4n+1)^{1/4}} \left(\frac{m\omega_0}{\hbar} \right)^{1/4}. \tag{3.112}$$

A comparison of the exact and WKB wave functions at the origin is given in Table 3.1. The accuracy of the WKB approximation is impressive. The agreement with the exact results is surprisingly good considering that WKB is a semi-classical approach.

The way that the WKB connection procedure cures the problem at a turning point can be illustrated with the harmonic oscillator by comparing the exact solution to the WKB solution. This is shown in Figure 3.9.

Table 3.1 Comparison of the wave function at the origin for the harmonic oscillator using the exact solution and the WKB approximation. We have suppressed a factor $(m\omega/\hbar)^{1/4}$ for each entry.

$2n$	exact	WKB
0	0.7511	0.7979
2	−0.5311	−0.5336
4	0.4600	0.4607
6	−0.4199	−0.4202
8	0.3928	0.3929

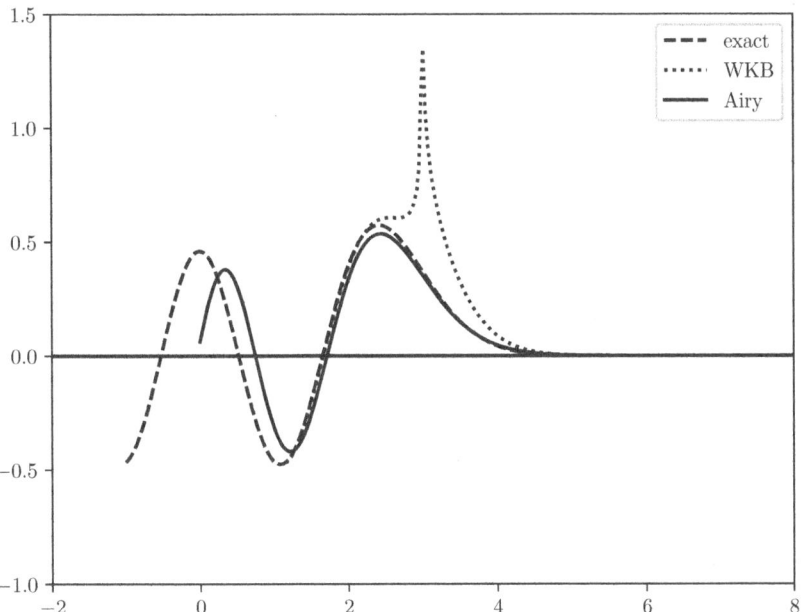

Figure 3.9 The connection formulae for WKB match the behavior of the WKB wave function on the two sides of the turning points. At the turning points, the WKB wave function diverges, preventing straightforward comparison of the function and its derivative at the two sides. The wave function for the $n = 4$ state of the simple harmonic oscillator is shown here in units of $a = \sqrt{m\omega/\hbar}$. The Airy function provides the connection between the two pieces of the WKB wave function, which is seen to be a very good approximation of the exact result, except near the turning point.

3.10 Barrier Penetration with the WKB Method

We wish to use WKB to discuss "leakage" through a barrier. Consider a "thick" barrier, as shown in Figure 3.10. Region I, to the left, $x < x_1$, is classically-allowed. Region II, $x_1 < x < x_2$ is classically forbidden. Finally, region III, $x > x_2$ is again classically allowed. We

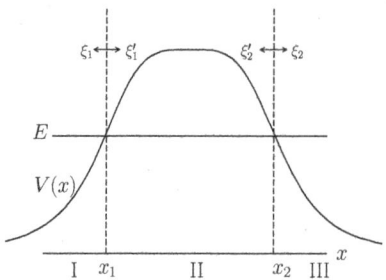

Figure 3.10 Representation of barrier penetration in WKB approximation. The turning points at x_1 and x_2 are indicated, as well as the orientation of the connection functions ξ_1, ξ_1', ξ_2, and ξ_2'.

assume $V(x)$ and E are such that WKB is applicable. For a wave proceeding from the left, at the interface at x_1 we go from using ξ_1 to ξ_1', while at x_2 we go from using ξ_2' to ξ_2. Physically, we have particles incident from the left upon the barrier; most are reflected, but a small probability exists for penetration through the barrier from I to III. Here, for the classically-allowed regions, we define

$$\xi_1 = \int_x^{x_1} k(x')dx'; \quad \xi_2 = \int_{x_2}^x k(x')dx', \tag{3.113}$$

while in the classically forbidden region we define

$$\xi_1' = \int_{x_1}^x \kappa(x')dx'; \quad \xi_2' = \int_x^{x_2} \kappa(x')dx'. \tag{3.114}$$

To find the transmission coefficient, we start in region III with a wave going to the right, with a phase chosen for convenience:

$$\psi_{III} \cong \frac{A}{\sqrt{k(x)}}e^{i(\xi_2 - \pi/4)}$$

$$= \frac{A}{\sqrt{k(x)}}[\cos(\xi_2 - \pi/4) + i\sin(\xi_2 - \pi/4)]. \tag{3.115}$$

We now use the WKB connection formula and find ψ in region II. From the sin term we know there is a contribution to ψ_{II}

$$-i\frac{Ae^{\xi_2'}}{\sqrt{\kappa}}. \tag{3.116}$$

We cannot determine how much there is of $e^{-\xi_2'}$ because there could be non-leading pieces from the sin term along with the known piece from the cos. However, for a thick barrier, such a term will be insignificant. Now we need to see what this tells us about the connection at x_1. We can set

$$\xi_1 + \xi_2 = \Theta \equiv \int_{x_1}^{x_2} \kappa(x)dx \tag{3.117}$$

and write

$$\psi_{II} \cong -i\frac{Ae^{\Theta - \xi_1'}}{\sqrt{\kappa}}. \tag{3.118}$$

Next, we make the connection to the classically allowed wave functions

$$\psi_I \cong -i\frac{Ae^{\Theta}}{\sqrt{k}}\left(2\cos(\xi_1 - \pi/4)\right)$$

$$= \frac{Ae^{\Theta}}{\sqrt{k(x)}}\left(e^{-i(\xi_1 + \pi/4)} - e^{i(\xi_1 + \pi/4)}\right). \tag{3.119}$$

The first term is the incoming wave and the second is the reflected wave. The transmission coefficient is defined as the ratio of the particle current emerging on the right to the incident current of the left of the barrier.

$$\mathbf{j} = \mathfrak{Re}\left(\psi^* \frac{\hbar}{im}\nabla\psi\right) = \frac{\hbar\mathbf{k}}{m}|\psi|^2 \tag{3.120}$$

Since

$$|\psi_{WKB}|^2 \propto \frac{1}{k(x)}, \tag{3.121}$$

we see that the transmission coefficient is defined as

$$T = \frac{j_{\text{outgoing}}}{j_{\text{incident}}} = \frac{k_{III}|\psi_{III}|^2}{k_1|\psi_I^{\text{inc}}|^2}. \tag{3.122}$$

Thus, the WKB approximation for the transmission coefficient is

$$T_{thick}^{WKB} \cong e^{-2\Theta} = \exp\left(-2\int_{x_1}^{x_2}\kappa(x)dx\right). \tag{3.123}$$

The barrier penetration formula has had a long and honorable history of application – Gamow-Gurney-Condon (1928-9) on α decay, Fowler and Nordheim (1929) for cold field emission from metals, μ-mesic catalysis of nuclear fusion (Jackson, 1957), etc.

3.11 Symmetrical Double-Well Potential

Some atomic and molecular systems exhibit closely degenerate states whose separation can be understood in terms of tunneling. As a model for the phenomenon, consider a symmetrical potential in one dimension, $V(-x) = V(x)$, with a shape as sketched in Figure 3.11.

Eventually we shall use WKB to find the energy levels, but first consider the extreme situation in which the central lump is infinitely high ($V_{\max} \to +\infty$). See Figure 3.12.

Figure 3.11 A symmetric double-well potential.

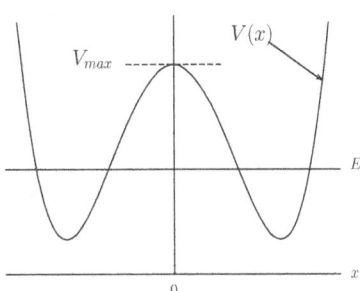

The particle is either in one well or the other, i.e. $\psi = \psi_n^{(L)}$ or $\psi = \psi_n^{(R)}$. But if we consider the combined potential (even though $V_{max} \to +\infty$), we know that solutions can be expressed as either even or odd in x. Thus,

$$\psi_n^{(\pm)}(x) = \frac{1}{\sqrt{2}}\left(\psi_n^{(L)} \pm \psi_n^{(R)}\right). \tag{3.124}$$

The energy levels E_n, $n = 0, 1, 2 \ldots$ are doubly degenerate.

Now we consider the circumstance in which V_{max} is finite, but large. Since the potential is even in x, the wave functions are either even or odd. We expect near degeneracy. (The even states will be below the odd states because they have the "smoother" wave functions and so less kinetic energy.)

We apply our WKB quantization procedure to the even (odd) states. See Figure 3.13. Define

$$\Theta_B = \int_{-x_1}^{x_1} \kappa(x')dx' \tag{3.125}$$

as the "phase" integral of the whole central barrier. Then within the barrier, on the interval $0 < x < x_1$, we have

$$\psi_{\mathrm{I}} = \frac{A}{\sqrt{\kappa(x)}} \frac{\cosh}{\sinh}(\Theta_B/2 - \xi_1') = \frac{A}{2\sqrt{\kappa(x)}}\left(e^{\xi_1'-\Theta_B/2} \pm e^{-\xi_1'+\Theta_B/2}\right). \tag{3.126}$$

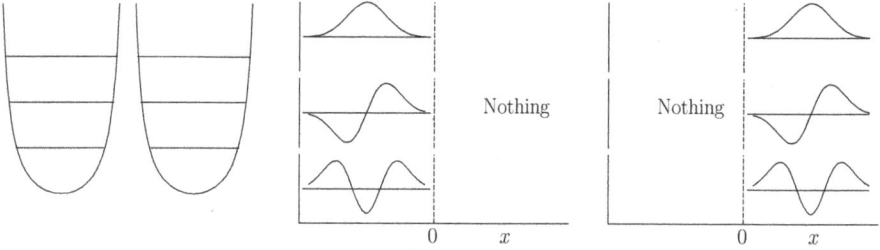

Figure 3.12 The extreme case of the symmetric double well in which the two wells are completely separated by an infinitely high barrier. There are solutions in which only one or the other of the two wells is occupied.

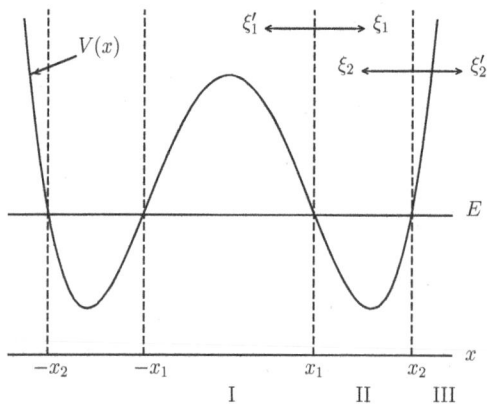

Figure 3.13 The double-well problem labeled by the classical turning points appropriate for a WKB treatment. In the regions I and III, classically the motion is forbidden and the wave function has exponential behavior. In region II, classically the motion is allowed and the wave function has oscillatory behavior.

Here we keep both exponentials because we have a symmetry property to lock onto. The solution in region II must connect properly with both regions I and III. The connection rules between regions I and II give us one version of ψ_{II}:

$$\psi_{\mathrm{II}}^{(1)} = \frac{A}{\sqrt{k(x)}} \left[-\frac{1}{2} e^{-\Theta_B/2} \sin(\xi_1 - \pi/4) \pm e^{\Theta_B/2} \cos(\xi_1 - \pi/4) \right]. \tag{3.127}$$

If instead we consider region III and demand a decaying exponential there, we will find in region II the solution is

$$\psi_{\mathrm{II}}^{(2)} = \frac{A'}{\sqrt{\kappa(x)}} \cos(\xi_2 - \pi/4). \tag{3.128}$$

Rewrite $\psi_{\mathrm{II}}^{(1)}$ as follows: put $\pm A e^{\Theta_B/2} = A''$. Then

$$\psi_{\mathrm{II}}^{(1)} = \frac{A''}{\sqrt{k(x)}} \left[\cos(\xi_1 - \pi/4) \mp \frac{1}{2} e^{-\Theta_B} \sin(\xi_1 - \pi/4) \right]. \tag{3.129}$$

Now define $\tan \alpha = e^{-\Theta_B}/2$ and find

$$\psi_{\mathrm{II}}^{(1)} = \frac{A'''}{\sqrt{k(x)}} \cos(\xi_1 - \pi/4 \pm \alpha). \tag{3.130}$$

With

$$\Theta_{12} = \int_{x_1}^{x_2} k(x') dx' \tag{3.131}$$

we have $\xi_1 + \xi_2 = \Theta_{12}$. Hence

$$\psi_{\mathrm{II}}^{(2)} = \frac{A'}{\sqrt{k(x)}} \cos(-\pi/4 + \xi_1 - \Theta_{12} + \pi/2)$$

$$= \psi_{\mathrm{II}}^{(1)} = \frac{A'''}{\sqrt{k(x)}} \cos(\xi_1 - \pi/4 \pm \alpha). \tag{3.132}$$

We thus have the eigenvalue condition

$$\Theta_{12} - \frac{\pi}{2} = n\pi \mp \alpha$$

$$\int_{x_1}^{x_2} k(x') dx' = (n + \frac{1}{2})\pi \mp \tan^{-1}(e^{-\Theta_B}/2), \tag{3.133}$$

where the upper sign pertains to the even wave function and the lower to the odd wave function. We see that if we neglect the barrier term, the eigenvalue condition is just that of the single side of the symmetric potential. For a thick barrier the angle α is small and we have

$$\int_{x_1}^{x_2} k(x') dx' = (n + \frac{1}{2})\pi \mp \frac{1}{2} e^{-\Theta_B}. \tag{3.134}$$

Now

$$F(E) = \int_{x_1}^{x_2} k(x') dx' \tag{3.135}$$

is a function of E because the turning points x_1 and x_2, as well as k are functions of E.

So the solutions at E_{odd} for the odd wave function and at E_{even} occur for

$$F(E_{\text{odd}}) = (n + \frac{1}{2})\pi + \frac{1}{2}e^{-\Theta_B}$$

$$F(E_{\text{even}}) = (n + \frac{1}{2})\pi - \frac{1}{2}e^{-\Theta_B} \tag{3.136}$$

so that the splitting ΔE between the even and odd states for a given n is

$$\Delta E \frac{dF}{dE} = e^{-\Theta_B}. \tag{3.137}$$

On the other hand,

$$\frac{dF}{dE} = \pi \frac{dn}{dE}. \tag{3.138}$$

In the harmonic oscillator approximation, $E_n = \hbar\omega_{\text{cl}}(n + 1/2)$, so $dE/dn = \hbar\omega_{\text{cl}}$. This frequency is that of the oscillation in the well on one side. This gives us an approximation for the energy splitting

$$\Delta E = \frac{\hbar\omega_{\text{cl}}}{\pi} e^{-\Theta_B}. \tag{3.139}$$

See Figure 3.14.

To interpret this result in terms of tunneling, put a particle in the right-hand well at $t = 0$:

$$\psi(x, 0) = \frac{1}{\sqrt{2}}\psi_{\text{even}}(x, 0) + \frac{1}{\sqrt{2}}\psi_{\text{odd}}(x, 0). \tag{3.140}$$

What happens for $t > 0$?

$$\psi(x, t) = \frac{1}{\sqrt{2}}\psi_{\text{even}}(x, 0)e^{-i(\omega_n - \Delta E/(2\hbar))t} + \frac{1}{\sqrt{2}}\psi_{\text{odd}}(x, 0)e^{-i(\omega_n + \Delta E/(2\hbar))t}$$

$$= e^{-i(\omega_n - \Delta E/(2\hbar))t}\frac{1}{\sqrt{2}}\left[\psi_{\text{even}}(x, 0) + \psi_{\text{odd}}(x, 0)e^{-i\Delta Et/\hbar}\right] \tag{3.141}$$

When $\Delta Et/\hbar = \pi$, the particle is primarily in the left well. The tunneling time is

$$\tau_{\text{tunnel}} = \frac{\hbar\pi}{\Delta E} = \frac{\pi^2 e^{\Theta_B}}{\omega_{\text{cl}}} = \frac{\pi}{2}e^{\Theta_B}\tau_{\text{cl}}, \tag{3.142}$$

where τ is the oscillation period for the similar harmonic oscillator. So the particle "knocks on the door" $\pi e^{\Theta_B}/2$ times on average before getting through. See Figure 3.15.

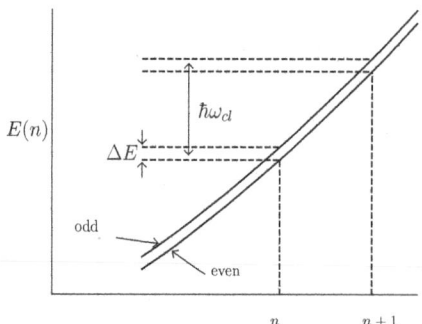

Figure 3.14 The odd solutions to the double-well problem lie higher than the even solutions for the same n by ΔE, which is given in the WKB approximation by $\Delta E = e^{-\Theta_B}\hbar\omega_{\text{cl}}/\pi$, where ω_{cl} is the natural frequency of the well and Θ_B is the WKB thickness of the barrier.

Figure 3.15 A particle placed in the right-hand well will oscillate back and forth between the two wells with a tunneling time $\tau = (\pi/2)e^{\Theta_B}\tau_{cl}$. So for a thick barrier, the tunneling time is very long compared the time for oscillation within one well.

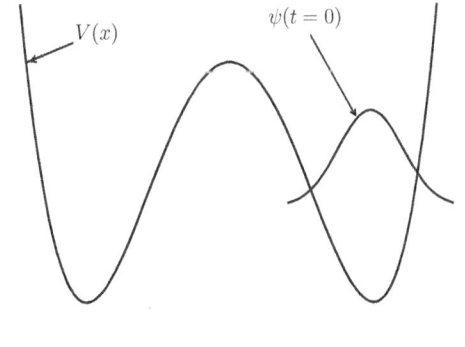

3.12 Application of the WKB Method to Ammonia Molecule

The ammonia molecule, NH_3, is pyramidal, with a base that is an equilateral triangle of hydrogen atoms with the nitrogen situated directly above the center of the triangle. See Figure 3.16.

The vibration due to a single well results in a frequency near 950 cm^{-1}, so this would be the separation between the two lowest modes. However, these two levels are themselves split because of the near degeneracy of the even and odd solutions to the two-well problem. The splitting of the lower level is 0.8 cm^{-1}, while the splitting of the next level is 36 cm^{-1}. Thus, for the lower level we find

$$e^{\Theta_B} = \frac{950}{\pi(0.8)} = 378 \qquad \text{or } \Theta_B \simeq 6. \tag{3.143}$$

For the second level

$$e^{\theta_B} = \frac{950}{\pi(36)} = 378 \qquad \text{or } \Theta_B \simeq 2 \tag{3.144}$$

in agreement with the expectation that the barrier is much thinner at the higher level. The splitting of the lower level, corresponding to a frequency of 23.87 GHz or a wavelength of 1.25 cm, was utilized in the invention of the maser by Charles Townes and collaborators in 1954. (J.P. Gordon, H.J. Zeiger, and C.H. Townes, *Phys. Rev.* **95**, 282 (1954).)

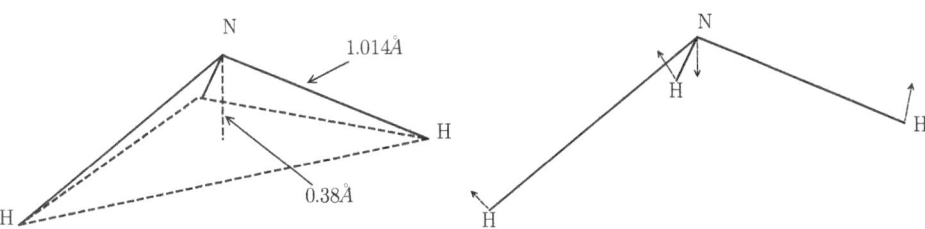

Figure 3.16 The ammonia molecule and its fundamental vibrational mode. Left: The nitrogen atom lies 0.38 Å above the plane of the hydrogen atoms. The angle between N-H bonds is 107°. Right: In the squashing mode, the nitrogen moves towards the plane of the hydrogen atoms while they move outward and upwards.

3.13   References and Suggested Reading

For the definitive mathematical treatment of one-dimensional WKB:

N. Fröman and P.O. Fröman, *JWKB Approximation: Contributions to the Theory*, North-Holland, Amsterdam (1965).

Important papers in the systematic treatment of WKB:

R. E. Langer, Phys. Rev. **51**, 669 (1937).
W. H. Furry, Phys. Rev. **71**, 360 (1947).

For the method of steepest descents and stationary phase:

Jeffreys and Jeffreys *Methods of Mathematical Physics*, pp. 501–511.
Dennery and Kryzwicki, *Mathematics for Physicists*, pp. 87–94.

For integral representations of special functions, see the sections on Bessel functions, etc. in:

Whittaker and Watson, *A Course of Modern Analysis*.
Jeffreys and Jeffreys, *Methods of Mathematical Physics*.
Dennery and Kryzwicki, *Mathematics for Physicists*.
Sommerfeld, *Partial Differential Equations in Physics*.

Problems

3.1 **(a)** Use the WKB quantization rule to estimate the number of bound states in the one-dimensional attractive potential $V(x) = -V_0/(\cosh x/a)^2$, as a function of $2mV_0a^2/\hbar^2$.

(b) From the result of (a), estimate the minimum strength V_0 such that there are just two bound states.

(c) The exact energy expression (Landau and Lifshitz, *Quantum Mechanics*, 2nd edition, pp.72–73) is

$$E_n = -\frac{\hbar^2}{2ma^2}\left[\frac{1}{2}\sqrt{1 + \frac{8mV_0a^2}{\hbar^2}} - (n + \frac{1}{2})\right]^2 ; \quad n = 0, 1, 2, ... \qquad (3.145)$$

Compare your result from (b) with the exact V_0 needed to give just two bound states. Is there one bound state regardless of how small V_0 is? Compare with WKB expectations and discuss any significant difference.

3.2 For a potential $V(x)$ that is infinitely repulsive for $x < 0$, but finite in magnitude for all finite $x > 0$, obtain the appropriate WKB quantization rule. This situation applies to r times the radial wave function for s-states in a spherically symmetric potential $V(r)$. Then $x = r$ is restricted to the range $0 < x < \infty$ and the wave function vanishes at $x = 0$. (Hint: consider the relationship of the energy levels for this potential with those of a potential that is equal to the present one for $x > 0$ and is even in x.)

3.3 Suppose that the potential in Problem 3.2 is $V = V_0x/a$ for $x > 0$, with $V_0/a = 1.2$ GeV/fm and the reduced mass in the Schrödinger equation is $m = 0.8$ GeV/c^2. This applies approximately to the "charmonium" system, a quark-antiquark pair bound by a linear potential conjectured to be relevant from considerations of quantum chromodynamics (QCD).

Determine the spacing in GeV between the first and second s-wave energy levels, using the WKB quantization rule of Problem 3.2. Compare with experiment (the spacing between the so-called J/ψ and the ψ'). See http://pdg.lbl.gov or the original discoveries, Aubert et al., *Phys. Rev.* **33**, 1404 (1974), Augustin et al., *Phys. Rev. Letters* **33**, 1406 (1974); Abrams et al., *Phys. Rev. Letters* **33**, 1453 (1974).

3.4* For a particle of mass m in a central field $V(r)$, the effective potential for radial motion is

$$V_{\text{eff}}(r) = V(r) + \frac{L^2\hbar^2}{2mr^2}, \tag{3.146}$$

where L is the magnitude of the angular momentum in units of $\hbar$. For large enough L, the centrifugal barrier keeps the particle away from the origin. Then the conventional WKB quantization rule applies.

(a) For an attractive Coulomb potential $V(r) = -e^2/r$, show that the WKB quantization condition can be written

$$n_r + \frac{1}{2} = \frac{1}{\pi} \int_{u_1}^{u_2} \frac{du}{u^2} \sqrt{-\epsilon + 2u - L^2 u^2}, \tag{3.147}$$

where $n_r = 0, 1, 2, ...$ is the radial quantum number, $u = a_0/r$ and $\epsilon = -2a_0 E/e^2$ is the binding energy in Rydberg units. The turning points where $k(r) = 0$ are u_1 and u_2.

(b) Evaluate the integral in (a) and determine ϵ in terms of n_r and L. What value of L makes the WKB answer agree with the exact result?

3.5* **(a)** Use the WKB approximation to calculate the transmission coefficient of the potential barrier

$$V(r) = 0, \qquad \text{for } r < R$$
$$V(r) = zZe^2/r, \qquad \text{for } r > R, \tag{3.148}$$

as shown in Figure 3.17, for a particle of mass m and kinetic energy $E < zZe^2/R$. This corresponds to the Coulomb barrier penetration for particles of charge ze, R being the nuclear radius. Of course the case of interest is $z = 2$, alpha decay. Express your answer in the form

Figure 3.17 Simple model for barrier penetration inside a nucleus showing Coulomb repulsion outside an attractive nuclear force.

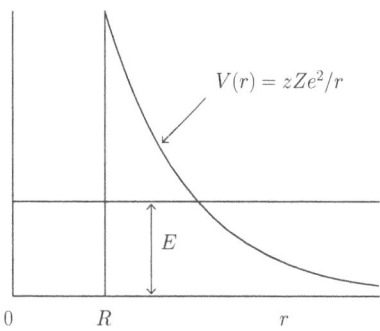

$$T \approx e^{-A/E^{1/2}+BR^{1/2}} \tag{3.149}$$

and determine the constants A and B in terms of the charges, mass, and other physical constants.

(b) Use uncertainty principle arguments to estimate the speed of an alpha particle ($m_\alpha = 4m_p$) inside the nucleus, and hence the typical rate at which the alpha particle strikes the nuclear surface. From this result and the answer from part (a), write down an approximate formula for the half-life of a radioactive alpha emitter. Evaluate the constant A and B in part (a) assuming the energy E is in MeV and the radius R in fm.

(c) Make a log-log plot of half-life in seconds versus energy in MeV with at least 20 decades vertically in half-life and from 3 to 10 MeV in energy (but still a log scale). Use the data on the Po (Z = 84) and Pu (Z = 94) alpha emitters in Tables 3.2 and 3.3 and put them on the plot, together with the curves from part (b).

Sources of data:

Table 3.2 Alpha decays of Po isotopes, with the alpha energy and half-life. If there are other decay channels, the $\log_{10}$ of the alpha branching ratio is given.

Po isotope	Half-life	Energy(MeV)	$-\log_{10}BR_\alpha$
192	3.4×10^{-2}s	7.170	
193	0.36s	6.940	
194	0.41s	6.960	
195	4.5s	6.609	
196	5.5s	6.520	
197	56s	6.281	0.4
198	1.76m	6.182	0.15
199	5.2m	5.952	0.9
200	11.5m	5.863	0.8
201	15.3m	5.683	1.8
202	44.7m	5.588	1.7
203	36.7m	5.384	3.0
204	3.53h	5.377	2.2
205	1.80h	5.220	2.9
206	8.80d	5.224	1.3
207	5.8h	5.115	3.7
208	2.9y	5.114	
209	102y	4.882	
210	138d	5.304	
211	.516s	7.450	
212	$3. \times 10^{-7}$s	8.784	
213	4.2×10^{-6}s	8.375	
214	1.6×10^{-4}s	7.687	
215	1.78×10^{-3}s	7.386	
216	0.15s	6.779	
217	10.s	6.539	
218	3.11m	6.022	

Table 3.3 Alpha decays of Pu isotopes, with the alpha energy and half-life. If there are other decay channels, the $\log_{10}$ of the alpha branching ratio is given.

Pu isotope	Half-life	Energy(MeV)	$-\log_{10} BR_{\alpha}$
232	34.1m	6.600	0.9
233	20.9m	6.300	2.9
234	8.8h	6.202	1.4
235	25.3m	5.850	4.7
236	2.85y	5.768	0.16
237	45.1d	5.333	4.7
238	87.74y	5.499	0.15
239	2.41×10^4y	5.157	0.14
240	6.56×10^3y	5.168	0.14
241	14.35y	4.869	4.7
242	3.76×10^5y	4.901	0.1
242	8.08×10^7y	4.589	0.1

Lederer and Shirley, *Table of Isotopes*, 7th edition.
Westmeier and Merklin, "Catalog of Alpha Particles from Radioactive Decay", Physik Daten/Physics Data No. 29-1 (1985).

3.6 Electrons in a metal can be treated approximately as free electrons. At low temperatures, they fill the Fermi sea to an energy E_F above the bottom of the conduction band. The surface of the sea is an energy ϕ below zero, as shown on the left in Figure 3.18. The depth of the Fermi sea below zero, ϕ, is called the work function of the metal.

When a strong electric field F is applied to the metal, the potential outside the metal becomes $V(z) = -eFz$, as shown on the right in Figure 3.18. (We ignore the image potential, $-e^2/4z$.)

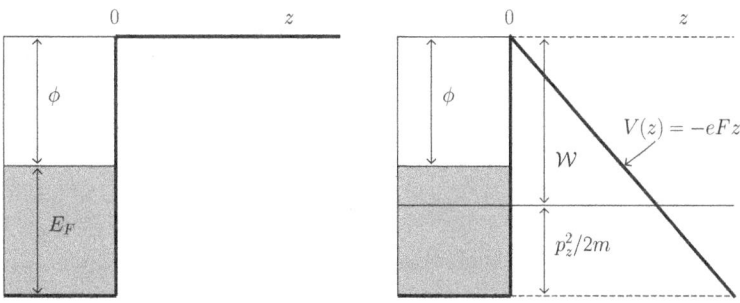

Figure 3.18 Simple model for field emission. The conduction band is filled up to the Fermi level E_F. The work function ϕ is the energy required to lift an electron from the Fermi level to the top of the conduction band. An electric field is applied and electrons with momentum p_z can tunnel through in field emission, escaping from the barrier of maximum height $\mathcal{W}$.

(a) For a typical electron in the sea, at an energy $\mathcal{W}$ below zero, use the WKB approximation to calculate the probability, $P(\mathcal{W})$, that it will leak through the barrier. Note that in terms of quantities already defined, $\mathcal{W} = \phi + E_F - p_z^2/2m$, where p_z is the component of the electron's momentum perpendicular to the surface of the metal. Remember also that the electrons fill the Fermi sphere in momentum space up to a maximum momentum $P_F = (2mE_F)^{1/2}$, related to the density of electrons.

(b) Use the result of part (a) and the fact that the electric current incident on the surface from inside is $-ep_z/m$ for an electron of momentum p_z, to integrate over the appropriate part of the Fermi distribution at zero temperature to obtain the cold-field emission formula of Fowler and Nordheim (1928):

$$ J = -\frac{e^3 F^2}{16\pi^2 \hbar \phi} \exp\left(-\frac{4}{3} \sqrt{\frac{2m}{\hbar^2}} \frac{\phi^{3/2}}{eF} \right). \tag{3.150} $$

Actually, Fowler and Nordheim had a factor of order unity out front, coming from a slightly fancier treatment of the barrier penetration. (Reference: Good, Jr. and Muller, "Field Emission", in Encyclopedia of Physics (Handbuch der Physik), Vol. XXI, Springer (1956), pp. 176–231.)

Cold field emission is important in electron and ion field emission spectroscopy, in particular the scanning tunneling microscope.

4

Rotations, Angular Momentum, and Central Force Motion

In many physical circumstances, the potential is spherically symmetric, that is, invariant under rotations. Thus, any solution to its Schrödinger equation, when rotated, remains a solution. The spherical harmonics, $Y_{\ell m}(\theta, \phi)$, provide the means to express such solutions since under rotations a single $Y_{\ell m}(\theta, \phi)$ can be expressed as a linear combination of $Y_{\ell, m'}(\theta, \phi)$ with $-\ell \leq m' \leq \ell$. Every rotation can be expressed in terms of the three Euler angles, α, β, γ. Since L^2 and L_z commute, we can establish a basis of states in which these are diagonalized. Because rotations leave the total angular momentum unchanged, every rotation can be represented by its matrix elements $D^j_{m'm}(\alpha, \beta, \gamma)$ between states $|jm\rangle$ and $|jm'\rangle$. The Clebsch-Gordan coefficients provide a multiplication table for representations of the rotation group. Many physical situations can be analyzed simply by exploiting rotational invariance as expressed through the Wigner-Eckart theorem. Armed with the spherical harmonics, we can reduce three-dimensional problems with rotational symmetry to more tractable one-dimensional problems. Of special significance is the Coulomb problem.

4.1 Infinitesimal Rotations

Earlier, we discussed the relationship of invariance of the Lagrangian under certain transformation to conservation laws.

(a) Time translation $\leftrightarrow$ Conservation of Energy
(b) Spatial translation $\leftrightarrow$ Conservation of Momentum

We now return to the third example: rotations.

(c) Rotations $\leftrightarrow$ Conservation of Angular Momentum

For an infinitesimal rotation $\delta\omega$, the transformations of position and velocity are given by

$$\mathbf{x} \to \mathbf{x} + \delta\omega \times \mathbf{x} = \mathbf{x}'$$
$$\mathbf{v} \to \mathbf{v} + \delta\omega \times \mathbf{v} = \mathbf{v}'. \tag{4.1}$$

See Figure 4.1. We seek the corresponding small unitary transformation

John David Jackson: A Course in Quantum Mechanics, First Edition. Robert N. Cahn.
© 2024 John Wiley & Sons, Inc. Published 2024 by John Wiley & Sons, Inc.
Companion website: www.wiley.com/go/Jackson/QuantumMechanics

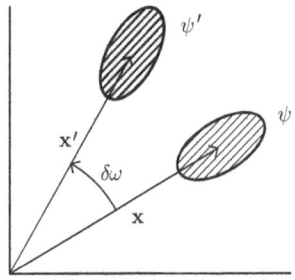

Figure 4.1 An infinitesimal rotation of $\delta\omega$ moves the state $|\psi\rangle$ to a new state $|\psi'\rangle$.

$$\delta U = 1 - i\mathcal{O}{\cdot}\delta\omega. \tag{4.2}$$

In quantum mechanics, consider the Schrödinger wave function $\psi_\alpha(\mathbf{x}) = \langle\mathbf{x}|\alpha\rangle$. The rotated state is produced by a unitary transformation,

$$\psi'_\alpha(\mathbf{x}) = \delta U\psi_\alpha(\mathbf{x}) = \psi_\alpha(\mathbf{x} - \delta\omega\times\mathbf{x}). \tag{4.3}$$

Now examine the Taylor series expansion

$$\delta U\psi(\mathbf{x}) = \psi(\mathbf{x}) - (\delta\omega\times\mathbf{x})\cdot\nabla\psi(\mathbf{x}) + \cdots$$
$$= \psi(\mathbf{x}) - i\frac{\delta\omega}{\hbar}\cdot\left(\frac{\hbar}{i}\mathbf{x}\times\nabla\right)\psi(\mathbf{x}) + \cdots$$
$$= \left(1 - i\frac{\delta\omega}{\hbar}\cdot\mathbf{L}\right)\psi(\mathbf{x}) + \cdots \tag{4.4}$$

Thus, we identify

$$\delta U = 1 - \frac{i}{\hbar}\delta\omega\cdot\mathbf{L}, \tag{4.5}$$

where $\mathbf{L}$ is the operator for orbital angular momentum.

We can generalize this by considering a many-particle wave function $\psi(\mathbf{x}_1, \mathbf{x}_2, \cdots, \mathbf{x}_N)$ for which again

$$\delta U = 1 - \frac{i}{\hbar}\delta\omega\cdot\mathbf{L}, \tag{4.6}$$

where

$$\mathbf{L} = \sum_j \mathbf{L}_j. \tag{4.7}$$

A second and fundamental generalization is required to account for half-integral angular momenta, for which there is no classical counterpart. Here we must postulate a rotation operator $\mathbf{J}$ that acts on all states, with integral or half-integral angular momenta so that

$$\delta U = 1 - i\delta\omega\cdot\mathbf{J}. \tag{4.8}$$

Here, and mostly in the future, we suppress the $\hbar$ and speak of $\mathbf{J}$ as the angular momentum.

Many operators have very specific properties under rotations. Operators S that are invariant under rotations are termed scalar operators:

$$[S, \mathbf{J}] = 0. \tag{4.9}$$

For example, $\mathbf{x}_1 \cdot \mathbf{x}_2, \mathbf{A} \cdot \mathbf{p}$, and $\mathbf{E} \cdot \mathbf{r}$ are scalar operators.

Vector operators are those that transform just as coordinates transform. Classically, this is described by

$$\mathbf{V} \to \mathbf{V}' = \mathbf{V} + \delta\boldsymbol{\omega} \times \mathbf{V}. \tag{4.10}$$

Quantum mechanically, we expect that the relation will preserve the classical interpretation, namely expectation values with respect to the rotated states will be the expectation values of $\mathbf{V}'$ with respect to the original states, i.e.

$$\langle \beta \left| \mathbf{V}' \right| \alpha \rangle = \langle \beta' \left| \mathbf{V} \right| \alpha' \rangle = \langle \beta \left| U^{\dagger} \mathbf{V} U \right| \alpha \rangle \tag{4.11}$$

or $\mathbf{V}' = U^{\dagger}\mathbf{V}U$. For an infinitesimal rotation, $\delta U = 1 - i\delta\boldsymbol{\omega} \cdot \mathbf{J}$, we have

$$\mathbf{V}' = (1 + i\delta\boldsymbol{\omega} \cdot \mathbf{J})\mathbf{V}(1 - i\delta\boldsymbol{\omega} \cdot \mathbf{J}) = \mathbf{V} + i[\delta\boldsymbol{\omega} \cdot \mathbf{J}, \mathbf{V}]. \tag{4.12}$$

We can thus identify

$$\delta\boldsymbol{\omega} \times \mathbf{V} = i[\delta\boldsymbol{\omega} \cdot \mathbf{J}, \mathbf{V}], \tag{4.13}$$

which we can rewrite as

$$[\mathbf{a} \cdot \mathbf{J}, \mathbf{b} \cdot \mathbf{V}] = i(\mathbf{a} \times \mathbf{b}) \cdot \mathbf{V} \tag{4.14}$$

or

$$[\mathbf{J}_i, \mathbf{V}_j] = i\epsilon_{ijk}V_k. \tag{4.15}$$

Multiplying both sides by $\epsilon_{ijk'}$ and summing over i, j

$$\mathbf{J} \times \mathbf{V} + \mathbf{V} \times \mathbf{J} = 2i\mathbf{V}. \tag{4.16}$$

Since $\mathbf{J}$ itself transforms like a vector under rotations, we get

$$[J_{\alpha}, J_{\beta}] = i\epsilon_{\alpha\beta\gamma}J_{\gamma} \tag{4.17}$$

or

$$\mathbf{J} \times \mathbf{J} = i\mathbf{J}. \tag{4.18}$$

Were $\mathbf{J}$ simply a constant vector rather than an operator, the cross product would, of course, vanish. But the different components of $\mathbf{J}$ do not commute with each other. The cross product of vector operators needs to be understood in terms of Eq. 4.14. The commutation relations of the components of $\mathbf{L}$ among themselves must be completely analogous, and similarly for the components of $\mathbf{S}$, which generate rotations of spin.

Now we compute the commutator of $J^2 = \mathbf{J} \cdot \mathbf{J}$ with $\mathbf{V}$:

$$[J^2, \mathbf{b} \cdot \mathbf{V}] = [\mathbf{J} \cdot \mathbf{J}, \mathbf{b} \cdot \mathbf{V}] = \mathbf{J} \cdot [\mathbf{J}, \mathbf{b} \cdot \mathbf{V}] + [\mathbf{J}, \mathbf{b} \cdot \mathbf{V}] \cdot \mathbf{J}$$
$$= i\mathbf{J} \cdot (\mathbf{b} \times \mathbf{V}) + i(\mathbf{b} \times \mathbf{V}) \cdot \mathbf{J} = i\mathbf{b} \cdot (\mathbf{V} \times \mathbf{J} - \mathbf{J} \times \mathbf{V}) \tag{4.19}$$

or

$$[J^2, \mathbf{V}] = i(\mathbf{V} \times \mathbf{J} - \mathbf{J} \times \mathbf{V}). \tag{4.20}$$

In particular,

$$[J^2, \mathbf{J}] = 0. \tag{4.21}$$

4.2 Construction of Irreducible Representations

Since J_z commutes with J^2, they can be diagonalized simultaneously. First note that since J_z is self-adjoint, if we set $|\beta\rangle = J_z |\alpha\rangle$ then

$$\langle \alpha | J_z^2 | \alpha \rangle = \langle \beta | \beta \rangle \geq 0. \tag{4.22}$$

It follows that the spectrum (that is, the set of eigenvalues) of J^2 is non-negative and moreover

$$\langle \alpha | J^2 | \alpha \rangle \geq \langle \alpha | J_z^2 | \alpha \rangle. \tag{4.23}$$

Thus among the states with a particular eigenvalue for J^2, the square of any eigenvalue of J_z must be no larger than the eigenvalue of J^2. We now make the traditional assignments

$$J_\pm = J_x \pm i J_y, \tag{4.24}$$

so $J_\pm^\dagger = J_\mp$. We then find the commutation relations

$$[J_z, J_\pm] = \pm J_\pm; \qquad [J_\pm, J_\mp] = \pm 2 J_z; \qquad [J^2, J_\pm] = 0. \tag{4.25}$$

The key relations we need are expressions for J^2 in terms of $J_\pm$ and J_z:

$$\begin{aligned}
J^2 &= \frac{1}{2}(J_+ J_- + J_- J_+) + J_z^2 \\
&= J_+ J_- + \frac{1}{2}[J_-, J_+] + J_z^2 \\
&= J_+ J_- + J_z(J_z - 1) \\
&= J_- J_+ + J_z(J_z + 1),
\end{aligned} \tag{4.26}$$

which can be written

$$\begin{aligned}
J_- J_+ &= J^2 - J_z(J_z + 1) \\
J_+ J_- &= J^2 - J_z(J_z - 1).
\end{aligned} \tag{4.27}$$

Now suppose we have a simultaneous eigenstate $|\lambda, \mu\rangle$ of J^2 and J_z such that

$$\begin{aligned}
J^2 |\lambda, \mu\rangle &= \lambda |\lambda, \mu\rangle \\
J_z |\lambda, \mu\rangle &= \mu |\lambda, \mu\rangle,
\end{aligned} \tag{4.28}$$

where λ is real and non-negative and μ is real and bounded by $\mu^2 \leq \lambda$.

Now consider $J_\pm J^2 |\lambda, \mu\rangle$. Since $[J^2, J_\pm] = 0$, this is still an eigenstate of J^2 with eigenvalue λ. What can we say about J_z? Consider

$$\begin{aligned}
J_z J_\pm |\lambda, \mu\rangle &= (J_\pm J_z + [J_z, J_\pm]) |\lambda, \mu\rangle = (\mu J_\pm \pm J_\pm) |\lambda, \mu\rangle \\
&= (\mu \pm 1)(J_\pm |\lambda, \mu\rangle).
\end{aligned} \tag{4.29}$$

Thus, we have established that $J_\pm |\lambda, \mu\rangle$ is an eigenstate of J_z with eigenvalue $\mu \pm 1$. Using J_+, we can create eigenstates with higher and higher values of μ. This must terminate because $|\mu|^2 \leq \lambda$. Let the maximum value of μ be $\mu_{\max}$ and consider $|\lambda, \mu_{\max}\rangle$. We know that $J_+ |\lambda, \mu_{\max}\rangle$ must vanish. Now

$$\left\langle \lambda, \mu_{\text{max}} \left| J_- J_+ \right| \lambda, \mu_{\text{max}} \right\rangle = \left\langle \lambda, \mu_{\text{max}} \left| J^2 \right| \lambda, \mu_{\text{max}} \right\rangle - \left\langle \lambda, \mu_{\text{max}} \left| J_z (J_z + 1) \right| \lambda, \mu_{\text{max}} \right\rangle$$

$$0 = \lambda - \mu_{\text{max}}(\mu_{\text{max}} + 1). \tag{4.30}$$

It is useful to give μ_{max} a name: $\mu_{\text{max}} = j$, where j is real and non-negative. Then

$$\lambda = j(j + 1). \tag{4.31}$$

Similarly, there is a smallest (negative) μ value. Call it $-j'$. Then we have $j_- |\lambda, -j'\rangle = 0$ and we find

$$0 = \lambda - (-j')(-j' - 1); \quad \lambda = j'(j' + 1), \tag{4.32}$$

so $j' = j$. The values of μ run from $-j$ to j, a total of $2j + 1$ values. Since $2j + 1$ is an integer, j itself must be an integer or half-integer.

The Hilbert space for angular momentum decomposes into a set of invariant subspaces, one for each value of j, with $2j + 1$ linearly independent vectors in it. The set of $2j + 1$ vectors may be chosen as the eigenvectors of J_z for that value of j. We can choose a set of orthonormal vectors of J^2 and J_z, which we denote by $|jm\rangle$ with

$$J^2 |jm\rangle = j(j + 1) |jm\rangle; \qquad J_z |jm\rangle = m |jm\rangle. \tag{4.33}$$

where $m = -j, -j + 1, \ldots, j - 1, j$. It is clear that these states are orthogonal since we can compare the action of the self-adjoint operators J^2 and J_z acting to the left and right between states with different values of j or m:

$$\langle j', m' | J^2 | jm \rangle = j'(j' + 1) \langle j', m' | jm \rangle = j(j + 1) \langle j', m' | jm \rangle$$

$$\langle j', m' | J_z | jm \rangle = m' \langle j', m' | jm \rangle = m \langle j', m' | jm \rangle. \tag{4.34}$$

However, it still remains to normalize the states. Let $J_+ |jm\rangle = c_m |j, m + 1\rangle$. Now

$$\langle jm | J_- J_+ | jm \rangle = \langle jm | J^2 - J_z(J_z + 1) | jm \rangle = j(j + 1) - m(m + 1) = |c_m|^2 \tag{4.35}$$

The phase of c_m is undetermined and must be set by convention. The universal choice is the Condon and Shortley phase convention

$$J_+ |jm\rangle = \sqrt{j(j + 1) - m(m + 1)} \, |j, m + 1\rangle$$

$$J_- |jm\rangle = \sqrt{j(j + 1) - m(m - 1)} \, |j, m - 1\rangle. \tag{4.36}$$

It is easy to see that J_+ annihilates the state $|j, j\rangle$, while J_- annihilates $|j, -j\rangle$.

With these results in hand, we can display explicit representations of the operators J^2, J_x, J_y, J_z in the $|jm\rangle$ basis. From

$$\langle m' | J^2 | m \rangle = j(j + 1) \, \delta_{m'm}$$

$$\langle m' | J_z | m \rangle = m \, \delta_{m'm}$$

$$\langle m' | J_\pm | m \rangle = \sqrt{(j+1) - m(m \pm 1)} \, \delta_{m', m\pm 1} \tag{4.37}$$

we find

$$\langle m' | J_x | m \rangle = \frac{1}{2} \sqrt{(j+1) - m(m + 1)} \, \delta_{m', m+1} + \frac{1}{2} \sqrt{(j+1) - m(m - 1)} \, \delta_{m', m-1}$$

$$\langle m' | J_y | m \rangle = \frac{1}{2i} \sqrt{(j+1) - m(m + 1)} \, \delta_{m', m+1} - \frac{1}{2i} \sqrt{(j+1) - m(m - 1)} \, \delta_{m', m-1}. \tag{4.38}$$

For the fundamental case of $j = 1/2$, the matrices are

$$J^2 = \begin{pmatrix} \frac{3}{4} & 0 \\ 0 & \frac{3}{4} \end{pmatrix}, \quad J_x = \begin{pmatrix} 0 & \frac{1}{2} \\ \frac{1}{2} & 0 \end{pmatrix}, \quad J_y = \begin{pmatrix} 0 & -\frac{i}{2} \\ \frac{i}{2} & 0 \end{pmatrix}, \quad J_z = \begin{pmatrix} \frac{1}{2} & 0 \\ 0 & -\frac{1}{2} \end{pmatrix}. \tag{4.39}$$

With these matrices, we can use explicit column vectors for the eigenkets

$$|1/2, 1/2\rangle \rightarrow \begin{pmatrix} 1 \\ 0 \end{pmatrix}; \qquad |1/2, -1/2\rangle = \begin{pmatrix} 0 \\ 1 \end{pmatrix}. \tag{4.40}$$

These matrices are related to the Pauli spin matrices σ_i:

$$\mathbf{J} = \frac{1}{2}\boldsymbol{\sigma}. \tag{4.41}$$

The case of angular momentum 1/2 is so special that it has its own notation: $\mathbf{J} \rightarrow \mathbf{s} = \frac{1}{2}\boldsymbol{\sigma}$, with

$$\sigma_x = \begin{pmatrix} 0 & 1 \\ 1 & 0 \end{pmatrix}, \quad \sigma_y = \begin{pmatrix} 0 & -i \\ i & 0 \end{pmatrix}, \quad \sigma_z = \begin{pmatrix} 1 & 0 \\ 0 & -1 \end{pmatrix} \tag{4.42}$$

and we denote the column vectors by

$$\alpha = \begin{pmatrix} 1 \\ 0 \end{pmatrix}; \qquad \beta = \begin{pmatrix} 0 \\ 1 \end{pmatrix}. \tag{4.43}$$

There are special relations that apply for the Pauli matrices without analogs for $j \neq 1/2$:

$$\sigma_x^2 = \sigma_y^2 = \sigma_z^2 = I$$
$$\sigma_x \sigma_y = i\sigma_z, \quad \text{etc.}$$
$$\sigma_i \sigma_j + \sigma_j \sigma_i = 2\delta_{ij}$$
$$\boldsymbol{\sigma} \cdot \mathbf{a}\, \boldsymbol{\sigma} \cdot \mathbf{b} = \mathbf{a} \cdot \mathbf{b} + i\boldsymbol{\sigma} \cdot (\mathbf{a} \times \mathbf{b}). \tag{4.44}$$

It is convenient to define, as well,

$$\sigma_+ = \frac{1}{2}(\sigma_x + i\sigma_y) = \begin{pmatrix} 0 & 1 \\ 0 & 0 \end{pmatrix} \tag{4.45}$$

$$\sigma_- = \frac{1}{2}(\sigma_x - i\sigma_y) = \begin{pmatrix} 0 & 0 \\ 1 & 0 \end{pmatrix}. \tag{4.46}$$

Within the subspace of spin-1/2, we can use the special properties of the 2 x 2 Pauli matrices to simplify the unitary operation describing a finite rotation. For a rotation by an angle θ around the direction defined by the unit vector $\hat{\mathbf{n}}$, the unitary operator is

$$U = e^{-i\theta\hat{\mathbf{n}}\cdot\mathbf{J}}. \tag{4.47}$$

For systems with $j = 1/2$ we have

$$U = e^{-i\frac{\theta}{2}\hat{\mathbf{n}}\cdot\boldsymbol{\sigma}} = 1 - i\theta/2(\hat{\mathbf{n}}\cdot\boldsymbol{\sigma}) + (-i\theta/2)^2\frac{1}{2!}(\hat{\mathbf{n}}\cdot\boldsymbol{\sigma})^2 + (-i\theta/2)^3\frac{1}{3!}(\hat{\mathbf{n}}\cdot\boldsymbol{\sigma})^3 \cdots \quad (4.48)$$

Since $(\hat{\mathbf{n}}\cdot\boldsymbol{\sigma})^{2n} = I$ and $(\hat{\mathbf{n}}\cdot\boldsymbol{\sigma})^{2n+1} = \hat{\mathbf{n}}\cdot\boldsymbol{\sigma}$ we see that

$$e^{-i\frac{\theta}{2}\hat{\mathbf{n}}\cdot\boldsymbol{\sigma}} = \cos(\theta/2) - i\sin(\theta/2)\hat{\mathbf{n}}\cdot\boldsymbol{\sigma}. \quad (4.49)$$

This is a convenient representation for practical use.

4.3 Coordinate Representation of Angular Momentum Eigenvectors

In the Schrödinger picture we often make use of coordinate representatives of angular momentum eigenvectors. Let us consider how to construct them. Among the $2j + 1$ kets $|jm\rangle$, the "top" and "bottom" kets are special. We have

$$J_+ |j, m = j\rangle = 0. \quad (4.50)$$

Let us suppose that $\mathbf{J} = -i\mathbf{r} \times \nabla$. In spherical coordinates, this becomes

$$\mathbf{J} = -i\hat{\boldsymbol{\phi}}\frac{\partial}{\partial\theta} + i\hat{\boldsymbol{\theta}}\frac{1}{\sin\theta}\frac{\partial}{\partial\phi}, \quad (4.51)$$

from which we can find

$$J_+ = e^{i\phi}\left(\frac{\partial}{\partial\theta} + i\cot\theta\frac{\partial}{\partial\phi}\right)$$

$$J_- = e^{-i\phi}\left(-\frac{\partial}{\partial\theta} + i\cot\theta\frac{\partial}{\partial\phi}\right)$$

$$J_z = -i\frac{\partial}{\partial\phi}$$

$$J^2 = -\frac{\partial^2}{\partial\theta^2} - \cot\theta\frac{\partial}{\partial\theta} - \frac{1}{\sin^2\theta}\frac{\partial^2}{\partial\phi^2}. \quad (4.52)$$

If we indicate the coordinate representation of $|jm\rangle$ by $\psi_{jm}(\theta,\phi)$, we have the defining equation

$$\left(\frac{\partial}{\partial\theta} + i\cot\theta\frac{\partial}{\partial\phi}\right)\psi_{jj}(\theta,\phi) = 0 \quad (4.53)$$

and also

$$-i\frac{\partial}{\partial\phi}\psi_{jj} = j\psi_{jj}. \quad (4.54)$$

With the substitution $\psi_{jj} = e^{ij\phi}\Theta(\theta)$ we find

$$\left(\frac{\partial}{\partial\theta} - j\cot\theta\right)\Theta(\theta) = 0, \quad (4.55)$$

whose solution is

$$\Theta(\theta) = const. \times (\sin\theta)^j \quad (4.56)$$

so that

$$\psi_{jj}(\theta, \phi) = const. \times e^{ij\phi}(\sin \theta)^j = const. \times \left(\frac{x + iy}{r}\right)^j. \tag{4.57}$$

If j is an integer ℓ, we can see that $\psi_{\ell\ell}(\theta, \phi)$ is none other than $Y_{\ell\ell}$ and we can use $(J_-)^{\ell-m}$ to get $Y_{\ell,m}$, the well-known spherical harmonics. Suppose, on the contrary, $j = 1/2$. We would then find

$$\psi_{1/2,1/2} \propto \sqrt{\sin \theta} e^{i\phi/2}. \tag{4.58}$$

Not only is this a dubious function because $\sin \theta$ can be negative as well as positive, but if we try to form $\psi_{1/2,-1/2}$ we find

$$\psi_{1/2,-1/2} \propto J_-\psi_{1/2,1/2} \propto e^{-i\phi}\left(\frac{\partial}{\partial \theta} - j \cot \theta\right)\sqrt{\sin \theta} e^{i\phi/2}$$

$$\propto -\frac{\cos \theta}{\sqrt{\sin \theta}}, \tag{4.59}$$

which is singular already at $\theta = 0$. While this function is integrable for $0 \le \theta \le \epsilon$, it would be a point source of probability current. The conclusion is that there are no coordinate representations for half-integral singular momenta. We can draw the same conclusion by considering the transformation of a Schrödinger wave function under rotations.

$$\psi(\mathbf{x}) \rightarrow \psi'(\mathbf{x}) = U(R)\psi(\mathbf{x}) = \psi(R^{-1}\mathbf{x}). \tag{4.60}$$

Now rotating $\mathbf{x}$ by 2π about any axis leaves $\mathbf{x}$ unchanged and so $\psi(\mathbf{x})$ must be similarly unchanged by such a rotation. But we know for $j = 1/2$

$$U = \cos \theta/2 - i\boldsymbol{\sigma} \cdot \hat{\mathbf{n}} \sin \theta/2, \tag{4.61}$$

so if $\theta = 2\pi$ then $U = -I$, contradicting the requirement of a Schrödinger wave function. Indeed, quite generally for j half-integral $U(2\pi) = -I$, though this cannot be realized with a co-ordinate space representation.

4.4 Observation of Sign Change for Rotation by 2π

Lest the reader imagine that the relation $U(2\pi) = -I$ is a purely mathematical construct, we examine the experiment of S. A. Werner, et al., *Phys. Rev. Lett.* **35**, 1053 (1975). The principle features of the experiment are shown in Figure 4.2. An unpolarized beam of neutrons whose momenta correspond to a wavelength $\lambda = 0.1445$ fm impinges on a thin slab of crystalline silicon at A. The beam is split by Bragg reflection. The beam proceeding from A to C passes for a distance ℓ through a region of magnetic field B, perpendicular to the path. At points B and C, the beams again undergo Bragg reflection and are recombined in the third slab of silicon at D.

If the spatial wave amplitude associated with the neutron entering the interferometer is I_0, the amplitude transmitted and refracted beams have amplitudes tI_0 and $i(rI_0)$, where t and r are relatively real and $t^2 + r^2 = 1$. See Problem 4.7. The wave amplitude in segment BD is $i(rtI_0)$, while in the absence of the magnetic field in segment CD it is $-r^2I_0$. At D there is interference and at C_2 the wave amplitude is $i(rt^2I_0) - i(r^3I_0)$, while at C_3 it is $-2r^2tI_0$. These amplitudes are modified by the application of the magnetic field.

Figure 4.2 Schematic representation of the apparatus demonstrating that a rotation of the neutron's spin by 4π is necessary to complete a full oscillation. The shaded region represents the presence of a magnetic field. Bragg reflection occurs at the points A, B, C, and D. Rates are measured at C_2 and C_3. Adapted from Werner, et al., 1975.

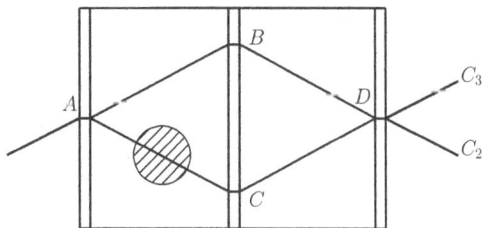

The interaction Hamiltonian arising from the magnetic field is

$$H = -\boldsymbol{\mu} \cdot \mathbf{B}, \tag{4.62}$$

where

$$\boldsymbol{\mu} = g_N \mu_N \boldsymbol{\sigma}. \tag{4.63}$$

Here $\mathbf{s} = (1/2)\boldsymbol{\sigma}$ is the neutron's spin, $\mu_N = e\hbar/(2m_p c) = 3.15 \times 10^{-8}$ eV/T in SI units, and g_N is the neutron's magnetic moment, -1.913, measured in nuclear magnetons, μ_N. (Despite the notation, g_N is not the neutron's g-factor, which is twice as large since the neutron has spin 1/2.)

The neutron passes through the length of the magnetic field ℓ in a time $T = \ell/v$, where v is the neutron's velocity. The rotation of the neutron's spin is then governed by

$$\exp(-i\boldsymbol{\mu} \cdot \mathbf{B} T/\hbar) = \exp\left(-i\frac{2g_N\mu_N B\ell}{\hbar v}\frac{1}{2}\boldsymbol{\sigma} \cdot \hat{\mathbf{B}}\right), \tag{4.64}$$

where we have displayed rotation operator explicitly so the resulting rotation is apparent:

$$\theta = \frac{2g_N\mu_N B\ell}{\hbar v}. \tag{4.65}$$

Because the neutron beams undergo Bragg scattering, it is appropriate to use in place of the neutron's velocity its corresponding de Broglie wavelength defined through

$$\lambda k = 2\pi = \lambda(p/\hbar); \qquad \lambda = 2\pi\hbar/p = 2\pi\hbar/(mv). \tag{4.66}$$

Thus, we can write the rotation angle θ about the axis defined by the direction of the magnetic field as

$$
\begin{aligned}
\theta/(B\ell\lambda) &= 4\pi \frac{g_N\mu_N m_N}{h^2} \\
&= \frac{4\pi \cdot 1.913 \times 3.15 \times 10^{-8}\text{eV/T} \cdot 939 \times 10^6 \text{eV/c}^2}{(4.136 \times 10^{-15}\text{eV-s})^2} \\
&= 4.61 \times 10^{-2} \, (\text{G cm Å})^{-1},
\end{aligned} \tag{4.67}
$$

where we have used gauss, centimeters, and Angstroms in keeping with the experimental paper. We anticipate that a full rotation of 4π required to return a spin-1/2 particle to its original phase will require a combination of the strength of the magnetic field and its length traversed

$$B\ell = \frac{4\pi}{0.0461\lambda} = \frac{272}{\lambda} \tag{4.68}$$

with B in Gauss, ℓ in centimeters, and λ in Angstroms.

The operator $\exp(-i\theta\boldsymbol{\sigma}\cdot\mathbf{B}/2)$ will act on the neutron spin. Since the beam is unpolarized, we can imagine it to be an incoherent sum of spin-up and spin-down components in equal proportions, with up and down defined relative to the direction of the magnetic field. The full wave function is the product of the spatial wave function and the spin wave function. If we have a spin-up neutron entering the magnetic field, at the end we have

$$\exp(-i\theta\frac{1}{2}\boldsymbol{\sigma}\cdot\hat{\mathbf{B}})\begin{pmatrix}1\\0\end{pmatrix} = \left[\cos(\theta/2)I - i\sin(\theta/2)\hat{\mathbf{B}}\cdot\boldsymbol{\sigma}\right]\begin{pmatrix}1\\0\end{pmatrix} = \begin{pmatrix}e^{-i\theta/2}\\0\end{pmatrix}. \tag{4.69}$$

Similarly

$$\exp(-i\theta\frac{1}{2}\boldsymbol{\sigma}\cdot\hat{\mathbf{B}})\begin{pmatrix}0\\1\end{pmatrix} = \left[\cos(\theta/2)I - i\sin(\theta/2)\hat{\mathbf{B}}\cdot\boldsymbol{\sigma}\right]\begin{pmatrix}0\\1\end{pmatrix} = \begin{pmatrix}0\\e^{i\theta/2}\end{pmatrix}. \tag{4.70}$$

Thus, the presence of the magnetic field inserts a factor $e^{\pm i\theta/2}$ into the amplitude for the path through segment AC. The resulting wave amplitudes are

$$C_3 : -r^2t(1 + e^{\pm i\theta/2})I_0 \tag{4.71}$$

$$C_2 : irt^2I_0 - ir^3e^{\pm i\theta/2}I_0. \tag{4.72}$$

The corresponding intensities are

$$C_3 : 2r^4t^2(1 + \cos(\theta/2))|I_0|^2 \tag{4.73}$$

$$C_2 : \left[r^2 - 2r^4t^2(1 + \cos(\theta/2))\right]|I_0|^2. \tag{4.74}$$

The quantity of interest is difference between the rates:

$$[r^2 - 4r^4t^2(1 + \cos(\theta/2))]|I_0|^2. \tag{4.75}$$

The data of Werner, et al. are shown in Figure 4.3.

The effective length of the path through the magnetic field was determined to be 2.7 cm. From the graph, we read that the full period of the oscillation as a function of the magnetic field was near 62 gauss. Together then,

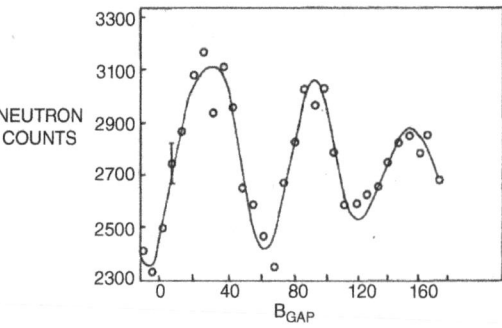

Figure 4.3 Data showing the difference between the counters at C_2 and C_3. Werner, et al., 1975 / American Physical Society.

$$B\ell\lambda = 2.7 \times 62 \times 1.445\,\text{G cm Å} = 242\,\text{G cm Å} \tag{4.76}$$

in pretty good agreement with the expectation calculated above.

4.5 Euler Angles, Wigner d-functions

From classical mechanics, we know that the orientation of a rigid body with respect to some standard orientation can be specified by three angles, called the Euler angles (α, β, γ).

Three successive rotations are required. The unitary operator is thus

$$U = U_3 U_2 U_1. \tag{4.77}$$

There are many conventions for choosing these. This is ours:

- Rotation 1: Positive rotation of α about the laboratory z axis, taking the y-axis fixed to the body to a new direction y''. Similarly, the x direction fixed to the body is now x''.
- Rotation 2: Positive rotation by β around the direction y''. This leaves y'' fixed but moves z to z' and x'' to x'''.
- Rotation 3: Positive rotation by γ about the z' direction, x''' to the final position x' and y'' to y' so that finally the body's x,y, and z directions now point to x', y' and z'. See Figure 4.4.

Thus,

$$U = e^{-i\gamma J_{z'}} e^{-i\beta J_{y''}} e^{-i\alpha J_z}. \tag{4.78}$$

This is not too convenient because it involves components of $\mathbf{J}$ along the y'' and z', rather than x, y, z. We recognize that the operator $J_{y''}$ acts on the rotated states $U_1\,|jm\rangle$ just the way J_y acts or $|jm\rangle$. Thus,

$$\left\langle \beta \left| U_1^\dagger J_{y''} U_1 \right| \alpha \right\rangle = \left\langle \beta \left| J_y \right| \alpha \right\rangle, \tag{4.79}$$

so

$$J_{y''} = U_1 J_y U_1^\dagger. \tag{4.80}$$

Similarly,

$$J_{z'} = U_2 U_1 J_z U_1^\dagger U_2^\dagger. \tag{4.81}$$

Figure 4.4 The Euler angles, α, β, γ define a rotation.

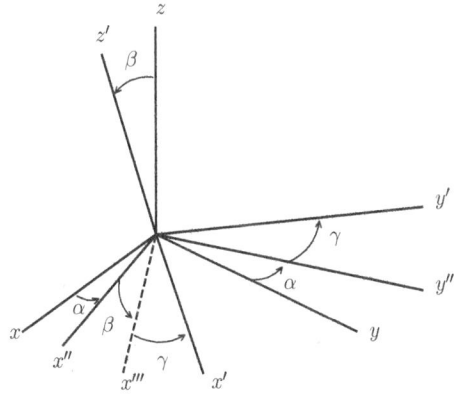

It follows that

$$U_3 = e^{-i\gamma J_{z'}} = U_2 U_1 e^{-i\gamma J_z} U_1^\dagger U_2^\dagger \tag{4.82}$$

and

$$U_2 = e^{-i\beta J_{y''}} = U_1 e^{-i\beta J_y} U_1^\dagger. \tag{4.83}$$

Altogether then,

$$U(\alpha, \beta, \gamma) = U_3 U_2 U_1 = U_2 U_1 e^{-i\gamma J_z} U_1^\dagger U_2^\dagger U_2 U_1 \tag{4.84}$$

$$= U_2 U_1 e^{-i\gamma J_z} = U_1 e^{-i\beta J_y} U_1^\dagger U_1 e^{-i\gamma J_z} \tag{4.85}$$

$$= e^{-i\alpha J_z} e^{-i\beta J_y} e^{-i\gamma J_z}. \tag{4.86}$$

The order of appearance of the angles is opposite to the original form, and only components with respect to the "laboratory" axes occur!

Armed with this expression, we are in position to determine the action of any rotation R, specified by α, β, γ, on our basis states, $|jm\rangle$. We indicate the result of rotating the basis states by

$$|jm\rangle_R = U(R) |jm\rangle. \tag{4.87}$$

Because J^2 commutes with J_i, it commutes as well with $U(R)$. Thus the rotated state is still an eigenstate of J^2, with the same j as the original. But J_z does not commute with $U(R)$ in general (if $\beta \neq 0$). Thus $|jm\rangle_R$ is a linear combination of states with different m values, but the same j:

$$|jm\rangle_R = \sum_{m'} |jm'\rangle \langle jm' |U(R)| jm\rangle. \tag{4.88}$$

The expansion coefficients are defined to be Wigner D-functions or Rotation Matrices.

$$D^j_{m'm}(R) = \langle jm' |U(R)| jm\rangle. \tag{4.89}$$

We have

$$D^j_{m'm}(R) = \langle jm' |e^{-i\alpha J_z} e^{-i\beta J_y} e^{-i\gamma J_z}| jm\rangle$$
$$= e^{-i\alpha m'} \langle jm' |e^{-i\beta J_y}| jm\rangle e^{-i\gamma m}$$
$$\equiv e^{-i\alpha m'} d^j_{m'm}(\beta) e^{-i\gamma m}. \tag{4.90}$$

Explicit expressions for $d^j_{m'm}(\beta)$ were first obtained by Wigner. A derivation of the general result is given by Sakurai and Napolitano, *Modern Quantum Mechanics*, 3rd edition, pp. 221–223, using the technique invented by Schwinger. We shall content ourselves with the cases $j = 1/2, 1$.

We can write $iJ_y = \frac{1}{2}(J_+ - J_-)$, so

$$e^{-i\beta J_y} = e^{-\frac{\beta}{2}(J_+ - J_-)} \tag{4.91}$$

and

$$d^j_{m'm}(\beta) = \sum_{k=0}^{\infty} \frac{1}{k!}\left(-\frac{\beta}{2}\right)^k \langle jm' | (J_+ - J_-)^k | jm\rangle. \tag{4.92}$$

Note that with the Condon and Shortley phase conventions, $d^j_{m'm}(\beta)$ is real.
For $j = 1/2$, we have directly $e^{-i\beta J_y} = \cos(\beta/2) - i\sigma_y \sin(\beta/2)$, so

$$d^{1/2}_{m'm}(\beta) = \begin{pmatrix} \cos\frac{\beta}{2} & -\sin\frac{\beta}{2} \\ \sin\frac{\beta}{2} & \cos\frac{\beta}{2} \end{pmatrix}. \tag{4.93}$$

For $j = 1$, we use the identity

$$e^{-i\beta J_y} = 1 - iJ_y \sin\beta - (1 - \cos\beta)J_y^2, \tag{4.94}$$

which can be derived from the results of Problem 4.1. Using the explicit results for the matrices of J_y and J_y^2,

$$d^1_{m'm}(\beta) = \begin{pmatrix} \frac{1}{2}(1 + \cos\beta) & -\frac{1}{\sqrt{2}}\sin\beta & \frac{1}{2}(1 - \cos\beta) \\ \frac{1}{\sqrt{2}}\sin\beta & \cos\beta & -\frac{1}{\sqrt{2}}\sin\beta \\ \frac{1}{2}(1 - \cos\beta) & \frac{1}{\sqrt{2}}\sin\beta & \frac{1}{2}(1 + \cos\beta) \end{pmatrix}. \tag{4.95}$$

The rotational transformations enjoy the properties of a group. We already used this aspect without articulating it in finding $U(R) = U_3(\gamma)U_2(\beta)U_1(\alpha)$. If $R = (\alpha, \beta, \gamma)$, the inverse rotation is $R^{-1} = (-\gamma, -\beta, -\alpha)$. We have $U(R^{-1}) = U^{-1}(R) = U^\dagger(R)$. Consider

$$\langle jm' | U^\dagger(R) | jm\rangle = \langle jm' | U(R^{-1}) | jm\rangle = \langle jm | U(R) | jm'\rangle^*. \tag{4.96}$$

We thus have

$$D^j_{m'm}(-\gamma, -\beta, -\alpha) = D^j_{mm'}(\alpha, \beta, \gamma)^*$$
$$e^{im'\gamma}d^j_{m'm}(-\beta)e^{im\alpha} = \left[e^{-im\alpha}d^j_{mm'}(\beta)e^{-im'\gamma}\right]^*$$
$$d^j_{m'm}(-\beta) = d^j_{mm'}(\beta). \tag{4.97}$$

The group property $U(R_2 R_1) = U(R_2)U(R_1)$ can be used to deduce the following symmetries:

$$d^j_{m'm}(\beta) = d^j_{-m,-m'}(\beta) = (-1)^{m'-m}d^j_{mm'}(\beta) = (-1)^{m'-m}d^j_{-m',-m}(\beta)$$
$$d^j_{m'm}(\pi - \beta) = (-1)^{j+m'}d^j_{m',-m}(\beta). \tag{4.98}$$

See Problem 4.6.

Unitarity leads to additional relations between the $D_{m'm}^j$ functions. Evidently we have

$$D_{m'm}^j(R_1 R_2) = \sum_{m''} D_{m'm''}^j(R_1) D_{m''m}^j(R_2). \tag{4.99}$$

If $R_1 = R$ and $R_2 = R^{-1}$, we have

$$\delta_{m'm} = \sum_{m''} D_{m'm''}^j(R) D_{m''m}^j(R^{-1}) = \sum_{m''} D_{m'm''}^j(R) D_{mm''}^{j*}(R). \tag{4.100}$$

Similarly,

$$\delta_{m'm} = \sum_{m''} D_{m''m}^{j*}(R) D_{m''m}^j(R). \tag{4.101}$$

These correspond to $UU^\dagger = I$ and $U^\dagger U = I$.

The classic question about the relationship between the projection of **J** along an arbitrary axis $\hat{n}$, if we know that the state has a definite J_z eigenvalue along the lab's z-axis, can be answered with the rotation matrices. Let the polar angles of $\hat{n}$ be (θ, ϕ). Then we choose R to be $\alpha = \phi$ and $\beta = \theta$. Any choice for γ defines the same direction, $\hat{n}$. Popular choices are $\gamma = -\phi$ and $\phi = 0$.

Now $\hat{n} \cdot \mathbf{J} = J_{z'}$. Hence the probability amplitude for a value m' for $J_{z'}$ is

$$A_{m'm} = {}_R\langle jm' | jm \rangle = \sum_{m''} D_{m''m'}^{j*}(R) \langle jm'' | jm \rangle = D_{mm'}^{j*}(R). \tag{4.102}$$

Thus the probability that a state with $J_z = m$ will be observed as $J_{z'} = m'$ is $d_{mm'}^j(\theta)^2$. As an example, consider a deuteron beam ($J = 1$) with $m = 1$ along the lab's z-axis. Then if the angular momentum is measured instead along $\hat{z}'$ with $\hat{z}' \cdot \hat{z} = \cos\theta$, then the probabilities for $J_{z'}$ are

$$P_{11}(\theta) = |d_{11}^1(\theta)|^2 = \cos^4(\theta/2) \tag{4.103}$$

$$P_{01}(\theta) = |d_{10}^1(\theta)|^2 = \frac{1}{2}\sin^2(\theta) \tag{4.104}$$

$$P_{-11}(\theta) = |d_{1-1}^1(\theta)|^2 = \sin^4(\theta/2). \tag{4.105}$$

4.6 Application to Nuclear Magnetic Resonance

The phenomena of atomic, molecular, and nuclear magnetic resonance have a vast literature. The molecular beam resonance technique for measuring nuclear magnetic moments was developed by I. I. Rabi and collaborators in 1938. Nuclear magnetic resonance in bulk materials was discovered by Felix Bloch and Edward Purcell independently in the late 1940s. It has widespread application in chemistry, biology, material science, and medical science.

Figure 4.5 shows a schematic representation of the apparatus of Rabi, et al. (*Phys. Rev.* **55**, 526 (1939)). Neutral atoms are heated in the oven and emerge at various angles and with various energies. A narrow slit is placed just in front of the central magnet, C. Magnet A has a field gradient, making it effectively a Stern-Gerlach device. Magnet A thus exerts a force on the atom that depends on the orientation of the magnetic moment of the nucleus, supposing that there is no net magnetic moment from the electrons (which, if non-zero, would be much greater than that of the nucleus).

Some combination of the energy, initial direction of the atom, and the orientation of the magnetic moment will be just right to pass the atom through the narrow slot, providing a beam with a uniquely defined orientation of the moment. Magnet B has its gradient just opposite that of magnet A and, if the orientation of the magnetic moment is unchanged, the atom arrives at the detector.

The Hamiltonian for the interaction of a magnetic moment μ with a magnetic field $\mathbf{B}$ is

$$H = -\boldsymbol{\mu} \cdot \mathbf{B}. \tag{4.106}$$

The moment is $\boldsymbol{\mu} = \gamma \hbar \mathbf{J}$, where γ is the gyromagnetic ratio. (For particles, atoms, and molecules, $\gamma = g\mu_0 \equiv ge\hbar/2mc$, where g is the Landé g-factor.) With $\mathbf{B} = B_0\hat{\mathbf{z}}$,

$$H = -\gamma B_0 \hbar J_z, \tag{4.107}$$

so it is natural to define $\omega_0 = \gamma B_0$. Thus, we have $2j + 1$ nuclear magnetic levels spaced by $\hbar\omega_0$. The time dependence of an arbitrary state $|\xi, t\rangle$ is

$$|\xi, t\rangle = e^{-iHt/\hbar} |\xi, 0\rangle = e^{i\omega_0 t J_z} |\xi, 0\rangle. \tag{4.108}$$

This corresponds to a clockwise rotation around the z-axis at a frequency ω_0 (assuming $\omega_0 > 0$).

In addition to the static field $\mathbf{B}_0$, we add a time-dependent (rotating) magnetic field $\mathbf{B}_1$ at a frequency ω. See Figure 4.6. This radio-frequency field has components

$$B_x = B_1 \cos \omega t$$
$$B_y = -B_1 \sin \omega t. \tag{4.109}$$

This magnetic field rotates clockwise at frequency ω and has constant magnitude. We expect some sort of resonant behavior when $\omega \approx \omega_0$, because then the rotating magnetic moment

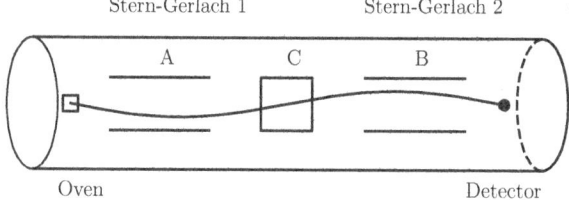

Stern-Gerlach 1   Stern-Gerlach 2

A   C   B

Oven   Detector

Figure 4.5 Schematic representation of the apparatus of I. I. Rabi and collaborators used to measure nuclear magnetic moments.

Figure 4.6 The configuration of the magnetic fields in region C of the apparatus of Rabi, et al. There is a static field $\mathbf{B}_0$ defining the z direction. A field of strength B_1 rotates clockwise with angular frequency ω.

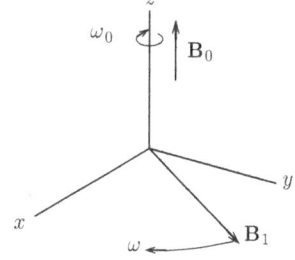

will be in phase with the combined field and there will be a torque tending to flip **J**. A deft application of the rotation group reduces the problem to one that is straightforward.

The Hamiltonian is now

$$H = -\hbar\omega_0 J_z - \hbar\Gamma(J_x \cos\omega t - J_y \sin\omega t), \tag{4.110}$$

where $\Gamma = \omega_0 B_1/B_0$. Generally, $B_1 \ll B_0$ and Γ is much smaller than ω_0.

Evidently, if we go to a rotating frame with angular speed ω around the z-axis, the magnetic field appears time-independent. We achieve the equivalent by considering the transformed state vector

$$|t\rangle_1 = e^{-i\omega t J_z} |t\rangle. \tag{4.111}$$

This should be viewed as $U^\dagger |t\rangle$ and corresponds to the passive mode - changing the axes, not the state. Now we need the Schrödinger equation satisfied by $|t\rangle_1$:

$$i\hbar\frac{\partial}{\partial t} |t\rangle_1 = i\hbar(-i\omega J_z)e^{-i\omega t J_z} |t\rangle + e^{-i\omega t J_z} i\hbar\frac{\partial}{\partial t} |t\rangle$$

$$= i\hbar(-i\omega J_z) |t\rangle_1 + e^{-i\omega t J_z} H e^{i\omega t J_z} |t\rangle_1. \tag{4.112}$$

Thus we can identify an effective Hamiltonian,

$$H_{\text{eff}} = \hbar\omega J_z + e^{-i\omega t J_z} H e^{i\omega t J_z}, \tag{4.113}$$

and write

$$i\hbar\frac{\partial}{\partial t} |t\rangle_1 = H_{\text{eff}} |t\rangle_1. \tag{4.114}$$

We can evaluate H_{eff} explicitly:

$$H_{\text{eff}} = \hbar\omega J_z + e^{-i\omega t J_z}\left[-\hbar\omega_0 J_z - \hbar\Gamma(J_x \cos\omega t - J_y \sin\omega t)\right]e^{i\omega t J_z}$$

$$= -\hbar(\omega_0 - \omega)J_z - \hbar\Gamma\cos\omega t\, e^{-i\omega t J_z} J_x e^{i\omega t J_z} + \hbar\Gamma\sin\omega t\, e^{-i\omega t J_z} J_y e^{i\omega t J_z}. \tag{4.115}$$

We recognize the rotations of the operators J_x and J_y as corresponding to a rotation by $\theta = -\omega t$, so

$$H_{\text{eff}} = -\hbar(\omega_0 - \omega)J_z - \hbar\Gamma\cos\omega t\, [J_x \cos\omega t + J_y \sin\omega t] \tag{4.116}$$

$$+ \hbar\Gamma\sin\omega t\, [-J_x \sin\omega t + j_y \cos\omega t]$$

$$= -\hbar[(\omega_0 - \omega)J_z + \Gamma J_x]. \tag{4.117}$$

We define

$$\nu = \sqrt{(\omega_0 - \omega)^2 + \Gamma^2} \tag{4.118}$$

and

$$\cos\alpha = \frac{\omega_0 - \omega}{\nu}; \qquad \sin\alpha = \frac{\Gamma}{\nu}, \tag{4.119}$$

then

$$H_{\text{eff}} = -\hbar\nu(J_z \cos\alpha + J_x \sin\alpha). \tag{4.120}$$

We now have a time-independent Hamiltonian, as expected, with the "magnetic field" tipped at an angle α with respect to the z-axis. See Figure 4.7.

Figure 4.7 The effective magnetic field of the effective Hamiltonian in the frame rotating at angular frequency ω, the frequency of the applied field B_1 transverse to the static field B_0.

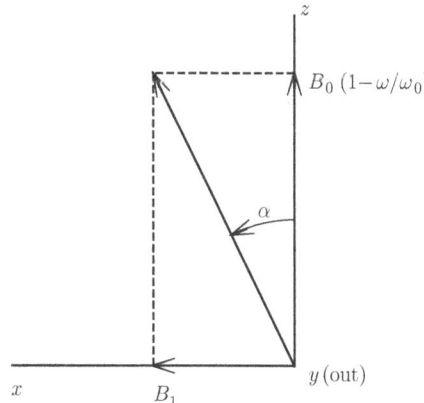

We now make a second transformation, suggested by this expression:

$$|t\rangle_2 = e^{i\alpha J_y} |t\rangle_1 .$$
(4.121)

Then

$$i\hbar \frac{\partial}{\partial t} |t\rangle_2 = e^{i\alpha J_y} H_{\text{eff}} e^{-i\alpha J_y} |t\rangle_2 ,$$
(4.122)

providing us with a new effective Hamiltonian

$$H_2 = -\hbar \nu J_z.$$
(4.123)

In terms of this Hamiltonian, we have a simple time-dependence:

$$|t\rangle_2 = e^{i\nu t J_z} |0\rangle_2 .$$
(4.124)

We now proceed to display the unitary time evolution operator $U(t)$ that solves the original problem. We want

$$|t\rangle = e^{i\omega t J_z} |t\rangle_1 = e^{i\omega t J_z} e^{-i\alpha J_y} |t\rangle_2$$
$$= e^{i\omega t J_z} e^{-i\alpha J_y} e^{i\nu t J_z} |0\rangle_2.$$
(4.125)

But

$$|0\rangle_2 = e^{i\alpha J_y} |0\rangle_1 = e^{i\alpha J_y} |0\rangle .$$
(4.126)

Altogether, if $|t\rangle = U(t) |0\rangle$, then

$$U(t) = e^{i\omega t J_z} e^{-i\alpha J_y} e^{i\nu t J_z} e^{i\alpha J_y} .$$
(4.127)

This solves the problem in terms of a product of rotation operators.

What we really want is the probability that a nucleus in the state $J_z = m$ entering the region C will emerge in the state $J_z = m'$. Let us write the amplitude for this transition as

$$A_{m'm}(t) = \langle jm'|t\rangle = \langle jm'|U(t)|jm\rangle$$
$$= e^{i\omega tm'}\langle jm'|e^{-i\alpha J_y}e^{i\nu t J_z}e^{i\alpha J_y}|jm\rangle. \tag{4.128}$$

The general result is obtained by inserting a complete set of states $|jm''\rangle\langle jm''|$ so that

$$A_{m'm}(t) = e^{i\omega tm'}\sum_{m''}d^j_{m'm''}(\alpha)d^j_{mm''}(\alpha)e^{i\nu tm''}, \tag{4.129}$$

but we examine only the case $j = 1/2$. Quite generally

$$e^{-i\alpha J_y}J_z e^{i\alpha J_y} = \cos\alpha J_z + \sin\alpha J_x. \tag{4.130}$$

Writing

$$\cos\theta J_z + \sin\theta J_x = \hat{n}\cdot J; \qquad \hat{n} = \cos\theta\,\hat{z} + \sin\theta\,\hat{x}, \tag{4.131}$$

we have for $j = 1/2$,

$$e^{-i\alpha J_y}e^{i\nu t J_z}e^{i\alpha J_y} = e^{i\nu t\hat{n}\cdot\frac{1}{2}\sigma}$$
$$= \cos(\nu t/2) + i\sin(\nu t/2)\hat{n}\cdot\sigma, \tag{4.132}$$

and thus

$$e^{i\omega t\frac{1}{2}\sigma_z}e^{-i\alpha J_y}e^{i\nu t J_z}e^{i\alpha J_y}$$
$$= \begin{pmatrix} e^{i\omega t/2}(\cos\nu t/2 + i\cos\alpha\sin\nu t/2) & ie^{i\omega t/2}\sin\alpha\sin\nu t/2 \\ ie^{-i\omega t/2}\sin\alpha\sin\nu t/2 & e^{-i\omega t/2}(\cos\nu t/2 - i\cos\alpha\sin\nu t/2) \end{pmatrix}. \tag{4.133}$$

From this we find the transition probabilities directly

$$P_{-\frac{1}{2},\frac{1}{2}}(t) = P_{\frac{1}{2},-\frac{1}{2}}(t) = \sin^2\alpha\sin^2\nu t/2$$
$$P_{\frac{1}{2},\frac{1}{2}}(t) = P_{-\frac{1}{2},-\frac{1}{2}}(t) = \cos^2\nu t/2 + \cos^2\alpha\sin^2\nu t/2 = 1 - P_{-\frac{1}{2},\frac{1}{2}}(t). \tag{4.134}$$

In terms of the physical parameters, $\omega_0 - \omega, \Gamma, t$, we have

$$P_{-\frac{1}{2},\frac{1}{2}}(t) = \frac{\Gamma^2}{(\omega_0 - \omega)^2 + \Gamma^2}\sin^2\left[\frac{1}{2}\sqrt{(\omega_0 - \omega)^2 + \Gamma^2}\,t\right]. \tag{4.135}$$

This transition probability is shown for three values of Γt in Figure 4.8. Recall that in the experiment of Rabi, et al., a flip of the nuclear spin results in the atom failing to reach the detector, and thus a reduction in the observed signal. An example of the data from Rabi, et al. is shown in Figure 4.9.

The resonant frequency is related to the energy spacing associated with the change of one unit in J_z. One defines the g-factor so that g times the nuclear magneton, $\mu_N = e\hbar/(2m_N c)$ times the magnetic field, is the energy interval. Thus,

$$h\nu = \hbar\omega = g\mu_N B. \tag{4.136}$$

The Rabi, et al. experiment determined the g-factor, not the total nuclear moment, $\mu = Jg\mu_N$. From Figure 4.9, we see the resonant frequency given as 7.76×10^6 s^{-1}, with a field about 1940 gauss. In these units, the constants are

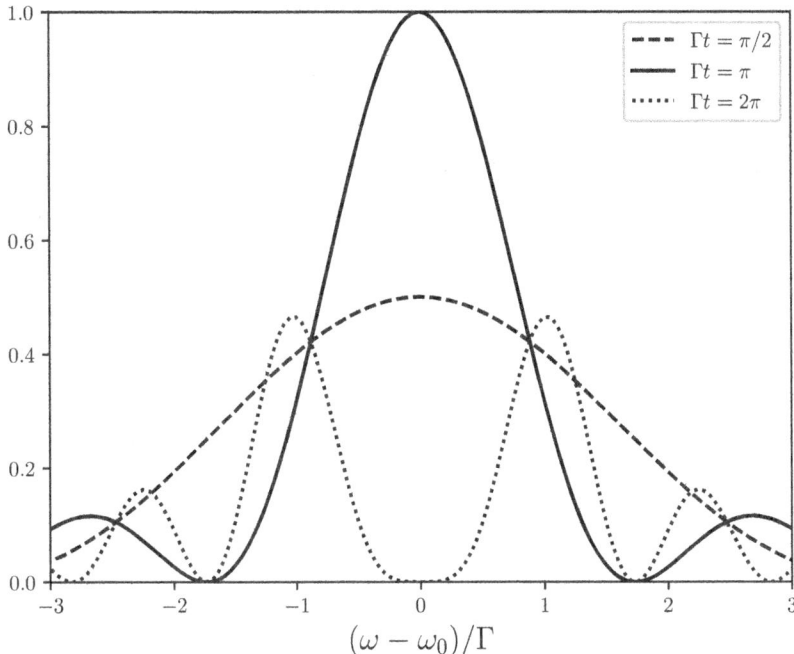

Figure 4.8 The probability that the m value for a $j = 1/2$ will flip in an apparatus like the one used by Rabi, et al. as a function of $(\omega - \omega_0)/\Gamma$ for three values of Γt. Here $\Gamma = \omega_0 B_1/B_0$, where ω_0 is the natural rotation frequency in the static field B_0, and B_1 is the strength of the transverse oscillating field. The nuclei are exposed to the fields in section C of the apparatus for a time t.

Figure 4.9 Data from Rabi, et al., 1939 / American Physical Society for the F^{19} nucleus. From the frequency $\nu = 7.76$ MHz and the field at resonance near 1940 gauss, the g factor is determined to be 5.24.

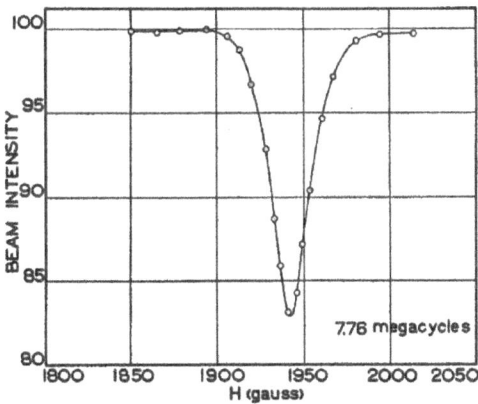

$$\mu_N = 5.05 \times 10^{-24} \text{ erg gauss}^{-1} \tag{4.137}$$

$$h = 6.63 \times 10^{-27} \text{erg sec.} \tag{4.138}$$

Thus,

$$g = \frac{\nu}{B} \cdot \frac{h}{\mu_N} = \frac{\nu}{B} \cdot 1.31 \times 10^{-3} \text{gauss sec.} \tag{4.139}$$

For this particular measurement, $g = 5.24$. The spin of F^{19}, known independently to be 1/2, gives a total magnetic moment of $2.62 \, \mu_N$.

4.7 Addition of Angular Momenta

Many physical situations involve a total angular momentum **J** being built as a sum of different contributions. As an example, consider an electron with orbital angular momentum **L** and spin **s**:

$$\mathbf{J} = \mathbf{L} + \mathbf{s}, \tag{4.140}$$

or more generally, several electrons in an atom

$$\mathbf{j}_k = \mathbf{L}_k + \mathbf{s}_k; \quad \mathbf{j} = \sum_k \mathbf{j}_k \tag{4.141}$$

or

$$\mathbf{L} = \sum_k \mathbf{L}_k; \quad \mathbf{S} = \sum_k \mathbf{s}_k; \quad \mathbf{J} = \mathbf{L} + \mathbf{S}. \tag{4.142}$$

The common feature is that the different angular momentum operators commute. The basic structure is that we have two independent generators of rotations - angular momenta – $\mathbf{J}_1, \mathbf{J}_2$ obeying

$$[J_{1\alpha}, J_{1\beta}] = i\epsilon_{\alpha\beta\gamma}J_{1\gamma}$$

$$[J_{2\alpha}, J_{2\beta}] = i\epsilon_{\alpha\beta\gamma}J_{2\gamma}$$

$$[J_{1\alpha}, J_{2\beta}] = 0. \tag{4.143}$$

We consider the space that is the direct product of the spaces for $\mathbf{J}_1$ and $\mathbf{J}_2$, that is, all the states that can be made from

$$|j_1 m_1 j_2 m_2\rangle = |j_1 m_1\rangle \otimes |j_2 m_2\rangle. \tag{4.144}$$

Each of these states is an eigenket of J_1^2, J_{1z} and J_2^2, J_{2z}. Formally, the operators on this product space are $\mathbf{J}_1 \otimes I$ and $I \otimes \mathbf{J}_2$, and thus the sum is

$$\mathbf{J} = \mathbf{J}_1 \otimes I_2 + I_1 \otimes \mathbf{J}_2, \tag{4.145}$$

but we shall write simply $\mathbf{J} = \mathbf{J}_1 + \mathbf{J}_2$, where each $\mathbf{J}_i$ acts only on its factor in the product space. Now it is clear that the sum $\mathbf{J}$ itself obeys

$$[J_\alpha, J_\beta] = i\epsilon_{\alpha\beta\gamma}J_\gamma, \tag{4.146}$$

so we must be able to diagonalize J^2 and J_z on these states with fixed J_1^2 and J_2^2. We thus write,

$$|(j_1 j_2)jm\rangle = \sum_{m_1 m_2} |j_1 m_1 j_2 m_2\rangle \langle j_1 m_1 j_2 m_2 |(j_1 j_2)jm\rangle. \tag{4.147}$$

The coefficients, which we will abbreviate as $\langle j_1 m_1 j_2 m_2 | jm\rangle$, are called Clebsch-Gordan coefficients or vector-addition coefficients. We thus have two distinct bases for the same space:

$|j, m\rangle$ and $|j_1 m_1\rangle |j_2 m_2\rangle$. They are related by the expression above and by its inverse,

$$|j_1 m_1 j_2 m_2\rangle = \sum_m |(j_1 j_2)jm\rangle \langle jm|j_1 m_1 j_2 m_2\rangle. \tag{4.148}$$

With Condon and Shortley convention, the Clebsch-Gordan coefficients are real. Thus,

$$\langle jm|j_1 m_1 j_2 m_2\rangle = \langle j_1 m_1 j_2 m_2|jm\rangle. \tag{4.149}$$

We can establish a number of basic properties of the Clebsch-Gordan coefficients:

(a)

$$\langle jm|j_1 m_1 j_2 m_2\rangle = 0 \tag{4.150}$$

unless $m = m_1 + m_2$, as can be verified by applying $J_z = J_{1z} + J_{2z}$ to both sides of the defining relation.

(b)

$$\sum_{m_1 m_2} \langle j'm'|j_1 m_1 j_2 m_2\rangle \langle j_1 m_1 j_2 m_2|jm\rangle = \delta_{jj'}\delta_{mm'}, \tag{4.151}$$

which follows from evaluating $\delta_{jj'}\delta_{mm'} = \langle j'm'|jm\rangle$.

(c)

$$\sum_{jm} \langle j_1 m_1 j_2 m_2|jm\rangle \langle jm|j_1 m_1' j_2 m_2'\rangle = \delta_{m_1 m_1'}\delta_{m_2 m_2'} \tag{4.152}$$

as we see from examining $\delta_{m_1 m_1'}\delta_{m_2 m_2'} = \langle j_1 m_1' j_2 m_2'|j_1 m_1 j_2 m_2\rangle$.

We can construct the Clebsch-Gordan coefficients recursively. The highest value of m occurs for $m_1 = j_1$ and $m_2 = j_2$. Thus the highest possible value for j is $j = j_1 + j_2$:

$$|jm = j\rangle = |j_1 m_1 = j_1\rangle |j_2 m_2 = j_2\rangle. \tag{4.153}$$

To construct $|jm = j - 1\rangle$, we use the lowering operator $J_- = J_{1-} + J_{2-}$.

$$J_- |jj\rangle = \sqrt{j(j+1) - j(j-1)}\, |jj-1\rangle = (J_{1-}|j_1 j_1\rangle)\,|j_2 j_2\rangle + |j_1 j_1\rangle J_{2-}\,|j_2 j_2\rangle$$

$$= \sqrt{j_1(j_1-1)}\, |j_1 j_1 - 1\rangle |j_2 j_2\rangle + \sqrt{j_2(j_2-1)}\, |j_1 j_1\rangle |j_2 j_2 - 1\rangle. \tag{4.154}$$

In this way, we have found two Clebsch-Gordan coefficients:

$$\langle jj - 1|j_1 j_1 - 1, j_2 j_2\rangle = \frac{\sqrt{j_1(j_1-1)}}{\sqrt{j(j+1) - j(j-1)}}$$

$$\langle jj - 1|j_1 j_1, j_2 j_2 - 1\rangle = \frac{\sqrt{j_2(j_2-1)}}{\sqrt{j(j+1) - j(j-1)}}. \tag{4.155}$$

We could continue with the lowering operator to find all the states $|jm\rangle$. The space of states with $J_z = j_1 + j_2 - 1$ is two-dimensional, so there is a state that is orthogonal to $|jj - 1\rangle$. Clearly, since it has $J_z = j_1 + j_2 - 1$, it must have $J = j_1 + j_2 - 1$. Using the lowering operator, we could find all the states $|j - 1m\rangle$.

What is the lowest value of j we obtain from j_1 and j_2? There is only one way to achieve $m = j_1 + j_2$, but there can be many ways to achieve other values of m. What is the greatest number of different ways some value of m can be obtained? Each value of m is $m_1 + m_2$. So if, for example, $j_1 < j_2$, there are at most $2j_1 + 1$ different ways of achieving a value of m. But each value of j needs to achieve that value of m. So there are $2j_< + 1$ different values of j, where $j_<$ is the minimum of j_1 and j_2. Thus the lowest j is $j_> + j_< - 2j_< = j_> - j_< = |j_1 - j_2|$.

4.8 Integration Over the Rotation Group

If we perform the same rotation R on the two spaces $|j_1 m_1\rangle$ and $|j_2 m_2\rangle$, we can indicate the action by

$$U(R) = U_1(R)U_2(R). \tag{4.156}$$

Now the matrix elements of $U(R)$ can be related to those of $U_1(R)$ and $U_2(R)$ by inserting a complete set of states in the $|j_1 m_1\rangle\,|j_2 m_2\rangle$ basis and the $|j_1' m_1'\rangle\,|j_2' m_2'\rangle$ basis:

$$
\begin{aligned}
D_{m'm}^{j}(R) &= \langle j_1 j_2 j m' | U(R) | j_1 j_2 j m \rangle = \langle j_1 j_2 j m' | U_1(R)U_2(R) | j_1 j_2 j m \rangle \\
&= \sum_{\substack{m_1' m_2' \\ m_1 m_2}} \langle jm' | j_1 m_1' j_2 m_2' \rangle \langle j_1 m_1' j_2 m_2' | U_1(R)U_2(R) | j_1 m_1 j_2 m_2 \rangle \langle j_1 m_1 j_2 m_2 | jm \rangle \\
&= \sum_{\substack{m_1' m_2' \\ m_1 m_2}} \langle jm' | j_1 m_1' j_2 m_2' \rangle D_{m_1' m_1}^{j_1}(R) D_{m_2' m_2}^{j_2}(R) \langle j_1 m_1 j_2 m_2 | jm \rangle.
\end{aligned} \tag{4.157}
$$

The inverse relation is

$$D_{m_1' m_1}^{j_1}(R)D_{m_2' m_2}^{j_2}(R) = \sum_{jmm'} \langle j_1 m_1' j_2 m_2' | jm' \rangle D_{m'm}^{j}(R) \langle jm | j_1 m_1 j_2 m_2 \rangle. \tag{4.158}$$

A powerful technique is integration over the rotation group. To do this, we want to give "equal weight" to each rotation. Because we can think of the first two Euler angles as specifying the axis for the final rotation, it is natural to suppose that the integration element will be proportional to $(d\alpha d \cos \beta)d\gamma$, where the first two factors are familiar from integrating over a sphere. We would like to define dR so that $\int dR = 1$. Because we know that for half-integral representations we need to go to 4π in some angle, we choose

$$dR = \frac{1}{32\pi^2} \sin \beta d\beta d\alpha d\gamma$$

$$0 \leq \alpha \leq 4\pi; \quad 0 \leq \beta \leq \pi; \quad 0 \leq \gamma \leq 4\pi. \tag{4.159}$$

We could have chosen either α or γ to be limited to 2π and changed the overall factor to $1/16\pi^2$. Just as when integrating over the sphere, we are free to redefine the axes and use the new θ and ϕ for the integration. We can perform a fixed rotation, R_0 and then $d(R_0 R) = dR$. This is analogous, as well, to using $x' = x + a$ as the integration variable. Now the group property tells us that

$$D_{m'm}^{j}(R_0 R) = \sum_{m''} D_{m'm''}^{j}(R_0)D_{m''m}^{j}(R). \tag{4.160}$$

We can use this to prove a very useful result. Consider

$$\int dR\, D^j_{m'm}(R) = \int dR\, D^j_{m'm}(R_0 R) = \sum_{m''} D^j_{m'm''}(R_0) \int dR\, D^j_{m''m}(R), \tag{4.161}$$

where we used the invariance of the integration volume $d(R_0 R) = dR$. But the result cannot depend on the arbitrary rotation R_0. This is possible only if $j = 0$ and, of course, then $m = m' = 0$. We see then that

$$\int dR\, D^j_{m'm}(R) = \delta_{m'0}\delta_{m0}\delta_{j0}. \tag{4.162}$$

The normalization of dR gives the unit coefficient for the right-hand side.

The D^j's are a unitary representation of the rotation group, so

$$D^j_{m'm}(R^{-1}) = D^{j*}_{mm'}(R). \tag{4.163}$$

Now consider

$$\int dR\, D^{j_1*}_{m'_1 m_1}(R) D^{j_2}_{m'_2 m_2}(R) = \int dR\, D^{j_1}_{m_1 m'_1}(R^{-1}) D^{j_2}_{m'_2 m_2}(R). \tag{4.164}$$

We can conclude several things about this integral without actually evaluating it completely. First, the product of the D^j factors must be some linear combination of single D^j functions with $|j_1 - j_2| \le j \le j_1 + j_2$. But the integral $\int dR D^j$ vanishes unless $j = 0$, so we know there is a factor δ_{j_1,j_2} in the answer. Moreover, the integrand has as its α dependence $e^{i\alpha(m'_1 - m'_2)}$, so there is a factor $\delta_{m'_1 m'_2}$. The same consideration for γ requires a factor $\delta_{m_1 - m_2}$. Thus, we can write

$$\int dR\, D^{j_1}_{m_1 m'_1}(R^{-1}) D^{j_2}_{m'_2 m_2}(R) = A\delta_{j_1 j_2}\delta_{m'_1 m'_2}\delta_{m_1 m_2}. \tag{4.165}$$

Now set $j_1 = j_2$ and multiply both sides by $\delta_{m'_1 m'_2}$ and sum over m'_2:

$$\sum_{m'_2} \int dR\, D^{j_1}_{m_1 m'_2}(R^{-1}) D^{j_1}_{m'_2 m_2}(R) = \int dR\, D^{j_1}_{m_1 m_2}(I) = \delta_{m_1 m_2}$$

$$= (2j_1 + 1)A\delta_{m_1 m_2}, \tag{4.166}$$

so $A = 1/(2j_1 + 1)$ and the identity is

$$\int dR\, D^{j_1*}_{m'_1 m_1}(R) D^{j_2}_{m'_2 m_2}(R) = \int dR\, D^{j_1}_{m_1 m'_1}(R^{-1}) D^{j_2}_{m'_2 m_2}(R)$$

$$= \frac{\delta_{j_1 j_2}\delta_{m'_1 m'_2}\delta_{m_1 m_2}}{2j_1 + 1}. \tag{4.167}$$

4.9   Gaunt Integral

These results enable us to calculate the Gaunt integral, which is used over and over again in molecular, atomic, and nuclear physics.

$$\int dR \, D^{j_1}_{m'_1 m_1}(R) D^{j_2}_{m'_2 m_2}(R) D^{j_3*}_{m'_3, m_3}(R)$$

$$= \int dR \sum_{jmm'} \langle j_1 m'_1 j_2 m'_2 | jm' \rangle D^{j}_{m'm}(R) D^{j_3*}_{m'_3, m_3}(R) \langle jm | j_1 m_1 j_2 m_2 \rangle$$

$$= \frac{\langle j_1 m'_1 j_2 m'_2 | j_3 m'_3 \rangle \langle j_3 m_3 | j_1 m_1 j_2 m_2 \rangle}{2j_3 + 1}. \qquad (4.168)$$

Using the relation (see Problem 4.6)

$$D^{j*}_{m'm}(R) = (-1)^{m'-m} D^{j}_{-m'-m}(R), \qquad (4.169)$$

we write alternatively

$$\int dR \, D^{j_1}_{m'_1 m_1}(R) D^{j_2}_{m'_2 m_2}(R) D^{j_3}_{m'_3, m_3}(R)$$

$$= (-1)^{m'_3 - m_3} \frac{\langle j_1 m'_1 j_2 m'_2 | j_3 - m'_3 \rangle \langle j_3 - m_3 | j_1 m_1 j_2 m_2 \rangle}{2j_3 + 1}. \qquad (4.170)$$

But we can write the phase just as well as

$$(-1)^{(j_3-m_3)-(j_3-m'_3)} = (-1)^{(j_3-m_3)+(j_3-m'_3)} \qquad (4.171)$$

since $j_3 - m_3$ is necessarily an integer. Following tradition, we introduce a factor $(-1)^{j_1-j_2-m_3}$ and write

$$\begin{pmatrix} j_1 & j_2 & j_3 \\ m_1 & m_2 & m_3 \end{pmatrix} = \frac{(-1)^{j_1-j_2-m_3}}{\sqrt{2j_3+1}} \langle j_1 m_1 j_2 m_2 | j_3 - m_3 \rangle. \qquad (4.172)$$

Because the integrand in Eq. (4.170) is symmetric in 1-2-3, it is clear that this factor is symmetric under the same interchange, up to a possible overall factor of -1. With this definition, the Wigner 3-j symbol thus defined is invariant under even permutations of its columns and acquires a factor -1 if $j_1 + j_2 + j_3$ is odd and the permutation is odd. The proof of this must refer to the fundamental definition of the Clebsch-Gordan coefficients and the Condon and Shortley sign conventions. The proof is given, for example, in Wigner's *Group Theory and Its Application to the Quantum Mechanics of Atomic Spectra*, pp. 291-292. This symmetrical form is quite useful and evocative. Note that it implies relations for Clebsch-Gordan coefficients under the interchange $j_1 \leftrightarrow j_3$, which would be otherwise difficult to prove. See Appendix B.3 for further discussion of the Wigner 3-j symbol and related structures.

We now have the elegant relation

$$\int dR\, D^{j_1}_{m'_1 m_1}(R) D^{j_2}_{m'_2 m_2}(R) D^{j_3}_{m'_3,m_3}(R)$$

$$= \begin{pmatrix} j_1 & j_2 & j_3 \\ m'_1 & m'_2 & m'_3 \end{pmatrix} \begin{pmatrix} j_1 & j_2 & j_3 \\ m_1 & m_2 & m_3 \end{pmatrix}. \quad (4.173)$$

The rotation functions reduce to spherical harmonics when one m value is zero.

$$D^{\ell}_{m0}(\alpha,\beta,0) = \sqrt{\frac{4\pi}{2\ell+1}} Y^*_{\ell m}(\beta,\alpha) \quad (4.174)$$

were β plays the role usually assigned to θ and α replaces ϕ. Using Eq. (4.158), we have

$$Y_{\ell_1 m_1}(\theta,\phi) Y_{\ell_2 m_2}(\theta,\phi)$$

$$= \sum_{LM} \sqrt{\frac{(2\ell_1+1)(2\ell_2+1)}{4\pi(2L+1)}} \langle L0|\ell_1 0\ell_2 0\rangle \langle \ell_1 m_1 \ell_2 m_2|LM\rangle Y_{LM}(\theta,\phi). \quad (4.175)$$

The Gaunt integral for spherical harmonics follows directly from this, with $dR \to d\Omega/4\pi$

$$\int d\Omega\, Y_{\ell_1 m_1}(\theta,\phi) Y_{\ell_2 m_2}(\theta,\phi) Y^*_{\ell_3,m_3}(\theta,\phi)$$

$$= \sqrt{\frac{(2\ell_1+1)(2\ell_2+1)}{4\pi(2\ell_3+1)}} \langle \ell_3 0|\ell_1 0\ell_2 0\rangle \langle \ell_1 m_1 \ell_2 m_2|\ell_3 m_3\rangle. \quad (4.176)$$

We can write a more symmetrical form with the Wigner 3-j symbol:

$$\int d\Omega\, Y_{\ell_1 m_1}(\theta,\phi) Y_{\ell_2 m_2}(\theta,\phi) Y_{\ell_3,m_3}(\theta,\phi)$$

$$= \sqrt{\frac{(2\ell_1+1)(2\ell_2+1)(2\ell_3+1)}{4\pi}} \begin{pmatrix} \ell_1 & \ell_2 & \ell_3 \\ 0 & 0 & 0 \end{pmatrix} \begin{pmatrix} \ell_1 & \ell_2 & \ell_3 \\ m_1 & m_2 & m_3 \end{pmatrix}. \quad (4.177)$$

4.10 Tensor Operators

Many physical operators have definite transformation properties under rotations, for example the E1 operator, $\mathbf{d} = \sum_j e_j \mathbf{r}_j$ transforming as a vector, $\mathbf{s}_1 \cdot \mathbf{s}_2$ transforming as a scalar, and $x_i x_j - \frac{1}{3} r^2 \delta_{ij}$ transforming as $L = 2$.

In classical physics, a rotation is defined by $x'_i = \sum_j a_{ij} x_j$, where a_{ij} is an orthogonal matrix in order to preserve the length of $\mathbf{x}$ ($\sum_i a_{ij} a_{ik} = \delta_{jk}$, a_{ij} real). For an mth rank tensor $M_{ij...p}$, its transformation under rotations is

$$M'_{ij...p} = \sum_{i'j'...p'} a_{ii'} a_{jj'} ... a_{pp'} M_{i'j'...p'}. \quad (4.178)$$

A first rank tensor is called a vector and transforms like a coordinate vector.

Higher rank Cartesian tensors are reducible. For example, a second rank tensor M_{ij} has $3 \times 3 = 9$ elements. These decompose into five independent components of a traceless symmetric tensor

$$M_{ij}^{(s)} = \frac{1}{2}(M_{ij} + M_{ji}) - \frac{1}{3}\delta_{ij}(\sum_k M_{kk}), \tag{4.179}$$

an antisymmetric tensor with three independent elements

$$M_{ij}^{(a)} = \frac{1}{2}(M_{ij} - M_{ji}), \tag{4.180}$$

and a single tensor proportional to the trace

$$M_{ij}^{(t)} = \delta_{ij}\text{Tr}\, M = \delta_{ij}\sum_k M_{kk}. \tag{4.181}$$

These three components are termed irreducible: under rotations, each element becomes a linear combination of the other elements in its component and, in addition, no component can be broken into smaller pieces with that same property. This decomposition of the Cartesian tensor M_{ij} is the analog of the angular momentum product $(J = 1) \times (J = 1) = (J = 2) + (J = 1) + (J = 0)$.

A vector operator transforms as

$$\mathbf{V'} = U^\dagger \mathbf{V} U. \tag{4.182}$$

For example, a rotation by the angle α around the z-axis gives

$$V_x' = V_x \cos\alpha - V_y \sin\alpha$$
$$V_y' = V_x \sin\alpha + V_y \cos\alpha.$$
$$V_z' = V_z \tag{4.183}$$

We can define a basis for $\mathbf{V}$ that, rather than being Cartesian, is analogous to $|jm\rangle$ for $j = 1$. Let

$$V_1 = -\frac{1}{\sqrt{2}}(V_x + iV_y); \quad V_{-1} = \frac{1}{\sqrt{2}}(V_x - iV_y); \quad V_0 = V_z. \tag{4.184}$$

Then

$$V_1' = -\frac{1}{\sqrt{2}}[V_x \cos\alpha - V_y \sin\alpha + i(V_x \sin\alpha + V_y \cos\alpha)]$$

$$= -\frac{1}{\sqrt{2}}(V_x(\cos\alpha + i\sin\alpha) + V_y(-\sin\alpha + i\cos\alpha) = e^{i\alpha}V_1. \tag{4.185}$$

Similarly,

$$V_{-1}' = e^{-i\alpha}V_{-1}. \tag{4.186}$$

We see that the complex conjugate of $D_{m'm}^1(\alpha, 0, 0)$ enters.

The extension to arbitrary order k (k integral) is accomplished by introducing $2k + 1$ operators $T_q^{(k)}$, $-k \le q \le k$ that transform under rotations as irreducible representations of order k of the rotation group. To infer the transformation properties of such an operator, imagine that it acts on a vacuum state. The result must be something that transforms as a state with $j = k$ and $m = q$:

$$T_q^{(k)} |0\rangle = |kq\rangle. \tag{4.187}$$

Now

$$U(R)T_q^{(k)}U(R^{-1}) |0\rangle = U(R)T_q^{(k)} |0\rangle = U(R) |kq\rangle = \sum_{q'} |kq'\rangle \langle kq' |U(R)| kq\rangle$$

$$= \sum_{q'} T_{q'}^{(k)} |0\rangle \langle kq' |U(R)| kq\rangle = \sum_{q'} T_{q'}^{(k)} |0\rangle D_{q'q}^k(R). \tag{4.188}$$

Knowing that it is the conjugated $U(R)T_q^{(k)}U(R^{-1})$ that must have a proper form, we surmise that

$$U(R)T_q^{(k)}U(R^{-1}) = \sum_{q'} T_{q'}^{(k)}D_{q'q}^k(R). \tag{4.189}$$

The definition above is general and satisfactory, but often one sees the tensor operators defined by their commutation properties with $\mathbf{J}$. These are found by means of infinitesimal rotations in the by-now-familiar way. We write

$$U(R) = 1 - i\delta\omega\hat{\mathbf{n}} \cdot \mathbf{J} + \dots \tag{4.190}$$

Then

$$UT_q^{(k)}U^\dagger = T_q^{(k)} - i\delta\omega[\hat{\mathbf{n}} \cdot \mathbf{J}, T_q^{(k)}] + \dots \tag{4.191}$$

The right-hand side of Eq. (4.189) to leading order is

$$\sum_{q'} \langle kq' |1 - i\delta\omega\hat{\mathbf{n}} \cdot \mathbf{J}| kq\rangle T_{q'}^{(k)} = T_q^{(k)} - i\delta\omega \sum_{q'} T_{q'}^{(k)} \langle kq' |\hat{\mathbf{n}} \cdot \mathbf{J}| kq\rangle. \tag{4.192}$$

Comparing with the righthand side of Eq. (4.191), we see

$$[\hat{\mathbf{n}} \cdot \mathbf{J}, T_q^{(k)}] = \sum_{q'} T_{q'}^{(k)} \langle kq' |\hat{\mathbf{n}} \cdot \mathbf{J}| kq\rangle. \tag{4.193}$$

Taking successively $\hat{\mathbf{n}}$ as $\hat{\mathbf{z}}$, $(\hat{\mathbf{x}} \pm i\hat{\mathbf{y}})/\sqrt{2}$, we obtain the commutation relations

$$[J_z, T_q^{(k)}] = qT_q^{(k)}$$
$$[J_\pm, T_q^{(k)}] = \sqrt{k(k+1) - q(q \pm 1)}\, T_{q\pm1}^{(k)}. \tag{4.194}$$

These are the generalization of our previous results for vector and scalar operators and can serve as the defining relations for $T_q^{(k)}$. We see that the generators of rotations act by commutation on $T_q^{(k)}$ the way they act directly on the state $|kq\rangle$.

New tensor operators can be created from existing ones simply by combining them in a fashion analogous to that used to combine states of well-defined angular momentum, that is, with Clebsch-Gordan coefficients. Let $A_{q_1}^{(k_1)}$ and $B_{q_2}^{(k_2)}$ be tensor operators. Then

$$C_q^{(k)} = \sum_{q_1,q_2} \langle kq|k_1 q_1 k_2 q_2 \rangle A_{q_1}^{(k_1)} B_{q_2}^{(k_2)} \qquad (4.195)$$

is a tensor operator of rank k. The proof is a straightforward exercise. Note that nothing has been said about the commutativity of the two initial operators. The tensorial character is independent of that matter.

Familiar examples include operators like $\mathbf{L \cdot s}$ and $\mathbf{s}_1 \cdot \mathbf{s}_2$. Such scalar operators are especially important, so consider $A^{(k)}$ and $B^{(k)}$ and define

$$A^{(k)} \bullet B^{(k)} = (-1)^k \sqrt{2k+1} \sum_{q_1 q_2} \langle 00|kq_1 kq_2 \rangle A_{q_1}^{(k)} B_{q_2}^{(k)}. \qquad (4.196)$$

Using

$$\langle 00|jmj'm' \rangle = \frac{(-1)^{j-m'}}{\sqrt{2j+1}} \delta_{j'j} \delta_{m'm}, \qquad (4.197)$$

we have

$$A^{(k)} \bullet B^{(k)} = \sum_q (-1)^q A_q^{(k)} B_{-q}^{(k)}. \qquad (4.198)$$

The contraction of two second-rank tensors into a scalar occurs in many physical situations. The nuclear quadrupole moment Q_{ij} couples to the gradients of the electrostatic field of the orbiting electrons, $\partial^2 \Phi_{\text{electrons}}/\partial x_i \partial x_j$. The second-rank coordinate tensor, $3x_i x_j - r^2 \delta_{ij}$, contracted with the second-rank spin tensor, $(3/2)(\sigma_{1i}\sigma_{2j} + \sigma_{1j}\sigma_{2i}) - \sigma_1 \cdot \sigma_2 \delta_{ij}$, is relevant to the tensor force in the nucleon-nucleon system, proportional to

$$S_{12} = 3\sigma_1 \cdot \hat{\mathbf{r}} \sigma_2 \cdot \hat{\mathbf{r}} - \sigma_1 \cdot \sigma_2. \qquad (4.199)$$

4.11 Wigner-Eckart Theorem

The celebrated Wigner-Eckart theorem concerns matrix elements of tensor operators taken between states of definite J^2 and J_z. The theorem states that

$$\left\langle \alpha' j'm' \left| T_q^{(k)} \right| \alpha jm \right\rangle = \langle j'm'|jmkq \rangle \langle \alpha' j' \| T^{(k)} \| \alpha j \rangle \qquad (4.200)$$

where the final factor, the reduced matrix element, does not depend on m, m', or q. The Wigner-Eckart theorem breaks a matrix element in the angular momentum basis into a product of a geometric factor, the Clebsch-Gordan coefficient, and a factor incorporating the physics.

This theorem can be proved in a variety of ways. Let us first demonstrate a simple point. Let $T_q^{(k)}$ be a rank k tensor operator. Then

$$\left\langle \alpha j'm' \left| T_q^{(k)} \right| 0 \right\rangle = A\delta_{jk} \delta_{m'q}. \qquad (4.201)$$

To see this, consider

$$\left\langle \alpha j'm' \left| J^2 T_q^{(k)} \right| 0 \right\rangle = j'(j'+1) \left\langle \alpha j'm' \left| T_q^{(k)} \right| 0 \right\rangle, \tag{4.202}$$

where J^2 acted to the left. On the other hand, $[J^2, T_q^{(k)}] = k(k+1)T_q^{(k)}$, so acting to the right

$$\left\langle \alpha j'm' \left| J^2 T_q^{(k)} \right| 0 \right\rangle = k(k+1) \left\langle \alpha j'm' \left| T_q^{(k)} \right| 0 \right\rangle. \tag{4.203}$$

So the matrix element vanishes unless $k = j'$. Similarly, using J_z, we find $m' = q$ if the matrix element is non-vanishing. Thus

$$\left\langle \alpha j'm' \left| T_q^{(k)} \right| 0 \right\rangle = A \delta_{jk} \delta_{m'q}. \tag{4.204}$$

In fact, A does not depend on m. This is established by considering

$$\left\langle \alpha j'm \left| [J_+, T_{m-1}^{(k)}] \right| 0 \right\rangle = \sqrt{j(j+1) - m(m-1)} \left\langle \alpha j'm \left| T_m^{(k)} \right| 0 \right\rangle, \tag{4.205}$$

where we used the commutator. Undoing the commutator and acting to both the left and right,

$$\left\langle \alpha j'm \left| [J_+, T_{m-1}^{(k)}] \right| 0 \right\rangle = \sqrt{j(j+1) - m(m-1)} \left\langle \alpha j'm - 1 \left| T_{m-1}^{(k)} \right| 0 \right\rangle. \tag{4.206}$$

Thus $\left\langle \alpha j'm \left| T_m^{(k)} \right| 0 \right\rangle$ is independent of m.

Now let us imagine the state $|\alpha jm\rangle$ as the result of a tensor operator acting on the vacuum $|0\rangle$:

$$A_m^j |0\rangle = |\alpha jm\rangle, \tag{4.207}$$

so

$$\left\langle \alpha' j'm' \left| T_q^{(k)} \right| \alpha jm \right\rangle = \left\langle \alpha' j'm' \left| T_q^{(k)} A_m^j \right| 0 \right\rangle. \tag{4.208}$$

We can invert Eq. (4.195) to write

$$T_q^{(k)} A_m^j = \sum_{j'',m''} C_{m''}^{(j'')} \left\langle jmkq | j''m'' \right\rangle \tag{4.209}$$

where $C_{m''}^{(j'')}$ are tensor operators. Now in

$$\sum_{j''m''} \left\langle \alpha' j'm' \left| C_{m''}^{(j'')} \right| 0 \right\rangle \tag{4.210}$$

there will be contributions only for $j'' = j', m'' = m'$. The value there we write as

$$\left\langle \alpha' j' \| T^{(k)} \| \alpha j \right\rangle, \tag{4.211}$$

noting that we have shown above that this value is independent of m'. Now we can write

$$\left\langle \alpha' j'm' \left| T_q^{(k)} \right| \alpha jm \right\rangle = \left\langle j'm' | jmkq \right\rangle \left\langle \alpha' j' \| T^{(k)} \| \alpha j \right\rangle. \tag{4.212}$$

Different authors write the expression with additional overall factors, such as $1/\sqrt{2j+1}$. In any event, the reduced matrix element can be determined by evaluating the left-hand side for any single choice of m, m', q, for example $m = m' = q = 0$ so

$$\left\langle \alpha' j' \| T^{(k)} \| \alpha j \right\rangle = \frac{\left\langle \alpha' j'0 \left| T_0^{(k)} \right| \alpha j0 \right\rangle}{\left\langle j'0 | j0k0 \right\rangle}. \tag{4.213}$$

4.12 Applications of the Wigner-Eckart Theorem

4.12.1 Selection Rules

The Wigner-Eckart theorem is a powerful tool, which we will invoke repeatedly. While knowing the reduced matrix element determines the full matrix element for all values of the ms, much of the usefulness of the Wigner-Eckart theorem is not dependent on the reduced matrix element. The Clebsch-Gordan coefficient itself controls which matrix elements vanish and which do not. Such determinations are called selection rules. The Wigner-Eckart theorem thus gives the selection rules that derive from the rotation group. Frequently, we use the Wigner-Eckart theorem together with a simpler principle based on parity symmetry. The parity operator takes every position and momentum into their opposites: $\mathbf{r} \to -\mathbf{r}; \mathbf{p} \to -\mathbf{p}$. Whether this symmetry is a real symmetry depends on whether the interaction in question obeys this symmetry. It was generally believed until 1956 that parity was a good symmetry for all interactions. In that year, T. D. Lee and C. N. Yang pointed out that while there was firm evidence that parity was conserved in strong and electromagnetic interactions, there was no such evidence for weak interactions, and they proposed tests of parity conservation. Early in 1957, reports came from several experiments that showed convincingly that parity symmetry was violated in weak interactions.

Electromagnetism does obey parity symmetry and radiative transitions – absorption and emission – reflect this. Laporte's rule, derived from observations of atomic spectra 1924, showed that only certain energetically allowed (E1) radiative transitions occur, ones between "even" (gerade) and "odd" (ungerade) terms. Laporte's rule was what stimulated Wigner to formulate the concept of parity. Even states are ones whose wave function obeys $\psi(\mathbf{x}) = \psi(-\mathbf{x})$, while for odd states $\psi(\mathbf{x}) = -\psi(\mathbf{x})$. For hydrogen, states with even angular momentum are even and those with odd angular momentum are odd. As we shall develop in great detail later, electromagnetic transitions are governed by interactions of the form $\mathbf{E} \cdot \mathbf{r}$ or $\mathbf{B} \cdot \boldsymbol{\mu}$ and by similar operators containing additional factors of $\mathbf{r}$. These are tensor operators susceptible to treatment with the Wigner-Eckart theorem. The first two are labeled E1 and M1, and the subsequent ones E2, M2, etc. Under the rotation group, $\mathbf{r}$ is odd, while magnetic moments, which can arise from orbital motion and are thus proportional to $\mathbf{r} \times \mathbf{p}$, are even. More generally, under parity, the EL operator has parity $(-1)^L$ while the ML operator has parity $(-1)^{L+1}$.

A system like a nucleus or an atom can have a complicated distribution of charge, current, and magnetization, leading to various static electromagnetic moments. We can add to our list E0, representing charge itself. By parity conservation, these static moments are possible only for EL with L even or ML for L odd. The following is a list of some sample nuclei and their observed moments:

Proton: E0, M1 ($j = 1/2$)
Deuteron: E0, M1, E2 ($j = 1$)
Alpha particle: E0 ($j = 0$)
^{7}Li: E0, M1, E2, M3 moment ($j = 3/2$)
$^{165}_{67}$Ho: E0, M1, E2, M3(?),E4 ($j = 7/2$)

We see that the pattern conforms to the Wigner-Eckart theorem because the Clebsch-Gordan coefficient is non-zero only if the triangular inequality is satisfied with (j, j, L) for the static moment EL (L even) or ML (L odd) so $0 \le L \le 2j$. Thus,

$$j = 0 \qquad\qquad\qquad L = 0$$

$$j = 1/2 \qquad\qquad\qquad L = 0, 1$$

$$j = 1 \qquad\qquad\qquad L = 0, 1, 2$$

$$j = 3/2 \qquad\qquad\qquad L = 0, 1, 2, 3$$

$$\cdot$$

$$j = 7/2 \qquad\qquad\qquad L = 0, 1, 2, 3, 4, 5, 6, 7. \tag{4.214}$$

Radiative and beta-decay transitions are governed by matrix elements of tensor operators

$$\left\langle \alpha' j' m' \left| T_q^{(L)} \right| \alpha j m \right\rangle, \tag{4.215}$$

which are non-vanishing only if

$$m' = m + q; \qquad |j' - j| \le L \le j + j'. \tag{4.216}$$

For example, if $j' = j = 1/2$, only $L = 0, 1$ are allowed. For electromagnetic transitions, $L = 0$ is not allowed and there is only dipole radiation. For $j' = 5/2, j = 3/2$, we can have $L = 1, 2, 3, 4$, i.e. dipole, quadrupole, octopole, 2^4-pole. Usually, the lowest allowed multipole dominates, although in nuclei there are often mixtures of multipoles, e.g. $M1$ and $E2$.

4.12.2 Relative Intensities of Spectral Lines within a Multiplet

Closely spaced spectral lines occur as a result of fine structure, hyperfine structure, or applied fields (Zeeman and Stark effects). Then the spatial wave functions are essentially the same for the initial group of states and for the final group. The relative intensities of the transitions are governed by angular momentum couplings. For relative intensities of allowed electric dipole transitions between magnetic substates in the Zeeman effect, say, all that enters is the square of the Clebsch-Gordan coefficient:

$$\Gamma_{\text{fi}} \propto |\langle j'm'|jm1q\rangle|^2. \tag{4.217}$$

4.12.3 Angular Distributions and Polarizations in Dipole Radiation

The dipole radiation operator is a scalar product of the photon's polarization and the system's dipole moment operator, **p**. This may be either the E1 or M1 operator. The scalar operator is thus

$$\epsilon^{(\lambda)} \cdot \mathbf{p}, \tag{4.218}$$

where $\epsilon^{(\lambda)}$ is a polarization vector of a certain type (circular or linear polarization). We omit the superscript for a while, but will reinstate it later, when needed. We organize this scalar product into a form useful for considering transitions between states of different J_z. Thus,

$$\epsilon \cdot \mathbf{p} = \frac{1}{\sqrt{2}}(\epsilon_x - i\epsilon_y)\frac{1}{\sqrt{2}}(p_x + ip_y) + \epsilon_z p_z + \frac{1}{\sqrt{2}}(\epsilon_x + i\epsilon_y)\frac{1}{\sqrt{2}}(p_x - ip_y). \tag{4.219}$$

Remembering that $(p_x \pm ip_y)$ connects states with $m' = m \pm 1$, while p_z connects states with $m' = m$, we see that

$$\langle \alpha' j'm + 1 | \boldsymbol{\epsilon} \cdot \mathbf{p} | \alpha jm \rangle = \frac{1}{\sqrt{2}} (\epsilon_x - i\epsilon_y) \left\langle \alpha' j'm + 1 \left| \frac{1}{\sqrt{2}} (p_x + ip_y) \right| \alpha jm \right\rangle$$

$$\langle \alpha' j'm | \boldsymbol{\epsilon} \cdot \mathbf{p} | \alpha jm \rangle = \epsilon_z \langle \alpha' j'm | p_z | \alpha jm \rangle$$

$$\langle \alpha' j'm - 1 | \boldsymbol{\epsilon} \cdot \mathbf{p} | \alpha jm \rangle = \frac{1}{\sqrt{2}} (\epsilon_x + i\epsilon_y) \left\langle \alpha' j'm - 1 \left| \frac{1}{\sqrt{2}} (p_x - ip_y) \right| \alpha jm \right\rangle. \qquad (4.220)$$

These relations are independent of the choice of polarization vectors, $\boldsymbol{\epsilon}$. The angular dependence does, of course, depend on the state of polarization. Consider first linear polarization. If the photon has direction $\hat{\mathbf{k}}$, defined by spherical angles θ, ϕ, the standard linear polarization vectors are chosen to be the unit vectors $\hat{\boldsymbol{\theta}}$ and $\hat{\boldsymbol{\phi}}$ in the directions of increasing θ and ϕ. See Figure 4.10. These are sometimes denoted by $\epsilon_\parallel = \hat{\boldsymbol{\theta}}$ and $\epsilon_\perp = \hat{\boldsymbol{\phi}}$, where parallel and perpendicular refer to the plane containing $\hat{\mathbf{k}}$ and the z-axis. We see that

$$\hat{\boldsymbol{\theta}} = \hat{\mathbf{x}} \cos\theta \cos\phi + \hat{\mathbf{y}} \cos\theta \sin\phi - \hat{\mathbf{z}} \sin\theta$$

$$\hat{\boldsymbol{\phi}} = -\hat{\mathbf{x}} \sin\phi + \hat{\mathbf{y}} \cos\phi. \qquad (4.221)$$

For $\Delta m = \pm 1$ transitions, the needed polarization factors are for $\epsilon_\parallel = \hat{\boldsymbol{\theta}}$

$$\frac{1}{\sqrt{2}} (\epsilon_x \mp i\epsilon_y) = \frac{1}{\sqrt{2}} \cos\theta \, e^{\mp i\phi}, \qquad (4.222)$$

while for $\epsilon_\perp = \hat{\boldsymbol{\phi}}$

$$\frac{1}{\sqrt{2}} (\epsilon_x \mp i\epsilon_y) = \mp \frac{i}{\sqrt{2}} e^{\mp i\phi}. \qquad (4.223)$$

For $\Delta m = 0$ transitions, the needed polarization factors are for $\epsilon_\parallel = \hat{\boldsymbol{\theta}}$

$$\epsilon_z = -\sin\theta, \qquad (4.224)$$

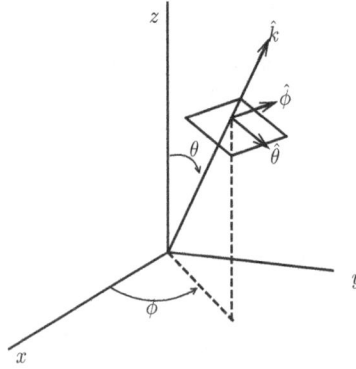

Figure 4.10 For a photon emitted in the direction $\hat{\mathbf{k}}$, the linear polarization directions can be chosen to be $\hat{\boldsymbol{\theta}}$ and $\hat{\boldsymbol{\phi}}$, unit vectors in the direction of increasing θ and increasing ϕ. An alternative notation is $\epsilon_\parallel = \hat{\boldsymbol{\theta}}$ and $\epsilon_\perp = \hat{\boldsymbol{\phi}}$, where parallel and perpendicular refer to the plane containing $\hat{\mathbf{k}}$ and the z-axis.

and for $\epsilon_\perp = \hat{\phi}$

$$\epsilon_z = 0. \tag{4.225}$$

For circular polarization, we form the complex combinations $\epsilon^{(\lambda)} = \frac{1}{\sqrt{2}}(\hat{\theta} \pm i\hat{\phi})$ to describe helicity ± 1 photons (left and right circular polarization, respectively). Because the vector is complex, we must be careful. In the matrix element for emission of radiation, it is actually $\epsilon^* \cdot \mathbf{p}$ that occurs (the photon is in the final state and is associated with a bra on the left in the matrix element). For $\lambda = +1$ and $\Delta m = +1$, we want

$$\frac{1}{\sqrt{2}}(\epsilon_x^{\lambda=+1} - i\epsilon_y^{\lambda=+1}) = \frac{1}{\sqrt{2}}\left(\frac{1}{\sqrt{2}}\right)\left((\hat{\theta}_x - i\hat{\theta}_y) - i(\hat{\phi}_x - i\hat{\phi}_y)\right)$$

$$= \frac{1}{2}e^{-i\phi}(\cos\theta - 1). \tag{4.226}$$

We can understand the $(1 - \cos\theta)$ behavior as follows. The emitted photon carries off $J_z = -1$ when $\Delta m = +1$. For a photon with helicity $\lambda = +1$, the amplitude will vanish at $\theta = 0$ and maximize at $\theta = \pi$.

For $\lambda = -1$, but $\Delta m = +1$, the polarization factor is obtained merely by changing the sign of the i between $(\hat{\theta}_x - i\hat{\theta}_y)$ and $(\hat{\phi}_x - i\hat{\phi}_y)$ above. Thus,

$$\frac{1}{\sqrt{2}}(\epsilon_x^{\lambda=-1} - i\epsilon_y^{\lambda=-1}) = \frac{1}{2}e^{-i\phi}(\cos\theta + 1). \tag{4.227}$$

Similarly, we find

$$\lambda = +1, \Delta m = -1 : \quad \frac{1}{2}e^{i\phi}(\cos\theta + 1)$$

$$\lambda = -1, \Delta m = -1 : \quad \frac{1}{2}e^{i\phi}(\cos\theta - 1)$$

$$\lambda = +1, \Delta m = 0 : \quad -\frac{1}{\sqrt{2}}\sin\theta. \tag{4.228}$$

Up to a phase of ± 1, these various factors are equal to $e^{-i\Delta m\phi}d_{\lambda,\Delta m}^1(\theta)$.

The angular distribution of the emitted radiation for a given Δm and state of polarization is obtained by taking the absolute squares of the appropriate matrix elements.

For linear polarization, we have

$$\Delta m = \pm 1 : \quad \frac{dw_\parallel}{d\Omega} = \frac{1}{2}\cos^2\theta \left|\langle\alpha'j'm \pm 1|\frac{1}{\sqrt{2}}(p_x \pm ip_y)|\alpha jm\rangle\right|^2$$

$$\frac{dw_\perp}{d\Omega} = \frac{1}{2}\left|\langle\alpha'j'm \pm 1|\frac{1}{\sqrt{2}}(p_x \pm ip_y)|\alpha jm\rangle\right|^2$$

$$\Delta m = 0 : \quad \frac{dw_\parallel}{d\Omega} = \sin^2\theta \left|\langle\alpha'j'm|p_z|\alpha jm\rangle\right|^2$$

$$\frac{dw_\perp}{d\Omega} = 0 \tag{4.229}$$

The sum of the parallel and perpendicular polarizations gives the angular distribution when the polarization is unobserved. For M1 transitions, the roles of **E** and **B** are interchanged and so the angular distributions for parallel and perpendicular polarizations are reversed.

For circular polarization, we find

$$\Delta m = \pm 1; \lambda = \pm 1 : \frac{dw}{d\Omega} = \frac{1}{4}(1 - \cos\theta)^2 \left| \langle \alpha' j'm \pm 1 | \frac{1}{\sqrt{2}}(p_x \pm ip_y)|\alpha jm \rangle \right|^2$$

$$\Delta m = \pm 1; \lambda = \mp 1 : \frac{dw}{d\Omega} = \frac{1}{4}(1 + \cos\theta)^2 \left| \langle \alpha' j'm \pm 1 | \frac{1}{\sqrt{2}}(p_x \pm ip_y)|\alpha jm \rangle \right|^2$$

$$\Delta m = 0; \lambda = \pm 1 : \frac{dw}{d\Omega} = \frac{1}{2}\sin^2\theta \left| \langle \alpha' j'm|p_z|\alpha jm \rangle \right|^2. \tag{4.230}$$

Summing over the polarizations gives the same sum as in the linear case, of course.

The Wigner-Eckart theorem can be used to relate the matrix elements for $\Delta m = 0$ to those for $\Delta \pm 1$:

$$\langle \alpha' j'm \pm 1 | \frac{1}{\sqrt{2}}(p_x \pm ip_y)|\alpha jm \rangle \propto \langle j'm \pm 1|jm; 1, \pm 1 \rangle$$

$$\langle \alpha' j'm|p_z|\alpha jm \rangle \propto \langle j'm|jm10 \rangle. \tag{4.231}$$

Thus when summing over polarizations, the intensities $\mathcal{I}_{m'm}$ obey

$$\mathcal{I}_{m \pm 1, m} = \text{const.} \cdot \frac{1}{2}(1 + \cos^2\theta)|\langle j'm \pm 1|jm; 1, \pm 1 \rangle|^2$$

$$\mathcal{I}_{m, m} = \text{const.} \cdot \sin^2\theta|\langle j'm|jm; 1, 0 \rangle|^2. \tag{4.232}$$

4.13 Two-Body Central Force Motion

Our analysis of the rotation group enables us to turn three-dimensional problems with rotational symmetry into one-dimensional problems. If two particles interact via a central conservative force described by a potential energy $V(|\mathbf{r}_1 - \mathbf{r}_2|)$, the Hamiltonian

$$H = \frac{p_1^2}{2m_1} + \frac{p_2^2}{2m_2} + V(|\mathbf{r}_1 - \mathbf{r}_2|) \tag{4.233}$$

can be transformed by introducing center-of-mass coordinates, relative coordinates, and a reduced mass: $(\mathbf{r}_1, \mathbf{p}_1, \mathbf{r}_2, \mathbf{p}_2, m_1, m_2) \to (\mathbf{R}, \mathbf{P}, M, \mathbf{r}, \mathbf{p}, m)$, with the result

$$H = \frac{P^2}{2M} + \frac{p^2}{2m} + V(r). \tag{4.234}$$

The commuting variables are $H, \mathbf{P}, \mathbf{L}_{\text{cm}}, \mathbf{L} = \mathbf{r} \times \mathbf{p}$. The internal motion is now governed by a one-particle Hamiltonian for a "reduced particle" of mass $m = m_1 m_2/(m_1 + m_2)$. The complete wave function describing the translation of the system as a whole and internal motion of a definite energy E is

$$\psi(\mathbf{r}, \mathbf{R}, t) - e^{i\mathbf{P}\cdot\mathbf{R}/\hbar}\psi_E(\mathbf{r})e^{-i(\frac{P^2}{2M}+E)t/\hbar}. \tag{4.235}$$

The internal wave function satisfies the Schrödinger equation

$$\left[\frac{p^2}{2m} + V(r)\right]\psi_E(\mathbf{r}) = E\psi_E(\mathbf{r}). \tag{4.236}$$

Because L^2 and L_z commute with H and with each other, we can have simultaneous eigenstates of the three operators. It is therefore useful to separate the radial and angular variables. To do this, we perform some vector gymnastics:

$$\begin{aligned}
(\mathbf{r} \times \nabla) \cdot (\mathbf{r} \times \nabla) &= \epsilon_{ijk}\epsilon_{imn}r_j\partial_k r_m\partial_n \\
&= (\delta_{jm}\delta_{kn} - \delta_{jn}\delta_{km})r_j\partial_k r_m\partial_n \\
&= (\delta_{jm}\delta_{kn} - \delta_{jn}\delta_{km})(r_j r_m\partial_k\partial_n + \delta_{km}r_j\partial_n) \\
&= r^2\nabla^2 - r_j r_m\partial_j\partial_m - 2\mathbf{r}\cdot\nabla. \tag{4.237}
\end{aligned}$$

But

$$(\mathbf{r}\cdot\nabla)^2 = r_j r_m\partial_j\partial_m + \mathbf{r}\cdot\nabla \tag{4.238}$$

so that

$$\begin{aligned}
(\mathbf{r} \times \nabla) \cdot (\mathbf{r} \times \nabla) &= r^2\nabla^2 - (\mathbf{r}\cdot\nabla)^2 - \mathbf{r}\cdot\nabla \\
\nabla^2 &= \frac{1}{r^2}\left[(\mathbf{r}\times\nabla)\cdot(\mathbf{r}\times\nabla) + \left(r\frac{\partial}{\partial r}\right)^2 + r\frac{\partial}{\partial r}\right] \\
&= \frac{(\mathbf{r}\times\nabla)\cdot(\mathbf{r}\times\nabla)}{r^2} + \frac{1}{r}\frac{\partial^2}{\partial r^2}r. \tag{4.239}
\end{aligned}$$

This enables us to write

$$\frac{p^2}{2m} = -\frac{\hbar^2}{2m}\frac{1}{r}\frac{\partial^2}{\partial r^2}r + \frac{\hbar^2 L^2}{2mr^2}. \tag{4.240}$$

On occasion, one needs

$$L^2 = -\left[\frac{1}{\sin\theta}\frac{\partial}{\partial\theta}\left(\sin\theta\frac{\partial}{\partial\theta}\right) + \frac{1}{\sin^2\theta}\frac{\partial^2}{\partial\phi^2}\right]. \tag{4.241}$$

Separation of variables leads to solutions of the form

$$\psi_E(\mathbf{r}) = \frac{u(r)}{r}Y_{\ell m}(\theta, \phi), \tag{4.242}$$

where $u(r)$ satisfies a standard one-dimensional Schrödinger equation

$$\left[-\frac{\hbar^2}{2m}\frac{d^2}{dr^2} + V(r) + \frac{\ell(\ell+1)\hbar^2}{2mr^2}\right]u(r) = Eu(r) \tag{4.243}$$

with the boundary condition that the wave function be square integrable at the origin, i.e. $\int_0^\epsilon dr|u(r)|^2 < \infty$. Combining the potential and the angular momentum barrier gives an effective potential, as shown in Figure 4.11.

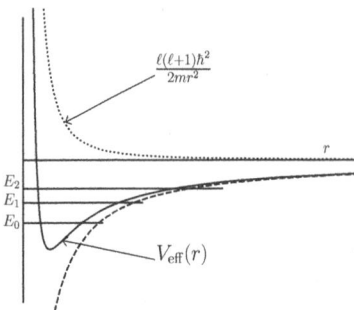

Figure 4.11 Combining the potential and the angular momentum barrier yields an effective potential for the radial wave function. Here there is an attractive potential shown as dashed, while the repulsive angular momentum barrier is shown as dotted. There may be no bound states, a finite number of bound states, or an infinite number of bound states.

Several observations are immediate.

(1) Parity commutes with H. The energy eigenstates, if not degenerate, are eigenstates of parity with parity $(-1)^\ell$.

(2) Since there are observables that commute with H but not with each other (L_x, L_y, L_z), the energy levels are degenerate unless $L = 0$. This is made explicit in the appearance of $Y_{\ell,m}$.

(3) Behavior near the origin: provided $r^2 V(r) \to 0$ as $r \to 0$, this is governed by

$$\left[\frac{d^2}{dr^2} - \frac{\ell(\ell+1)}{r^2} \right] u(r) = 0. \tag{4.244}$$

The solutions are proportional to $r^{\ell+1}$ or $r^{-\ell}$. For the solution to be square integrable, $u(r) \propto r^{\ell+1}$ as $r \to 0$.

(4) Behavior at infinity for $E < 0$: If $V(r) \to 0$ as $r \to \infty$, then at large r the differential equation is

$$\left[\frac{d^2}{dr^2} + \frac{2mE}{\hbar^2} \right] u(r) = 0. \tag{4.245}$$

For $E < 0$, $u(r) \propto e^{\pm \alpha r}$, where $\alpha^2 = -2mE/\hbar^2$. For a localized state with finite norm, we must choose the decaying exponential, i.e.

$$\lim_{r \to \infty} U(r) \propto e^{-\alpha r}. \tag{4.246}$$

The two boundary conditions at $r = 0$ and $r \to \infty$ define the energy eigenvalue problem. See Figure 4.12.

(5) Behavior at $r \to \infty$ for $E > 0$: Assume $rV(r) \to 0$ as $r \to \infty$. Let $k^2 = 2mE/\hbar^2$. Then the solution at large r is

$$u(r) = A \sin kr + B \cos kr. \tag{4.247}$$

In the circumstance of scattering when the angular momentum is ℓ, this is generally written as

$$u(r) \propto \sin(kr + \delta - \ell\pi/2). \tag{4.248}$$

For $V(r) \propto 1/r$ as $r \to \infty$, as in the Coulomb problem, we find $u \propto \sin(kr + \delta - \epsilon \ln(2kr))$, with $\epsilon = z_1 z_2 e^2 / \hbar v$, corresponding to a classical hyperbolic orbit.

Figure 4.12 Radial wave functions for (upper) $\ell = 2, n_r = 0$ and (lower) $\ell = 1, n_r = 1$, where n_r is the number of nodes in the radial wave function.

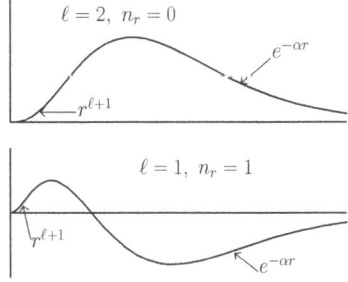

4.14 The Coulomb Problem

Finding the bound states of the Coulomb potential can be achieved in a variety of ways. Here we demonstrate the power of complex variable theory to solve differential equations. We define the Coulomb problem with an infinitely massive positive charge $+Ze$ at the center and a (spinless) electron with charge $-e$ at a radial distance r. The radial wave equation for angular momentum ℓ can be written in dimensionless form if we introduce

$$x = r/a; \qquad a = \frac{\hbar^2}{me^2 Z} = \frac{a_0}{Z}$$

$$\epsilon = -\frac{E}{\text{Ryd}}; \qquad \text{Ryd} = \frac{Z^2 me^4}{2\hbar^2} = \frac{Z^2 e^2}{2a_0}, \tag{4.249}$$

then the radial Schrödinger equation is

$$\frac{d^2 u}{dx^2} + \left[\frac{2}{x} - \frac{\ell(\ell+1)}{x^2} - \epsilon\right] u = 0. \tag{4.250}$$

We know the behavior near $x = 0$, so we set $u(x) = x^{\ell+1} v(x)$ and find

$$x\frac{d^2 v}{dx^2} + 2(\ell+1)\frac{dv}{dx} + (2 - \epsilon x)v = 0. \tag{4.251}$$

Since this is only linear in x but second order in d/dx, a solution in momentum space is attractive:

$$v(x) = \int_{-\infty}^{\infty} dq\, e^{iqx}\phi(q). \tag{4.252}$$

We see that the replacements needed in the differential equation are $\frac{d}{dx} \to iq,\ x \to i\frac{d}{dq}$. The result is

$$i\frac{d}{dq}(-q^2\phi) + 2(\ell+1)iq\phi + 2\phi - i\epsilon\frac{d\phi}{dq} = 0. \tag{4.253}$$

Reorganizing,

$$\frac{d\phi}{dq} + \frac{2(i - \ell q)}{q^2 + \epsilon}\phi = 0. \tag{4.254}$$

The solution to this first-order equation is simply

$$\phi(q) = A\exp\left[-2\int \frac{(i - \ell q)}{q^2 + \epsilon}dq\right]. \tag{4.255}$$

For bound states $E < 0$ and $\epsilon > 0$, so we can write $\epsilon = \alpha^2$. Then, using partial fractions,

$$\frac{(i - \ell q)}{q^2 + \alpha^2} = \frac{1}{2}\left(\frac{1}{\alpha} - \ell\right)\frac{1}{q - i\alpha} - \frac{1}{2}\left(\frac{1}{\alpha} + \ell\right)\frac{1}{q + i\alpha},$$

$$-2\int \frac{(i - \ell q)}{q^2 + \alpha^2}\,dq = \left(\ell - \frac{1}{\alpha}\right)\ln(q - i\alpha) + \left(\ell + \frac{1}{\alpha}\right)\ln(q + i\alpha). \tag{4.256}$$

We thus obtain

$$\phi(q) = A(q^2 + \alpha^2)^\ell \left(\frac{q + i\alpha}{q - i\alpha}\right)^{1/\alpha}. \tag{4.257}$$

and

$$u(x) = Ax^{\ell+1}\int_{-\infty}^{\infty} dq\, e^{iqx}(q^2 + \alpha^2)^\ell \left(\frac{q + i\alpha}{q - i\alpha}\right)^{1/\alpha}. \tag{4.258}$$

Now we know from general considerations of second order differential equations of this sort, that while the small x behavior is either $x^{\ell+1}$ or $x^{-\ell}$, the large x behavior is $\exp(\pm\alpha x)$. The eigenvalue problem in this context is the determination of the values for α that lead to $\exp(-\alpha x)$ with $x^{\ell+1}$ fixed at low x.

We consider three distinct classes:

(a) $0 < 1/\alpha < \infty$, $1/\alpha$ not an integer.
(b) $1/\alpha$ an integer less than $\ell + 1$.
(c) $1/\alpha$ an integer greater than ℓ.

Consider (b) first.

$$(q^2 + \alpha^2)^\ell \left(\frac{q + i\alpha}{q - i\alpha}\right)^{1/\alpha} = (q + i\alpha)^{\ell+1/\alpha}(q - i\alpha)^{\ell-1/\alpha}. \tag{4.259}$$

Now both $\ell + 1/\alpha$ and $\ell - 1/\alpha$ are non-negative integers, so the integrand has no poles in the complex q plane. Because of the $\exp(iqx)$, we can close the contour that stretches along the real axis from $-\infty$ to $+\infty$ with a semi-circle at very large $|q|$ in the upper half-plane. See Figure 4.13. Thus the entire integral vanishes.

In case (a), $(q + i\alpha)^{\ell+1/\alpha}$ and $(q - i\alpha)^{\ell-1/\alpha}$ have branch points at $-i\alpha$ and $+i\alpha$, respectively. See Figure 4.14 and Appendix B.1. Since $1/\alpha$ is not an integer, we see that the integrals I_{L} and I_{R} on opposite sides of the cut in the upper half-plane differ by a phase $e^{2\pi i/\alpha}$, and thus

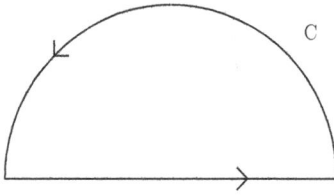

Figure 4.13 For case (b), there are no singularities in the upper half-plane and the contour can be closed there. The total contour integral is therefore zero. The contribution at large $|q|$ in the upper half-plane vanishes. As a result, the integral along the real axis vanishes.

cannot cancel. Look then just at I_R, and put $q = i\alpha + it/x$. Then we find that

$$I_R \propto x^{\ell+1} \int_0^\infty dt \frac{e^{-\alpha x}}{x} e^{-t} \left(2\alpha + \frac{t}{x}\right)^{\ell+1/\alpha} \left(\frac{t}{x}\right)^{\ell-1/\alpha}. \tag{4.260}$$

In the limit $x \to 0$, we see that $I_R \propto x^{-\ell}$, which is inadmissible behavior for a bound-state wave function. So finally we turn to case (c).

We put $1/\alpha = \ell + 1 + n_r$ with $n_r = 0, 1, 2 \ldots$ Then consider

$$v(x) = A \int_\infty^\infty dq\, e^{iqx} \frac{(q + i\alpha)^{2\ell+1+n_r}}{(q - i\alpha)^{n_r+1}}. \tag{4.261}$$

Thanks to the factor e^{iqx}, the contour can be closed in the upper half-plane. There is a pole at $q = i\alpha$. See Figure 4.15. Let us put $q = -i\alpha + iz/x$. Then we have

$$v(x) = A \frac{i(-1)^\ell e^{\alpha x}}{x^{2\ell+1}} \oint dz\, e^{-z} \frac{z^{2\ell+1+n_r}}{(z - 2\alpha x)^{n_r+1}}. \tag{4.262}$$

Cauchy's theorem tells us

$$\oint dz \frac{f(z)}{(z - z_0)^{n+1}} = \frac{2\pi i}{n!} \frac{d^n f}{dz^n}\bigg|_{z=z_0}. \tag{4.263}$$

Thus we have ultimately,

$$v(x) = \frac{i(-1)^\ell e^{\alpha x}}{x^{2\ell+1}} \times \frac{2\pi i}{n_r!} \left[\frac{d^{n_r}}{dz^{n_r}} \left(z^{2\ell+1+n_r} e^{-z}\right)\right]_{z=2ax}. \tag{4.264}$$

The derivatives will create a polynomial whose lowest power of z is $z^{2\ell+1}$, so v is well-behaved as $x \to 0$. We also see that at large x, v has the general behavior $e^{-\alpha x}$, as desired. We introduce the associated Laguerre polynomials with the definition used by Magnus and Oberhettinger,

Figure 4.14 For case (a), there are branch points at $q = \pm i\alpha$. The values of $(q - i\alpha)^{1/\alpha}$ and $(q + i\alpha)^{1/\alpha}$ must be specified with respect to the branch points. Here $(q - i\alpha)^{1/\alpha} = |q - i\alpha|^{1/\alpha} e^{i\theta/\alpha}$, where θ is measured counterclockwise from the direction joining $i\alpha$ and the origin. The values of $(q - i\alpha)^{\ell-1/\alpha}$ on opposite sides of the cut, indicated by R and L, differ by a phase $e^{2\pi i/\alpha}$ because the values of θ differ by 2π on the two sides. As a consequence, there cannot be cancellation between the integrals I_R and I_L on the two sides. The resulting function will have unacceptable behavior as $x \to 0$.

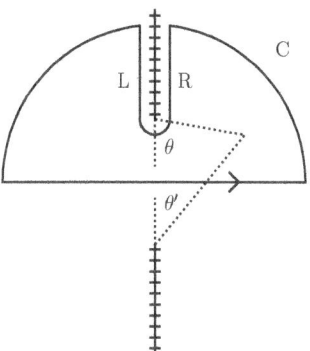

Figure 4.15 For case (c), there is a pole at $+i\alpha$. Because there is no branch cut and because the integrand vanishes at infinity in the upper half-plane, the contour can be closed. The pole makes the unique contribution to the integral.

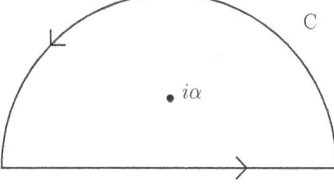

Formulas and Theorems for the Functions of Mathematical Physics, and the Bateman Project *Higher Transcendental Functions* Vol 2, p. 188)

$$L_n^\beta(z) = \frac{e^z z^{-\beta}}{n!} \frac{d^n}{dz^n} \left(e^{-z} z^{n+\beta} \right)$$

$$= \frac{e^z z^{-\beta}}{n!} \left[e^{-z} \frac{d^n}{dz^n} z^{\beta+n} + n \frac{d}{dz} e^{-z} \frac{d^{n-1}}{dz^{n-1}} z^{\beta+n} + \frac{n(n-1)}{2} \frac{d^2}{dz^2} e^{-z} \frac{d^{n-2}}{dz^{n-2}} z^{\beta+n} \cdots \right]$$

$$= \left[\frac{(\beta+n)!}{n!\beta!} - \frac{(\beta+n)!}{(\beta+1)!(n-1)!} z + \frac{(\beta+n)!}{2(\beta+2)!(n-2)!} z^2 + \cdots + (-1)^n z^n \right]$$

$$= \frac{(\beta+n)!}{n!\beta!} \left[1 - \frac{n}{\beta+1} z + \frac{1}{2!} \frac{n(n-1)}{(\beta+1)(\beta+2)} z^2 + \cdots + (-1)^m \frac{n!\beta!}{(\beta+n)!} z^n \right]. \quad (4.265)$$

Other authors have definitions that differ from this. With our definition,

$$u(x) \propto x^{\ell+1} e^{-\alpha x} L_{n_r}^{2\ell+1}(2\alpha x) \quad (4.266)$$

with a normalization to be determined later.

The energy eigenvalues are

$$\epsilon = \alpha^2 = \frac{1}{(n_r + 1 + \ell)^2} = \frac{1}{n^2}. \quad (4.267)$$

where we make the customary definition $n = n_r + 1 + \ell$ as the principal quantum number. From Eq. (4.265), we find

$$L_{n_r}^{2\ell+1}(2x/n) \propto 1 - \frac{n_r}{2\ell+2} \frac{2x}{n} + \frac{1}{2!} \frac{n_r(n_r+1)}{(2\ell+2)(2\ell+3)} \left(\frac{2x}{x} \right)^2$$

$$- \frac{1}{3!} \frac{n_r(n_r-1)(n_r-2)}{(2\ell+2)(2\ell+3)(2\ell+4)} \left(\frac{2x}{n} \right)^3 + \cdots \quad (4.268)$$

If we write $u_{n=n_r+\ell+1,\ell}$, we have the unnormalized radial wave functions

$$u_{n,\ell}(r) \propto r^{\ell+1} e^{-r/n} \left[1 - \frac{n_r}{2\ell+2} \frac{2r}{n} + \frac{1}{2!} \frac{n_r(n_r+1)}{(2\ell+2)(2\ell+3)} \left(\frac{2r}{x} \right)^2 \right.$$

$$\left. - \frac{1}{3!} \frac{n_r(n_r-1)(n_r-2)}{(2\ell+2)(2\ell+3)(2\ell+4)} \left(\frac{2r}{n} \right)^3 + \cdots \right]. \quad (4.269)$$

It is straightforward to get the normalization from

$$\int_0^\infty dr\, u_{n,\ell}^2 = 1 \quad (4.270)$$

to obtain the radial wave functions for hydrogen with r in units of a_0

$$u_{1,0} = 2re^{-r}$$

$$u_{2,0} = \frac{r}{\sqrt{2}}e^{-r/2}\left(1 - \frac{r}{2}\right),$$

$$u_{2,1} = \frac{r^2}{2\sqrt{6}}e^{-r/2},$$

$$u_{3,0} = \frac{2r}{3\sqrt{3}}e^{-r/3}\left(1 - \frac{2}{3}r + \frac{2}{27}r^2\right),$$

$$u_{3,1} = \frac{8r^2}{27\sqrt{6}}e^{-r/3}\left(1 - \frac{r}{6}\right),$$

$$u_{3,2} = \frac{4r^3}{81\sqrt{30}}e^{-r/3}$$

$$u_{4,0} = \frac{r}{4}e^{-r/4}\left(1 - \frac{3}{4}r + \frac{1}{8}r^2 - \frac{1}{192}r^3\right)$$

$$u_{4,1} = \frac{r^2}{16}\sqrt{\frac{5}{3}}e^{-r/4}\left(1 - \frac{1}{4}r + \frac{1}{80}r^2\right)$$

$$u_{4,2} = \frac{r^3}{64\sqrt{5}}e^{-r/4}\left(1 - \frac{1}{12}r\right)$$

$$u_{4,3} = \frac{r^4}{768\sqrt{35}}e^{-r/4} \tag{4.271}$$

in agreement with Bethe and Salpeter, *Quantum Mechanics of One- and Two-Electron Atoms*, p. 15, except that we give the radial wave functions with the extra power of r. The full wave function includes an appropriate spherical harmonic.

4.15 Patterns of Bound States

For fixed orbital angular moment ℓ, if there are bound states then the one with the lowest energy will have no radial nodes ($n_r = 0$). The next state will have one radial node, etc. Because the centrifugal barrier increases the repulsion as ℓ increases, we can conclude that the lowest state for each ℓ lies higher than that for $\ell' = \ell - 1$ and lower than for $\ell'' = \ell + 1$. The Grotrian diagram shows energy levels broken out according to ℓ and will therefore appear as shown in Figure 4.16.

The pattern above the "ground state" of each ℓ depends on the exact shape of $V(r)$. To get some intuition on how the levels behave, one can

(a) Look at some standard problems to see how the shape of $V(r)$ affects the pattern of levels.
(b) Use approximation techniques:
  (i) A one-dimensional simple harmonic oscillator approximation around the minimum in $V_{\text{eff}}(r)$.
  (ii) WKB approximation:

$$(n_r + 1/2)\pi = \int_{x_1}^{x_2} \sqrt{\frac{2m}{\hbar^2}(E - V(r)) - \frac{(\ell + 1/2)^2}{r^2}}\, dr \tag{4.272}$$

where we use the Langer replacement $\ell(\ell + 1) \rightarrow (\ell + 1/2)^2$ (Langer, R., *Phys. Rev.* **51**, 669 (1937)), which turns to work better for bound states.

The Grotrian diagrams for the three-dimensional simple harmonic oscillator and for the Coulomb problem are shown in Figure 4.17.

(a) The way the lowest energy states for each ℓ progress with increasing ℓ is understood in terms of $[V_{\mathrm{eff}}(r)]_{\min}$ vs ℓ.

(b) The relative energies of the lowest p-wave state and the second lowest s-wave state in the two cases can be understood in terms of the increase of $V(r)$ at large r for the simple harmonic oscillator. The $n_r = 1, \ell = 0$ state goes out further in r.

(c) The accidental degeneracies in both these problems are unique to the specific forms of $V(r)$. There are special symmetries that can be identified as being the cause. The classical Coulomb problem has the energy depending upon the length of the semi-major axis, but not on the eccentricity of the ellipse.

(d) A potential that is attractive, but falls off faster than $1/r$, for example a screened Coulomb potential for a valence electron in an atom, will have (in Coulomb-potential notation) the

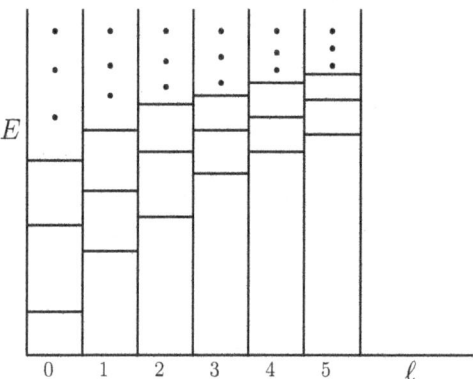

Figure 4.16 The Grotrian diagram shows the bound state energies spread out according to the angular momentum.

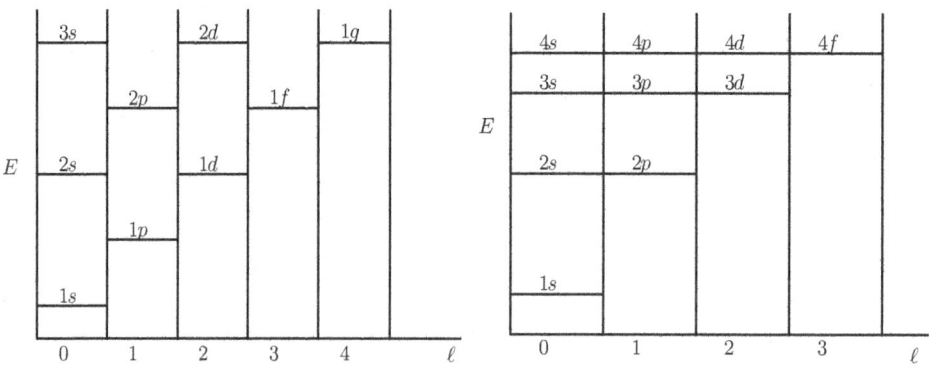

Figure 4.17 The Grotrian diagrams for the simple harmonic oscillator (left) and the Coulomb problem (right). Note that it is conventional in the Coulomb problem to label the first two degenerate states 2s and 2p, rather than 2s and 1p, as the labels are given for the simple harmonic oscillator. The advantage of this seemingly irregular notation is that it provides a reminder of the degeneracy pattern, e.g. in the non-relativistic hydrogen atom 3s, 3p, and 3d are degenerate.

2s level below the 2p, and the 3p below the 3d. This phenomenon is called the effect of penetrating orbits because lower ℓ values penetrate to smaller r.

4.16 Hellmann-Feynman Theorem

Suppose H depends on some parameter ξ. Then the energy E_n in some eigenstate $\psi_{n,\xi}(\mathbf{x})$ depends on ξ. Thus we have

$$E_n(\xi) = \int d^3x\, \psi_{n,\xi}^*(\mathbf{x}) H \psi_{n,\xi}(\mathbf{x}), \tag{4.273}$$

where we have assumed that $\psi_{n,\xi}$ is normalized. Now consider

$$\frac{\partial E_n(\xi)}{\partial \xi} = \frac{\partial}{\partial \xi} \int d^3x\, \psi_{n,\xi}^*(\mathbf{x}) H \psi_{n,\xi}(\mathbf{x}). \tag{4.274}$$

Let us write

$$\frac{\partial}{\partial \xi} \psi_{n,\xi} \equiv \psi_{n,\xi'} \tag{4.275}$$

so that

$$\begin{aligned}
\frac{\partial E_n(\xi)}{\partial \xi} &= \int d^3x\, \psi_{n,\xi}^*(\mathbf{x}) \frac{\partial H}{\partial \xi} \psi_{n,\xi}(\mathbf{x}) + \int d^3x\, \psi_{n,\xi'}^* H \psi_{n,\xi} + \int d^3x\, \psi_{n,\xi}^* H \psi_{n,\xi'} \\
&= \int d^3x\, \psi_{n,\xi}^*(\mathbf{x}) \frac{\partial H}{\partial \xi} \psi_{n,\xi}(\mathbf{x}) + E_n(\xi) \frac{\partial}{\partial \xi} \int d^3x\, \psi_{n,\xi}^* \psi_{n,\xi}.
\end{aligned} \tag{4.276}$$

But since the state is normalized, we have

$$\frac{\partial E_n(\xi)}{\partial \xi} = \int d^3x\, \psi_{n,\xi}^*(\mathbf{x}) \frac{\partial H}{\partial \xi} \psi_{n,\xi}(\mathbf{x}). \tag{4.277}$$

While this is called the Hellmann-Feynman theorem, it was known long before Feynman's undergraduate thesis and his publication of 1939. Hellmann published the equation in a textbook and used it in various molecular physics problems. Hellmann himself apparently knew that the relation was in Pauli's 1933 *Handbuch der Physik* article. It appears, as well, in a 1932 paper by Güttinger.

We apply this theorem to central field motion with a state of angular momentum ℓ and energy $E = E(n_r, \ell)$, where n_r is the number of radial nodes. In terms of the radial Schrödinger equation, we have

$$H = -\frac{\hbar^2}{2m} \frac{d^2}{dr^2} + V(r) + \frac{\ell(\ell+1)\hbar^2}{2mr^2}. \tag{4.278}$$

If we think of H as a function of ℓ, we can compute

$$\frac{\partial H}{\partial \ell} = (2\ell + 1) \frac{\hbar^2}{2mr^2}. \tag{4.279}$$

We apply the Hellmann-Feynman theorem to obtain

$$\left\langle \frac{1}{r^2} \right\rangle = \frac{2m}{(2\ell+1)\hbar^2} \frac{\partial E(n_r, \ell)}{\partial \ell}. \tag{4.280}$$

For hydrogen-like states, with $n = n_r + \ell + 1$, we have

$$E(n_r, \ell) = -\frac{Z^2 e^2}{2a_0} \frac{1}{(n_r + \ell + 1)^2} \tag{4.281}$$

and

$$\frac{\partial E(n_r, \ell)}{\partial \ell} = \frac{Z^2 e^2}{a_0} \frac{1}{n^3}, \tag{4.282}$$

so

$$\left\langle \frac{1}{r^2} \right\rangle = \frac{Z^2}{a_0^2} \frac{1}{(\ell + 1/2)n^3}. \tag{4.283}$$

This result is particularly welcome because it cannot be obtained by the methods developed in Problems 4.12 and 4.13, which determine many other radial moments.

For the isotropic oscillator,

$$E(n_r, \ell) = (2n_r + \ell + 3/2)\hbar\omega; \qquad \left\langle \frac{1}{r^2} \right\rangle = \frac{1}{\ell + 1/2} \frac{m\omega}{\hbar}. \tag{4.284}$$

References and Suggested Reading

For the quantum theory of angular momentum:

Rose, *Elementary Theory of Angular Momentum*.
Edmonds, *Angular Momentum in Quantum Mechanics*.

Two references on molecular beams are:

N. F. Ramsey, in *Experimental Nuclear Physics*, Vol. 1, editor E. Segrè, (1953), pp. 358 ff.
N. F. Ramsey, *Molecular Beams*, Oxford University Press, (1956, 1985), especially Chapters 1, 5.

For nuclear magnetic resonance:

Abragam, A., *The Principles of Nuclear Magnetism*. Oxford University Press (1961).
Abragam, A. and Goldman, M., *Nuclear Magnetism: Order and Disorder*. Oxford University Press (1982).
Gottfried, K., *Quantum Mechanics: Fundamentals*, pp. 427–434. Our treatment is close to that presented there.
Slichter, C. P., *Principles of Magnetic Resonance*, 2nd edition. Springer (1978).

Problems

4.1 For $j = 1$, use the results of the raising and lowering operators to construct explicitly the 3×3 matrices for $J^2, J_x, J_y, J_z, J_x^2, J_y^2, J_z^2$ and J_x^3, J_y^3, J_z^3. What about higher powers?

4.2 **(a)** From the commutation relations of a scalar operator S with J^2 and J_z, show that the matrix elements of S with respect to the states $|\alpha jm\rangle$ have the form

$$\langle \alpha' j'm' |S| \alpha jm\rangle = \delta_{m'm}\delta_{jj'} \langle \alpha' j||S||\alpha j\rangle, \tag{4.285}$$

where $\langle \alpha' j||S||\alpha j\rangle$ is called a reduced matrix element and is independent of m.

(b) Write out the commutation relations of the components of a vector operator $\mathbf{V}$ with J_z and $J_\pm$. Define $V_\pm = V_x \pm iV_y$ and use them and V_z as the three independent components of the vector. From these commutation relations, deduce the selection rules for the magnetic quantum number for the matrix elements

$$\langle \alpha' j'm' |V_z| \alpha jm\rangle \text{ and } \langle \alpha' j'm' \left|V_\pm\right| \alpha jm\rangle. \tag{4.286}$$

(c) From the commutation relations of a vector operator $\mathbf{V}$ with J^2 and J_i, show that the double commutator of $\mathbf{V}$ with J^2 has the form

$$[J^2, [J^2, \mathbf{V}]] = 2\mathbf{V}J^2 + 2J^2\mathbf{V} - 4\mathbf{J}(\mathbf{J} \cdot \mathbf{V}). \tag{4.287}$$

(d) Using the double commutator of part (c), show that the restricted set of matrix elements of $\mathbf{V}$ between states of the same j has the form

$$\langle \alpha' jm' |\mathbf{V}| \alpha jm\rangle = \frac{\langle jm' |\mathbf{J}| jm\rangle}{j(j+1)} \langle \alpha' j||\mathbf{J} \cdot \mathbf{V}||\alpha j\rangle. \tag{4.288}$$

This relation is a special case of the Wigner-Eckart theorem and provides justification for the vector model of atomic physics. It shows that the expectation value of any vector operator $\mathbf{V}$ in a state $|\alpha jm\rangle$ is proportional to m and points in the z-direction.

4.3 The classical generators of the infinitesimal rotations, defined from $\delta\mathbf{x} = \delta\mathbf{\Omega}\times\mathbf{x}$, have the 3×3 matrix representations

$$S_1 = \begin{pmatrix} 0 & 0 & 0 \\ 0 & 0 & -1 \\ 0 & 1 & 0 \end{pmatrix}, \quad S_2 = \begin{pmatrix} 0 & 0 & 1 \\ 0 & 0 & 0 \\ -1 & 0 & 0 \end{pmatrix}, \quad S_3 = \begin{pmatrix} 0 & -1 & 0 \\ 1 & 0 & 0 \\ 0 & 0 & 0 \end{pmatrix}. \tag{4.289}$$

These anti-Hermitian matrices are the space parts of the rotation generators of special relativity. See, for example, Jackson, *Classical Electrodynamics*, 3rd edition, p. 546.

The matrix $i\mathbf{S}$ must be unitarily equivalent to the (j, m) representatives of the angular momentum operator $\mathbf{J}$ for $j = 1$ of Problem 4.1. Thus

$$\mathbf{J} = U^\dagger(i\mathbf{S})U. \tag{4.290}$$

Determine explicitly the form of the 3×3 matrix U, up to an overall phase. What is the significance of the columns of U?

4.4 A spin-1/2 system with magnetic moment $\boldsymbol{\mu} = \mu_0\boldsymbol{\sigma}$ and Hamiltonian $H = -\boldsymbol{\mu} \cdot \mathbf{B}$ is located in a uniform, static magnetic field B_0 in the positive z-direction. For the time interval, $0 < t < T$, an additional uniform, constant field B_1 is applied in the positive x-direction. During this period, the system is again in a uniform, static field, but of a different magnitude and direction (z') from the initial one. At and before $t = 0$, the system is in the $m = 1/2$ state with respect to the z-axis.

(a) At $t = 0^+$, what are the amplitudes for finding the system with spin projections $m' = \pm 1/2$ with respect to the z' direction?

(b) What is the time development of the energy eigenstates with respect to the z' direction during the time interval $0 < t < T$?

(c) What is the probability amplitude at $t = T$ of observing the system in the spin state $m = -1/2$ along the original z-axis? At times $t > T$?

Express your answers in terms of the angle θ between the z and z' axes and the frequency $\omega_0 = \mu_0 B_0/\hbar$.

4.5* A system of angular momentum **J** and magnetic moment $\mu = \gamma\hbar\mathbf{J}$ interacts with an external magnetic induction $\mathbf{B} = \mathbf{B}_0 + \mathbf{B}_1(t)$, where $\mathbf{B}_0$ is a static uniform field in the z direction and $\mathbf{B}_1$ is a harmonic field with components $B_x = B_1\cos\omega t$ and $B_y = -B_1\sin\omega t$.

(a) Using the Heisenberg equations of motion for **J**, show by direct calculation that the z-component of **J** at time t is

$$J_z(t) = J_z(0)\left[1 - 2\frac{\Gamma^2}{\nu^2}\sin^2\frac{\nu t}{2}\right]$$
$$+ \frac{\Gamma}{\nu}\left[\frac{\Delta\omega}{\nu}J_x(0)(1 - \cos\nu t) - J_y(0)\sin\nu t\right], \tag{4.291}$$

where $\Delta\omega = \omega_0 - \omega, \Gamma = \omega B_1/B_0, \nu = \sqrt{(\Delta\omega)^2 + \Gamma^2}, \omega_0 = \gamma B_0$. (HINT: One way to find a solution is to obtain an uncoupled second order differential equation for dJ_z/dt. Another is to follow the sequence of transformations to rotating frames for the Schrödinger equation. Still another is to observe that there are few functional forms for the time dependence and expand all components of **J** in those forms. Here you are asked to use the first approach.)

(b) Assume that at $t = 0$, the system has only a net magnetic moment in the z direction. Write down an expression for $\mu_z(t)$ and graph it as a function of time for resonance ($\Delta\omega = 0$) and slightly off resonance ($\Delta\omega = \pm\Gamma$).

(c) Use the unitary transformation $U(t)$ that gives the time evolution of the states in the Schrödinger picture. In this description,

$$\langle J_z(t)\rangle = \langle t|J_z|t\rangle = \langle 0|U^\dagger(t)J_zU(t)|0\rangle. \tag{4.292}$$

Show explicitly by carrying out the indicated unitary transformation that the result is equivalent to the Heisenberg operator expression of part (a). (HINT: Make repeated use of the relation between the rotated vector operator **V**' and the original operator **V**: $\mathbf{V}' = U^\dagger\mathbf{V}U$.)

4.6 Use unitarity and the group properties of rotations to establish the following symmetry properties of the Wigner d-functions:

(1) $d_{m'm}(-\beta) = d^j_{mm'}(\beta)$.
(2) $d^j_{m'm}(\pi - \beta) = (-1)^{j-m}d^j_{-mm'}(\beta)$.
(3) $d^j_{m'm}(\beta) = (-1)^{m'-m}d^j_{-m'-m}(\beta)$.
(4) $d^j_{m'm}(\beta) = d^j_{-m-m'}(\beta)$.

(a) From the unitarity of $U(0, \beta, 0)$ and the reality of the d-functions, prove relation (1).

(b) We must have $\exp(-i\alpha J_y) \, |jm\rangle = c_m \, |j - m\rangle$. (Why?) Use $[J_y, U(0, \pi, 0)] = 0$ to prove $c_m = (-1)^{j-m}$ or $d^j_{m'm}(\pi) = (-1)^{j-m}\delta_{m',-m}$.

(c) Use the result of (b) and the group property of rotations to derive relations (2) and (3).

(d) Use $U(0, \beta, 0) = U(-\pi, -\beta, \pi)$ and relation (1) to establish relation (4).

4.7 Referring to Figure 4.2, suppose the amplitude incident at A is set to unity and suppose that at each of the three layers of silicon the amplitude for transmission is T while that for refraction is R. Then, because there is no absorption of neutrons and we suppose that all neutrons are either transmitted or refracted along a unique Bragg angle, we must have $|T|^2 + |R|^2$. Now consider not just the beams shown in the figure, but also the continuations of transmitted beams at B and C. Now show that the sum of the intensities of the four beams before the slab at D does add up to one. Next, consider the four beams present after the third slab: the two from the transmitted beams at B and C and the two beams involving interference measured at C_2 and C_3. Show that conservation of the number of neutrons requires that the R and T be relatively imaginary.

4.8 Use the angular momentum operator $\mathbf{J}$ as the tensor operator in the Wigner-Eckart theorem to show that

$$\langle jm'|jm1q\rangle = \frac{\langle jm' \, |J_q| \, jm\rangle}{\sqrt{j(j+1)}}, \tag{4.293}$$

where $J_{\pm 1} = \mp(J_x \pm iJ_y)/\sqrt{2}$ and $J_0 = J_z$. (Hint: to get the normalization, evaluate the absolute square of the matrix element of J_q, summed over m and q.)

Use this result to establish the final relation in Problem 4.2, now using the Wigner-Eckart theorem, including the connection between $\langle \alpha' j||\mathbf{J} \cdot \mathbf{V}||\alpha j\rangle$ and the Wigner-Eckart reduced matrix element.

4.9 (a) Given that $T^{(2)}_0 = (JJ)_0 = 3J_zJ_z - J^2$ is the $q = 0$ member of a rank-two tensor operator $(JJ)_q$, write down all the other components.

(b) Apply the Wigner-Eckart theorem to $(JJ)_q$ and show that the Clebsch-Gordan coefficients for $j_1 = j$, $j_2 = 2$, and $j = j$ can be expressed as

$$\langle jm'|jm2q\rangle = \frac{\langle jm' \, |(JJ)_q| \, jm\rangle}{\sqrt{(2j-1)j(j+1)(2j+3)}}. \tag{4.294}$$

(c) An electron moving in a central-field potential is in a state characterized by $\ell, s = 1/2, j, m$ and a radial quantum number. The quadrupole moment operator (a second rank tensor operator) is normalized so $Q_0 = 3z^2 - r^2$. Calculate the expectation values of the quadrupole moment operator for arbitrary j and m. Express your answer in terms of the expectation value $\langle R^2 \rangle$ and j and m (not ℓ) insofar as possible. The quadrupole moment Q of a state is defined as the expectation value of Q_0 in a state with $m = j$. What is Q?

4.10 Two spin-1/2 particles with relative coordinates vector $\mathbf{r} = \mathbf{r}_1 - \mathbf{r}_2$ have a total angular momentum operator $\mathbf{J} = \mathbf{L} + (\boldsymbol{\sigma}_1 + \boldsymbol{\sigma}_2)/2$. The interaction between the articles involves a function of $|\mathbf{r}|$ times the tensor force operator, $S_{12} = 3\sigma_1 \cdot \hat{\mathbf{r}}\sigma_2 \cdot \hat{\mathbf{r}} - \sigma_1 \cdot \sigma_2$. The operator S_{12} can be viewed as the scalar product of two second-rank tensor operators, one in coordinate space and one in spin space.

(a) Starting with the $q = 0$ member, $A_0^{(2)} = 3z^2 - r^2$, use the commutation relation for vector and tensor operators to find (by brute force) explicit expressions for the $q = \pm1, \pm2$ members of the spatial second rank tensor.

(b) Repeat the process for the spin operator, starting with $B_0^{(2)} = 3\sigma_{1z}\sigma_{2z} - \sigma_1 \cdot \sigma_2$.

(c) Form the scalar product $A^{(k)} \cdot B^{(k)} = \sum_q (-1)^q A_q^{(k)} B_{-q}^{(k)}$ and establish its relation to S_{12}.

4.11* The deuteron is a bound state of neutron and proton with unit total angular momentum and even parity. An important part of the binding is provided by a tensor force of the form $V_T(r)S_{12}$. There is also an attractive central force, $V_C(r) = V_1(r) + \sigma_1 \cdot \sigma_2 V_2(r)$. A convenient basis of states is the simultaneous eigenstates of $L^2, S^2, J^2,$ and J_z, where $\mathbf{S} = (\sigma_1 + \sigma_2)/2$ is the total spin ($S = 0, 1$). Consider the matrix elements of S_{12}:

$$\langle L'S'J'M' |S_{12}| LSJM \rangle. \tag{4.295}$$

(a) From the fact that S_{12} is rotation- and reflection-invariant, but is the scalar product of two second-rank tensors, one involving the spatial coordinates and the other the spin operators, deduce the selection rules for the above matrix elements, i.e. for what values of the primed and unprimed quantum numbers is it non-vanishing.

(b) If $L = 0, S = 1$, find explicitly (as numbers) all the non-vanishing matrix elements of S_{12}.

(c) The fact that the deuteron has $J = 1$ says something about the spin character of the neutron-proton force. What is it? (Ignore a possible tensor force.) The fact that the magnetic moment $\mu = +0.8574\mu_N$ is approximately the sum of the proton's and neutron's moments says something about the dominant angular momentum in the ground state What is it? What does the existence of a quadrupole moment ($Q/e = 2.71 \times 10^{-27}$ cm^2) imply about the deuteron's wave function? Is the tensor force relevant?

4.12* When the bound state in central-field motion has angular momentum greater than zero, the radial wave equation includes the kinetic energy associated with the orbital motion. A generalized virial theorem for central-field motion can be obtained by multiplying the radial Schrödinger equation,

$$\frac{d^2u}{dr^2} + W(r)u = 0, \tag{4.296}$$

where

$$W(r) = \frac{2m}{\hbar^2}\left(E - V(r) - \frac{\ell(\ell+1)\hbar^2}{2mr^2}\right) \tag{4.297}$$

by $r^q du/dr$ and integrating from zero to infinity.

(a) For $\ell \neq 0$ and localized bound states, show that the generalized virial theorem reads

$$\frac{1}{2}q(q-1)(q-2)\langle r^{q-3}\rangle + 2q\langle r^{q-1}W(r)\rangle + \langle r^q\frac{dW}{dr}\rangle = 0. \tag{4.298}$$

What conditions apply to the values of q and the behavior of $W(r)$ at the origin for fixed ℓ? What value of q yields the usual virial theorem?

(b) For $\ell = 0$, discuss the modifications, if any, and any different conditions on q or $W(r)$ at the origin.

(c) For $\ell = 0$ and $q = 0$, show that the theorem yields the relation

$$|\Psi(0)|^2 = \frac{m}{2\pi\hbar^2}\langle\frac{dV}{dr}\rangle, \tag{4.299}$$

a result used in charmonium spectroscopy to estimate the e^+e^- widths of the $J^{CP} = 1^{--}$ states.

4.13 (a) Specialize the result of the previous problem, part (a), to the attractive Coulomb potential $V(r) = -Ze^2/r$.

(b) Evaluate the expression for $q = 0, 1, 2, 3$ and thereby determine the averages of r^{-1}, r, r^{-3} and r^2 for a hydrogen-like state of arbitrary ℓ and n. Compare your results with the table of Bethe and Salpeter, *Quantum Mechanics of One- and Two-Electron Atoms*, p. 17. Note that the average value of r^{-2} does not emerge. It is obtained using the Hellmann-Feynman theorem in Section 4.16.

5

Time-Independent Perturbation Theory

The modification of energy levels and wave functions of systems subjected to static external fields or to small, previously neglected interactions is the subject of time-independent perturbation theory. This technique is ubiquitous. Here we use it to determine the Stark and Zeeman effects and the leading corrections to the hydrogen spectrum. We consider the "unperturbed" problem to be solved. The zeroth order Hamiltonian is H_0, with eigenvectors and energies $|n^{(0)}\rangle, E_n^{0)}$. The situation of interest is the addition of a small time-independent Hamiltonian λH_1, where λ is inserted for bookkeeping purposes.

5.1   Time-Independent Perturbation Expansion

We are interested in the eigenvalue problem

$$(H_0 + \lambda H_1)\,|n\rangle = E_n\,|n\rangle \tag{5.1}$$

with λ a small, dimensionless parameter. Our approach will be that of successive approximations. We assume here that the eigenkets $|n^{(0)}>$ are non-degenerate in energy. We expand

$$|n> = |n^{(0)}> +\lambda|n^{(1)}> +\lambda^2|n^{(2)}> + \cdots \tag{5.2}$$

$$E_n = E_n^{(0)} + \lambda E_n^{(1)} + \lambda^2 E_n^{(2)} + \cdots \tag{5.3}$$

Then we have

$$(H_0 + \lambda H_1)(|n^{(0)}> +\lambda|n^{(1)}> + \cdots)$$
$$= (E_n^{(0)} + \lambda E_n^{(1)} + \cdots)(|n^{(0)}> +\lambda|n^{(1)}> + \cdots). \tag{5.4}$$

We compare the left- and right-hand sides:

$$\text{LHS} := H_0|n^{(0)}> +\lambda[H_0|n^{(1)}> +H_1|n^{(0)}>] + \lambda^2[H_0|n^{(2)}> +H_1|n^{(1)}>] + \cdots$$

$$\text{RHS} := E_n^{(0)}|n^{(0)}> +\lambda[E_n^{(0)}|n^{(1)}> +E_n^{(1)}|n^{(0)}>]$$
$$+ \lambda^2[E_n^{(0)}|n^{(2)}> +E_n^{(1)}|n^{(1)}> +E_n^{(2)}|n^{(0)}>] + \cdots \tag{5.5}$$

John David Jackson: A Course in Quantum Mechanics, First Edition. Robert N. Cahn.
© 2024 John Wiley & Sons, Inc. Published 2024 by John Wiley & Sons, Inc.
Companion website: www.wiley.com/go/Jackson/QuantumMechanics

Equating powers of λ through λ^2

$$H_0|n^{(0)}> = E_n^{(0)}|n^{(0)}>$$

$$H_0|n^{(1)}> + H_1|n^{(0)}> = E_n^{(0)}|n^{(1)}> + E_n^{(1)}|n^{(0)}>$$

$$H_0|n^{(2)}> + H_1|n^{(1)}> = E_n^{(0)}|n^{(2)}> + E_n^{(1)}|n^{(1)}> + E_n^{(2)}|n^{(0)}> . \tag{5.6}$$

As a preliminary, we note that we can assume $|n^{(1)}>, |n^{(2)}>, ...$ are orthogonal to $|n^{(0)}>$ and $\langle n^{(0)}|n^{(0)}\rangle = 1$. This is true because if λH_1 did cause an addition to $|n^{(0)}>$, we could absorb it into the definition of the normalized unperturbed ket $|n^{(0)}>$. Now act with $\langle n^{(0)}|$ on the first of the three equations to get

$$\langle n^{(0)}|H_1|n^{(0)}\rangle + \langle n^{(0)}|H_0|n^{(1)}\rangle = E_n^{(1)} + E^{(0)}\langle n^{(0)}|n^{(1)}\rangle. \tag{5.7}$$

Note that the cancellation of the second term on each side does not depend on $\langle n^{(0)}|n^{(1)}\rangle = 0$. We thus find the first order energy shift:

$$\langle n^{(0)}|H_1|n^{(0)}\rangle = E_n^{(1)}. \tag{5.8}$$

Now act on both sides of Eq. (5.5) instead with $\langle m^{(0)}| \neq \langle n^{(0)}|$:

$$\langle m^{(0)}|H_1|n^{(0)}\rangle + \langle m^{(0)}|H_0|n^{(1)}\rangle = E_n^{(1)}\langle m^{(0)}|n^{(0)}\rangle + E_n^{(0)}\langle m^{(0)}|n^{(1)}\rangle \tag{5.9}$$

$$\langle m^{(0)}|H_1|n^{(0)}\rangle + E_m^{(0)}\langle m^{(0)}|n^{(1)}\rangle = E_n^{(0)}\langle m^{(0)}|n^{(1)}\rangle. \tag{5.10}$$

Thus,

$$\langle m^{(0)}|n^{(1)}\rangle = \frac{\langle m^{(0)}|H_1|n^{(0)}\rangle}{E_n^{(0)} - E_m^{(0)}}. \tag{5.11}$$

So we have determined

$$|n^{(1)}> = \sum_m{}'|m^{(0)}> \frac{\langle m^{(0)}|H_1|n^{(0)}\rangle}{E_n^{(0)} - E_m^{(0)}} \tag{5.12}$$

where the prime means omit $m = n$ from the sum. Note that the non-singular nature of the sum depends upon there being no degeneracy. This result is the first-order correction to the state vector.

We now look at the third line of Eq. (5.6) and evaluate the matrix element with $\langle n^{(0)}|$.

$$\langle n^{(0)}|H_0|n^{(2)}\rangle + \langle n^{(0)}|H_1|n^{(1)}\rangle = E_n^{(0)}\langle n^{(0)}|n^{(2)}\rangle + E_n^{(1)}\langle n^{(0)}|n^{(1)}\rangle + E_n^{(2)}. \tag{5.13}$$

The first term on the left cancels with the first term on the right, while the second term on the right vanishes by orthogonality. Thus,

$$E_n^{(2)} = \langle n^{(0)}|H_1|n^{(1)}\rangle$$
$$= \sum_m{}' \frac{\langle n^{(0)}|H_1|m^{(0)}\rangle\langle m^{(0)}|H_1|n^{(0)}\rangle}{E_n^{(0)} - E_m^{(0)}}. \tag{5.14}$$

The relations derived above are of great utility as we explore below.

5.2   Interlude: Spectra and History

The intertwined history of atomic physics and quantum mechanics is presented in Max Born's classic *Atomic Physics*. In the late nineteenth century, high-resolution atomic spectroscopy provided an enormous wealth of data, which could not be completely understood prior to the development of quantum mechanics starting in 1925. Certain results were critical in that development. Pieter Zeeman's studies just prior to the start of the twentieth century showed that spectral lines were modified in the presence of a magnetic field. The "normal Zeeman effect" showed single lines being split into triplets, which could be understood in terms of a spinless electron, with a component of the orbital angular momentum quantized along the direction of the magnetic field. The orbiting electron provided a magnetic moment

$$\mu_B L_z, \tag{5.15}$$

where

$$\mu_B = \frac{e\hbar}{2mc} = 5.788 \times 10^{-5} \, \text{eV/T} \tag{5.16}$$

is the Bohr magneton. However the "anomalous Zeeman effect," in which the single line was split into several components, defied this simple explanation. Even spectra taken in the absence of a magnetic field, for example the famous doublet in sodium at 5890-5896 Å, could not be explained in the spinless-electron picture. The resolution of these puzzles was provided by the proposal of Goudsmit and Uhlenbeck, that the electron has an intrinsic spin of one-half. The picture of the interaction of the electron with electric and magnetic fields had to be modified to include the effect of spin.

While the proper starting point for the behavior of an electron in electric and magnetic fields is the Dirac equation, an adequate description was deduced prior to Dirac's inspiration. The classical Hamiltonian for a spinless particle of charge $-e$ and mass m in static electric and magnetic fields is given by

$$H = \frac{1}{2m}\left(\mathbf{p} + \frac{e}{c}\mathbf{A}\right)^2 - e\Phi, \tag{5.17}$$

where $\mathbf{p}$ is the canonical momentum, $\mathbf{A}$ is the vector potential, and Φ is the electric potential. For a constant magnetic field, we can take $\mathbf{A} = \frac{1}{2}(\mathbf{B} \times \mathbf{r})$. We then find, with $\mathbf{L} = \mathbf{r} \times \mathbf{p}/\hbar$ as the dimensionless angular momentum,

$$\mathbf{p} \cdot \mathbf{A} + \mathbf{A} \cdot \mathbf{p} = \mathbf{p} \cdot (\mathbf{B} \times \mathbf{r}) = (\mathbf{r} \times \mathbf{p}) \cdot \mathbf{B} = \hbar \mathbf{L} \cdot \mathbf{B}, \tag{5.18}$$

since $[p_i, r_j] = 0$ if $i \neq j$. Thus in this instance, the first-order magnetic interaction is

$$H' = \frac{e\hbar}{2mc}\mathbf{L} \cdot \mathbf{B} = -\boldsymbol{\mu}_\ell \cdot \mathbf{B}, \tag{5.19}$$

where

$$\boldsymbol{\mu}_\ell = -\frac{e\hbar}{2mc}\mathbf{L} = -\mu_B \mathbf{L} \tag{5.20}$$

is the magnetic moment due to the orbital motion of the negatively-charged electron.

For a particle with an intrinsic spin, the natural extension is the addition of a term $H' = -\boldsymbol{\mu} \cdot \mathbf{B}$, where $\boldsymbol{\mu}$ is the magnetic moment of the charged particle. The magnetic moment $\boldsymbol{\mu}$ is necessarily proportional to the spin, $\mathbf{s}$, and the Bohr magneton, $e\hbar/2mc$. The coefficient g_e in

$$\boldsymbol{\mu}_e = -g_e \mu_B \mathbf{s} \tag{5.21}$$

is the g-factor of the electron and the minus sign is the consequence of the negative charge of the electron. The proposal of Goudsmit and Uhlenbeck was that the electron had spin 1/2 and to obtain the measured Zeeman g_e must be taken to be very nearly 2. Thus the Hamiltonian for the electron (with charge -e !) in an electric field and a constant magnetic field should be, dropping the second order term in A,

$$H = \frac{p^2}{2m} - e\Phi + \mu_B(\mathbf{L} + 2\mathbf{s}) \cdot \mathbf{B}. \tag{5.22}$$

The primitive version of quantum theory available to Goudsmit and Uhlenbeck was enough to establish the result of Problem 4.2(d):

$$\langle \alpha' jm' |\mathbf{V}| \alpha jm \rangle = \frac{\langle jm' |\mathbf{J}| jm \rangle}{j(j+1)} \langle \alpha' j||\mathbf{J} \cdot \mathbf{V}||\alpha j\rangle. \tag{5.23}$$

Thus, the splitting due to a magnetic field in the z direction can be computed as

$$\mu_B B \langle jm | L_z + 2S_z | jm \rangle = \mu_B B \langle jm |J_z| jm \rangle \frac{\langle jm |(\mathbf{L} + 2\mathbf{S}) \cdot \mathbf{J}| jm \rangle}{j(j+1)}$$

$$= \mu_B Bm \left[1 + \frac{j(j+1) - \ell(\ell+1) + s(s+1)}{2j(j+1)} \right]$$

$$\equiv \mu_B mBg, \tag{5.24}$$

where g is the Landé g-factor for the atom.

Another piece was required to explain the spin-orbit interaction, which arises because an electron in motion in an electric field $\mathbf{E}$ senses a magnetic field

$$\mathbf{B} = -\frac{\mathbf{v}}{c} \times \mathbf{E}. \tag{5.25}$$

An electron in a central field with potential energy $-e\Phi = V(r)$ will sense a magnetic field

$$\mathbf{B} = \frac{\mathbf{v}}{ec} \times \frac{\mathbf{r}}{r} \frac{dV}{dr} = -\frac{\hbar \mathbf{L}}{ecmr} \frac{dV}{dr} \tag{5.26}$$

so that we would be tempted to add a term

$$H_{so} \doteq g_e \mu_B \mathbf{s} \cdot \mathbf{L} \frac{\hbar}{ecmr} \frac{dV}{dr} = g_e \frac{\hbar^2}{2m^2 c^2} \mathbf{s} \cdot \mathbf{L} \frac{1}{r} \frac{dV}{dr} \qquad (wrong!). \tag{5.27}$$

This, however, does not give the right value for splittings like that between 2p states in hydrogen, which have $j = 1/2$ and $j = 3/2$. The correct term, as first shown by L. H. Thomas, is

$$H_{so} = (g_e - 1)\frac{\hbar^2}{2m^2 c^2} \mathbf{s} \cdot \mathbf{L} \frac{1}{r} \frac{dV}{dr}. \tag{5.28}$$

The mistake in the first form is due to the failure to consider the rotating frame of the spinning electron. See Jackson, *Classical Electrodynamics*, 3rd edition, pp. 548-552. With g_e nearly exactly 2, the change decreases the spin-orbit interaction by 1/2, sometimes called the Thomas factor.

As we shall see later, all this is confirmed directly by the Dirac equation. Armed with these pieces of the Hamiltonian, we can compute to good precision the spectrum of hydrogen and its modification by external electric and magnetic fields.

5.3 Fine Structure of Hydrogen

The exact solution of the relativistic effects in the hydrogen atom can be found via the Dirac equation, but the effects are so small that they can be evaluated accurately with first-order perturbation theory. There are two effects of order v^2/c^2:

(a) Relativistic corrections to the kinetic energy.
(b) Spin-orbit interaction.

5.3.1 Relativistic Corrections to the Kinetic Energy

The kinetic energy is defined by

$$T = E - mc^2 = \sqrt{p^2c^2 + m^2c^4} - mc^2 = \frac{p^2}{2m} - \frac{1}{8}\frac{p^4}{m^3c^2} + \cdots \tag{5.29}$$

Thus, our perturbation is

$$H_1 = -\frac{1}{8}\frac{p^4}{m^3c^2} = -\frac{1}{2mc^2}T^2 = -\frac{1}{2mc^2}(H_0 - V)^2$$

$$= -\frac{1}{2mc^2}(H_0^2 - H_0V - VH_0 + V^2). \tag{5.30}$$

Slightly generalizing to hydrogenic systems, we take $V = -Ze^2/r$, we get

$$\langle n\ell m\,|H_1|\,n\ell m\rangle = -\frac{1}{2mc^2}\left[E_{n\ell}^2 + 2E_{n\ell}\left\langle n\ell m\left|\frac{Ze^2}{r}\right|n\ell m\right\rangle\right.$$

$$\left. + Z^2e^4\left\langle n\ell m\left|\frac{1}{r^2}\right|n\ell m\right\rangle\right]. \tag{5.31}$$

From the virial theorem (see Problems 1.7 and 4.12),

$$\left\langle\frac{Ze^2}{r}\right\rangle = -2E_{n\ell} = -2\left(-\frac{Z^2e^2}{2a_0n^2}\right). \tag{5.32}$$

while from the Hellmann-Feynman theorem proved above,

$$\left\langle\frac{1}{r^2}\right\rangle = \frac{Z^2}{a_0^2}\frac{1}{(\ell+1/2)n^3}; \quad Z^2e^4\left\langle\frac{1}{r^2}\right\rangle = 4E_{n\ell}^2\left(\frac{n}{\ell+1/2}\right). \tag{5.33}$$

Altogether,

$$\langle H_1\rangle = -\frac{1}{2mc^2}E_{n\ell}^2\left[1 - 4 + \frac{4n}{\ell+1/2}\right] = \frac{2E_{n\ell}^2}{mc^2}\left(\frac{3}{4} - \frac{n}{\ell+1/2}\right) \tag{5.34}$$

$$= Z^2\alpha^2\left(\frac{e^2}{2a_0}\right)\cdot\frac{1}{n^4}\left(\frac{3}{4} - \frac{n}{\ell+1/2}\right). \tag{5.35}$$

The presence of the factor α^2 shows the v^2/c^2 scale of the result. This is the relativistic kinetic energy contribution.

5.3.2 Spin-orbit Interaction

We take $V = -Ze^2/r$, and since the electron's g-factor is 2 up to radiative corrections, $g - 1 \approx 1$, so

$$H_{so} = \frac{Ze^2}{2m^2c^2} \frac{\hbar^2}{r^3} \mathbf{s} \cdot \mathbf{L}. \tag{5.36}$$

We now must consider the spin degrees of freedom. We have $\mathbf{J} = \mathbf{s} + \mathbf{L}$ so that $\mathbf{s} \cdot \mathbf{L} = (J^2 - L^2 - s^2)/2$ as an operator expression. If we suppose that we have eigenstates of J^2, L^2, s^2, J_z, then we have

$$\langle \mathbf{L} \cdot \mathbf{s} \rangle = \frac{1}{2}\left[j(j+1) - \ell(\ell+1) - \frac{3}{4} \right]. \tag{5.37}$$

With $j = \ell \pm 1/2$, $\langle \mathbf{L} \cdot \mathbf{s} \rangle = \ell/2$ for $j = \ell + 1/2$ and $\langle \mathbf{L} \cdot \mathbf{s} \rangle = -(\ell+1)/2$ for $j = \ell - 1/2$.

The first-order energy shift is

$$\Delta E_{L \cdot s} = \frac{Ze^2 \hbar^2}{2m^2c^2} \langle \mathbf{s} \cdot \mathbf{L} \rangle \left\langle \frac{1}{r^3} \right\rangle. \tag{5.38}$$

From the virial theorem (See Problems 4.12 and 4.13),

$$\left\langle \frac{1}{r^3} \right\rangle = \frac{Z^3}{n^3 \ell (\ell + 1/2)(\ell + 1) a_0^3}. \tag{5.39}$$

Thus,

$$\Delta E_{L \cdot s} = Z^4 \alpha^2 \cdot \frac{e^2}{2a_0} \cdot \frac{\langle \mathbf{s} \cdot \mathbf{L} \rangle}{n^3 \ell (\ell + 1/2)(\ell + 1)}. \tag{5.40}$$

We see that this is the same order and same power of Z as the relativistic correction to the kinetic energy. Combining the two effects,

$$\Delta E = Z^4 \alpha^2 \cdot \frac{e^2}{2a_0} \left[\frac{3}{4n^4} + \cdot \frac{\langle \mathbf{L} \cdot \mathbf{s} \rangle}{n^3 \ell (\ell + 1/2)(\ell + 1)} - \frac{1}{n^3 (\ell + 1/2)} \right]. \tag{5.41}$$

Inserting $\langle \mathbf{L} \cdot \mathbf{s} \rangle$ and simplifying, we obtain

$$\Delta E = Z^4 \alpha^2 \cdot \frac{e^2}{2a_0} \left[\frac{3}{4n^4} - \frac{1}{n^3 (j + 1/2)} \right] \tag{5.42}$$

for both values of $j = \ell \pm 1/2$. For the $n = 2$ $P_{1/2}$ and $P_{3/2}$ states, we have energy shifts

$$\Delta E(2^2 P_{1/2}) = Z^4 \alpha^2 \left(\frac{3}{64} - \frac{1}{8} \right) \text{ Ryd.}$$

$$\Delta E(2^2 P_{3/2}) = Z^4 \alpha^2 \left(\frac{3}{64} - \frac{1}{16} \right) \text{ Ryd.} \tag{5.43}$$

Some convenient conversions are:

$$\frac{e^2}{2a_0} = 1 \text{ Ryd.} = 13.605693 \text{ eV} = 1.097373 \times 10^5 \text{ cm}^{-1} = 3.289842 \times 10^{15} \text{ Hz.} \tag{5.44}$$

The splitting between the p-states is

$$E(2^2P_{3/2}) - E(2^2P_{1/2}) = \frac{Z^4\alpha^2}{16}\,\text{Ryd.} = Z^4 \times 10.950\ \text{GHz}. \tag{5.45}$$

Our formula fortuitously gives the correct result for $\ell = 0$, even though $\langle 1/r^3 \rangle = \infty$ and $\langle \mathbf{L}\cdot\mathbf{s} \rangle = 0$. Dirac relativistic theory explains the result for $\ell = 0$ through the Darwin term. The Dirac equation will also explain the surprising result that there is dependence only on j, making the $2^2S_{1/2}$ degenerate with the $2^2P_{1/2}$, a degeneracy lifted by the Lamb shift.

5.4 Stark Effect in Ground-State Hydrogen

If a hydrogen atom is subjected to a uniform electric field $\mathcal{E}$ in the z-direction, the potential energy along the z-axis would appear as shown in Figure 5.1. Evidently, any bound state will now be unstable and not have a well defined energy. However, the barrier is very thick unless $n \gg 1$. Recall that the ground-state energy $e^2/(2a_0)$ is 13.6 eV while the Bohr radius a_0 is 0.528×10^{-8} cm. The scale of the internal electric field is $e/a_0^2 \approx 5 \times 10^9$ V/cm, much greater than any imaginable external field. The distance to the peak value of the potential is $z_0 = a_0\sqrt{(e^2/a_0)/\mathcal{E}}$, which is many Bohr radii. For example, if $\mathcal{E} = 10^3$ V/cm, z_0 is about $2000\,a_0$. A hydrogenic state with this as the radius would have $n \approx 50$. Thus stability issues arise only for so-called "Rydberg atoms," which we will not consider here.

For the ground state, or indeed any non-degenerate state that is an eigenstate of parity, the lowest order energy shift caused by $H_1 = -e\mathcal{E}z$ vanishes. We thus must go at least to second order. The needed matrix elements are

$$\langle n\ell m\,|H_1|\,100\rangle = -e\mathcal{E}\,\langle n\ell m\,|z|\,100\rangle. \tag{5.46}$$

Because $z \propto Y_{10}$, only $m = 0$, $\ell = 1$ states contribute to the sum. We thus have

$$\Delta E_1^{(2)} = -e^2\mathcal{E}^2 \sum_n{}' \frac{|\langle n10\,|z|\,100\rangle|^2}{E_n - E_1} \tag{5.47}$$

where $E_n = -e^2/(2a_0n^2)$. The sum actually includes the continuum as well as the bound states. We make several observations:

Figure 5.1 The potential energy for an electron in hydrogen in the presence of a uniform electric field, $\mathcal{E}$. The strength of the electric field is enormously exaggerated. For a practicable electric field, the linear potential would appear nearly flat, so tunneling would be extremely improbable except for large values of n.

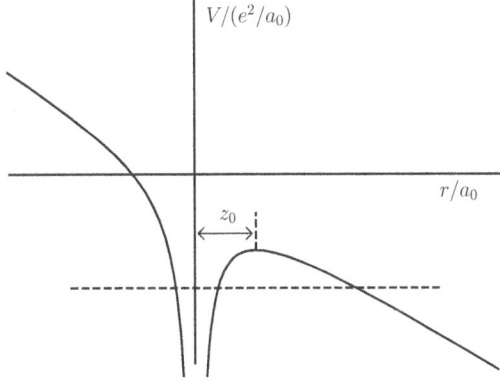

(i) The sum is positive since $E_n - E_1 > 0$, and thus $\Delta E_1^{(2)} < 0$ and $\Delta E_1^{(2)} \propto \mathcal{E}^2$.

(ii) An upper bound for the sum is obtained by noting that the smallest value of $E_n - E_1$ is $(3/8)e^2/a_0$. Thus,

$$\sum_n{}' \frac{|\langle n10\,|z|\,100\rangle\,|^2}{E_n - E_1} < \frac{8a_0}{3e^2} \sum_n{}' |\langle n10\,|z|\,100\rangle\,|^2 < \frac{8a_0}{3e^2} |\langle 100\,|z^2|\,100\rangle\,|. \qquad (5.48)$$

Here we used the completeness of the sum over the states $|n\ell m\rangle$, including only those states that would contribute to the matrix element. Now for 1s hydrogen $\langle r^2 \rangle = 3a_0^2$, so $\langle z^2 \rangle = a_0^2$, and so

$$|\Delta E^{(2)}| < \frac{8}{3}\mathcal{E}^2 a_0^3. \qquad (5.49)$$

(iii) A lower bound can be obtained using the average value of $E_n - E_1$.

   This follows from a simple inequality: if a_n and b_n are positive numbers, then

$$\sum_n a_n b_n \sum_n \frac{a_n}{b_n} \geq \left(\sum_n a_n\right)^2. \qquad (5.50)$$

This is proved with the Cauchy-Schwarz inequality

$$(\mathbf{A} \cdot \mathbf{B})^2 < |\mathbf{A}|^2 |\mathbf{B}|^2 \qquad (5.51)$$

taking $\mathbf{A}_n = a_n^{1/2} b_n^{1/2}$, $\mathbf{B}_n = a_n^{1/2} b_n^{-1/2}$.

   Now take

$$a_n = |\langle n10\,|z|\,100\rangle\,|^2$$
$$b_n = E_n - E_1. \qquad (5.52)$$

   Using the inequality above,

$$\sum_n{}' \frac{|\langle n10\,|z|\,100\rangle\,|^2}{E_n - E_1} \geq \frac{(\sum_n{}' |\langle n10\,|z|\,100\rangle\,|^2)^2}{\sum_n{}' |\langle n10\,|z|\,100\rangle\,|^2(E_n - E_1)}. \qquad (5.53)$$

The numerator on the right-hand side is evaluated using completeness of the intermediate states with the result a_0^4. The denominator on the right-hand side can be evaluated with the celebrated Thomas-Reiche-Kuhn sum rule, which we now demonstrate. Let

$$H = \frac{p^2}{2m} + V(r). \qquad (5.54)$$

From the fundamental commutation relation,

$$\frac{1}{i\hbar}\langle n\,|zp_z|\,n\rangle - \frac{1}{i\hbar}\langle n\,|p_z z|\,n\rangle = 1. \qquad (5.55)$$

But $p_z = m\dot{z} = \frac{m}{i\hbar}[z, H]$, so

$$\frac{m}{\hbar^2} \sum_{n'} [\langle n\,|z|\,n'\rangle\langle n'\,|Hz - zH|\,n\rangle - \langle n\,|Hz - zH|\,n'\rangle\langle n'\,|z|\,n\rangle] = 1, \qquad (5.56)$$

and finally

$$\frac{2m}{\hbar^2} \sum_{n'} (E_{n'} - E_n)|\langle n'\,|z|\,n\rangle\,|^2 = 1. \qquad (5.57)$$

An individual piece of this sum

$$f_{n,n'} = \frac{2m}{\hbar^2}(E_{n'} - E_n)|\langle n'|z|n\rangle|^2 \tag{5.58}$$

is called the oscillator strength for the transition $n \rightarrow n'$, in analogy with a classical one-dimensional oscillator. This makes quantum mechanical results for dielectric constants or energy loss look like the classical expressions. See, for example, Jackson, *Classical Electrodynamics*, 3rd edition, pp. 310, 635. The Thomas-Reiche-Kuhn sum rule for the oscillator strengths reads

$$\sum_{n'} f_{n,n'} = 1. \tag{5.59}$$

If there is more than one final state possible, the oscillator's"strength" is "spread out," but the total "strength" is unity.

Consider a one-dimensional harmonic oscillator initially in a state $n \neq 0$. Dipole transitions are possible to states $n' = n \pm 1$. Looking up the expression for x in terms of a and $a^\dagger$, we find $f_{n+1,n} = n+1, f_{n-1,n} = -n$. The sum is unity, as required. Note that there is an algebraic sign attached to f_{ij} from the sign of $E_j - E_i$.

Applying the Thomas-Reiche-Kuhn sum rule for $n = 1$ to Eq. (5.53), we have the bound

$$\sum_n{}' \frac{|\langle n\ell m|z|100\rangle|^2}{E_n - E_1} > a_0^4 \frac{2m}{\hbar^2} = \frac{2a_0^3}{e^2}. \tag{5.60}$$

Altogether, we have as limits on the Stark effect on $n = 1$ hydrogen,

$$-\frac{8}{3}\mathcal{E}^2 a_0^3 < \Delta E_1^{(2)} < -2\mathcal{E}^2 a_0^3. \tag{5.61}$$

(iv) The exact coefficient is 9/4, found from solving the Schrödinger equation in parabolic coordinates (Bethe and Salpeter, *Quantum Mechanics of One- and Two-Electron Atoms*, pp. 228 ff) or by a trick (see Merzbacher, *Quantum Mechanics* pp. 423-424).

5.5 Perturbation Theory with Degeneracy

We consider the Hamiltonian $H = H_0 + H_1$, where H_0 has degeneracy, i.e. we suppose that the nth energy level is N-fold degenerate.

$$H_0|n_i^{(0)}\rangle = E_n^{(0)}|n_i^{(0)}\rangle; \qquad i = 1, 2, ... N. \tag{5.62}$$

The states $|n_i^{(0)}\rangle$ are assumed to be orthonormal. If the nth energy level is far removed from those of the other eigenstates, we can appeal to the expansion

$$|n_i^{(1)}\rangle = \sum_m{}' |m^{(0)}\rangle \frac{\langle m^{(0)}|H_1|n_i^{(0)}\rangle}{E_n^{(0)} - E_m^{(0)}} \tag{5.63}$$

to argue that for sufficiently small H_1, we can ignore all but the N degenerate states. This reduces the problem to one that is straight-forward: diagonalizing the Hamiltonian H_1 on an N-dimensional space.

If H_1 is not diagonal with respect to these states, the eigenvalue problem $(H_0 + H_1)\, |n\rangle = E_n\, |n\rangle$ will be solved by linear combinations of the N eigenstates of H_0. There will be N eigenstates, of course, with the degeneracy partially or totally lifted. We therefore write

$$|n\rangle = \sum_{i=1}^{N} a_i\, |i^{(0)}\rangle, \tag{5.64}$$

where we write $|i^{(0)}\rangle$ for $|n_i^{(0)}\rangle$, since now we consider exclusively the N degenerate states at the nth level. Insert this into the Schrödinger equation and find

$$\sum_{i}^{N} a_i \left[(E_n^{(0)} - E_n)\, |i^{(0)}\rangle + H_1\, |i^{(0)}\rangle \right] = 0. \tag{5.65}$$

Now $H_1\, |i^{(0)}\rangle$ is some new ket that we can expand in terms of the N degenerate kets (assuming we can neglect the states separated in energy from those under consideration):

$$H_1\, |i^{(0)}\rangle = \sum_{j=1}^{N} |j^{(0)}\rangle \langle j^{(0)} |H_1| i^{(0)}\rangle. \tag{5.66}$$

The Schrödinger equation then becomes

$$\sum_{i}^{N} a_i \left[(E_n^{(0)} - E_n)\, |i^{(0)}\rangle + \sum_{j=1}^{N} |j^{(0)}\rangle \langle j^{(0)} |H_1| i^{(0)}\rangle \right] = 0. \tag{5.67}$$

Acting with $\langle k^{(0)}|$ on the left,

$$\left[(E_n^{(0)} - E_n) + \langle k^{(0)} |H_1| k^{(0)}\rangle \right] a_k + \sum_{\substack{i=1 \\ i \neq k}}^{N} \langle k^{(0)} |H_1| i^{(0)}\rangle\, a_i = 0. \tag{5.68}$$

This set of N linear equations can be written succinctly with $E_n^{(0)} - E_n = \epsilon$, $\langle k^{(0)} |H_1| i^{(0)}\rangle = h_{ki}$ as

$$\sum_{i=1}^{N} (h_{ki} - \delta_{ki}\epsilon) a_i = 0. \tag{5.69}$$

There are solutions only when ϵ satisfies

$$\det(h_{ki} - \delta_{ki}\epsilon) = 0. \tag{5.70}$$

This eigenvalue equation yields N eigenvalues for ϵ, called $\epsilon_\alpha, \alpha = 1, 2, \ldots N$. Corresponding to each eigenvalue, there is an eigenket

$$|\xi^{(\alpha)}\rangle = \sum_{i=1}^{N} |i^{(0)}\rangle\, a_i^{(\alpha)}, \tag{5.71}$$

where the $a_i^{(\alpha)}$ have been found in the usual way from the N linear equations. These linear combinations of the original set of N degenerate states are special, in that they have only diagonal matrix elements for the perturbation H_1. This means that the first-order energy shifts are just as if there were no degeneracy

$$\epsilon_\alpha = \langle \xi^{(\alpha)} |H_1| \xi^{(\alpha)}\rangle. \tag{5.72}$$

Moreover, the second-order shift can be calculated in the usual way

$$\Delta E_\alpha^{(2)} = \sum_m{}' \frac{|\langle m^{(0)} |H_1| \xi^{(\alpha)} \rangle|^2}{E_n^{(0)} - E_m^{(0)}}, \tag{5.73}$$

provided the prime on the sum over m is interpreted to exclude all the N originally degenerate states.

5.6 Linear Stark Effect in Hydrogen

The Stark effect in $n = 2$ hydrogen is a particularly apt example. If we ignore the relativistic corrections calculated above, without the applied electric field the energy level $n = 2$, with $E = -e^2/(8a_0)$, is four-fold degenerate: the 2s state with $m = 0$ and the three 2p states with $m = -1, 0, 1$. Now, any linear combination of the four state vectors is also an eigenstate of the non-relativistic H_0. We can ask which linear combinations will be states of definite energy in the presence of H_1. That is, which linear combinations will be eigenstates of $H_0 + H_1$, where $H_1 = -e\mathcal{E}z$? It is clear that H_1 transforms under rotations just as $Y_{10} \propto \cos\theta$ does. Thus, it can connect only states with the same value of m. Here, that means the mixing is only between 2s and 2p with $m = 0$. To compute the non-vanishing matrix element $\lambda = \langle 2p_0 |-e\mathcal{E}z| 2s_0 \rangle$, we use the standard hydrogenic wave functions, Eq. (4.271).

$$\psi_{2s} = \frac{1}{\sqrt{2}} e^{-r/2}(1 - r/2) \frac{1}{\sqrt{4\pi}}$$

$$\psi_{2p_0} = \frac{1}{2\sqrt{6}} e^{-r/2} r \sqrt{\frac{3}{4\pi}} \cos\theta. \tag{5.74}$$

Here r is measured in units of the Bohr radius, a_0. Including H_1, we find

$$\lambda = -e\mathcal{E} \frac{a_0}{2\sqrt{12}} \int_0^\infty r^2 dr\, e^{-r} r^2 (1 - \frac{r}{2}) \int d\Omega \frac{1}{\sqrt{4\pi}} \sqrt{\frac{3}{4\pi}} \cos^2\theta$$

$$= -e\mathcal{E} \frac{a_0}{2\sqrt{12}} \cdot 24(1 - \frac{5}{2}) \cdot \frac{1}{\sqrt{3}} = +3e\mathcal{E}a_0. \tag{5.75}$$

We have only the simple 2×2 matrix

$$\det(H_1 - \epsilon I) = \begin{vmatrix} -\epsilon & \lambda \\ \lambda & -\epsilon \end{vmatrix} = 0. \tag{5.76}$$

Thus,

$$\epsilon = \pm\lambda = \pm 3e\mathcal{E}a_0. \tag{5.77}$$

The eigenstates are simply

$$\epsilon = \pm\lambda : \qquad\qquad \psi_\pm = \frac{1}{\sqrt{2}}(\psi_{2s_0} \pm \psi_{2p_0})$$

$$\epsilon = 0 \qquad\qquad \psi_0^{(1)} = \psi_{2p_1}; \qquad \psi_0^{(2)} = \psi_{2p_{-1}}. \tag{5.78}$$

The energy ϵ, as a function of the field $\mathcal{E}$, is shown schematically in Figure 5.2.

The $2p \to 1s$ radiative transitions will be split into three lines. The polarization properties of the photons are related to Δm. Thus, the split lines ψ_+, ψ_- will have different polarization character ($\epsilon_\parallel$) from the unshifted lines $\Delta m = \pm 1$ ($\epsilon_\perp$).

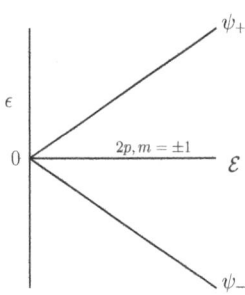

Figure 5.2 The energy of the four $n = 2$ states in hydrogen (ignoring spin, which plays no role here) when a uniform electric field $\mathcal{E}$ is applied. This linear Stark effect arises because of the degeneracy of the $n = 2$ states in hydrogen. Here the relativistic corrections and Lamb shift are ignored.

For $n = 3$, the electric field will couple the $m = 0$ states for $\ell = 0, 1, 2$. Hence we will get a 3×3 matrix. But it will also couple $m = 1$ for $\ell = 1$ and $\ell = 2$. Similarly, it will couple the $m = -1$ states. Only the $m = \pm 2$ states from $\ell = 2$ remain unmixed. The $m = 0$ matrix will have the form

$$\begin{pmatrix} 0 & \lambda^* & 0 \\ \lambda & 0 & \mu^* \\ 0 & \mu & 0 \end{pmatrix}, \tag{5.79}$$

where the columns and rows represent $3s_0, 3p_0, 3d_0$. The matrices for both $m = 1$ and $m = -1$ are simply of the form

$$\begin{pmatrix} 0 & \nu^* \\ \nu & 0 \end{pmatrix}. \tag{5.80}$$

5.7 Perturbation Theory with Near Degeneracy

The actual $n = 2$ levels in hydrogen, including spin and the Lamb shift, illustrate a practical situation in which energy level spacings are small, but not zero. See Figure 5.3. The effect depends on how strong the perturbation is compared to the pre-existing splittings. In particular, the $2p_{3/2} - 2p_{1/2}$ splitting is $\Delta = \frac{\alpha^2}{32} \frac{e^2}{a_0} = 4.527 \times 10^{-5}$ eV, while the Stark effect splitting, ignoring spin, is $\lambda = 3e\mathcal{E}a_0 = 1.587 \times 10^{-8} \mathcal{E}$ (V/cm) eV. If $\mathcal{E}$ is much less than 3 kV/cm, the Stark splitting is small compared to the fine structure. Then the shift of the $2p_{3/2}$ state will actually be quadratic, not linear. Only the degenerate (without the Lamb shift) $2s_{1/2}$ and $2p_{1/2}$ states will show a linear Stark effect.

If you wish to explore the level behavior as a function of $\mathcal{E}$, you must solve a matrix diagonalization with the pre-existing energy perturbations as part of the degenerate state perturbation, that is, take

$$H_1 = H_1(\text{rel. mass}) + H_1(\text{spin-orbit}) + H_1(\text{Lamb}) - e\mathcal{E}z. \tag{5.81}$$

The first three pieces are diagonal in the basis (j, ℓ, s, j_z). There are eight states to contend with: four with $\ell = 1, j = 3/2$, two with $\ell = 1, j = 1/2$, and two with $\ell = 0, j = 1/2$.

For orientation, note that although the Stark effect perturbation $H_1 = -e\mathcal{E}z$ breaks the spherical symmetry, and so will not respect L^2 or J^2, it is azimuthally symmetric and so will be diagonal in J_z. Recalling that it is an odd-parity operator, we see that $-e\mathcal{E}z$ will have non-vanishing matrix elements only for the states $s_{1/2} - p_{1/2}$ (with $m = \pm 1/2$) and $s_{1/2} - p_{3/2}$ (with $m = \pm 1/2$). The states $p_{3/2}, m = \pm 3/2$ do not enter into the mixing (They are like the

Figure 5.3 The energies of the $n = 2$ states in hydrogen in the absence of an external electric field, but with the relativistic corrections and Lamb shift included.

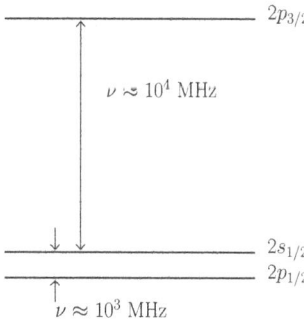

$m = \pm 1$ states in the spinless approximation.). We need the decomposition for $|\ell, s, j, m\rangle$ in a mixed representation with a spherical harmonic times the spin state indicated by $|1/2, m_s\rangle$:

$$|0, 1/2, 1/2, m\rangle = Y_{00}\,|1/2, m\rangle$$

$$|1, 1/2, 1/2, m\rangle = -\sqrt{\frac{\frac{3}{2} - m}{3}}\,Y_{1,m-1/2}\,|1/2, 1/2\rangle + \sqrt{\frac{\frac{3}{2} + m}{3}}\,Y_{1,m+1/2}\,|1/2, -1/2\rangle$$

$$|1, 1/2, 3/2, m\rangle = \sqrt{\frac{\frac{3}{2} + m}{3}}\,Y_{1,m-1/2}\,|1/2, 1/2\rangle + \sqrt{\frac{\frac{3}{2} - m}{3}}\,Y_{1,m+1/2}\,|1/2, -1/2\rangle. \quad (5.82)$$

With these, we find the two non-vanishing matrix elements,

$$2s_{1/2} - 2p_{1/2}\,:\quad \langle 0, 1/2, 1/2, m\,|-e\mathcal{E}z|\,1, 1/2, 1/2, m\rangle = \mp\frac{1}{\sqrt{3}}\lambda \quad (m = \pm 1/2)$$

$$2s_{1/2} - 2p_{3/2}\,:\quad \langle 0, 1/2, 1/2, m\,|-e\mathcal{E}z|\,1, 1/2, 3/2, m\rangle = \sqrt{\frac{2}{3}}\lambda \quad (m = \pm 1/2),$$

$$(5.83)$$

where $\lambda = 3e\mathcal{E}a_0$ is the matrix element from the spinless approximation.

We now display the combined perturbation matrix, choosing the position of the degenerate $2s_{1/2} - 2p_{1/2}$ level as zero energy. The ordering of the rows and columns is chosen to obtain a block-diagonal form.

$$\begin{pmatrix}
0 & -\frac{1}{\sqrt{3}}\lambda & \sqrt{\frac{2}{3}}\lambda & & & & & \\
-\frac{1}{\sqrt{3}}\lambda & 0 & 0 & & & & & \\
\sqrt{\frac{2}{3}}\lambda & 0 & \Delta & & & & & \\
& & & 0 & \frac{1}{\sqrt{3}}\lambda & \sqrt{\frac{2}{3}}\lambda & & \\
& & & \frac{1}{\sqrt{3}}\lambda & 0 & 0 & & \\
& & & \sqrt{\frac{2}{3}}\lambda & 0 & \Delta & & \\
& & & & & & \Delta & 0 \\
& & & & & & 0 & \Delta
\end{pmatrix} \qquad (5.84)$$

Since the sign of λ is not relevant, as seen below, there is only one 3×3 matrix to diagonalize. The eigenvalue problem $\det(H_{ij} - \epsilon\delta_{ij}) = 0$ becomes

$$
\begin{vmatrix}
-\epsilon & -\frac{1}{\sqrt{3}}\lambda & \sqrt{\frac{2}{3}}\lambda \\
-\frac{1}{\sqrt{3}}\lambda & -\epsilon & 0 \\
\sqrt{\frac{2}{3}}\lambda & 0 & \Delta - \epsilon
\end{vmatrix} = 0
\tag{5.85}
$$

or $\epsilon^3 - \Delta\epsilon^2 - \lambda^2\epsilon + \frac{\Delta}{3}\lambda^2 = 0$. This is a cubic equation for ϵ in Δ (spin-orbit) and λ (Stark). Before discussing the solution in general, we examine limiting situations.

(i) $\Delta = 0$: Then $\epsilon^3 - \lambda^2\epsilon = 0$ and $\epsilon = 0, +\lambda, -\lambda$. This is the spinless case, recovered.
(ii) $\Delta \gg \lambda$: $\epsilon = \Delta$ (for the $p_{3/2}$ states), $\epsilon = \pm\lambda/\sqrt{3}$ for the degenerate $s_{1/2}, p_{1/2}$ states.

For the general solution, write $z = \epsilon/\Delta$, $x = \lambda/\Delta$. Then we have

$$
z^3 - z^2 - zx^2 + \frac{x^2}{3} = 0.
\tag{5.86}
$$

A change of variables, $z = \frac{1}{3}(1-y)$, gives the standard form for finding the analytic solutions to a cubic equation

$$
y^3 - 3py + 2 = 0; \qquad p = 1 + 3x^2.
\tag{5.87}
$$

Figure 5.4 shows the results for ϵ/Δ versus λ/Δ. The dashed lines originating at $\epsilon/\Delta = 1/3$ are the spinless approximation. The solid curves are the roots of the cubic equation. Note that for small λ/Δ, the Stark splittings for the degenerate $s_{1/2}$ and $p_{1/2}$ states are linear (with slopes $\pm1/\sqrt{3}$ as large as the spinless), while the non-degenerate $p_{3/2}$ states show a quadratic Stark effect. Only when λ/Δ is large does the mixing of the $s_{1/2}$ state into the $p_{3/2}$ become large enough to convert the shift into an approximately linear one. Finding the eigenstates is left to the reader.

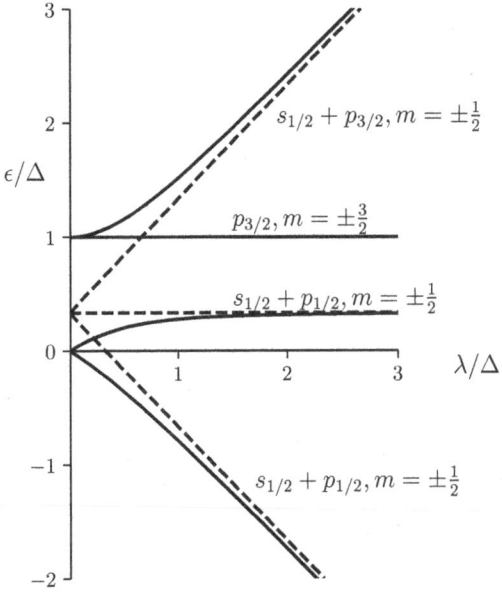

Figure 5.4 Stark effect in $n = 2$ hydrogen. The energy levels ϵ, divided by the relativistic splitting Δ between $2p_{3/2}$ and $2p_{1/2}$, are shown as a function of $\lambda = 3e\mathcal{E}a_0$, the strength of the Stark effect for electric field $\mathcal{E}$, divided by Δ. The $m = \pm3/2$ states are unmixed, while the $p_{3/2}, m = \pm1/2$ states mix with $s_{1/2}$. The $s_{1/2}$ states also mix with $p_{1/2}$. The dashed lines show the Stark effect if the relativistic splitting is ignored. At low fields, the Stark effect for $p_{3/2}$ is quadratic, since there appears to be no degeneracy. For larger fields, the relativistic splitting seems small and there is an approach to the linear Stark effect as expected.

5.8 Zeeman and Paschen-Back Effects in Hydrogen

The effect of an external magnetic field on the s-wave states of hydrogen is easily determined. From Eq. (5.22), we see that the $m_s = 1/2$ and $m_s = -1/2$ states are split by

$$2\mu_B B = 2 \times 5.788 \times 10^{-5} B(\mathrm{T})\,\mathrm{eV} = 0.933\,B(\mathrm{T})\mathrm{cm}^{-1}. \tag{5.88}$$

For p-wave states, $j = 1/2$ and $j = 3/2$, we must consider both the spin-orbit interaction and the magnetic effect. For very intense fields, the spin-orbit interaction can be neglected and the shift of a state with quantum numbers m_ℓ, m_s is simply

$$\Delta E = \mu_B(m_\ell + 2m_s)B. \tag{5.89}$$

Now taking the z axis along the direction of the magnetic field, rotations about the z axis remain a good symmetry and the magnetic field can connect only states with the same $m = j_z$. There is mixing between the $m = 1/2$ states for $j = 1/2$ and $j = 3/2$ and similarly for the $m = -1/2$ states. At the level of first-order perturbation theory, the radial wave functions are unaffected and orthogonality then restricts mixing to states with common n and ℓ. Thus, there are only two 2 x 2 mixing matrices to consider. See Problem 5.5.

References and Suggested Reading

For the definitive treatment of hydrogen:
   Bethe and Salpeter, *Quantum Mechanics of One- and Two-Electron Atoms*, pp. 205-241.

Problems

5.1   A particle of mass m in a one-dimensional harmonic oscillator potential of frequency ω_0 is subject to a cubic perturbation

$$H_1 = \lambda m\omega_0^2 x^3/d \tag{5.90}$$

where λ is a small number and d is a length. Write this perturbation in terms of creation and destruction operators and use abstract operator methods to find the first non-vanishing energy level shifts for the ground state and first excited state.

5.2   The Coulomb potential in a hydrogen-like atom is modified at short distances because of the finite size of the nucleus. If it is assumed that the total charge Ze is distributed uniformly through a sphere of radius R, the potential is

$$V(r) = -\frac{3}{2}\frac{Ze^2}{R}\left(1 - \frac{r^2}{3R^2}\right); \qquad\qquad 0 \le r \le R$$

$$= -\frac{Ze^2}{r}; \qquad\qquad R \le r. \tag{5.91}$$

(a)   Treating the difference between the point-Coulomb potential and the one for a finite nucleus as a perturbation, calculate the first- order energy change for the ground state of a hydrogen-like atom consisting of a nucleus of total charge Ze and a particle of mass m and charge $-e$, without any assumptions about the relative sizes of the atom and nucleus.

(b) For $Z = 82$ and $R = 8 \times 10^{-13}$ cm (lead nucleus), find the ratio of the energy shift to the unperturbed binding energy for an electronic atom (81-fold ionized lead!) and a muonic atom ($M_\mu = 206.77m_e$). Also calculate the ratio of the mean radius of the unperturbed atom to the nuclear radius in both instances.

(c) Is the result for the muonic lead atom reliable? Can you develop a better answer for the lowest state of the muon in the presence of a lead nucleus?

5.3 The hyperfine interaction in atomic hydrogen between magnetic moments of the electron and proton is described (for $\ell = 0$ states) by the effective Hamiltonian

$$H_{\text{eff}} = -\frac{8\pi}{3}\boldsymbol{\mu}_e \cdot \boldsymbol{\mu}_p \, \delta(\mathbf{r}). \tag{5.92}$$

See Jackson, *Classical Electrodynamics*, 3rd edition, pp. 190-191. Because of this interaction, the 1s ground state is split slightly.

(a) Consider states of total angular momentum $\mathbf{F} = \mathbf{s} + \mathbf{I}$, where $\mathbf{s}$ is the electron spin and $\mathbf{I}$ is the proton spin. Use lowest order perturbation theory to evaluate the splitting between the $F = 0$ and $F = 1$ states.

(b) Which is the true ground state of the hydrogen atom? Evaluate the splitting numerically in electron volts and also find the wavelength of electromagnetic radiation corresponding to a transition between the two states.

5.4* Consider a hydrogen atom in the 1s electronic state, as in the previous problem, but with an external magnetic field $\mathbf{B}$ applied in the positive z-direction. The spin and magnetic moment of the proton cannot be ignored.

(a) Show that the interaction Hamiltonian for the spins takes the form

$$H_1 = a\mathbf{I} \cdot \mathbf{s} - k_1 s_z - k_2 I_z. \tag{5.93}$$

Find explicit expressions for a, k_1, k_2 and specify clearly their signs.

(b) Use degenerate state perturbation theory with the four states in the (F, m_F) basis to determine the energy levels as a function of the magnetic field strength. This is the Breit-Rabi formula.

(c) Make an accurate plot of the energy levels (relative to the unperturbed spinless ground state) in units of a as a function of $(k_2 - k_1)/a$. Use the physical values for a, μ_e, and μ_p to find the proportionality of the abscissa with magnetic field in kilogauss. Make sure your plot goes out to at least $2 \text{ kG} = 0.2$ T.

(d) Write out the eigenstates of the diagonalized Hamiltonian for arbitrary magnetic field strength. Compare the limiting forms for $B = 0$ and B very large.

5.5* Consider the $n = 2$ p-wave levels in hydrogen. Let the spin-orbit splitting between the $j = 3/2$ and $j = 1/2$ $n = 2$ p-wave states in hydrogen be Δ ($\approx 11{,}000 \text{ cm}^{-1}$). A magnetic field B with $\mu_B B/\Delta = a$ is applied in the z direction. Determine the energy eigenstates $J_z = \pm 3/2, \pm 1/2$ and their eigenenergies. Graph the resulting energies in units of Δ as a function of a.

6

Atomic Structure

The hydrogen atom is the perfect testing ground for quantum mechanics. In the preceding chapter, we showed how its fine structure and behavior in applied electric and magnetic fields are explained by quantum mechanics. However, the claim of quantum mechanics is much greater. As a comprehensive theory, it addresses all of atomic physics - and more. Here the focus is on the multi-electron atoms. The fundamental principle of antisymmetry of electronic wave-functions takes the center stage. Unlike the situation for hydrogen, it is approximate solutions that are sought. A particularly powerful approach is that of Russell-Saunders or L-S coupling, where the atom's state is described by the values of the squares of the total angular momentum, the orbital angular momentum, and the spin angular momentum. L-S coupling is an approximation, and deviations from it can be important, especially in high-Z atoms. The Thomas-Fermi model is a remarkably successful description of many-electron atoms, which treats the electrons as essentially free with a density determined by the effective potential. The Hartree and Hartree-Fock equations provide means of calculating atomic structure self-consistently.

6.1 Parity

An important discrete symmetry operation is that of space inversion, or parity. Consider a system of particles with coordinates $\mathbf{r}_1, \mathbf{r}_2, \ldots, \mathbf{r}_n$. The active version of the parity operation P is to consider a Schrödinger representative $\Psi(\mathbf{r}_1, \mathbf{r}_2, \ldots, \mathbf{r}_N)$ and produce a new state:

$$\Psi_P(\mathbf{r}_1, \mathbf{r}_2, \ldots, \mathbf{r}_N) = \Psi(-\mathbf{r}_1, -\mathbf{r}_2, \ldots, -\mathbf{r}_N). \tag{6.1}$$

Thus, we imagine reflecting the coordinates of each particle through the common origin. The value of the wave function at the space-inverted point $P' = (-\mathbf{r}_1, -\mathbf{r}_2, \ldots, -\mathbf{r}_N)$ is not, in general, the same as the value at the original point. Then $\Psi_P \neq \Psi$.

It is not just the coordinates that are transformed, but any polar vector, e.g. $\mathbf{p}_i$. Thus for momentum-space wave functions, $P\Psi(\{\mathbf{p}_i\}) = \Psi(\{-\mathbf{p}_i\})$. Note that $P^2 = I$ is an operator equation. The eigenvalues of P are ± 1. $P = 1$ is called even parity and $P = -1$ is called odd parity. Operators transform under P as $A_P = PAP^{-1}$. If $A = A(\mathbf{r}_i, \mathbf{p}_i)$, then $A_P(\mathbf{r}_i, \mathbf{p}_i) = A(-\mathbf{r}_i, -\mathbf{p}_i)$. Thus under parity, while $\mathbf{r} \to -\mathbf{r}; \mathbf{p} \to -\mathbf{p}$, we have $\mathbf{L} \to \mathbf{L}$ because $\mathbf{L} = \mathbf{r} \times \mathbf{p}$ is an axial, not polar, vector.

John David Jackson: A Course in Quantum Mechanics, First Edition. Robert N. Cahn.
© 2024 John Wiley & Sons, Inc. Published 2024 by John Wiley & Sons, Inc.
Companion website: www.wiley.com/go/Jackson/QuantumMechanics

In polar coordinates, $\mathbf{r}$ corresponds to (r, θ, ϕ) and $-\mathbf{r}$ to $(r, \theta' = \pi - \theta; \phi' = \pi + \phi)$. Now consider

$$Y_{\ell,m} \propto (L_-)^{\ell-m} \left(\frac{x + iy}{r} \right)^{\ell}. \tag{6.2}$$

Now applying the parity operator, $PL_-P^{-1} = L_-$, but $P(x + iy) = -(x + iy)$, so $Y_{\ell,m}(\theta, \phi) \rightarrow (-1)^{\ell} Y_{\ell,m}(\theta, \phi)$.

Note that the parity operator is related to total angular momentum of a system only for a single spinless particle or two-particle internal motion. For many-electron atoms, you must examine what happens for each electron. In the independent particle approximation $P = (-1)^{\Sigma_i \ell_i}$, ℓ_i is the orbital angular momentum of each electron. Note, as well, that parity is not to be confused with particle exchange. While parity takes $(\mathbf{r}_1, \mathbf{r}_2)$ to $(-\mathbf{r}_1, -\mathbf{r}_2)$, in the two-body case, particle exchange changes $(\mathbf{r}_1, \mathbf{r}_2)$ to $(\mathbf{r}_2, \mathbf{r}_1)$. For a two-body system, these may be the same since there is only one true variable: $\mathbf{r} = \mathbf{r}_1 - \mathbf{r}_2$. However, in other instances, this is not so. For example, in a helium atom with the nucleus at the origin, the positions of the electrons $\mathbf{r}_1$ and $\mathbf{r}_2$ are independent and under parity $\psi(\mathbf{r}_1, \mathbf{r}_2) \rightarrow \psi(-\mathbf{r}_1, -\mathbf{r}_2)$ while under particle exchange $\psi(\mathbf{r}_1, \mathbf{r}_2) \rightarrow \psi(\mathbf{r}_2, \mathbf{r}_1)$, which are two quite different things.

Many Hamiltonians describe particles interacting via two-body forces that depend only on the relative distances between the pairs, i.e. $V_{ij} = V(|\mathbf{r}_i - \mathbf{r}_j|)$. They may depend, too, on operators like $\sigma_1 \cdot \mathbf{r}_{12}\, \sigma_2 \cdot \mathbf{r}_{12}$. Such Hamiltonians commute with the parity operator. That is, $H_P = PHP^{-1} = H$. Since P and H commute, we can entertain the possibility of simultaneous eigenstates. In particular, if $[P, H] = 0$ and $|E'\rangle$ is a non-degenerate eigenstate of energy, then $|E'\rangle$ is also an eigenstate of parity. This is easily shown. Let $P |E'\rangle = |E'\rangle_P$.

$$PH |E'\rangle = E'P |E'\rangle = E' |E'\rangle_P$$

$$PHP^{-1}P |E'\rangle = H |E'\rangle_P = E' |E'\rangle_P. \tag{6.3}$$

Thus $|E'\rangle_P$ is also an eigenstate with the same energy. If there is no degeneracy, then $|E'\rangle_P = c |E\rangle$. But $P^2 = I$, so $c^2 = 1$. Thus $c = \pm 1$.

If there is degeneracy, we must suppose that P on one state $|E', \alpha'\rangle$ is a linear combination of the states degenerate with it:

$$P |E', \alpha'\rangle = \sum_{\alpha''}^{k} c(\alpha', \alpha'') |E', \alpha''\rangle. \tag{6.4}$$

Since $P^2 = 1$, we have

$$|E', \alpha'\rangle = \sum_{\alpha'\alpha''} c(\alpha', \alpha'')c(\alpha'', \alpha''') |E', \alpha'''\rangle. \tag{6.5}$$

So the $c(\alpha, \alpha')$ are elements of a $k \times k$ matrix C with $C^2 = 1$. So after diagonalization, each element is of the form

$$c(\beta, \beta') = c_{\beta}\delta_{\beta,\beta'} \tag{6.6}$$

with $c_{\beta} = \pm 1$.

In the non-relativistic hydrogen atom, we have degeneracy at every level except 1s. At every other level there are energy eigenstates that are not parity eigenstates, for example linear combinations of s and p states. However, it is natural to group the states by their orbital angular momentum, which thus separates them into groups with common parity values. In the three-dimensional simple harmonic oscillator, the energy eigenstates can be labeled by three integers (n_x, n_y, n_z), with $E = (n_x + n_y + n_z + 3/2)\hbar\omega$. The parity of a single oscillator is $(-1)^n$, so the parity of $|n_x, n_y, n_z\rangle$ is even if $n_x + n_y + n_z$ is even, and odd if it is odd. Thus the energy determines the parity and there is no need for diagonalization of P.

Operators between parity eigenstates vanish unless certain selection rules are obeyed. For example, electric and magnetic dipole transitions are governed by the matrix elements

$$\left\langle \alpha'', P'' \Big| \sum_j e_j \mathbf{r}_j \Big| \alpha', P' \right\rangle ; \qquad \left\langle \alpha'', P'' \Big| \sum_j \boldsymbol{\mu}_j \Big| \alpha', P' \right\rangle . \tag{6.7}$$

Let $\mathcal{O}$ be a general operator with well-defined behavior under parity: $\mathcal{O}_P = P\mathcal{O}P^{-1} = \pm\mathcal{O}$. Consider

$$\langle \alpha'', P'' |\mathcal{O}| \alpha', P' \rangle = \langle \alpha'', P'' |P^{-1}P\mathcal{O}P^{-1}P| \alpha', P' \rangle$$
$$= \pm P'P'' \langle \alpha'', P'' |\mathcal{O}| \alpha', P' \rangle . \tag{6.8}$$

Thus,

$$[1 - (\pm P'P'')]\langle \alpha'', P'' |\mathcal{O}| \alpha', P' \rangle = 0. \tag{6.9}$$

The matrix element must vanish unless the product of the parities of the two states is equal to the parity of the operator. Among the odd-parity operators are $\mathbf{r}$ and $\mathbf{p}$, which connect states of opposite parity. Even-parity operators, which connect states of equal parity, include $\boldsymbol{\mu}$ and $\boldsymbol{\sigma}_1 \cdot \boldsymbol{\sigma}_2$.

6.2 Identical Particles and the Pauli Exclusion Principle

The light atoms at the beginning of the periodic table provide evidence of the Pauli exclusion principle. We start with three concepts:

(a) hydrogenic energy levels (1s,2s,2p...)
(b) electrons have spin 1/2
(c) atomic number Z is the number of electrons

The lightest elements have features we seek to explain: the ionization potential, size, behavior in a Stern-Gerlach device, the g-factor (Eq. (5.24)) measured in a magnetic field.

Hydrogen -(1e): Ionization potential (I)=13.6 eV, size $\sim a_0$, Stern-Gerlach apparatus splits beam into two parts, $g = 2, J = 1/2$.

He - (2e): I=24.6 eV, size about $\frac{1}{2}a_0$. Stern-Gerlach apparatus - no splittings, so $J = 0$. No low-lying states with $J = 1$. Lowest $S = 1 = J$ state is only bound by 4.8 eV.

Lithium - (3e): I=5.4 eV, size about a_0. Stern-Gerlach - split into two beams, $g = 2, J = 1/2$. Li$^+$ ion - like helium but more tightly bound and small. Thus, Li has a helium-like core of $Z = 3$, plus a loosely bound electron.

Beryllium - (4e): I= 9.3 eV, size about a_0. Stern-Gerlach - no splitting, $J = 0$.

Boron - (5e): I= 8.3 eV, Stern-Gerlach - split into two beams, so $J = 1/2$, but $g = 2/3$, corresponding to orbital angular momentum (one electron in p state).

This pattern is understood, as we all know, in terms of an independent-electron, central-field approximation (upon which we do perturbation theory to get the details right) and the Pauli exclusion principle: "No two electrons can have the same four quantum numbers," where the quantum numbers are n, ℓ, m_ℓ, m_s in the central-field description.

We know that the independent-particle central-field description assigns each electron to an "orbital" (n, ℓ) with the Bohr principle quantum number for n and s,p,d,f,g... as a letter standing for $\ell = 0, 1, 2, 3, 4, ...$ Thus the configurations are:

$$H : \qquad\qquad (1s)$$
$$He : \qquad\qquad (1s)^2$$
$$Li : \qquad\qquad (1s)^2(2s)$$
$$Be : \qquad\qquad (1s)^2(2s)^2$$
$$B : \qquad\qquad (1s)^2(2s)^2(2p) \qquad\qquad (6.10)$$

and so on. The above properties all fit nicely into this description if we demand that $(1s)^2$ has $S = 0$, to conform to the exclusion principle.

The Pauli principle, as originally stated, was specific to the central-field, independent particle model of atoms, but when phrased in terms of antisymmetric state vectors has general applicability.

Consider a many-identical-particle system with "coordinates" $\xi_1 \xi_2, \xi_3, ... \xi_n$, ($\xi_i = \mathbf{x}_i, \mathbf{p}_i$, $\mathbf{s}_i, ..$). The Hamiltonian $H = H(\xi_1, \xi_2, ... \xi_n)$ must be symmetric with respect to all interchanges of pairs of coordinates (because the particles are postulated to be identical and indistinguishable), or equivalently, if $\mathcal{P}$ is an operator corresponding to some permutation of the coordinates, $\mathcal{P}H\mathcal{P}^{-1} = H$.

Although not mandatory, it is conceptually useful to build up the many-particle state vectors from products of single-particle kets. Let $|a_i\rangle, |b_i\rangle, |c_i\rangle, ...$ be the basis of the ith particle states. Then an n-particle state vector might be $|a_1, b_2, c_3 ...\rangle = |a_1\rangle |b_2\rangle |c_3\rangle ...$ If we apply the permutation operator corresponding to the interchange of particles 1 and 2, we get a different state vector, in general

$$\mathcal{P}_{12} |a_1, b_2, c_3 ...\rangle = |b_1, a_2, c_3 ...\rangle = |b_1\rangle |a_2\rangle |c_3\rangle ..., \qquad (6.11)$$

where we have chosen to keep the particle labels in the standard order and to interchange the states - it is the same either way. It might be that the action of $\mathcal{P}_{12}$ on the original ket gave back the same ket, apart from a sign change. We would then call that ket symmetric or antisymmetric under the exchange or permutation $\mathcal{P}_{12}$.

Because $H(\xi_i)$ is a symmetric function of the ξ_i, whatever symmetry character a ket has at time $t = 0$, it will not change that character with time. Once symmetric (antisymmetric), always symmetric (antisymmetric).

It is a remarkable fact of Nature that particles come in two basic kinds: half-integral intrinsic angular momentum and integral intrinsic angular momentum. State vectors for more than one identical system or particle are symmetric or antisymmetric under permutation of any pair of particles, depending on whether the intrinsic angular momentum is integral or half-integral.

At the level of non-relativistic quantum mechanics, this fact is just that. In relativistic quantum mechanics there is a spin-statistics theorem, based on Lorentz invariance and other basic, plausible assumptions, that requires that the exchange of integral-spin identical particles is symmetric while for half-integral identical particles it is antisymmetric. This was first proved by Pauli and subsequently proved with fewer and fewer assumptions. Half-integral spin particles are called fermions (or Fermi-Dirac particles) and integral-spin particles are called bosons (or Bose-Einstein particles), after Satyendra Bose, who first proposed the symmetry under exchange of like particles.

Compelling evidence for the connection between spin and statistics comes from molecular spectra. The spectra of diatomic molecules can be described using the Born-Oppenheimer approximation, which we sketch here.

The light electrons can quickly adapt to a displacement of the nuclei. It makes sense, then, to develop an effective potential as a function of the internuclear separation, R. There is some separation, R_0, on the scale of a_0 that minimizes this effective potential. Near the minimum, the potential is quadratic and we expect a simple harmonic oscillator behavior of the two nuclei. These states are labeled $\nu = 0, 1, 2, ...$ We can reduce the two-body system to a one-body problem with an effective mass $M = M_1 M_2/(M_1 + M_2)$, where M_1, M_2 are the masses of the nuclei. The electronic energies are all of the scale of atomic energies, a Rydberg $= 13.6$ eV. All the distances are of the scale of the Bohr radius. Thus, a sketch of the Hamiltonian is

$$\frac{p^2}{2M} + \frac{1}{2}A \cdot \text{Ryd}(R - R_0)^2/R_0^2. \tag{6.12}$$

with A of order unity. The coefficient of $\frac{1}{2}(R - R_0)^2$ must be $M\omega^2$, where ω is the natural molecular oscillation frequency. Thus,

$$M\omega^2 \sim \frac{\text{Ryd}}{a_0^2}, \tag{6.13}$$

but

$$\frac{1}{a_0^2} = 2 \, \text{Ryd} \, m_e/\hbar^2. \tag{6.14}$$

Thus, roughly,

$$M\omega^2 \sim (\text{Ryd})^2 m_e/\hbar^2; \qquad \frac{\hbar\omega}{\text{Ryd}} \sim \sqrt{\frac{m_e}{M}}. \tag{6.15}$$

If we take as a typical nuclear mass 10 times the mass of a nucleon, the reduced mass is about 10^4 times the mass of an electron, so the spacing of the vibrational levels is about $1/10$ eV. The molecule can rotate as well as vibrate. Taking as the moment of inertial of the molecule $\sim M a_0^2$, the rotational energy is

$$\sim \frac{L(L + 1)\hbar^2}{2M a_0^2} \sim L(L + 1)\frac{m_e}{M}\text{Ryd}. \tag{6.16}$$

Thus, the spacing of the rotational levels is smaller than the spacing of the vibrational levels by an additional factor of $\sqrt{m_e/M} \sim 10^{-2}$. At room temperature, $kT \sim 1/40$ eV so that few vibrational levels are populated, but many rotational levels are. See Figure 6.1.

Bose and Fermi statistics make their appearance in the molecular band spectra of homonuclear diatomic molecules. Transitions exhibit closely spaced lines that form "bands" under poor resolution. The intensities of closely spaced lines alternate in a definite pattern related to the nuclear intrinsic angular momentum and the symmetry of the wave functions. Let the two nuclear angular momenta each be I. Each nucleus has $(2I + 1)$ essentially degenerate

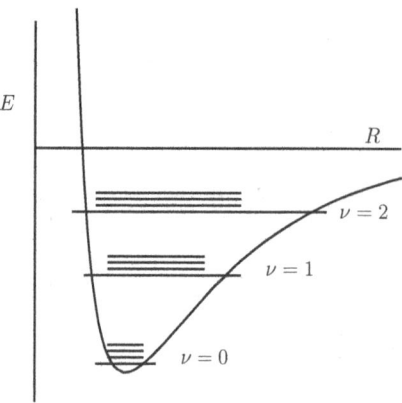

E

R

$\nu = 2$

$\nu = 1$

$\nu = 0$

Figure 6.1 A diatomic molecule can be understood in the Born-Oppenheimer approximation. The effective potential as a function of R, the interatomic separation, is shown. There are vibrational states, labeled by ν. Each of these has rotational levels, indicated without labeling. A separate figure could be drawn for each electronic state of the molecule.

states and thus there are $(2I + 1)^2$ states altogether. States of the form $|Im\rangle_1 |Im\rangle_2$ are necessarily even under interchange of the nuclei. States with values of $m \neq m'$ can be either even or odd under interchange: $|Im\rangle_1 |Im'\rangle_2 \pm |Im'\rangle_1 |Im\rangle_2$. Thus there are $(2I + 1)I$ states odd under interchange, but $(2I + 1)(I + 1)$ that are even. Now the state of the molecule can be written (we are not interested in the electronic state)

$$\psi(1, 2) = R_\nu(r)Y_{LM}(\theta, \phi)\chi(1, 2), \tag{6.17}$$

where $R_\nu(r)$ is the vibrational state and $\chi(1, 2)$ is the nuclear spin state for the two nuclei. Now suppose $\psi(1, 2)$ is symmetric. Then if L is even, so is χ and the degeneracy is $(I + 1)(2I + 1)$, while if L is odd the degeneracy is instead $I(2I + 1)$. If $\psi(1, 2)$ is antisymmetric, the argument is reversed: if L is even, χ is odd and the degeneracy is $I(2I + 1)$, while of L is odd χ is even and the degeneracy is $(I + 1)(2I + 1)$.

The most important electromagnetic transitions arise from electric dipole moments. But these are absent in low-lying states of homonuclear molecules like H_2, N_2, O_2, which explains why these molecules do not contribute significantly to the greenhouse effect responsible for global warming. Instead, the dominant transitions involve quadrupole moments and thus change the angular momentum by two units.

The energy levels of the rotational states are proportional to $L(L+1)$. The energy difference between L and $L + 2$ is proportional to $(L + 2)(L + 3) - L(L + 1) = 4L + 6$. For $L = 0, 2, ...,$ this is $6, 14, ...,$ while for $L = 1, 3, ...$ it is $10, 18, ...$ The two series will have different intensities because the number of particles in the initial state will be proportional to the degeneracy (since the energy splittings are small). In the case of even $\psi(1, 2)$, the odd-L degeneracy is proportional to I, while the even is proportional to $I + 1$, so the odd-L series will be less intense. For $\psi(1, 2)$ odd, this is reversed.

In fact, for C_2 (made from ^{12}C) and O_2 (made from ^{16}O), the lines with L odd are not seen. This leads to the conclusion that $I = 0$. For N_2 (made from ^{14}N), the intensity of the odd-L lines is about 1/2 that of the even-L lines, so $I = 1$. (Before the discovery of the neutron, the nitrogen nucleus was viewed as having 14 protons and 7 electrons, making it a fermion, in conflict with the spectroscopic data.) In contrast, for H_2 the L-odd lines are three times as strong as the L-even lines, so $I = 1/2$. All this is in agreement with the expectation that half-integral identical particles require antisymmetric wave functions, while identical integral-spin particles have symmetric wave functions.

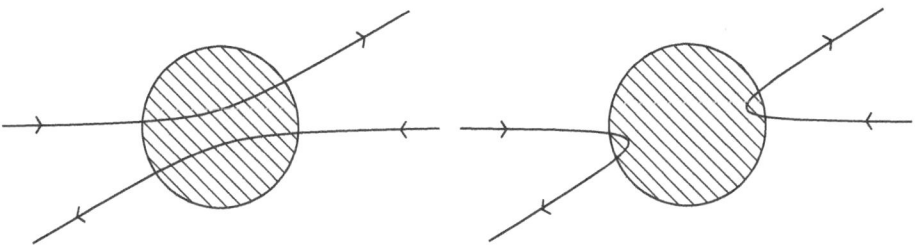

Figure 6.2 When identical particles scatter, the amplitudes scattering by θ and by $\pi - \theta$ must be added. Whether they are added with a relative plus sign or a relative minus depends on whether the spatial wave function is symmetric or antisymmetric.

The consequences of symmetry and antisymmetry under particle interchange are manifest in scattering. Identical particle scattering cannot distinguish between scattering through an angle θ and an angle $\pi - \theta$ and therefore the amplitudes for the two processes must be added before taking the square. See Figure 6.2. For spinless bosons, we have

$$\frac{d\sigma}{d\Omega} = \left| f(\theta) + f(\pi - \theta) \right|^2 = |f(\theta)|^2 + |f(\pi - \theta)|^2 + 2\Re[f^*(\theta)f(\pi - \theta)]. \tag{6.18}$$

For scattering of nuclei like ^{12}C at sufficiently low energy, the Coulomb repulsion keeps the nuclei far enough away from each other that only the Coulomb force is important. The first two terms in Eq. (6.18) give the classical result, in which two contributions to the cross-section are added. The complete answer includes the interference between the two contributing amplitudes given by quantum mechanics. Data for scattering of two ^{12}C nuclei are shown in Figure 6.3.

More generally, if the scattering is spin-independent and the scatters have spin I, then again the fraction of the $(2I+1)^2$ amplitudes that are even is $(I+1)/(2I+1)$, while a fraction $I/(2I+1)$ is antisymmetric. Then the cross-section is

$$\frac{d\sigma}{d\Omega} = \frac{I+1}{2I+1}|f(\theta) \pm f(\pi - \theta)|^2 + \frac{I}{2I+1}|f(\theta) \mp f(\pi - \theta)|^2$$

$$= |f(\theta)|^2 + |f(\pi - \theta)|^2 \pm \frac{2}{2I+1}\Re[f^*(\theta)f(\pi - \theta)], \tag{6.19}$$

where the upper sign is for bosons and the lower sign is for fermions. The classical result is given by the first two terms.

For $I = 0$, a symmetric case,

$$\frac{d\sigma}{d\Omega} = |f(\theta) + f(\pi - \theta)|^2 \tag{6.20}$$

and at $\theta = \pi/2$,

$$\frac{d\sigma}{d\Omega} = 2\left(\frac{d\sigma}{d\Omega}\right)_{\text{classical}}. \tag{6.21}$$

In contrast, for $I = 1/2$, an antisymmetric case

$$\frac{d\sigma}{d\Omega} = |f(\theta)|^2 + |f(\pi - \theta)|^2 - \Re[f^*(\theta)f(\pi - \theta)] \tag{6.22}$$

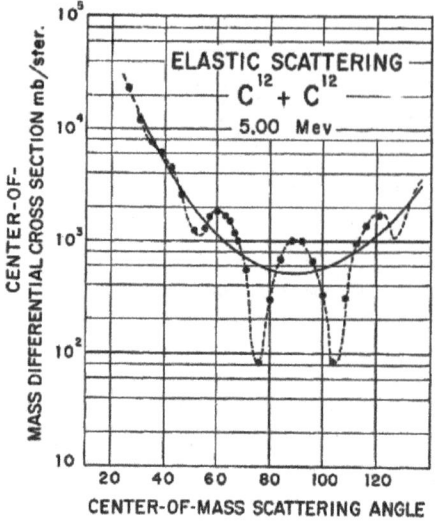

Figure 6.3 Data from Bromley, 1961 / American Physical Society for scattering of two ^{12}C nuclei at a center-of-mass energy 5 MeV. At this low energy, the nuclei feel only the electromagnetic force and the cross section is accurately predicted by the Mott formula (N. F. Mott, Proc. Roy. Soc. (London) **A126**, (1930)). The oscillations arise because the phase of $f(\theta)$ depends on θ.

and at $\theta = \pi/2$,

$$\frac{d\sigma}{d\Omega} = \frac{1}{2} \left(\frac{d\sigma}{d\Omega} \right)_{classical}. \tag{6.23}$$

6.3 Atoms

The ket $|a_1 b_2 c_3 \ldots h_n\rangle$ is specific in assigning particle 1 to the "state" a, particle 2 to the "state" b, and so on. In general, such a state possesses no particular permutation symmetry. If the particles are identical, this state will not have the proper behavior under interchange of the particles. We can create appropriate states for n identical bosons or n identical fermions by taking linear combinations of the $n!$ kets created by taking all possible permutations:

$$\psi_S(1, 2, \ldots, n) = \frac{1}{\sqrt{n!}} \sum_{\mathcal{P}} \mathcal{P} |a_1 b_2 c_3 \ldots h_n\rangle \qquad \text{[Bosons]}$$

$$\psi_A(1, 2, \ldots, n) = \frac{1}{\sqrt{n!}} \sum_{\mathcal{P}} (-1)^{\mathcal{P}} \mathcal{P} |a_1 b_2 c_3 \ldots h_n\rangle \qquad \text{[Fermions]} \tag{6.24}$$

where the sum is over all possible permutations $\mathcal{P}$ and $(-1)^{\mathcal{P}} = \pm 1$ for an even (odd) permutation of the coordinates. The totally antisymmetric kets can be written as a Slater determinant (J. C. Slater, 1929)

$$\psi_A = \frac{1}{\sqrt{n!}} \begin{vmatrix} |a_1\rangle & |a_2\rangle & |a_3\rangle & \cdots & |a_n\rangle \\ |b_1\rangle & |b_2\rangle & |b_3\rangle & \cdots & |b_n\rangle \\ \cdot & & & & \\ \cdot & & & & \\ |h_1\rangle & |h_2\rangle & |h_3\rangle & \cdots & |h_n\rangle \end{vmatrix}. \tag{6.25}$$

The individual kets $|a\rangle$, $|b\rangle$, ... have spatial and spin degrees of freedom specified. The Pauli exclusion principle is built in. The n electrons in an atom can be described by a family of these Slater determinants for each energy level. For example, a state of the helium atom corresponding to the configuration $(1s)(2s)$ is described by four Slater determinants:

$$\psi_1 = \frac{1}{\sqrt{2}} \begin{vmatrix} \psi_{1s}(1)\alpha_1 & \psi_{1s}(2)\alpha_2 \\ \psi_{2s}(1)\alpha_1 & \psi_{2s}(2)\alpha_2 \end{vmatrix}, \quad \psi_2 = \frac{1}{\sqrt{2}} \begin{vmatrix} \psi_{1s}(1)\alpha_1 & \psi_{1s}(2)\alpha_2 \\ \psi_{2s}(1)\beta_1 & \psi_{2s}(2)\beta_2 \end{vmatrix}$$

$$\psi_3 = \frac{1}{\sqrt{2}} \begin{vmatrix} \psi_{1s}(1)\beta_1 & \psi_{1s}(2)\beta_2 \\ \psi_{2s}(1)\alpha_1 & \psi_{2s}(2)\alpha_2 \end{vmatrix}, \quad \psi_4 = \frac{1}{\sqrt{2}} \begin{vmatrix} \psi_{1s}(1)\beta_1 & \psi_{1s}(2)\beta_2 \\ \psi_{2s}(1)\beta_1 & \psi_{2s}(2)\beta_2 \end{vmatrix} \tag{6.26}$$

where, as usual, α means spin up and β means spin down. Evidently, these determinantal wave functions are linear combinations of wave functions we can create by looking at the spatial and spin spaces separately. Consider $s_1 = 1/2, s_2 = 1/2$. With $\mathbf{S} = \mathbf{s}_1 + \mathbf{s}_2$, we have four states:

$$S = 1 \quad \chi_{1,1} = \alpha_1\alpha_2$$

$$\chi_{1,0} = \frac{1}{\sqrt{2}}(\alpha_1\beta_2 + \beta_1\alpha_2)$$

$$\chi_{1,-1} = \beta_1\beta_2$$

$$S = 0 \quad \chi_{0,0} = \frac{1}{\sqrt{2}}(\alpha_1\beta_2 - \beta_1\alpha_2). \tag{6.27}$$

The $S = 1$ spin states are symmetric under particle interchange; the $S = 0$ spin state is antisymmetric. Spatial wave functions of the sort $\psi_{1s}(1)\psi_{2s}(2)$ can be symmetrized or antisymmetrized to combine with the χ_{S,M_S} to give totally antisymmetric state vectors

$$S = 1: \quad \psi_{M_S} = \frac{1}{\sqrt{2}}[\psi_{1s}(\mathbf{r}_1)\psi_{2s}(\mathbf{r}_2) - \psi_{1s}(\mathbf{r}_2)\psi_{2s}(\mathbf{r}_1)]\chi_{1,M_S}$$

$$S = 0: \quad \psi_0 = \frac{1}{\sqrt{2}}[\psi_{1s}(\mathbf{r}_1)\psi_{2s}(\mathbf{r}_2) + \psi_{1s}(\mathbf{r}_2)\psi_{2s}(\mathbf{r}_1)]\chi_{0,0}. \tag{6.28}$$

Comparison shows that the Slater determinants are related as follows:

$$\psi_{M_S=1} = \psi_1; \qquad\qquad\qquad \psi_{M_S=-1} = \psi_4$$

$$\psi_{M_S=0} = \frac{1}{\sqrt{2}}(\psi_2 + \psi_3); \qquad\qquad \psi_0 = \frac{1}{\sqrt{2}}(\psi_2 - \psi_3). \tag{6.29}$$

A key point for Slater determinants involving individual kets $|n\ell\, m_\ell\, m_s\rangle$ is that each is an eigenket of $L_z = \sum_i \ell_{iz}$ and $S_z = \sum_i s_{iz}$, or $M_L = \sum_i m_{\ell i}, M_S = \sum_i m_{si}$ and thus of J_z, but not necessarily an eigenket of L^2, S^2, or J^2.

6.4 Helium Atom

The two-electron system of a helium-like atom affords an illustration of the Pauli exclusion principle and the factors that determine the energy levels of an atom. In atomic units (lengths in a_0, momenta in $\hbar/a_0$, energies in e^2/a_0), the Hamiltonian for a nucleus of charge Ze and two electrons is, in the non-relativistic limit,

$$H = \left(-\frac{1}{2}\nabla_1^2 - \frac{Z}{r_1}\right) + \left(-\frac{1}{2}\nabla_2^2 - \frac{Z}{r_2}\right) + \frac{1}{r_{12}}. \tag{6.30}$$

If we ignore the electron-electron interaction, the Hamiltonian is the sum of $H_1 + H_2$, and so has eigenkets that are products of single-particle kets. Since, at this level, there are only Coulomb interactions, the energy levels do not depend explicitly on spin. Imposition of the Pauli principle does, however, influence the electrostatic energies. Hence, the sequence of energies does depend on spin.

With $s_1 = 1/2, s_2 = 1/2$, we can have $S = 0, 1$. Helium states with $S = 0$ (singlet) are called parahelium and those with $S = 1$ are called orthohelium. The ground state configuration $(1s)^2$ has an antisymmetric spin state, $S = 0$, and a symmetric spatial state, since it is not possible to choose the symmetric spin state because the spatial part $(1s)^2$ cannot be made antisymmetric. If we consider a product wave function, we have

$$\psi_0(1, 2) = f(r_1)f(r_2)\chi_{0,0}. \tag{6.31}$$

Problem 6.1 asks for a variational estimate of the ground state energy, using hydrogen-like wave functions for $f(r)$,

$$f(r) = \frac{1}{\sqrt{\pi}}\left(\frac{Z_{\text{eff}}}{a_0}\right)^{3/2} e^{-Z_{\text{eff}}r/a_0} \tag{6.32}$$

where Z_{eff} is a variational parameter. You will find $Z_{\text{eff}} = Z - 5/16$ and an estimate of the total binding energy that is within a few percent of 79.0 eV, the experimental value. The term $-5/16$ represents the partial screening effect of one electron in the $1s$ shell on the other.

Better calculations of the helium-like ground state for $Z = 1, 2 \ldots$ are described in Bethe and Salpeter, *Quantum Mechanics of One- and Two-Electron Atoms*. Such variational calculations employ the above exponential form, times a polynomial in $s = r_1 + r_2$ even powers of $t = (r_1 - r_2)$ and $u = r_{12}$.

For the H^- ion, Chandrasekhar (Ap. J, **100**, 176 (1944)) employed

$$\psi = A(e^{-ar_1 - br_2} + e^{-br_1 - ar_2})(1 + cr_{12}) \tag{6.33}$$

With $c = 0$, he found $a = 1.039, b = 0.283$, and a binding energy of $I = 0.37$ eV (about half of the correct value, but at least bound). With c varied, he found $a = 1.75, b = 0.478$, $c = 0.312$, and $I = 0.705$ eV, compared to 0.7542 eV obtained with much more complex variational forms. Chandrasekhar's result is much better than an analogous three-parameter trial function that has an exponential of the form $e^{-\alpha(r_1 + r_2)}$ times a polynomial.

The first few bound states of helium are shown in Figure 6.4. Two striking features appear:

(1) The ground state has ~ 8 times the binding energy of the first excited state, i.e. the energy of excitation is 20 eV.
(2) Spectroscopy shows a parahelium spectrum and an orthohelium spectrum, but no crossovers. The $(1s)(2s)\,^3S_1$ is metastable.

The excited states occur when one electron is promoted to a higher orbital. The excited electron will be bound by roughly $1/(2n^2)$ in atomic units since the other electron will effectively shield one of the two nuclear electric charges. For $n = 2, 3$ ionization will require

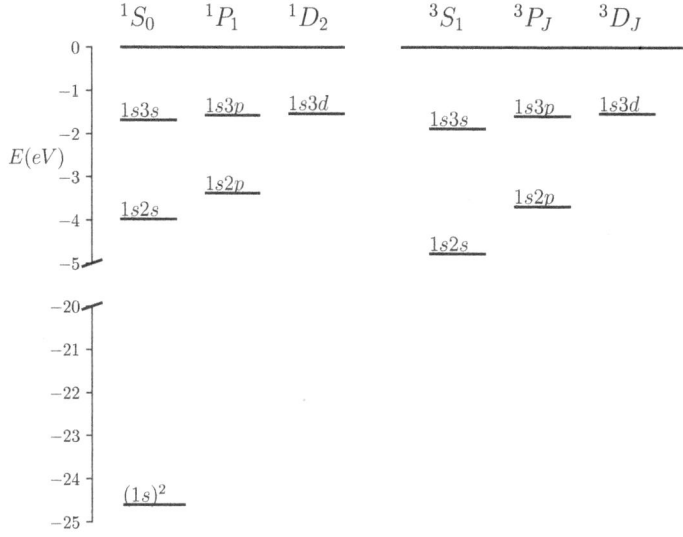

Figure 6.4 The low-lying states of helium. The parahelium ($S = 0$) states are on the left and the orthohelium ($S = 1$) on the right. There is no spin-triplet for $(1s)^2$ since that spatial state is intrinsically symmetric. For the other states shown, the spin-triplet lies lower than the spin-singlet.

Table 6.1 Energy levels of helium excited states demonstrating the stronger binding of triplet states. The antisymmetrized spatial wave functions of the spin-triplets have reduced electron-electron interaction.

(1s)(2s):	$^1S_0$	−3.98 eV
	$^3S_1$	−4.78 eV
(1s)(2p):	$^1P_1$	−3.38 eV
	^3P_J	−3.64 eV

about −3.4 eV to −1.5 eV, as is observed for the $(1s)(2s)$ or $(1s)(2p)$ and $(1s)(3s, 3p, 3d)$ configurations.

Table 6.1 shows the binding energies of the first excited states of helium. We see that the spin-triplet states lie lower than the corresponding singlet states. This is not the consequence of spin-spin interactions, but of the Pauli principle. Intuitively, we anticipate that the anti-symmetric wave function keeps the electrons away from each other, lowering the electrostatic energy. The generalized observation that electrons fill a level (n, ℓ) so as to maximize the total spin, S, is known as Hund's rule. The second part of Hund's rule asserts that among the states with the highest S, it is the state of highest L that has the lowest energy. These rules apply most reliably to the lowest energy state.

To estimate the singlet-triplet splitting, we treat the electron-electron interaction e^2/r_{12} as a perturbation. Consider excited states with occupied orbitals a and b (e.g. $a = 1s$, $b = 2p$). The triplet and singlet wave functions are

$$\psi^{(t)}(1,2) = \frac{1}{\sqrt{2}}[a(1)b(2) - b(1)a(2)]\chi_{1,M_S}$$

$$\psi^{(s)}(1,2) = \frac{1}{\sqrt{2}}[a(1)b(2) + b(1)a(2)]\chi_{0,0}. \qquad (6.34)$$

Thus, the energy difference between the state singlet state s and the triplet t is, in units of e^2/a_0 (sometimes called a Hartree, twice a Rydberg),

$$\begin{aligned}
E^{(s)} - E^{(t)} &= \frac{1}{2}\int d^3x_1\, d^3x_2\, [a(1)b(2) + b(1)a(2)]^* \frac{1}{r_{12}}[a(1)b(2) + b(1)a(2)] \\
&\quad - \frac{1}{2}\int d^3x_1\, d^3x_2\, [a(1)b(2) - b(1)a(2)]^* \frac{1}{r_{12}}[a(1)b(2) - b(1)a(2)] \\
&= 2\int d^3x_1\, d^3x_2\, [a(1)b(2)]^* \frac{1}{r_{12}}[b(1)a(2)]. \qquad (6.35)
\end{aligned}$$

This difference is the exchange energy. Note that

$$E_{\substack{s \\ t}} = \int d^3x_1\, d^3x_2\, \frac{1}{r_{12}}\left[|a(1)|^2|b(2)|^2 \pm a^*(1)b(1)b^*(2)a(2) \right], \qquad (6.36)$$

where the first term is the direct integral and the second is the exchange integral. An explicit example for $(1s)(2p)$ is given in Problem 6.6.

To ionize helium fully into an alpha particle and two free electrons requires 79.0 eV. Single ionization requires just 24.6 eV. Above 24.6 eV, there is a continuum of states, but embedded in the continuum are states where both electrons have been promoted above the ground state. The lowest of these is $(2s)^2$. A crude estimate of its binding is that the first electron is bound by $Z^2/2^2 = 1$ Rydberg, while the second is bound by $1/2^2$ Rydberg, altogether 17 eV, so 62 eV above the ground state, a slight overestimate. Higher states $(2s)(2p)$, ..., $(4s)(4p)$, ... can be excited, and above 65.4 eV there are states with one electron in the $(2s)$ and one electron free. Of course, the same pattern is repeated with electrons excited into states like $(3s)(4p)$ and $(3s)$ plus a free electron. Ultimately, we have fully ionized helium with two free electrons. The states like $(2s)^2$ above 24.6 eV can spontaneous decay to singly ionized helium $(1s)$ and are termed auto-ionizing. See Figure 6.5.

In Figure 6.6, we show the photo-absorption cross-section of helium in units of 10^{-17} cm^2 ($a_0^2 = 2.8 \times 10^{-17}$ cm^2) over a narrow range of energy (roughly 52 eV to 69 eV). The peak at 206 Å is due to the $(2s)(2p)$ state. At 195 Å, there is the $(2s)(3p)$ state, followed by an accumulation of states leading to the ionization threshold for the $(2s)$ state at 65.4 eV. At energies just above each peak, much less than an Angstrom lower in wave-length, there is a dip indicating interference.

The interference arises because there are two paths from the initial state of $(1s^2)$ helium and the incoming photon to the final state of a $(1s)$ helium ion and a continuum electron in a p-wave state. The first path is the direct interaction, whose amplitude is

$$\langle f(E)\,|T|\,i\rangle \qquad (6.37)$$

where $|f(E)\rangle$ is the final state whose electron has energy E equal to the incoming photon's energy less the 24.6 eV needed to singly ionize the helium. The electron is in a p-wave as a consequence of the "optically allowed" transition (discussed in a later chapter). The electromagnetic transition operator is indicated by T. This amplitude is a smooth function of E.

Figure 6.5 The full spectrum of helium in eV shows that ionization requires 24.6 eV. Above that, there is a continuum of He⁺ + e⁻. In addition, there are states like He(2s²) embedded in the continuum. These states are auto-ionizing and decay to continuum states. There is a series of such auto-ionizing states, 2s2p, etc.

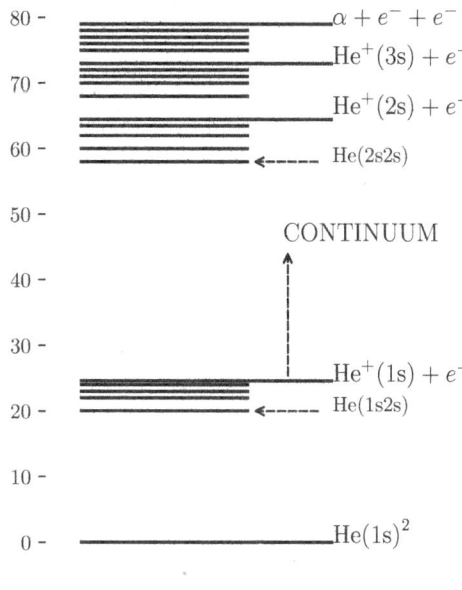

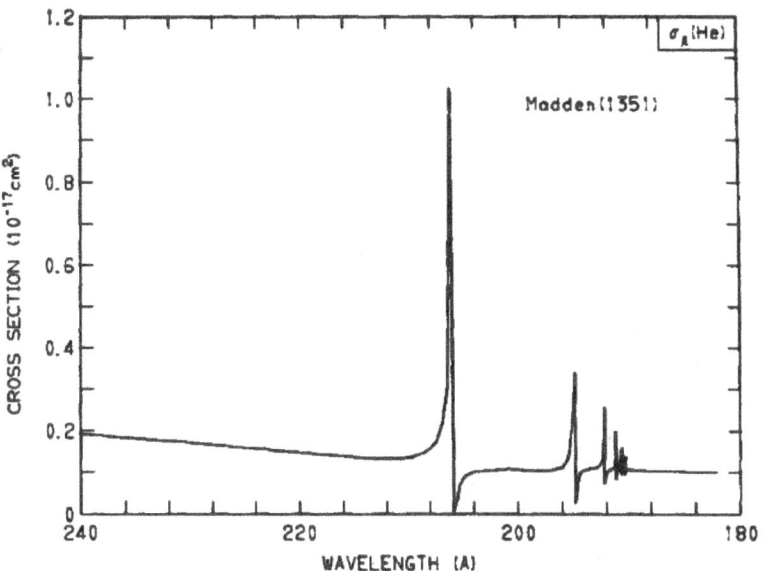

Figure 6.6 The photo-absorption cross section for helium taken from the compilation of Hudson and Kieffer, "Atomic Data and Nuclear Data Tables", **2**, 205 (1971). The prominent peaks correspond to various auto-ionizing states.

However, in the immediate neighborhood of one of the discrete (doubly-excited) states $|d\rangle$, there is another path to the final state with a second-order amplitude

$$\frac{\langle f(E)|V'|d\rangle\langle d|T|i\rangle}{E_d - E - i\Gamma/2}, \tag{6.38}$$

where

$$\Gamma \propto |\langle f|V'|d\rangle|^2 \tag{6.39}$$

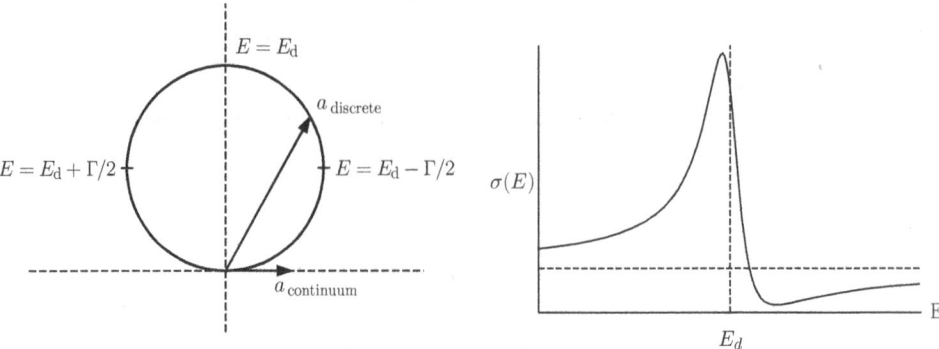

Figure 6.7 The two complex amplitudes, $a_{discrete}$ and $a_{continuum}$, for $\gamma + \text{He}(1s) \rightarrow \text{He}(1s)^+ + e^-$ near the threshold for excitation to a discrete auto-ionizing state. Because the continuum amplitude is taken to be real and positive, there is destructive interference after the peak.

is the width for decay of the discrete state to the continuum. The cross-section in the neighborhood of the discrete state is then

$$\sigma(E) \propto \left| \langle f(E) | T | i \rangle + \frac{\langle f(E) | V' | d \rangle \langle d | T | i \rangle}{E_d - E - i\Gamma/2} \right|^2. \tag{6.40}$$

The two amplitudes interfere. Near the discrete state, the first term is essentially constant. In a complex amplitude plane, we have the sum of two complex numbers, one of which, $a_{discrete}$, goes rapidly around a circle and interferes destructively with $a_{continuum}$. Because of the interference, the peak does not occur exactly at the resonance energy, E_d. In Figure 6.7, an example is shown with $a_{continuum}$ taken to be real and positive.

6.5 Periodic Table

A major triumph of quantum mechanics in the 1920s was the explanation in considerable detail of the periodic table of elements. The ingredients of the explanation are:

(a) the independent-electron central-field approximation for the motion of the electron around the nucleus
(b) the Pauli Principle (antisymmetric wave functions)

The central-field approximation assumes that the motion of any particular electron can be evaluated with a single-particle Schrödinger equation with a central potential $V(r)$ determined by the Coulomb interaction of the electron with the nucleus and the (averaged) charge cloud of the other electrons. If the nuclear charge is Ze and there are N electrons in the atom, then at short distances $V(r) \approx -Ze^2/r$, while at distances large compared to the orbits of the other electrons $V(r) \approx -e^2(Z - N + 1)/r$. Because the potential is not simply proportional to $1/r$ the accidental degeneracy of the hydrogen atom is absent. The single-particle energies depend on both ℓ and the number of nodes, $n_r = n - \ell - 1$, of the radial wave function. For fixed n, the smaller the value of ℓ, the lower the energy. This is displayed in Figure 6.8.

The Pauli principle limits the number of electrons in each (n, ℓ) group of states: $\ell = 0$ - two states, $\ell = 1$ - six states, $\ell = 2$ ten states. A glance at the periodic table shows how this

Figure 6.8 The Grotrian diagram for the central-field model of multi-electron atoms. The degeneracy of the hydrogen atom is lifted and the energy levels depend on ℓ as well as n. In particular, the 4s level is slightly below the 3d.

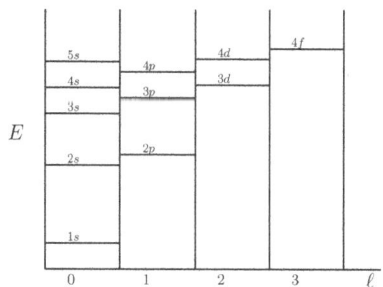

pattern is reflected. We saw in helium a very large difference between $(1s)^2$ and $(1s)(2s)$. This trend persists in low-Z elements, basically because $1/n^2$ differs a great deal for small n and the penetrating versus non-penetrating differences within a given n grouping are small. But eventually that effect becomes larger, at least for $\ell = 0$, than the $1/n^2$ difference.

The beginning of the periodic table is quite regular. The $1s$ shell is filled with H and He. The $2s$ shell is filled with Li and Be. The $2p$ follows with B, C, N, O, F, Ne. The $3s$ accounts for Na and Mg, followed by six more for the $3p$: Al, Si, P, S Cl, Ar. The a priori expectation would be to start filling the $3d$ shell. But instead, the penetrating orbit effect dominates for the $4s$. The next elements are K $((3p)^6(4s))$ and Ca $((3p)^6(4s)^2)$. The penetrating orbit effect lowers the $4s$ below the $3d$, but the $3d$ comes before the $4p$. This results in the ten elements of the first transition group from $Z = 21$ to $Z = 30$: Sc, Ti, V, Cr, Mn, Fe, Co, Ni, but Cu is $(4s)(3d)^{10}$ instead of $(4s)^2(3d)^9$, followed by Zn $(4s)^2(3d)^{10}$. Copper illustrates the preference for a closed shell as a particularly stable configuration.

The next eight elements ($Z = 31$ to $Z = 38$) fill the $4p$ and $5s$ shells. Then $Z = 39$ to $Z = 48$ are the second transition group, with the $4d$ shell being filled and some alternation between two electrons in the $5s$ and only one: e.g. Nb $((4d)^4(5s))$, Mo $((4d)^5(5s))$, Tc $((4d)^5(5s)^2)$, Ru $((4d)^7(5s))$, Rh $((4d)^8(5s))$, Pd $((4d)^{10})$, Ag $((4d)^{10}(5s))$, Cd$((4d)^{10}(5s)^2)$. The next group fills the $5p$ shell in an orderly fashion ($Z = 49 - 54$). After that, shells are filled as $[6s, 4f, 5d]$: $Z = 55 - 80$; $[6p]$:$Z = 81 - 86$; $[7s, 5f, 6d]$: $Z = 87 - 112$; $[7p]$: $Z = 113 - 118$.

Basic principles of chemistry follow, based on the nature of the last shell being filled. Examples are:

(a) Noble gases: Completed p shells -(Ne Ar, Kr, Xe, Rd).
(b) Alkalis: Next electron added after a completed p shell goes into an ns orbital - Na ($Z = 11$, $3s$), K ($Z = 19, 4s$), Rb ($Z = 37, 5s$), Cs ($Z = 55, 6s$), Fr (Z=87, $7s$).
(c) Halogens - one electron missing from a closed p shell: F($Z = 9, (2p)^{6-1}$), Cl ($Z = 17$, $(3p)^{6-1}$), Br ($Z = 35, (4p)^{6-1}$), I ($Z = 53, (5p)^{6-1}$), At ($Z = 85, (6p)^{6-1}$).
(d) Alkaline earths - closed $(ns)^2$ shells: Be and Mg are similar - next electron would be np. Ca, Sr, Ba are similar - next electron would be $(n - 1)d$. Zn, Cd, Hg are similar - closed $((n - 1)d)^{10}(ns)^2$ configurations.
(e) Noble metals - $((n - 1)d)^{10}(ns)$: Cu, Ag, Au.

6.6 Multiplet Structure, Russell-Saunders Coupling

The Pauli exclusion principle together with the independent-electron, central-field approximation gives an impressive explanation of the periodic table, and is thus a triumph of quantum mechanics. But quantum mechanics promised more than a qualitative description of nature: it claimed to be a complete description of atomic physics. While a full treatment of many-electron atoms was out of reach, an accessible challenge was a quantitative treatment of atomic spectra. This challenge was met in particular by the classic works by Eugene Wigner, *Group Theory and its Application to the Quantum Mechanics of Atomic Spectra* and Condon and Shortley's *The Theory of Atomic Spectra*. A major advance beyond these was made by Giulio Racah in a series of papers in The Physical Review in the 1940s.

In the independent-electron, central-field approximation, there are a certain number of states and a certain number of electrons outside closed shells. For example, in a configuration like $(3p)^3$, there are $2 \times (2\ell + 1)$ states for the $3p$ shell and there are three electrons to distribute among them. If the number of available states is N and the number of electrons is n, then the Pauli principle tells us that the actual number off allowed states is

$$D = \binom{N}{n}. \tag{6.41}$$

For example, for $(3p)^3$ there are six states among which we must chose three, so there are 20 possibilities. In the central-field, independent-electron approximation, all these states are degenerate. Some of the degeneracy is from rotational invariance of the isolated atom, but other degeneracy is the result of neglecting interactions. We already saw in helium with two electrons how the electron-electron Coulomb interaction entered and cause removal of the degeneracy. In order of decreasing importance, the neglected interactions are:

(1) Residual inter-electron interactions not included in the central-field approximation. Most important are the exchange contributions. See Problem 6.6 for a helium example.
(2) Spin-orbit interactions.
(3) Relativistic effects (velocity-dependent) and spin-spin coupling.
(4) Hyperfine interactions - electrons interacting with the nuclear electromagnetic multipole moments.
(5) Electron spin-spin interactions.

The relative significance of these effects is readily determined. The residual Coulomb interaction is of the form $e^2 / |\mathbf{r}_i - \mathbf{r}_j|$ and thus has the scale $e^2 / a_0 \sim$ Ryd. The spin-orbit interaction can be estimated

$$\sim \frac{\hbar^2}{m^2 c^2} \frac{1}{r} \frac{e^2}{r^2} \sim \frac{\hbar^2}{m^2 c^2} \frac{e^2}{a_0^3} \sim \frac{e^2}{a_0} \left(\frac{e^2}{\hbar c} \right)^2 \sim \alpha^2 \text{Ryd}. \tag{6.42}$$

The relativistic correction to the energy must be of order $(v/c)^2$ Ryd, thus of the same order as the spin-orbit corrections. However, the relativistic correction is determined entirely by the velocities of the individual electrons and is not, at lowest order, affected by how the one-particle states are joined to make the final eigenstates.

The hyperfine interaction of electrons with the magnetic moment of the nucleus is of the form

$$\frac{\mu_N \mu_e}{r^3} \sim \frac{e\hbar}{2m_c} \frac{e\hbar}{2m_N c} \frac{1}{a_0^3} \sim \alpha^2 \frac{m_e}{m_N} \text{ Ryd}, \tag{6.43}$$

thus three orders of magnitude smaller than the fine structure. See Problem 5.3.

In elements of small and modest Z, the fine-structure is generally sufficiently small that it can be ignored in the atomic dynamics and then treated in first-order perturbation theory.

The only interaction beyond the independent-particle central-field is the spin-independent Coulomb interaction. This leads to a natural description of the states of the electrons outside the closed shells in terms of total orbital angular momentum $\mathbf{L}$ and total spin-angular momentum $\mathbf{S}$. This is called Russell-Saunders or L-S coupling (the latter name having to do with the addition of $\mathbf{L}$ and $\mathbf{S}$ to get $\mathbf{J}$ for the fine-structure multiplets).

While the spin-orbit interaction is much less important than the electrostatic electron-electron interaction in hydrogen, that isn't always the case in very heavy atoms. If the spin-orbit interaction

$$H_{\text{so}} = \sum_i \frac{\hbar^2}{2m^2c^2} \mathbf{s}_i \cdot \boldsymbol{\ell}_i \frac{1}{r} \frac{dV}{dr} \tag{6.44}$$

is more important than the electrostatic, it is appropriate to take as a basis states that are made from electron eigenstates $|jlsm\rangle$, since is each is an eigenstate of $\mathbf{s}_i \cdot \boldsymbol{\ell}_i$. The various $\mathbf{j}_i$ can be combined to make $\mathbf{J} = \sum_i \mathbf{j}_i$. This is the j-j coupling scheme. In this basis, the electrostatic interaction is then computed perturbatively. We will examine the middle ground between these extremes below.

For Russell-Saunders coupling with n outer shell electrons with orbital and spin angular momenta $\boldsymbol{\ell}_i$ and $\mathbf{s}_i$, we define

$$\mathbf{L} = \sum_i \boldsymbol{\ell}_i, \qquad \mathbf{S} = \sum_i \mathbf{s}_i. \tag{6.45}$$

We then consider states of definite L^2, M_L and S^2, M_S. Because the electrostatic electron-electron interactions do not involve the spin, one might conclude that the states with different values of S would be degenerate. That would be a false conclusion, as we already saw for the singlet and triplet states of helium. Atomic energies depend on S as a consequence of the Pauli principle and the resulting different spatial symmetry character of the wave function depending on S.

6.6.1 Electrostatic Interaction for Two Electrons Generally

The total number of states for non-equivalent electrons is $D_{\text{non-eq}} = 2(2\ell_1 + 1) \times 2(2\ell_2 + 1) = 4(2\ell_1 + 1)(2\ell_2 + 1)$. For equivalent electrons ($n_1 = n_2, \ell_1 = \ell_2$), the total number of states is

$$D_{eq} = \binom{2(2\ell_1 + 1)}{2} = \frac{[2(2\ell_1 + 1)]!}{[2(2\ell_1 + 1) - 2]! 2!} = (2\ell_1 + 1)(4\ell_1 + 1). \tag{6.46}$$

Thus the number of possible states for equivalent electrons is less than one-half the number available for non-equivalent electrons. See Table 6.2.

We illustrate the influence of the Pauli principle by considering the shifts in energy levels caused by the electrostatic repulsion between two electrons, treated in first order perturbation theory. Suppose the one-electron spatial states are

$$\psi_a(1) = R_a(r_1) Y_{\ell_1 m_1}(\hat{\mathbf{x}}_1)$$
$$\psi_b(2) = R_b(r_2) Y_{\ell_2 m_2}(\hat{\mathbf{x}}_2). \tag{6.47}$$

Table 6.2 Comparison of the number of states available for two electrons with the same ℓ for equivalent and non-equivalent cases.

ℓ	$D_{\text{non-eq}}$	D_{eq}	$D_{\text{eq}}/D_{\text{non-eq}}$
0	4	1	0.250
1	36	15	0.420
2	100	45	0.450
3	196	91	0.464
4	324	153	0.472

We build up total L eigenfunctions that are explicitly even and odd under interchange as

$$\Phi_{LM}^{\ell_1,\ell_2,\pm}(\mathbf{x}_1,\mathbf{x}_2) = \frac{1}{\sqrt{2}}\sum_{m_1 m_2} \langle \ell_1 m_1 \ell_2 m_2 | L M_L\rangle \left[R_a(r_1) Y_{\ell_1 m_1}(\hat{\mathbf{x}}_1) R_b(r_2) Y_{\ell_2 m_2}(\hat{\mathbf{x}}_2) \right.$$

$$\left. \pm R_a(r_2) Y_{\ell_1 m_1}(\hat{\mathbf{x}}_2) R_b(r_1) Y_{\ell_2 m_2}(\hat{\mathbf{x}}_1) \right]. \quad (6.48)$$

To these we can attach the spin-singlet or spin-triplet state made from the two electron spins:

$$X_{00} = \frac{1}{\sqrt{2}}(\alpha_1\beta_2 - \beta_1\alpha_2); \quad X_{11} = \alpha_1\alpha_2, \text{ etc.} \quad (6.49)$$

The spin-singlet, odd under interchange, accompanies Φ^+ and the spin-triplet is paired to Φ^-.
If we introduce

$$\mathcal{Y}_{LM}(\hat{\mathbf{x}}_1,\hat{\mathbf{x}}_2) = \sum_{m_1 m_2} \langle \ell_1 m_1 \ell_2 m_2 | L M_L\rangle\, Y_{\ell_1 m_1}(\hat{\mathbf{x}}_1) Y_{\ell_2 m_2}(\hat{\mathbf{x}}_2), \quad (6.50)$$

we can write the spatial wave functions as

$$\Phi_{LM}^{\ell_1,\ell_2,\pm}(\mathbf{x}_1,\mathbf{x}_2) = \frac{1}{\sqrt{2}}\left[R_a(r_1)R_b(r_2)\mathcal{Y}_{LM}(\hat{\mathbf{x}}_1,\hat{\mathbf{x}}_2) \pm R_a(r_2)R_b(r_1)\mathcal{Y}_{LM}(\hat{\mathbf{x}}_2,\hat{\mathbf{x}}_1) \right]. \quad (6.51)$$

If we introduce symmetric and antisymmetric combinations of the radial and angular functions

$$\mathcal{Y}_{LM}^{\pm}(\hat{\mathbf{x}}_1,\hat{\mathbf{x}}_2) = \frac{1}{\sqrt{2}}\left[\mathcal{Y}_{LM}(\hat{\mathbf{x}}_1,\hat{\mathbf{x}}_2) \pm \mathcal{Y}_{LM}(\hat{\mathbf{x}}_2,\hat{\mathbf{x}}_1) \right]$$

$$R^{\pm}(r_1,r_2) = \frac{1}{\sqrt{2}}\left[R_a(r_1)R_b(x_2) \pm R_a(r_2)R_b(\dot{x}_1) \right], \quad (6.52)$$

we can write the spatial function even under $1 \leftrightarrow 2$ interchange as

$$\frac{1}{\sqrt{2}}\left[R^+\mathcal{Y}^+ + R^-\mathcal{Y}^- \right] \quad (6.53)$$

and the odd function as

$$\frac{1}{\sqrt{2}}\left[R^+\mathcal{Y}^- + R^-\mathcal{Y}^+ \right]. \quad (6.54)$$

We obtain states satisfying the Pauli principle by combining the interchange-odd spin singlet with the even spatial function and the interchange-even spin triplet with the odd spatial function.

$$\frac{1}{\sqrt{2}}\left[R^+\mathcal{Y}^+ + R^-\mathcal{Y}^-\right]X_{00}, \qquad \frac{1}{\sqrt{2}}\left[R^+\mathcal{Y}^- + R^-\mathcal{Y}^+\right]X_{1M_S}. \tag{6.55}$$

If the two electrons are equivalent, that is, they have the same values of n and ℓ, R^- vanishes. In addition, if $\ell_1 = \ell_2$, we see that

$$\mathcal{Y}_{LM}(\hat{\mathbf{x}}_2, \hat{\mathbf{x}}_1) = (-1)^L \mathcal{Y}_{LM}(\hat{\mathbf{x}}_1, \hat{\mathbf{x}}_2) \tag{6.56}$$

as a consequence of the symmetry of the Clebsch-Gordan coefficients

$$\langle \ell_2 m_2 \ell_1 m_1 | LM \rangle = (-1)^{\ell_1 + \ell_2 + L} \langle \ell_1 m_1 \ell_2 m_2 | LM \rangle. \tag{6.57}$$

As a result, if $\ell_1 = \ell_2$, as it does for the case of equivalent electrons, $\mathcal{Y}^-$ vanishes for even L and $\mathcal{Y}^+$ vanishes for odd L. If $\ell_1 = \ell_2$ but the electrons are inequivalent, for example $2p$ and $3p$, the surviving wave functions are

$$R^-\mathcal{Y}^+X_{1M_S} \quad (L \text{ even}); \qquad R^+\mathcal{Y}^+X_{00} \quad (L \text{ odd}). \tag{6.58}$$

Thus, every possible combination of L and S is permitted. However, if the electrons are equivalent, R^- vanishes and L-even is allowed only for the spin singlet and L-odd requires the spin triplet. For equivalent p electrons, this permits only $^1S_0, ^1D_2$ and ^3P_J, where $J = 0, 1, 2$.

Returning to the general case, because the interaction e^2/r_{12} is rotationally invariant and does not involve the electron spin, we need only calculate the first-order effect for diagonal elements with well-defined L, M_L, S and M_S.

$$\int d^3x_1 \, d^3x_2 \, \Phi_{LM}^{\ell_1\ell_2\pm}(\mathbf{x}_1, \mathbf{x}_2) \frac{1}{r_{12}} \Phi_{LM}^{\ell_1\ell_2\pm*}(\mathbf{x}_1, \mathbf{x}_2). \tag{6.59}$$

We expand $1/r_{12}$ as

$$\frac{1}{r_{12}} = \sum_{k=0}^{\infty} \frac{r_<^k}{r_>^{k+1}} P_k(\cos\theta_{12}) = \sum_{k=0}^{\infty} \frac{4\pi}{2k+1} \frac{r_<^k}{r_>^{k+1}} \sum_q Y_{kq}(1) Y_{kq}^*(2). \tag{6.60}$$

The angular integrals are of the form (see Eq. 4.177)

$$\int d\Omega \, Y_{\ell_1' m_1'}^*(\Omega) Y_{\ell_1 m_1}(\Omega) Y_{kq}(\Omega) = (-1)^{m_1'} \int d\Omega Y_{\ell_1' -m_1'}(\Omega) Y_{\ell_1 m_1}(\Omega) Y_{kq}(\Omega)$$

$$= (-1)^{m_1'} \sqrt{\frac{(2\ell_1'+1)(2\ell_1+1)(2k+1)}{4\pi}} \begin{pmatrix} \ell_1' & \ell_1 & k \\ 0 & 0 & 0 \end{pmatrix} \begin{pmatrix} \ell_1' & \ell_1 & k \\ -m_1' & m_1 & q \end{pmatrix}$$

$$\equiv \sqrt{\frac{2k+1}{4\pi}} c^k(\ell_1', m_1'; \ell_1, m_1) \quad \text{if } q = m_1' - m_1; \text{ else } = 0. \tag{6.61}$$

Defining the quantity $c^k(\ell', m'; \ell, m)$ is traditional in this setting; it is given in the 1935 classic *Theory of Atomic Spectra* by Condon and Shortley. Note that

$$c^k(\ell', m'; \ell, m) = (-1)^{m-m'} c^k(\ell, m; \ell' m'). \tag{6.62}$$

Next, consider

$$I = \left\langle n_1\ell_1 m_1; n_2\ell_2 m_2 \left| \frac{e^2}{r_{12}} \right| n_3\ell_3 m_3; n_4\ell_4 m_4 \right\rangle, \tag{6.63}$$

where electron 1 is associated with quantum numbers n_1, ℓ_1 and n_3, ℓ_3, while electron 2 is associated with n_2, ℓ_2 and n_4, ℓ_4.

The matrix element is

$$I = \int d^3x_1\, d^3x_2\, R_1(r_1)R_2(r_2)R_3(r_1)R_4(r_2)$$

$$\times e^2 \sum_{k,q} \frac{4\pi}{2k+1} \frac{r_<^k}{r_>^{k+1}} Y_{kq}(1)Y_{kq}^*(2)Y_{\ell_1 m_1}^*(1)Y_{\ell_3 m_3}(1)Y_{\ell_2 m_2}^*(2)Y_{\ell_4 m_4}(2). \tag{6.64}$$

Let us define the radial integral

$$R_{(12,34)}^k = e^2 \int r_1^2 dr_1\, r_2^2 dr_2\, R_1(r_1)R_2(r_2)R_3(r_1)R_4(r_2)\frac{r_<^k}{r_>^{k+1}}. \tag{6.65}$$

With this notation we have

$$I = \sum_k R_{(12,34)}^k c^k(\ell_1 m_1, \ell_3 m_3) c^k(\ell_4 m_4, \ell_2 m_2)\delta_{m_1+m_2,m_3+m_4}. \tag{6.66}$$

We need

$$\left\langle a\,\ell_1 m_1'; b\,\ell_2 m_2' \left| \frac{e^2}{r_{12}} \right| a\,\ell_1 m_1; b\,\ell_2 m_2 \right\rangle$$

$$\pm \left\langle a\,\ell_1 m_1'; b\,\ell_2 m_2' \left| \frac{e^2}{r_{12}} \right| b\,\ell_2 m_2; a\,\ell_1 m_1 \right\rangle \tag{6.67}$$

$$= \sum_k \left[R^k(ab;ab)c^k(\ell_1 m_1', \ell_1 m_1)c^k(\ell_2 m_2\ell_2 m_2') \right.$$

$$\left. \pm R^k(aa;bb)c^k(\ell_1 m_1', \ell_2 m_2)c^k(\ell_1 m_1\ell_2 m_2') \right]. \tag{6.68}$$

In the end, we need the matrix elements for the states of well-defined L and M for a specific power k:

$$\left\langle LM; \ell_1\ell_2 \left| \frac{e^2}{r_{12}} \right| LM; \ell_1\ell_2 \right\rangle = \sum_k \sum_{\substack{m_1',m_2'\\m_1,m_2}} \langle \ell_1 m_1 \ell_2 m_2|LM\rangle \langle \ell_1 m_1'\ell_2 m_2'|LM\rangle$$

$$\times \sum_k \left[R^k(ab;ab)c^k(\ell_1 m_1', \ell_1 m_1)c^k(\ell_2 m_2\ell_2 m_2') \right.$$

$$\left. \pm R^k(aa;bb)c^k(\ell_1 m_1', \ell_2 m_2)c^k(\ell_1 m_1\ell_2 m_2') \right]. \tag{6.69}$$

The sum over the magnetic quantum numbers is clearly independent of M. It is a function of ℓ_1, ℓ_2, L, and k alone and we may well expect that it can be represented in a more succinct fashion. The answer in a more general form is

$$J(\ell_1, \ell_2, \ell_1', \ell_2', L, k)$$

$$= \frac{1}{2L+1} \sum_{\substack{m_1' m_2' \\ m_1 m_2, M_L}} \langle \ell_1' m_1' \ell_2' m_2' | LM_L \rangle \langle \ell_1 m_1 \ell_2 m_2 | LM_L \rangle c^k(\ell_1' m_1' \ell_1 m_1) c^k(\ell_2 m_2 \ell_2' m_2')$$

$$= (-1)^{L+k} \sqrt{(2\ell_1 + 1)(2\ell_2 + 1)(2\ell_1' + 1)(2\ell_2' + 1)}$$

$$\times \begin{pmatrix} \ell_1' & \ell_1 & k \\ 0 & 0 & 0 \end{pmatrix} \begin{pmatrix} \ell_2 & \ell_2' & k \\ 0 & 0 & 0 \end{pmatrix} \begin{Bmatrix} \ell_1 & \ell_2 & L \\ \ell_2' & \ell_1' & k \end{Bmatrix}, \quad (6.70)$$

where the last factor is the Wigner 6-j symbol. See Appendix B.3 and Fano and Racah, *Irreducible Tensorial Sets*, p. 92., Eq. (16.17).

It is traditional to define

$$F^k(a,b) = \int_0^\infty r_1^2 dr_1 \int_0^\infty r_2^2 dr_2 \frac{r_<^k}{r_>^{k+1}} R_a(r_1)^2 R_b(r_2)^2$$

$$G^k(a,b) = \int_0^\infty r_1^2 dr_1 \int_0^\infty r_2^2 dr_2 \frac{r_<^k}{r_>^{k+1}} R_a(r_1) R_b(r_1) R_a(r_2) R_b(r_2) \quad (6.71)$$

in terms of which

$$\left\langle \frac{e^2}{r_{12}} \right\rangle = e^2 \sum_k \left[J(\ell_1, \ell_2, \ell_1, \ell_2, L, k) F^k(a,b) \pm (-1)^{\ell_1 + \ell_2 + L} \right.$$

$$\left. J(\ell_2, \ell_1, \ell_1, \ell_2, L, k) G^k(a,b) \right], \quad (6.72)$$

where the minus sign corresponds to an odd spatial function and the plus to an even spatial function. The coefficient of the second term in Eq. (6.72) has a factor of $(-1)^{\ell_1 + \ell_2 + L}$ relative to the first term from the opposite ordering of the two Clebsch-Gordan coefficients. Recall that $e^2 F$ and $e^2 G$ are energies on the scale of e^2/a_0.

The coefficient of F^0 is always unity, since when $k = 0$ we have

$$\begin{Bmatrix} \ell_1 & \ell_2 & L \\ \ell_2 & \ell_1 & 0 \end{Bmatrix} = \frac{(-1)^{\ell_1 + \ell_2 + L}}{\sqrt{(2\ell_1 + 1)(2\ell_2 + 1)}} \quad (6.73)$$

and $I(\ell_1, \ell_2, \ell_1, \ell_2, L, 0) = 1$.

If the two electrons have the same value of ℓ but are not necessarily equivalent (possibly differing values of n), we have

$$\left\langle \frac{e^2}{r_{12}} \right\rangle = e^2 \sum_k I(\ell, \ell, \ell, \ell, L, k) \left[F^k(a,b) \pm (-1)^L G^k(a,b) \right]. \quad (6.74)$$

For the case of two equivalent electrons with the same n and ℓ, we have $F(a,a) = G(a,a)$ and

$$\langle e^2/r_{12} \rangle = \sum_k I(\ell, \ell, \ell, \ell, L, k) [1 \pm (-1)^L] F^k(a,a). \quad (6.75)$$

The symmetric spatial function - the upper sign - requires the spin-singlet and we recover the result that the singlet goes with even values of L. Analogously, the spin-triplet demands odd L. The results for two equivalent p or d electrons are shown in Table 6.3.

Table 6.3 Theoretical splittings between terms for two equivalent electrons arising from their electrostatic interaction. The results from Eq. (6.74) are in agreement with the table in Bethe and Jackiw, *Intermediate Quantum Mechanics*, 3rd edition. There, the calculation is done using Slater determinants and sum rules derived from the invariance of the trace of a matrix under orthogonal transformations. Adapted from Bethe and Jackiw 2005 / Sarat Book House.

p^2		d^2		
^1S :	$\frac{2}{5}F^2$	^1S :	$\frac{2}{7}F^2$	$+\frac{2}{7}F^4$
^3P :	$-\frac{1}{5}F^2$	^3P :	$\frac{1}{7}F^2$	$-\frac{4}{21}F^4$
^1D :	$+\frac{1}{25}F^2$	^1D :	$-\frac{3}{49}F^2$	$+\frac{4}{49}F^4$
		^3F :	$-\frac{8}{49}F^2$	$-\frac{1}{49}F^4$
		^1G :	$\frac{4}{49}F^2$	$+\frac{1}{441}F^4$

As an example of inequivalent electrons, consider one s electron and one p electron. This necessarily has $L = 1$, so the terms are ^3P and ^1P. The triplet requires the spatially-odd wave function and the spin-singlet the spatially-even wave function. Inspection of Eqs. (6.70) and (6.72) shows that only $F^0 = 1$ and G^1 can contribute. The coefficient of G^1 is

$$\sqrt{3 \cdot 3} \begin{pmatrix} 1 & 0 & 1 \\ 0 & 0 & 0 \end{pmatrix} \begin{pmatrix} 1 & 0 & 1 \\ 0 & 0 & 0 \end{pmatrix} \begin{Bmatrix} 0 & 1 & 1 \\ 0 & 1 & 1 \end{Bmatrix} = \frac{1}{3}, \tag{6.76}$$

where we used Eq. (6.73) and the columns of a 6-j symbol can be freely permuted. The energy caused by the electrostatic interaction is thus

$$E(^3P) = F^0(ns, n'p) - \frac{1}{3}G^1(ns, n'p)$$

$$E(^1P) = F^0(ns, n'p) + \frac{1}{3}G^1(ns, n'p). \tag{6.77}$$

The F^k integrals are manifestly positive, though that is not the case for the G^k. However, G^k is found to be positive when calculated for specific cases (c.f. Condon and Shortley, *The Theory of Atomic Spectra*, p. 177). As a result, the ^3P state lies below the ^1P. This is to be expected since the anti-symmetric wave function in the spin-triplet tends to keep the two electrons away from each other, reducing the electrostatic energy. This is in accord, of course, with Hund's rule.

This program can be extended to the more complex case where there are more than two electrons outside the filled shells. There and here, the electrostatic electron-electron interaction leaves us with eigenstates of L^2 and S^2, but there is still degeneracy among the allowed J values for each L-S combination. That is lifted by the spin-orbit interaction.

6.7 Spin-Orbit Interaction

In Chapter 5, we calculated the effect of the spin-orbit interaction for hydrogenic atoms. It remains to determine how this interaction is manifested in multi-electron atoms. The starting point is given by the states of well-defined L^2 and S^2, including the electron-electron inter-action, as calculated above. That leaves degeneracy between the states with differing total angular momenta J constructed from the same L and S. The effective spin-orbit interaction is

$$H_{so} = \frac{\hbar^2}{2m^2c^2} \sum_i \left(\frac{1}{r}\frac{dV}{dr} \right)_i \boldsymbol{\ell}_i \cdot \mathbf{s}_i,$$ (6.78)

where $V(r)$ is the central potential.

Consider an atom with a number of filled shells, plus N electrons in an outer, unfilled shell. The Slater determinantal wave functions have permutations of the ket $|n_i \ell_i m_{\ell_i} m_{s_i}\rangle$, where the single set of quantum numbers stands for those of all the electrons. The matrix elements of H_{so} with respect to these states consists of two terms: $H_{so} = (H_{so})_{closed} + (H_{so})_{open}$.

The contribution from the closed shells is

$$\langle (H_{so})_{closed} \rangle = \frac{\hbar^2}{2m^2c^2} \sum_{i,closed} \left(\frac{1}{r}\frac{dV}{dr} \right)_i m_{\ell_i} m_{s_i}.$$ (6.79)

For each shell, $\frac{1}{r}\frac{dV}{dr}$ has the same value for each of its electrons, so the contribution from a single filled shell is proportional to $\sum_i m_{\ell_i} m_{s_i}$, where the sum is over the filled orbitals. But this sum clearly vanishes, since for every ℓ_i, m_{s_i} takes on both values $\pm 1/2$.

We restrict our consideration to N outer electrons falling in a single shell. Thus $\frac{1}{r}\frac{dV}{dr}$ has the same value for each electron and each electron has the same radial wave function, $R_{n\ell}(r)$. Although the spin-orbit interaction, by coupling $\mathbf{L}$ and $\mathbf{S}$, breaks the separate symmetries they generate, this perturbation is small compared to the interactions that do respect these symmetries and its effect can be reliably calculated from first-order perturbation theory. Thus, we need to evaluate the expectation value of the interaction for the state representing the unfilled shell. If we write the states as sums of Slater determinants like

$$\Psi(1, 2, \ldots N) = R(r_1)R(r_2) \cdots R(r_N) \begin{vmatrix} Y_{\ell m_1}(1)\chi_1 & & \cdot & & \cdot \\ \cdot & & Y_{\ell m_2}(2)\chi_2 & & \cdot & \cdot \\ \cdot & & & & \end{vmatrix},$$ (6.80)

we compute

$$\int \prod_i d^3 r_i \Psi(1, 2, \ldots N)^* H_{so} \Psi(1, 2, \ldots N) = (H_{so})_{eff} = \frac{\hbar^2}{2m^2c^2} \xi_{n\ell} \left\langle \sum_i \boldsymbol{\ell}_i \cdot \mathbf{s}_i \right\rangle$$ (6.81)

where

$$\xi_{n\ell} = \int_0^\infty dr\, r^2 |R_{n\ell}(r)|^2 \frac{1}{r}\frac{dV}{dr}.$$ (6.82)

The expectation of $\sum_i \boldsymbol{\ell}_i \cdot \mathbf{s}_i$ is to be evaluated over the angular and spin variables. Note that since $V(r)$ is monotonically increasing, $\frac{1}{r}\frac{dV}{dr}$ is non-negative and $\xi_{n\ell}$ is positive.

We want to determine the perturbation on well-defined states, that is, states with given values of J, L, and S. This is most efficiently carried out using the Wigner-Eckart theorem. Consider the diagonal element $\langle LSJM | \boldsymbol{\ell}_i \cdot \mathbf{s}_i | LSJM \rangle$. We expand in the $|LM_L SM_S\rangle$ basis and write $\boldsymbol{\ell}_i \cdot \mathbf{s}_i = \sum_q (-1)^q (\ell_i)_q (s_i)_{-q}$. In this way, we arrive at

$$\langle LSJM \left| \boldsymbol{\ell}_i \cdot \mathbf{s}_i \right| LSJM \rangle = \sum_{\substack{M'_L M'_S \\ M_L M_S}} \langle JM | LM'_L SM'_S \rangle \langle JM | LM_L SM_S \rangle$$

$$\times \sum_q (-1)^q \langle LM'_L \left| (\ell_i)_q \right| LM_L \rangle \langle SM'_S \left| (s_i)_{-q} \right| SM_S \rangle. \quad (6.83)$$

But now we apply the Wigner-Eckart theorem to the orbital and spin matrix elements separately. We use the relation proved in Problem 4.8, applied to S and L:

$$\langle LM'_L | LM_L 1q \rangle = \frac{\langle LM'_L \left| L_q \right| LM_L \rangle}{\sqrt{L(L+1)}}; \quad \langle SM'_S | LM_S 1q \rangle = \frac{\langle SM'_S \left| S_q \right| SM_S \rangle}{\sqrt{S(S+1)}}. \quad (6.84)$$

This enables us to relate the matrix elements of $\boldsymbol{\ell}$ and $\mathbf{s}$ to those of $\mathbf{L}$ and $\mathbf{S}$:

$$\langle LM'_L \left| (\ell_i)_q \right| LM_L \rangle = \langle LM'_L | LM_L 1q \rangle \langle L \left\|\ell_i\right\| L \rangle = \frac{\langle LM'_L \left| L_q \right| LM_L \rangle}{\sqrt{L(L+1)}} \langle L \left\|\ell_i\right\| L \rangle$$

$$\langle SM'_S \left| (s_i)_{-q} \right| SM_S \rangle = \langle SM'_S | LM_S 1q \rangle \langle S \left\|s_i\right\| S \rangle = \frac{\langle SM'_S \left| S_{-q} \right| SM_S \rangle}{\sqrt{S(S+1)}} \langle S \left\|s_i\right\| S \rangle. \quad (6.85)$$

We can now sew things back together using $\sum_q (-1)^q L_q S_q = \mathbf{L} \cdot \mathbf{S}$ and resum to get

$$\langle LSJM \left| \boldsymbol{\ell}_i \cdot \mathbf{s}_i \right| LSJM \rangle = \frac{\langle L \left\|\ell_i\right\| L \rangle}{\sqrt{L(L+1)}} \frac{\langle S \left\|s_i\right\| S \rangle}{\sqrt{S(S+1)}} \mathbf{L} \cdot \mathbf{S}, \quad (6.86)$$

where $\mathbf{L} \cdot \mathbf{S}$ is diagonal in the $|LSJM\rangle$ basis.

It is traditional to write

$$\langle \alpha LSJM \left| (H_{\text{so}})_{\text{eff}} \right| \alpha LSJM \rangle = \zeta(\alpha LS) \langle LSJM \left| \mathbf{L} \cdot \mathbf{S} \right| LSJM \rangle$$

$$= \zeta(\alpha LS) \frac{J(J+1) - L(L+1) - S(S+1)}{2} \quad (6.87)$$

so that the spin-orbit splitting is given by $\zeta(\alpha LS)$ times the factor depending on J.

An immediate consequence is that the levels are not equally spaced, but rather the spacing between the level J and the level $J - 1$ is proportional to J, the so-called Landé interval rule (A. Landé, Zeit. f. Phys., **15**, 189; **19**, 112 (1923)). The actual size of the splittings is determined by ζ_α, where α stands for (n, ℓ).

$$\zeta(\alpha LS) = \frac{\hbar^2}{2m^2c^2} \xi_\alpha \frac{\sum_i \langle L \left\|\ell_i\right\| L \rangle \langle S \left\|s_i\right\| S \rangle}{\sqrt{L(L+1)}\sqrt{S(S+1)}}. \quad (6.88)$$

If there is a single unpaired electron, the result for angular-spin factor is immediate because $\boldsymbol{\ell}_1 = \mathbf{L}, \mathbf{s}_1 = \mathbf{S}$. For a single electron in a p-shell, we have

$$\zeta_0 = \frac{\hbar^2}{2m^2c^2} \xi_{2p}. \quad (6.89)$$

Consider, the next simplest case: $2p^2$. Because the two electrons are identical, the Pauli principle restricts the possibilities to be products of spatial-even spin-odd or spatial-odd spin-even combinations. The spin-even case has $S = 1$ and $L = 1$: $^3P_{J=0,1,2}$. The spin-odd combinations are $^1D_2, {}^1S_0$. The latter two cannot have any spin-orbit coupling: J is unique and $\mathbf{L} \cdot \mathbf{S} = 0$. We choose $M_L = 1, M_S = 1$ for evaluating Eq. (6.81). The angular-spin wave function is a single Slater determinant:

$$\Psi(1,2) = \frac{1}{\sqrt{2}} \begin{vmatrix} Y_{11}(1)\alpha_1 & Y_{11}(2)\alpha_2 \\ Y_{10}(1)\alpha_1 & Y_{10}(2)\alpha_2 \end{vmatrix} = \frac{1}{\sqrt{2}} (Y_{11}(1)Y_{10}(2) - Y_{10}(1)Y_{11}(2))\,\chi_{1,1}(1,2). \quad (6.90)$$

In evaluating the expectation value of $\boldsymbol{\ell}_i \cdot \mathbf{s}_i$ with this state, only s_{i_z} contributes because this is an eigenstate of s_{i_z}. We thus find

$$\langle \Psi(1,2) | \boldsymbol{\ell}_1 \cdot \mathbf{s}_1 | \Psi(1,2) \rangle$$

$$= \frac{1}{2} \times \frac{1}{2} \langle Y_{11}(1)Y_{10}(2) - Y_{10}(1)Y_{11}(2) | \ell_{1z} | Y_{11}(1)Y_{10}(2) - Y_{10}(1)Y_{11}(2) \rangle$$

$$= \frac{1}{4} \quad (6.91)$$

with the identical result for $1 \to 2$.

Indicating by ξ_{2p} the value of Eq. (6.82) for the 2p state, we have

$$\left\langle \Psi(1,2) \left| \frac{\hbar^2}{2m^2c^2} \xi_{2p}(\boldsymbol{\ell}_1 \cdot \mathbf{s}_1 + \boldsymbol{\ell}_2 \cdot \mathbf{s}_2) \right| \Psi(1,2) \right\rangle = \frac{1}{2} \left(\frac{\hbar^2}{2m^2c^2} \right) \xi_{2p}. \quad (6.92)$$

This is the result for $J = 2, L = 1, S = 1$ for which $\mathbf{L} \cdot \mathbf{S} = 1$.

Thus we have

$$\zeta(^3P) = \frac{1}{4} \left(\frac{\hbar^2}{m^2c^2} \right) \xi_{2p} = \frac{1}{2}\zeta_0. \quad (6.93)$$

The splittings for $J = 0, 1$ follow from appending the corresponding values of $\mathbf{L} \cdot \mathbf{S}$.

Consider the more complex three-electron example, $(2p)^3$. The possible terms are $^4S, ^2D, ^2P$. The $L = 0$ state cannot have a spin-orbit effect. Let us analyze the state with $M_L = 2$ and $M_S = 1/2$. Clearly, this can only be $^2D_{5/2}$. We can construct it as a Slater determinant

$$\psi(1,2,3) = \frac{1}{\sqrt{6}} \begin{vmatrix} Y_{11}(1)\alpha_1 & \cdot & \cdot \\ \cdot & Y_{10}(2)\alpha_2 & \cdot \\ \cdot & \cdot & Y_{11}(3)\beta_3 \end{vmatrix}. \quad (6.94)$$

The other states of $^4S, ^2D, ^2P$ with various values of M_L and M_S may be made of more than one Slater determinant. With these as basis states, we know the spin-orbit interaction is diagonal and proportional to $\mathbf{L} \cdot \mathbf{S}$ and also to $\sum_i \boldsymbol{\ell}_i \cdot \mathbf{s}_i$. There are 20 Slater determinants for this system. With these as basis, the spin-orbit interaction is not diagonal. It is diagonal in the basis $|JLSJ_z\rangle$. We can calculate directly, indicating $Y_{1m}\alpha = m^+, Y_{1m}\beta = m^-$.

$$\langle \psi(1,2,3) | \ell_{1z} \cdot s_{1z} | \psi(1,2,3) \rangle$$

$$= \frac{1}{12} \langle (1^+, 0^+, 1^-) + (1^-, 1^+, 0^+) + (0^+, 1^-, 1^+)$$

$$- (0^+, 1^+, 1^-) - (1^-, 0^+, 1^+) - (1^+, 0^+, 1^-) |$$

$$(1^+, 0^+, 1^-) - (1^-, 1^+, 0^+) + 0(0^+, 1^-, 1^+) - 0(0^+, 1^+, 1^-)$$

$$+ (1^-, 0^+, 1^+) - (1^+, 0^+, 1^-) \rangle = \frac{1}{12}[1 - 1 + 0 + 0 - 1 + 1] = 0. \quad (6.95)$$

Table 6.4 Spin-orbit splittings for nitrogen, oxygen, and fluorine ions in cm^{-1} with the configuration $(2p)^3$. The calculation gives zero. The observed splittings are much smaller than those for $(2p)^2$. Data taken from https://www.pa.uky.edu/~peter/atomic. See van Hoof, P. A. M., 2018, Galaxies, 6, 63.

Atom	$^2D_{5/2}$	$^2D_{3/2}$	$^2P_{3/2}$	$^2P_{1/2}$
N I	19224.5	19233.2	28839.3	28838.9
O II	26810.5	26830.6	40468.0	40470.0
F III	34087.4	34123.2	51561.6	51560.4

Now the other parts of $\boldsymbol{\ell}_1 \cdot \mathbf{s}_1$ do not contribute. Moreover, by symmetry, the contributions of $\boldsymbol{\ell}_2 \cdot \mathbf{s}_2$ and $\boldsymbol{\ell}_3 \cdot \mathbf{s}_3$ also vanish. Thus for $^2D_{5/2}$, the spin-orbit interaction vanishes and therefore it vanishes for ^2D_J, generally.

Next consider the states with $M_L = 1, M_S = 1/2$. These can be ^2D or ^2P. Accordingly, there are two Slater determinants from which they must be made.

$$\psi_1(1,2,3) = \frac{1}{\sqrt{6}} \begin{vmatrix} Y_{11}(1)\alpha_1 & \cdot & \cdot \\ \cdot & Y_{10}(2)\alpha_2 & \cdot \\ \cdot & \cdot & Y_{10}(3)\beta_3 \end{vmatrix} \tag{6.96}$$

and

$$\psi_2(1,2,3) = \frac{1}{\sqrt{6}} \begin{vmatrix} Y_{1-1}(1)\alpha_1 & \cdot & \cdot \\ \cdot & Y_{11}(2)\alpha_2 & \cdot \\ \cdot & \cdot & Y_{11}(3)\beta_3 \end{vmatrix}. \tag{6.97}$$

Here we find

$$\left(\sum_i \ell_{iz} s_{iz}\right)_1 = \frac{1}{2}; \qquad \left(\sum_i \ell_{iz} s_{iz}\right)_2 = -\frac{1}{2} + \frac{1}{2} - \frac{1}{2} = -\frac{1}{2} \tag{6.98}$$

This enables us to deduce directly that the spin-orbit interaction vanishes as well for ^2P. To see this, recall that the trace of a matrix is invariant under a unitary transformation. Here we have a 2×2 matrix, and in the basis of ψ_1 and ψ_2 the trace of the interaction vanishes, so it does as well in the ^2D - ^2P basis. But we know that the spin-orbit interaction vanishes for ^2D, so it also vanishes for ^2P.

A representative collection of energy levels is shown in Table 6.4. The observed splittings are much smaller than those for $(2p)^2$ in comparable ions (see, for example, Problem 6.7), consistent with the theoretical predictions.

6.8 Intermediate Coupling

We have evaluated the effect of the spin-orbit coupling on the energy of a particular term $^{2S+1}L_J$. But the Russell-Saunders coupling scheme is an ansatz, not an exact result. If the spin-orbit interaction is more important than the electron-electron interaction, j-j coupling is more appropriate. In between the two extremes lies the case of intermediate coupling. Consider, for example, an atom with two valence electrons, with a ground state $(ns)^2 = {}^1S_0$ and an excited state $(ns\,np)$, which can be ${}^1P_1$ or ${}^3P_1$. A pertinent example is Hg. The Hamiltonian for the electrostatic and spin-orbit interactions for the two valence electrons is

$$H' = \frac{e^2}{r_{12}} + \frac{\hbar^2}{2m^2c^2}\xi(r_1)\boldsymbol{\ell}_1 \cdot \mathbf{s}_1 + \frac{\hbar^2}{2m^2c^2}\xi(r_2)\boldsymbol{\ell}_2 \cdot \mathbf{s}_2. \tag{6.99}$$

This interaction is invariant if we rotate both the space and spin parts, that is, under rotations by $\mathbf{J}$, but is not invariant under separate rotations by $\mathbf{L} = \boldsymbol{\ell}_1 + \boldsymbol{\ell}_2$ or $\mathbf{S} = \mathbf{s}_1 + \mathbf{s}_2$. Since $\mathbf{S}$ is not a good quantum number, there could be mixing between ${}^3P_1$ and ${}^1P_1$.

The famous resonant line at 2537 Å in mercury appears in the books as a ${}^3P_1$ to ${}^1S_0$ transition with a parity change. This should be forbidden as an E1 transition because it has $\Delta S = 1$ and as an M1 transition because it changes the parity. Comparing the allowed transition ${}^1P_1 \to {}^1S_0$ at 1850 Å to the forbidden transition at 2537 Å shows that the effective dipole matrix element for the latter is suppressed by only a factor of 0.17. (A. V. Smith and W. J. Alford, *Phys. Rev.* **A33**, 3172 (1986).) The reason for this is that Russell-Saunders coupling does not hold exactly for high-Z atoms like mercury. That such might be the case can be inferred from the term scheme. The fine structure splittings among the ${}^3P_J$ states are not negligible compared to the ${}^1P_1 - {}^3P_1$ electrostatic splitting.

If we consider the Russell-Saunders states as our basis, we have the electrostatic energy diagonal in this basis with the Slater determinants yielding

$$\Delta E({}^1P_1) = F_0 + G_1'$$
$$\Delta E({}^3P_J) = F_0 - G_1', \qquad J = 0, 1, 2, \tag{6.100}$$

where we have written G_1' for what was indicated previously as $G_1/3$, following the notation of Condon and Shortley, *Theory of Atomic Spectra*, in this section. Now let us write out the basis states explicitly for $m_J = 0$, taking care to form antisymmetric states. The spin-singlet wave function is antisymmetric, so we can write

$$\psi({}^1P_1) = \frac{1}{\sqrt{2}}[\alpha(1)\beta(2) - \alpha(2)\beta(1)]\frac{1}{\sqrt{2}}[s(1)p(2)_{m=0} + s(2)p(1)_{m=0}] \tag{6.101}$$

and, analogously,

$$\psi({}^3P_1) = \frac{1}{2}\alpha(1)\alpha(2)[s(1)p(2)_{m=-1} - s(2)p(1)_{m=-1}]$$
$$- \frac{1}{2}\beta(1)\beta(2)[s(1)p(2)_{m=1} - s(2)p(1)_{m=1}]. \tag{6.102}$$

Now act with the Hamiltonian on $\psi({}^1P_1)$, applying the appropriate Clebsch-Gordan coefficients.

$$\boldsymbol{\ell}_1 \cdot \mathbf{s}_1 \psi({}^1P_1) = \left[\frac{1}{2}(\ell_{1+}s_{1-} + \ell_{1-}s_{1+}) + \ell_{1z}s_{1z}\right]$$
$$\times \frac{1}{2}[\alpha(1)\beta(2) - \alpha(2)\beta(1)][s(1)p(2)_{m=0} + s(2)p(1)_{m=0}]$$
$$= \frac{1}{2\sqrt{2}}[\beta(1)\beta(2)s(2)p(1)_{m=1} - \alpha(1)\alpha(2)s(2)p(1)_{m=-1}] \tag{6.103}$$

$$\boldsymbol{\ell}_2 \cdot \mathbf{s}_2 \psi(^1P_1) = \left[\frac{1}{2}(\ell_{2+}s_{2-} + \ell_{2-}s_{2+}) + \ell_{1z}s_{1z}\right]$$

$$\times \frac{1}{2}(\alpha(1)\beta(2) - \alpha(2)\beta(1))(s(1)p(2)_{m=0} + s(2)p(1)_{m=0})$$

$$= \frac{1}{2\sqrt{2}}[-\beta(1)\beta(2)s(1)p(2)_{m=1} + \alpha(1)\alpha(2)s(1)p(2)_{m=-1}]. \tag{6.104}$$

Combining these results,

$$\langle {}^3P_1 | H' | {}^1P_1 \rangle = \frac{\zeta}{\sqrt{2}} \tag{6.105}$$

with

$$\zeta = \frac{\hbar^2}{2m^2c^2} \int_0^\infty dr\, r^2 p(r)^2 \frac{1}{r}\frac{dV}{dr}. \tag{6.106}$$

The full matrix for the Hamiltonian taken between the four states $^3P_{2,1,0}$ and $^1P_1$ is

$$\begin{pmatrix} \zeta/2 & 0 & 0 & 0 \\ 0 & -\zeta/2 & 0 & \zeta/\sqrt{2} \\ 0 & 0 & -\zeta & 0 \\ 0 & \zeta/\sqrt{2} & 0 & 0 \end{pmatrix}, \tag{6.107}$$

where the rows and columns are ordered $^3P_2, {}^3P_1, {}^3P_0, {}^1P_1$. The two $J = 1$ states mix. If we include the electrostatic interaction, the two-by-two mixing matrix is

$$\begin{pmatrix} -G_1' - \zeta/2 & \zeta/\sqrt{2} \\ \zeta/\sqrt{2} & G_1' \end{pmatrix}. \tag{6.108}$$

The eigenvalues are

$$\lambda_\pm = -\frac{\zeta}{4} \pm \sqrt{(G_1' + \zeta/4)^2 + \zeta^2/4}. \tag{6.109}$$

The energies of the eigenstates are

$$\Delta E(^3P_2) = F_0 - G_1' + \zeta/2$$

$$\Delta E(^1P_1') = F_0 - \zeta/4 + \sqrt{(G_1' + \zeta/4)^2 + \zeta^2/2}$$

$$\Delta E(^3P_1') = F_0 - \zeta/4 - \sqrt{(G_1' + \zeta/4)^2 + \zeta^2/2}$$

$$\Delta E(^3P_0) = F_0 - G_1' - \zeta. \tag{6.110}$$

If we write the eigenstates for $J = 1$ as $a\,|^3P_1\rangle + b\,|^1P_1\rangle$, then

$$\frac{a}{b} = -\frac{\sqrt{2}}{\zeta}\left((G_1' + \zeta/4) \pm \sqrt{(G_1' + \zeta/4)^2 + \zeta^2/2}\right). \tag{6.111}$$

Following Condon and Shortley, *The Theory of Atomic Spectra*, we characterize the strength of the spin-orbit interaction relative to the electron-electron interaction by

$$\chi = \frac{3}{4}\zeta/G_1'. \tag{6.112}$$

In order to display the mixing effect from pure L-S to pure j-j, we introduce yet another parameter,

$$\lambda = \frac{\chi}{1+\chi} \tag{6.113}$$

so that small $\lambda \ll 1$ corresponds to L-S coupling and $\lambda = 1$ is pure j-j coupling. At large χ, the splittings are proportional to χ, so we plot

$$\epsilon = \frac{\Delta E - F_0}{(1+\chi)G_1'} \tag{6.114}$$

as a function of λ. One finds that $\zeta/G_1' = 2/3$ gives a reasonable fit to the data, with $G_1' = 5290\,\text{cm}^{-1}$ and $\zeta = 3650\,\text{cm}^{-1}$. The eigenstates are then

$$|^3P_1'\rangle = 0.98\,|^3P_1\rangle - 0.19\,|^1P_1\rangle$$
$$|^1P_1'\rangle = 0.19\,|^3P_1\rangle + 0.98\,|^1P_1\rangle. \tag{6.115}$$

See Figure 6.9. The relative matrix elements of the electric dipole operator for the 2537 Å and 1850 Å lines are just the coefficients of the $^1P_1$ ket in each eigenket. The relative value of 0.19 corresponds closely to the experimental 0.17 from the measured oscillator strengths.

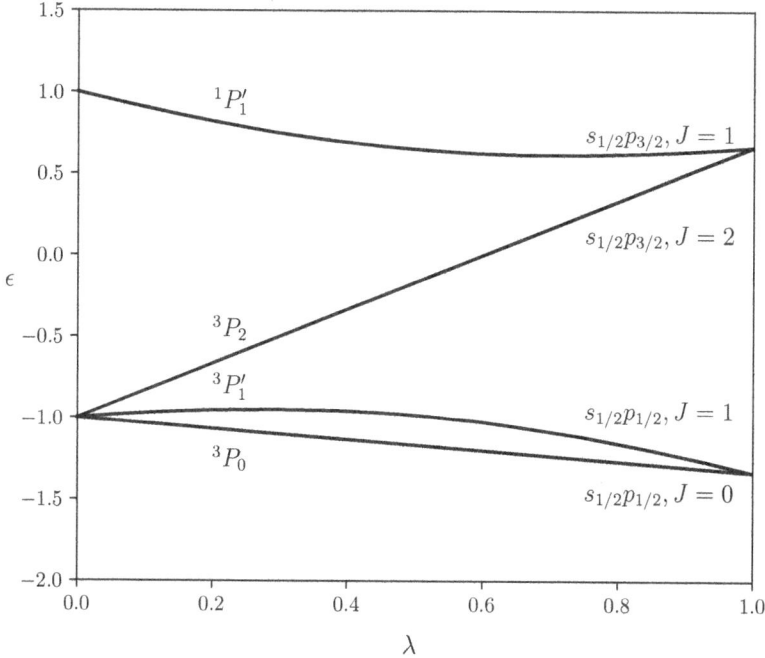

Figure 6.9 The ns-np system showing the transition from L-S coupling, where the electron-electron interaction dominates, to j-j coupling, where the spin-orbit interaction dominates. The abscissa λ is related to the ratio of the spin-orbit coupling to the electron-electron interaction by $\chi = (4/3)\zeta/G_1'$ and $\lambda = \chi/(1+\chi)$. Thus, $\lambda \ll 1$ is L-S coupling and the levels there are labeled in the L-S scheme. At the extreme $\lambda \simeq 1$, there is j-j coupling and the states are labeled by j values of the two electrons and by the total J. The value $\zeta/G_1' = 2/3$ corresponds to $\lambda = 1/3$ in the figure. Adapted from Condon and Shortley 1951 / Cambridge University Press.

6.9 Thomas-Fermi Atom

The Thomas-Fermi model is a semi-classical description of the many-electron atom, giving an approximation for the central-field potential and for the electron charge density as a function of the position. The idea is that the wavelength of most electrons is very short compared to atomic size. Locally (at fixed r), the states from the negative potential up to zero are filled with electrons that are treated as free particles whose kinetic energy is measured up from $T = 0$ at $E = -V(r)$. Thus, the Fermi gas has a local Fermi energy $E_F = -V(r)$. See Figure 6.10. Recall that for a box of volume $\mathcal{V}$ the density of states is: Figure

$$dN = 2 \times \frac{\mathcal{V}d^3k}{(2\pi)^3} \tag{6.116}$$

If we fill up the states with one particle in each state, up to a maximum wave number k_F, the density of electrons is

$$n = \frac{8\pi}{3} \cdot \frac{1}{8\pi^3} k_F^3 = \frac{k_F^3}{3\pi^2}. \tag{6.117}$$

Now we write $V(r) = -e\Phi(r)$, where $\Phi(r)$ is the electrostatic potential. With $e\Phi = E_F = \frac{\hbar^2 k_F^2}{2m}$, we find

$$n(\mathbf{x}) = \frac{1}{3\pi^2}\left(\frac{2me\Phi(\mathbf{x})}{\hbar^2}\right)^{3/2}. \tag{6.118}$$

The charge density caused by the electron cloud is thus

$$\rho_e(\mathbf{x}) = -en(\mathbf{x}). \tag{6.119}$$

Poisson's equation for the electrostatic potential will then be

$$\nabla^2\Phi = -4\pi\left[Ze\delta(\mathbf{x}) - en(\mathbf{x})\right] = -4\pi Ze\delta(\mathbf{x}) + \frac{4e}{3\pi}\left(\frac{2me\Phi(\mathbf{x})}{\hbar^2}\right)^{3/2}. \tag{6.120}$$

This is a non-linear equation for Φ, with boundary conditions $\Phi \to \frac{Ze}{r}$ as $r \to 0$, and $\Phi \to 0$ as $r \to \infty$, for a neutral atom. It is appropriate to put this in a dimensionless form and look

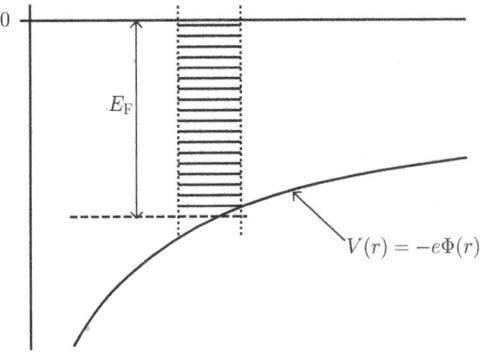

Figure 6.10 The Thomas-Fermi model for neutral atoms imagines that locally a sea of free particle states can be filled to a kinetic energy $E_F = \frac{\hbar^2 k_F^2}{2m} = e\Phi(r)$.

for a spherically symmetric solution. We write

$$\Phi(r) = \frac{Ze}{r}\chi(r) \tag{6.121}$$

and introduce a length parameter

$$b = \left(\frac{3\pi}{4}\right)^{2/3}\frac{a_0}{2Z^{1/3}} = 0.8853\, a_0/Z^{1/3}. \tag{6.122}$$

Then, with $x = r/b$, the differential equation becomes

$$\frac{d^2\chi}{dx^2} = \frac{1}{\sqrt{x}}\chi^{3/2}. \tag{6.123}$$

In terms of χ, the electron density is

$$n(\mathbf{x}) = \frac{Z}{4\pi b^3}\left(\frac{\chi}{x}\right)^{3/2}. \tag{6.124}$$

The boundary conditions are $\chi(0) = 1$, $\chi(\infty) = 0$. The non-linear equation can be integrated numerically if a value of $d\chi/dx$ at $x = 0$ is specified. If this is too negative, χ will fall to zero at a finite value of x. If it is not negative enough, χ will reach a minimum and then climb. The value of $d\chi/dx$ at $x = 0$, for which $\chi \to 0$ as $x \to \infty$, is near -1.588, as shown in Figure 6.11. The actual value is known to high precision: $-1.5880710226...$

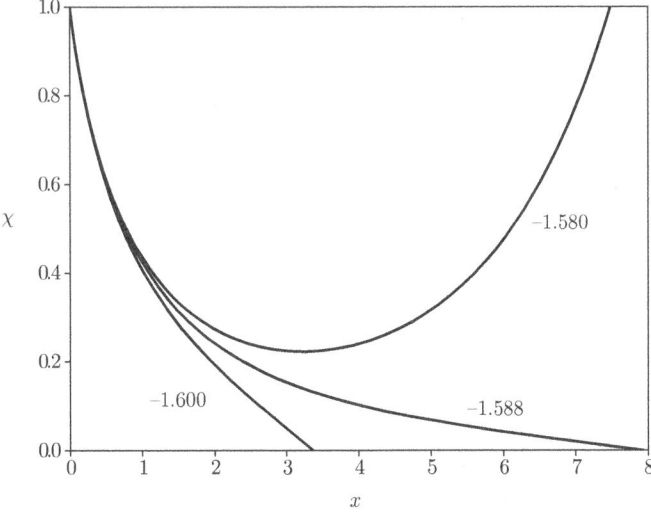

Figure 6.11 Numerical determination of the Thomas-Fermi function χ. The boundary conditions are $\chi(0) = 1$ and $\chi(\infty) = 0$. Trial solutions begin with an assumed value of $\chi'(0)$. The three curves are labeled by the value of $\chi'(0)$.

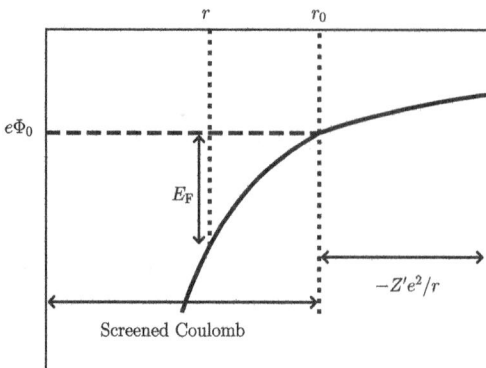

Figure 6.12 In the Thomas-Fermi model for ions, the function $\chi(x)$ falls to zero at some finite value r_0. The limiting Fermi energy, E_F at $r = bx$, is determined by $\Phi(x) - \Phi(x_0)$. Beyond r_0, the potential energy of an electron would be $-Z'e^2/r$, where $Z'e$ is the charge of the ion.

We can check that $n(\mathbf{x})$ indeed represents exactly Z electrons.

$$\int d^3r\, n(r) = b^3 \int d^3x\, \frac{Z}{4\pi b^3} \left(\frac{\chi}{x}\right)^{3/2} = Z \int_0^\infty dx\, x^{1/2} \chi^{3/2}$$

$$= Z \int_0^\infty dx\, x \frac{d^2\chi}{dx^2} = Z\left[x\frac{d\chi}{dx}\Big|_0^\infty - \int_0^\infty dx \frac{d\chi}{dx}\right] = Z \quad (6.125)$$

The Thomas-Fermi model can also describe positive ions. If the ion has positive charge $Z'e$, the potential for large r is $Z'e/r$ classically. In the Thomas-Fermi semi-classical description for positive ions, the electron density vanishes beyond some radius r_0, in contrast with the neutral atom picture, where the electron cloud extends to infinity. For the ion, we have at $r_0 = bx_0$, $\chi(x_0) = 0$, and $\Phi(r_0) \equiv \Phi_0 = Z'e/r_0$. The electrons at radius r fill up the Fermi sphere of energy $E_F = e(\Phi(r) - \Phi_0)$. See Figure 6.12. Thus,

$$n(r) = \frac{1}{3\pi^2}\left[\frac{2me}{\hbar^2}(\Phi(r) - \Phi_0)\right]^{3/2}. \quad (6.126)$$

We have $\Phi(r)-\Phi_0$ replacing $\Phi(r)$ in the neutral atom. The electron cloud in the Thomas-Fermi model of ions terminates when $\chi(x) = 0$. The charge of the electron cloud is

$$\int_0^{r_0} d^3r\, n(r) = b^3 \int_0^{x_0} d^3x\, \frac{Z}{4\pi b^3} \left(\frac{\chi}{x}\right)^{3/2} = Z \int_0^{x_0} dx\, x^{1/2} \chi^{3/2}$$

$$= Z \int_0^{x_0} dx\, x \frac{d^2\chi}{dx^2} = Z\left[x\frac{d\chi}{dx}\Big|_0^{x_0} - \int_0^{x_0} dx \frac{d\chi}{dx}\right]$$

$$= Z(x_0\chi'(x_0) + 1) = Z - Z'. \quad (6.127)$$

Thus, the fraction of electrons that are ionized is

$$Z'/Z = -x_0\chi'(x_0). \quad (6.128)$$

For a given value of Z'/Z, one integrates back from a trial value of x_0, using the slope at x_0, $\chi'(x_0) = -x_0 Z'/Z$. The value of x_0 is varied until $\chi(0)$ is sufficiently close to unity. Results are shown in Figure 6.13 and Table 6.5.

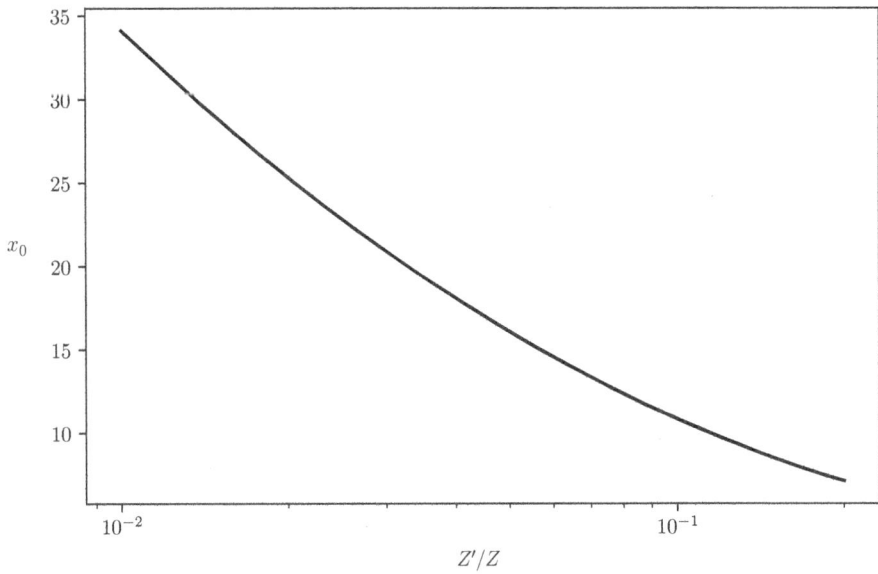

Figure 6.13 The value of x_0 where the Thomas-Fermi function $\chi(x)$ vanishes, as a function of Z'/Z, the fraction of ionization of the atom.

Table 6.5 For various values of Z'/Z the fraction of electrons ionized, the value of x_0 where the Thomas-Fermi function vanishes. The final column shows the radius at which the function vanishes for a singly ionized atom in units of the Bohr radius using $r_0/a_0 = 0.8853(Z'/Z)^{1/3}x_0$.

Z'/Z	x_0	r_0/a_0
0.01	34.3	6.53
0.02	25.2	6.05
0.03	20.8	5.72
0.05	16.1	5.25
0.08	12.5	4.76
0.10	11.0	4.51

Table 6.6 Representative ionic radii of alkali metals in angstroms as reported in the CRC Handbook of Physics. For the elements Na and on, when the Thomas-Fermi model might apply, the data are about a factor two below the predictions.

Z	11	19	37	55
Element	Na	K	Rb	Cs
Measured	0.97	1.33	1.47	1.67
Thomas-Fermi	2.44	2.75	3.07	3.24

The Thomas-Fermi model predictions can be compared to the radii of alkali metals as determined from crystal structures. The radii depend on the coordination number of the atom in the particular crystal. Representative results are shown in Table 6.6 in angstroms.

The detailed shape of the periodic table is determined by the entry of new orbital angular momenta as Z increases. The Thomas-Fermi model makes predictions for this pattern. If we consider $-e\Phi(r)$ as the potential for a single electron, we can consider the radial motion of an electron of angular momentum ℓ in the effective potential

$$V_{\text{eff}}(r) = -\frac{Ze^2}{r}\chi(r) + \frac{(\ell + 1/2)^2\hbar^2}{2mr^2}, \tag{6.129}$$

where we have used $(\ell + 1/2)^2$ in the spirit of the semi-classical approach. For a given Z and increasing ℓ, the potential appears as shown in Figure 6.14. Conversely, for a fixed ℓ, Z must reach some minimum value $Z_{\min}(\ell)$ in order for there to be a negative potential available for binding an electron with that ℓ value. The critical Z occurs when

$$\frac{Ze^2}{r}\chi(r) = \frac{(\ell + 1/2)^2\hbar^2}{2mr^2} \tag{6.130}$$

for some r. That is,

$$2Z\frac{r_0}{a_0}\chi(r) = (\ell + 1/2)^2 \tag{6.131}$$

$$2Z^{2/3}(0.8853)[x\chi(x)]_{\max} = (\ell + 1/2)^2. \tag{6.132}$$

From Figure 6.11, we find that the maximum of $x\chi(x)$ occurs near $x = 2.1$, where its value is 0.4865. Accordingly, we have

$$Z_{\min}(\ell) = 1.251(\ell + 1/2)^3. \tag{6.133}$$

This prediction works remarkably well, as shown in Table 6.7.

Fermi, in his 1927 article (Z. fur Physik, **48**,73 (1927)), used the model to estimate how many electrons there are of a given angular momentum as a function of Z. We can use the

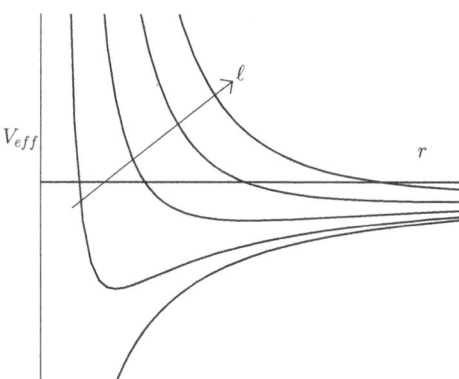

Figure 6.14 The effective potential for fixed Z and increasing ℓ obtained using the electrostatic potential from the Thomas-Fermi model. There is some maximum ℓ for which the potential is low enough to support a bound state. As Z increases, so do the values of ℓ for which bound states are possible.

Table 6.7 First appearance of a new angular momentum state as Z increases through the periodic table, comparing the prediction of the Thomas-Fermi model with the actual occurrence.

Orbital	ℓ	$Z_{\min}+1$	Z(first appearance)
s	0	1.15	1 (H)
p	1	5.22	5 (B)
d	2	20.6	21 (Sc)
f	3	54.6	57–58 (La-Ce)
g	4	115.	

WKB quantization rule with $V_{\text{eff}}(r)$ to estimate the number of levels of a given ℓ value. We have $\hbar^2 k_{\text{max}}^2/2m = -V_{\text{eff}}(r)$ and, generally from WKB, with $n = 1, 2, \ldots$

$$\int_{r_1}^{r_2} k(r)dr = (n - \tfrac{1}{2})\pi; \qquad k = \sqrt{\frac{2m}{\hbar^2}(E - V_{\text{eff}}(r))}. \tag{6.134}$$

We want to count the number of bound states N_ℓ to $E = 0$. For each ℓ, there are $2(2\ell + 1)$ such states available, so

$$\frac{N_\ell}{2(2\ell + 1)} = \frac{1}{2} + \frac{1}{\pi} \int_0^\infty \sqrt{\frac{2m}{\hbar^2}\left(e\Phi(x) - \frac{(\ell + 1/2)^2\hbar^2}{2mr^2}\right)}dr. \tag{6.135}$$

In terms of our dimensionless variable $x = r/b$ and the function $\chi(x)$, we have

$$N_\ell = 2\ell + 1 + \frac{2}{\pi}(2\ell + 1) \int_0^\infty \frac{dx}{x}\sqrt{\left(\frac{3\pi Z}{4}\right)^{2/3} x\chi(x) - (\ell + 1/2)^2}. \tag{6.136}$$

Apart from from the first term $(2\ell + 1)$, which comes from the $1/2$ in the WKB quantization, this expression is the one derived by Fermi in 1927. It is quoted by Gombás (Gombás, *Die Statistische Theorie des Atoms*, Springer Verlag (1949); *Statistische Behandlung des Atoms*, in Handbook of Physics, Vol. 36. Atoms II, Springer Verlag (1956), pp. 109 ff.) who gives the excellent approximation

$$\chi(x) = \exp(-1.44(\sqrt{x + 0.285} - \sqrt{0.285})), \tag{6.137}$$

which is good to within about 1% for $x < 5.5$. The integral of $(x\chi^3)^{1/2}$ is equal to 0.962 for this approximation, rather that unity.

Following Fermi, we compare the results of the Thomas-Fermi model with the actual occupation of the orbitals as Z moves through the periodic table. See Figure 6.15.

The agreement is quite impressive. The Thomas-Fermi model does a remarkable job representing multi-electron atoms, a job that is all the more impressive when we examine the challenges to a more direct approach in the next section.

6.10 Hartree-Fock Approximation

We give here only a very brief sketch of the Hartree-Fock method of calculating self-consistent potential and electron densities, energies, etc. in many-electron atoms and molecules.

The underlying assumption is that electrons move independently in a self-consistent central potential that is determined by quantum states of all the electrons. The equations governing the electronic motion are derived from a variational principle for the energy of the system. If the trial wave function is simply the product of single-electron orbitals, the resulting equations describe the Hartree approximation (1927). If the trial wave function is a Slater determinant of single-electron orbitals, the result is the Hartree-Fock approximation (due to Slater and Fock), which we shall describe.

The Hamiltonian of the system is taken to be the standard electrostatic Hamiltonian:

$$H = \sum_i H_i + \frac{1}{2}\sum_{i \neq j} \frac{e^2}{r_{ij}}; \qquad H_i = -\frac{\hbar^2}{2m}\nabla_i^2 - \frac{Ze^2}{r_i}. \tag{6.138}$$

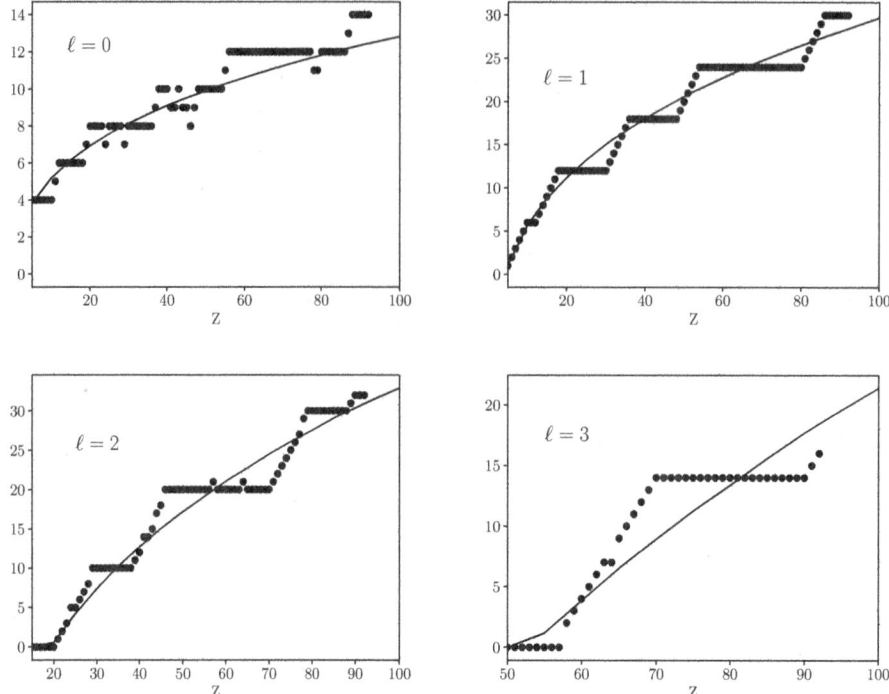

Figure 6.15 The occupancy of orbitals with $\ell = 0, 1, 2, 3$ as predicted by the Thomas-Fermi model shown as curves, compared with the actual occupancies of the elements.

We compute

$$E = \langle \Psi | H | \Psi \rangle \tag{6.139}$$

where

$$\Psi(1, 2, \ldots n) = \frac{1}{\sqrt{n!}} \begin{vmatrix} \phi_\alpha(1) & \cdot & \cdot & \cdot \\ \cdot & \phi_\beta(2) & \cdot & \cdot \\ \cdot & \cdot & \cdot & \cdot \\ \cdot & \cdot & \cdot & \phi_\zeta(n) \end{vmatrix}. \tag{6.140}$$

We assume that each $\phi_\alpha(i)$ is a normalized wave function of the spatial and spin degrees of freedom. Substitution gives

$$E = \sum_\alpha \langle \phi_\alpha(1) | H_1 | \phi_\alpha(1) \rangle + \frac{1}{2} \sum_{\alpha,\beta} \left\{ \left\langle \phi_\alpha(1)\phi_\beta(2) \left| \frac{e^2}{r_{12}} \right| \phi_\alpha(1)\phi_\beta(2) \right\rangle \right.$$

$$\left. - \left\langle \phi_\alpha(1)\phi_\beta(2) \left| \frac{e^2}{r_{12}} \right| \phi_\alpha(2)\phi_\beta(1) \right\rangle \right\}. \tag{6.141}$$

The second term, the exchange integral, is absent in the Hartree approximation. The $1/\sqrt{n!}$ is now gone and the $1/2$ for the e^2/r_{12} interaction compensates for each pair of electrons occurring twice.

Now we are ready to apply the variation principle to E. We use Lagrange multipliers to impose the constraints of orthonormality of the single-particle wave functions. The variational principle is thus

$$\delta\left(E - \sum_{\alpha,\beta} \langle \phi_\alpha | \phi_\beta \rangle \lambda_{\beta\alpha}\right) = 0 \tag{6.142}$$

with $\lambda_{\beta,\alpha} = \lambda_{\alpha,\beta}^*$. We now vary ϕ_α^* to obtain a differential equation for ϕ_α. Thus, we mean

$$\delta \langle \phi_\alpha | H_1 | \phi_\alpha \rangle = \langle \delta\phi_\alpha | H_1 | \phi_\alpha \rangle$$

$$\delta \langle \phi_\alpha | \phi_\alpha \rangle = \langle \delta\phi_\alpha | \phi_\alpha \rangle \tag{6.143}$$

$$\delta \left\langle \phi_\alpha(1)\phi_\beta(2) \left| \frac{e^2}{r_{12}} \right| \phi_\alpha(1)\phi_\beta(2) \right\rangle = \langle \delta\phi_\alpha | \tilde{V}_{\beta\beta}(1) | \phi_\alpha \rangle, \tag{6.144}$$

where we define

$$\tilde{V}_{\beta'\beta}(1) = \left\langle \phi_{\beta'}(2) \left| \frac{e^2}{r_{12}} \right| \phi_\beta(2) \right\rangle. \tag{6.145}$$

Then the exchange term variation is

$$\delta \left\langle \phi_\alpha(1)\phi_\beta(2) \left| \frac{e^2}{r_{12}} \right| \phi_\alpha(2)\phi_\beta(1) \right\rangle = \langle \delta\phi_\alpha | \tilde{V}_{\beta\alpha} | \phi_\beta \rangle. \tag{6.146}$$

The variation principle then reads as follows:

$$0 = \delta\left(E - \sum_{\alpha,\beta} \langle \phi_\alpha | \phi_\beta \rangle \lambda_{\beta\alpha}\right)$$

$$= \sum_\alpha \left\{ \langle \delta\phi_\alpha | H_1 | \phi_\alpha \rangle + \sum_\beta \langle \delta\phi_\alpha | \tilde{V}_{\beta\beta}(1) | \phi_\alpha \rangle \right.$$

$$\left. - \langle \delta\phi_\alpha | \tilde{V}_{\beta\alpha}(1) | \phi_\beta \rangle - \sum_\beta \langle \delta\phi_\alpha | \phi_\beta \rangle \lambda_{\beta\alpha} \right\}. \tag{6.147}$$

Since the ϕ_α^* are all independent, we must have, for each α, a Schrödinger-like Hartree-Fock equation:

$$H_1\phi_\alpha(1) + \sum_\beta \left(\tilde{V}_{\beta\beta}(1)\phi_\alpha(1) - \tilde{V}_{\beta,\alpha}(1)\phi_\beta(1) \right) = \sum_\beta \phi_\beta(1)\lambda_{\beta\alpha}. \tag{6.148}$$

The Lagrange-multiplier matrix $\lambda_{\alpha\beta}$ is Hermitian, so we can diagonalize it with some constant matrix, which is equivalent to choosing a new basis for the ϕ_α, say ψ_α. In this basis, we can write $\lambda_{\alpha\beta} = E_\alpha \delta_{\alpha\beta}$. Then the Hartree-Fock equation is

$$H_1\psi_\alpha(1) + \sum_\beta \left(V_{\beta\beta}(1)\psi_\alpha(1) - V_{\beta\alpha}(1)\psi_\beta(1) \right) = E_\alpha\psi_\alpha(1), \tag{6.149}$$

where now

$$V_{\beta'\beta}(1) = \left\langle \psi_{\beta'}(2) \left| \frac{e^2}{r_{12}} \right| \psi_\beta(2) \right\rangle. \tag{6.150}$$

The Hartree approximation is simpler, but less complete, than the Hartree-Fock method. Without the exchange term, one can take for the potential term its spherical average:

$$\sum_{\beta \neq \alpha} V_{\beta\beta}(1) \rightarrow V_{dir}^\alpha(r_1). \tag{6.151}$$

We look at this in more detail with $\psi_\beta = R_\beta(r_2)Y_\beta(2)\chi_\beta(2)$. We can evaluate the "direct potential"

$$\sum_{\beta\neq\alpha} \langle V_{\beta\beta}(1)\rangle_{angles} = \sum_{\beta\neq\alpha} \int d^3x_2 \left\langle \frac{e^2}{r_{12}} \right\rangle_{angles} |R_\beta(r_2)|^2 |Y_\beta(2)|^2. \tag{6.152}$$

Since

$$\frac{1}{r_{12}} = \sum_\ell \frac{r_<^\ell}{r_>^{\ell+1}} P_\ell(\hat{\mathbf{r}}_1 \cdot \hat{\mathbf{r}}_2), \tag{6.153}$$

we see that

$$\left\langle \frac{1}{r_{12}} \right\rangle_{angles} = \frac{1}{r_>} \tag{6.154}$$

and

$$V_{dir}^\alpha(r_1) = e^2 \int_0^\infty dr_2 \frac{r_2^2}{r_>} |R_\beta(r_2)|^2, \tag{6.155}$$

where $r_>$ is the greater of r_1 and r_2.

The scheme of calculation is now clear:

(a) Guess at $\sum_{\beta\neq\alpha} |R_\beta(r_2)|^2$, e.g. take from Thomas-Fermi $4\pi n(r_2)$.
(b) Compute the central potential with this guess.
(c) Solve for $\psi_\alpha^{(1)}(1)$ for all α and determine $E_\alpha^{(1)}$.
(d) Use $\psi_\alpha^{(1)}$ to compute new direct central potential.
(e) Repeat steps (c) and (d) until stability and sufficient accuracy have been obtained. A few iterations yields fairly good wave functions and values of E_α.

Hartree-Fock calculations are similar, but have the complication of the exchange terms. One scheme is to use the result of Problem 6.10 to make an approximation to the exchange term

$$V_{ex}(1) = -e^2 \left(\frac{3}{\pi} n(\mathbf{r}) \right)^{1/3}, \tag{6.156}$$

where

$$n(\mathbf{r}) = \sum_\beta \psi_\beta^* \psi_\beta. \tag{6.157}$$

See Bethe and Jackiw, *Intermediate Quantum Mechanics*, 3rd edition, pp. 72-74.

What is the significance of E_α in the Hartree-Fock equation? Compare E_α from the Hartree-Fock equation

$$E_\alpha = \langle \alpha |H_1| \alpha \rangle + \sum_\beta \langle \alpha |V_{\beta\beta}(1)| \alpha \rangle - \sum_\beta \langle \alpha |V_{\beta\alpha}(1)| \beta \rangle \tag{6.158}$$

to the original expression for the total energy

$$E = \sum_\alpha \langle \alpha |H_1| \alpha \rangle + \frac{1}{2} \sum_\alpha \sum_\beta \left[\langle \alpha |V_{\beta\beta}(1)| \alpha \rangle - \langle \alpha |V_{\beta\alpha}(1)| \beta \rangle \right]. \tag{6.159}$$

The electron-electron interaction has been counted twice in E_α, so $\sum_\alpha E_\alpha \neq E$. [We can define $E_\alpha' = (E_\alpha + \langle \alpha |H_1| \alpha \rangle)/2$, then $E = \sum_\alpha E_\alpha'$.]

If we assume that the wave functions of an ion obtained by removing one electron from the neutral atom are essentially the same as for the original atom, the E_α values approximate the removal energy of an electron from state α. This is known as Koopman's theorem. A comparison of data and Hartree-Fock predictions is given by Bethe and Jackiw, *Intermediate Quantum Mechanics*, 3rd edition, Chapter 5.)

References and Suggested Reading

The classic reference is Condon and Shortley, *The Theory of Atomic Spectra*, Cambridge Press, 1953.

Very extensive coverage is given in Bethe and Jackiw, *Intermediate Quantum Mechanics*, 3rd edition, pp. 118-175. In particular, see pp. 138-9 for a discussion of Hund's rule.

For the restricted range of hydrogen and helium, Bethe and Salpeter, *Quantum Mechanics of One- and Two-electron Atoms*.

For a derivation of Thomas precession from relativistic kinematics, see *Jackson, Classical Electrodynamics*, 3rd edition, pp. 548-553.

Problems

6.1 The ground state energy in helium-like atoms (two electrons around a nucleus of charge Ze) can be estimated well with a simple variational wave function

$$\Psi(\mathbf{r}_1, \mathbf{r}_2) = Ae^{-\beta(r_1+r_2)} \tag{6.160}$$

where β is a variational parameter.

(a) Find the best value for the ground state energy and the corresponding value of β. The kinetic energy and the electron-nucleus potential energy terms are simple. The only part requiring a little work is the term e^2/r_{12}.

(b) Find experimental results from helium-like atoms with $Z = 1, 2, 3, 4, 5$ and compare with your variational calculation.

6.2* Argon is a noble gas, with closed shells in its ground state. The orbital configuration is $(1s)^2(2s)^2(2p)^6(3s)^2(3p)^6$. Singly ionized argon has a ground state $...(3s)^2(3p)^5$ or $(3p)^{-1}$, that is, a hole in the $3p$ shell. The hole can be expected to have a negative spin-orbit energy.

(a) Consider the possible low-lying excited states of neutral argon obtained by exciting an electron from the $3p$ shell to higher orbitals. For each orbital configuration, determine the Russell-Saunders multiplets and parities. For each $^{2S+1}L_J$ state, determine the expected g-factor. How many low-lying odd (even) parity states do you expect? Are they grouped in any manner?

(b) From the data of Table 6.8 make an energy level diagram covering the interval from 90,000 cm^{-1} to 110,000 cm^{-1} above the ground state showing the first 14 excited states. Use the measured g-factors (from Zeeman splittings), J value data, and a knowledge of the magnitudes of fine-structure splitting to make as many Russell-Saunders assignments to the states as possible.

Table 6.8 Energy levels for neutral argon adapted from C. E. Moore, Atomic Energy Levels, NSRDS-NBS 35, reissued, December 1971. The column Config. gives an assignment for the $(3p)^5$ core and the promoted electron. The column Desig. shows an assignment for the sum of the j from the core and the ℓ from the promoted electron. The level is the energy above the ground state in cm^{-1}.

Config.	Desig.	J	Level	Obs. g
		0	0	
$^2P^o_{3/2}(4s)$	$4s[3/2]^o$	2	93144	1.506
		1	93751	1.404
$^2P^o_{1/2}(4s)$	$4s[1/2]^o$	0	94553	
		1	95399	1.102
$^2P^o_{3/2}(4p)$	$4p[1/2]$	1	104102	1.985
	$4p[5/2]$	3	105463	1.338
		2	105617	1.112
	$4p[3/2]$	1	106087	0.838
		2	106237	1.305
	$4p[1/2]$	0	107054	
$^2P^o_{1/2}(4p)$	$4p[3/2]$	1	107131	0.819
		2	107290	1.260
	$4p[1/2]$	1	107496	1.380
		0	108772	
$^2P^o_{3/2}(3d)$	$3d[1/2]^o$	0	111667	
		1	111818	
	$3d[7/2]^o$	4	112750	
		3	113020	
	$3d[3/2]^o$	2	112138	
		1	114147	
	$3d[5/2]^o$	2	113426	
		3	113716	
$^2P^o_{1/2}(3d)$	$3d[5/2]^o$	2	114641	
		3	114821	
	$3d[3/2]^o$	2	114305	
		1	115366	

(c) Repeat the arguments of part (a) for the ground state and low-lying excited states of ionized argon, given in Table 6.9. This situation is more complicated than neutral argon because the excited states often involve two equivalent p-shell holes and the excited electron. Try to explain, as much as possible, the properties of the ground state and excited states up to 140,000 cm^{-1}, taken from extracts of the NBS tables.

Table 6.9 Energy levels for singly-ionized argon adapted from C. E. Moore, Atomic Energy Levels, NSRDS-NBS 35, reissued, December 1971. The column Config. gives an assignment for the $(3p)^4$ core and the promoted electron. The column Desig. shows an assignment for $^{2S+1}L_J$. The level is the energy above the ground state in cm^{-1}.

Config.	Desig.	J	Level	Obs. g
$3s^2 3p^5$	$3p^5(^3P^o)$	3/2	0.0	
		1/2	1432	
$3s(3p)^6$	$3p^6(^2S)$	1/2	108772	
$3s^2 3p^4(^3P)3d$	$3d(^4D)$	7/2	132328	
		5/2	132482	
		3/2	132632	
		1/2	132738	
$3s^2 3p^4(^3P)4s$	$4s(^4P)$	5/2	134243	1.598
		3/2	135086	1.722
		1/2	135602	2.650
$3s^2 3p^4(^3P)4s$	$4s(^2P)$	3/2	138244	1.334
		1/2	139259	0.676
$3s^2 3p^4(^3P)3d$	$3d(^4F)$	9/2	142187	
		7/2	142718	
		5/2	143109	
		3/2	143372	
$3s^2 3p^4(^3P)3d$	$3d(^2P)$	1/2	144711	
		3/2	145669	
$3s^2 3p^4(^3P)3d$	$3d(^4P)$	1/2	147229	
		3/2	147504	
		5/2	147877	
$3s^2 3p^4(^1D)4s$	$3d(^2D)$	3/2	148621	0.803
		5/2	148843	1.202
$3s^2 3p^4(^3P)3d$	$3d(^2F)$	7/2	149180	
		5/2	150148	
$3s^2 3p^4(^3P)3d$	$3d(^2D)$	3/2	150476	
		5/2	151088	
$3s^2 3p^4(^3P)4p$	$4p(^4P^o)$	5/2	155044	1.599
		3/2	155352	1.720
		1/2	155709	2.638
$3s^2 3p^4(^3P)4p$	$4p(^4D^o)$	7/2	157234	1.427
		5/2	157674	1.334
		3/2	158168	1.199
		1/2	158429	0.00

(Continued)

Table 6.9 (Continued)

Config.	Desig.	J	Level	Obs. *g*
$3s^2 3p^4 (^3P)4p$	$4p(^2D^o)$	5/2	158731	1.241
		3/2	159394	0.918
$3s^2 3p^4 (^3P)4p$	$4p(^2P^o)$	1/2	159707	0.983
		3/2	160240	1.244
$3s^2 3p^4 (^3P)4p$	$4p(^4S^o)$	3/2	161049	1.987
$3s^2 3p^4 (^3P)4p$	$4p(^2S^o)$	1/2	161090	1.695

6.3* In the paper announcing the discovery of the element argon ($Z = 18$), extracted from atmospheric nitrogen (Lord Rayleigh and W. Ramsay, Phil. Trans **186A**, 187 (1895), reprinted in Vol IV of Lord Rayleigh's *Scientific Papers*, Dover reprint (1964)), Rayleigh and Ramsay state that spectroscopic observations by Mr. Crookes and Prof. Schuster on discharge tubes containing samples of the gas showed previously unknown prominent red lines and one intense violet line (at low pressure, where excitation dominated over ionization) and several strong blue lines (at atmospheric pressure, where ionization is more prominent). In 1895, the presence of both red light and blue light suggested to some people that there was more than one element present - Rayleigh and Ramsay were skeptical of this idea, and they were right.

In Table 6.10 are the wavelengths (in Angstroms) of some of the characteristic lines, taken from the CRC Handbook (including some prominent infrared lines not seen in 1895).

(a) From Table 6.8, identify the transitions responsible for the lines observed in neutral argon, listing the Russell-Saunders initial and final states, the total angular momentum change, and the parity change. Are there changes in the spin state? Are these transitions consistent with electric dipole radiation? List also the single-particle orbital transition in each case.

(b) Repeat part (a) for the lines listed for the atmospheric discharge.

6.4 A transition from a $p_{3/2}$ state to a $s_{1/2}$ state (the sodium D_2 line, for example) is split into six components by a weak uniform magnetic field. At right angles to the direction of the magnetic field ($\theta = \pi/2$), all six components are observed with a

Table 6.10 Spectral lines in angstroms of argon at low and atmospheric pressure from the CRC Handbook.

Low Pressure lines	Atmospheric pressure lines
6965	4727
7067	4736
7384	4765
8115	4806
9123	4880

Figure 6.16 Model for Problem 6.4.

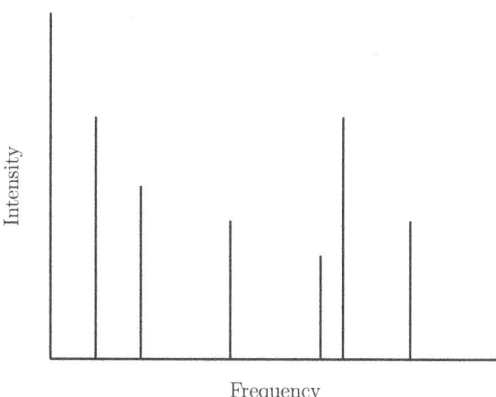

photon detector that accepts all polarizations, while the same detector placed along the field direction ($\theta = 0$) detects only four components (the central pair are absent). If a plane polarization-sensitive detector is placed to detect photons at right angles to the field, some components are observed to have linear polarization parallel to the field (π components) and some have linear polarization perpendicular to the field (σ or *senkrecht* components). Use the electric dipole polarization tensor and the Wigner-Eckart theorem, as discussed in Chapter 4, to discuss this physical problem.

(a) Determine the g-factors for the initial and final states, and so infer the arrangement of the six components in frequency with respect to the single line in the absence of the magnetic field. Why are there not $(2j + 1)(2j' + 1) = 8$ components?

(b) What are the relative intensities of the components for the polarization-insensitive detector at $\theta = 0$ and $\theta = \pi/2$? Make a diagram as shown in Figure 6.16, with the correct spacings and the sum of the intensities equal to unity.

(c) For the detector of linear polarization at $\theta = \pi/2$, repeat the calculation of part (b). Make a diagram in which the sum of the intensities of all the lines (both π and σ) is unity. Label the lines that are π and those that are σ.

6.5 Consider the Russell-Saunders description of three equivalent p-wave electrons (as occur in the nitrogen atom).

(a) By considering a determinantal wave function with $S = 3/2$ and $S_z = 3/2$, show that the only totally antisymmetric spatial wave function has the form

$$\psi_A(1, 2, 3) = \mathbf{r}_1 \cdot (\mathbf{r}_2 \times \mathbf{r}_3) f(r_1) f(r_2) f(r_3) \qquad (6.161)$$

corresponding to the $^4S_{3/2}$ state.

(b) Use the Slater technique of listing the possible m_ℓ and m_s (and M_L and M_S) values for the three electrons, consistent with the Pauli principle, to show that there are 20 possible magnetic substates and identify the Russell-Saunders multiplets.

(c) Express, as an explicit sum of determinants, the $^2P_{3/2}$ wave function with $M_J = 3/2$ and show that it can be written as

$$\psi(1, 2, 3) = \sum_P (-1)^P f(r_1) f(r_2) f(r_3) \mathbf{r}_1 \cdot \mathbf{r}_2 (x_3 + iy_3)[\alpha_1\beta_2 - \beta_1\alpha_2]\alpha_3, \qquad (6.162)$$

where P is a permutation of (1,2,3) and $(-1)^P$ is $+1$ for even permutations and -1 for odd permutations.

6.6 Consider the $1s2p$ singlet and triplet P states ($^1P_1$ and ^3P_J) of the helium-like atoms of nuclear charge Ze. Approximate the one-electron-wave functions by hydrogenic wave function corresponding to charge Z for the $1s$ electron and $Z-1$ for the $2p$ electron. (A variational calculation with effective charge parameters gives essentially this result.) The task here is to compute the electron-electron electrostatic splitting between the singlet and triplet.

(a) Compute the necessary integral in perturbation theory and find the energies of the $^1P_1$ and ^3P states for arbitrary Z. For helium, compare your results for the ionization energies of the two states with the experimental results, 27,182 cm^{-1} or 3.370 eV for the singlet state and 29,229 cm^{-1} or 3.624 eV for the triplet state.

(b) Use the data in Table 6.11 for helium-like atoms (from He I to F VIII, and even beyond) and plot the $1s2p$ singlet-triplet P state splitting in units of Z Rydbergs as a function of Z. On the same graph, show the result of your calculation in part (a). Comment on the agreement or disagreement. 1 Rydberg= 109,737 cm^{-1} = 13.6058 eV.

6.7 Use the data from Table 6.12 on the fine structure splittings for the ^3P_J ground state $(2p)^2$ multiplets of C I, N II,... Na VI from Table 6.12. Plot the fourth root of $(E(J = 1) - E(J = 0)) + (E(J = 2) - E(J = 0))$ in order to determine a value for Z_{eff} for the spin-orbit energy in the $2p$ shell. Use this Z_{eff} with

$$\frac{1}{r}\frac{dV}{dr} = \frac{Z_{\text{eff}}e^2}{r^3} \tag{6.163}$$

and a result from Problem 4.13 to calculate an estimate in cm^{-1} of the coefficient of $\mathbf{L}\cdot\mathbf{S}$ for the $(2p)^2$ atoms. Compare this estimate with the actual data used to find Z_{eff}.

Table 6.11 Data for the energy levels of helium-like elements through sodium for the states $^1P_1$ and ^3P_J (averaged over J). Data are from S. Bashkin and J. O. Stoner, Elsevier, 1975, supplemented by data from https://www.pa.uky.edu/ peter/new-page/levels.html as described in van Hoof, P. A. M., 2018, Galaxies, 6, 63.

Atom	$^1P_1$(cm^{-1})	$\langle ^3P_J\rangle$ (cm^{-1})	ΔE(Ryd)/Z
He I	171,129	169,081	0.00933
Li II	501,809	494,263	0.0229
Be III	997,466	983,368	0.0321
B IV	1,658,020	1,636,900	0.0385
C V	2,483,771	2,455,210	0.0434
N VI	3,473,790	3,438,450	0.0460
O VII	4,629,362	4,586,463	0.0489
F VIII	5,949,900	5,900,218	0.0505
Ne IX	7,346,600	7,379,212	0.0522
Na X	9,090,900	9,025,300	0.0543

Table 6.12 Data for $(2p)^2$ spin-orbit splittings, from
S. Bashkin and J. O. Stoner, Elsevier, 1975, supplemented by data
from https://www.pa.uky.edu/ peter/newpage/levels.html as
described in van Hoof, P. A. M., 2018, Galaxies, 6, 63.

Z	Atom	$E(^3P_1 - ^3P_0)$ cm^{-1}	$E(^3P_2 - ^3P_0)$ cm^{-1}	Ratio
6	C I	16.4	43.4	2.65
7	N II	48.7	130.8	2.69
8	O III	113.4	306.8	2.71
9	F IV	226	614	2.72
10	Ne V	414	1112	2.69
11	NA VI	698	1858	2.66

6.8 Consider the electron-electron electrostatic energy of two equivalent p-shell electrons.

  (a) Evaluate (in atomic units and in electron volts) the integral $F^2(2p, 2p)$, assuming a hydrogen-like wave function with an effective nuclear charge Z_{eff}.

  (b) Use Table 6.13 on the first and second excited states of the $(2p)^2$ atoms C I, N II,... Na VI and also the $(2p)^4 = (2p)^{-2}$ atoms (O I, F II,... Mg V). Tabulate the ratio of the $(^1S - ^1D)$ energy difference to the $(^1D - ^3P)$ energy difference for these atoms and compare with theory.

  Compute the $(^1S - ^1D)$ difference in electron volts and plot versus Z for the $(2p)^2$ and $(2p)^4$ atoms. What Z_{eff} can you infer? With that Z_{eff}, how does the calculated estimate from part (b) compare with the data?

6.9 **(a)** For the neutral Thomas-Fermi atom, prove that the total repulsive electron-electron interaction V_{ee} has an average value that is one-seventh the magnitude of the total attractive interaction of all the electrons with the nucleus V_{eZ}.

  (b) From the variational estimates of Problem 6.1 for helium ($Z = 2$), evaluate $-V_{ee}/V_{eZ}$ and compare with the Thomas-Fermi ratio of 1/7.

  (c) Calculate the average value of the kinetic energy for the Thomas-Fermi atom and combine with the potential energy terms to get the total (negative) energy of the atom. Using the known properties of the dimensionless solution of the Thomas-Fermi differential equation, express your answer as $E(Z) = -bZ^n$, where n is determined and b is in electron volts.

  Look up data (CRC Handbook of Chemistry and Physics has a useful table) for $Z = 2, 8, 16, 20$ on the total binding energy of each atom. Compare with the Thomas-Fermi estimate. If you don't like what you find, look up J. Schwinger, Phys. Rev. **A22**, 1827 (1980); **A24**, 2353 (1981).

6.10* For a free-electron gas at zero temperature (Fermi gas) with a density of n particles per unit volume, discuss the exchange Coulomb energy and the exchange charge density correlated with each electron.

  (a) Show by direct calculation that the exchange Coulomb energy per unit volume can be written in the various forms,

$$E_{ex} = -(e^2 k_F^4/4\pi^3) = -3e^2 k_F n/4\pi = -e^2(81/64\pi)^{1/3} n^{4/3}, \tag{6.164}$$

Table 6.13 Data for $(2p)^2$, $(2p)^4$, $(3p)^2$, and $(3p)^4$ atoms from S. Bashkin and J. Stoner, Elsevier, 1975, supplemented by data from https://www.pa.uky.edu/~peter/atomic as described in van Hoof, P. A. M., 2018, Galaxies, 6, 63.

$(2p)^2$	C I	N II	O III	F IV	Ne V	Na VI
^1S	21648	32689	43184	57344	63900	74394
^1D	10193	15316	20271	23241	30294	35478
^3P	0,16,43	0,49,131	0,113,306	0,226,614	0,414,1112	0,698,1858
$(^1S-^1D)/(^1D-^3P)$	1.124	1.134	1.130	1.121	1.109	1.097
$\Delta(^1S-^1D)$(eV)	1.42	2.15	2.84	3.51	4.17	4.82

$(2p)^4$	O I	F II	Ne III	Na IV	Mg V	
^1S	33792	44919	55747	66780	77712	
^1D	15868	20873	25841	31118	36348	
^3P	0,158,227	0,341,490	0,643,921	0,1106,1576	0,1782,2522	
$(^1S-^1D)/(^1D-^3P)$	1.130	1.152	1.157	1.146	1.138	
$\Delta(^1S-^1D)$(eV)	2.22	2.98	3.71	4.42	5.13	

$(3p)^2$	Si I	P II	S III	Cl IV		
^1S	15394	21576	27163	32550		
^1D	6299	8883	11320	13766		
^3P	0,77,223	0,164,469	0,299,833	0,492,1343		
$(^1S-^1D)/(^1D-^3P)$	1.444	1.429	1.400	1.365		
$\Delta(^1S-^1D)$(eV)	1.13	1.57	1.96	2.33		

$(3p)^4$	S I	Cl II	Ar III	K IV	Ca	
^1S	22181	27878	33267	38548	43847	
^1D	9239	11652	14010	16386	18831	
^3P	0,396,573	0,696,996	0,1112,1570	0,1671,2321	0,2404,3276	
$(^1S-^1D)/(^1D-^3P)$	1.401	1.394	1.375	1.352	1.328	
$\Delta(^1S-^1D)$(eV)	1.60	2.01	2.39	2.75	53.10	

where k_F is the Fermi wave number, related to n by $k_F = (3\pi^2 n)^{1/3}$.

(b) Show that an electron of a certain spin can be thought to have accompanying it an exchange particle density, of the same spin, given by

$$\rho_{Ex} = -(n/2)[3j_1(k_F r)/k_F r]^2, \tag{6.165}$$

where r is the distance away from the electron and $j_\ell(x)$ is the spherical Bessel function of order one. Discuss the value of ρ_{ex} at zero separation and also its volume integral. These results have application to the exchange energy in the Thomas-Fermi-Dirac atom (first incorporated by Dirac) and in energy calculations in condensed matter.

7

Time-Dependent Perturbation Theory and Scattering

Much that we wish quantum mechanics to describe involves change: emission or absorption of radiation, decay of particles or nuclei, collisions of particles. Time-dependent perturbation theory, developed by Dirac, provides the means to address these problems. So powerful is this technique that Fermi declared Dirac's basic result the "Golden Rule." We apply time-dependent perturbation theory in subsequent chapters in a variety of circumstances, but here primarily in the scattering of particles.

Scattering has been central to the exploration of fundamental particles and their interactions from the time of Rutherford, when the Geiger-Marsden experiment revealed the structure of the atom. Scattering by a weak potential or otherwise localized interaction can be treated by perturbation theory applicable to the positive-energy continuum. At low energies, it is convenient to analyze scattering in distinct channels of angular momentum. At very high energies, many angular momenta enter and eikonal techniques can be more effective. Resonance phenomena often provide key insights into dynamics and structure.

7.1 Time-dependent Perturbation Theory

7.1.1 Interaction Picture

The traditional approach to time-dependent perturbation theory, developed by Dirac in an early paper (1927-8), is called variation of parameters. One starts with

$$H = H_0 + H_1(t), \tag{7.1}$$

with $H_0 \psi_n = E_n \psi_n$. Then

$$\psi(t) = \sum_n a_n(t) \psi_n(\mathbf{x}) e^{-iE_n t/\hbar}, \tag{7.2}$$

If $H_1 = 0$, the $a_n(t)$ are constant in time. If $H_1 \neq 0$, the $a_n(t)$ evolve in time. For this approach, see Merzbacher, *Quantum Mechanics*, pp. 450–454 and Gasiorowicz, *Quantum Physics*, pp. 341–342.

An approach that is essentially the same, but more powerful and generalizable to quantum field theory, uses the interaction picture and unitary transformations. We consider $H = H_0 + H_1$, where H_1 is of finite duration or range and the initial and final states correspond to well-separated systems that are eigenstates of H_0. There may or may not be explicit time dependence in H_1.

John David Jackson: A Course in Quantum Mechanics, First Edition. Robert N. Cahn.
© 2024 John Wiley & Sons, Inc. Published 2024 by John Wiley & Sons, Inc.
Companion website: www.wiley.com/go/Jackson/QuantumMechanics

At early or late times (or large initial or final separations), the states of H_0 are relevant. In between, the full Hamiltonian is operative. Under its action, initial or "in" states evolve into final or "out" states:

$t = -\infty$				$t = +\infty$
(H_0, ψ_{in})	$\rightarrow$	(H, ψ)	$\rightarrow$	(H_0, ψ_{out}).

$$(7.3)$$

Assume for definiteness that $H_1 \neq 0$ only for $t_1 < t < t_2$. Then a unitary transformation will connect the "in" and "out" states:

$$|\alpha, \text{out}\rangle = U(t_2, t_1) |\alpha, \text{in}\rangle. \tag{7.4}$$

When $t_1 \rightarrow -\infty$ and $t_2 \rightarrow +\infty$, $U \rightarrow S$, the S-matrix of Heisenberg. As an example, $|\alpha, \text{in}\rangle$ might represent incoming particles with momenta $\mathbf{p}_1$ and $\mathbf{p}_2$. Then $|\alpha, \text{out}\rangle$ could represent a superposition of many states with momenta $\mathbf{p}_1'$ and $\mathbf{p}_2'$ with coefficients given by scattering amplitudes.

In the absence of H_1, the states in the Schrödinger picture have a "trivial" time dependence, i.e. each state of definite energy eigenvalue of H_0 has time dependence $\exp(-iE_n t/\hbar)$. It is convenient to go over to a picture where this time dependence is removed. The interaction picture has states

$$|\alpha, t\rangle_I = e^{iH_0 t/\hbar} |\alpha, t\rangle_S. \tag{7.5}$$

The equation of motion in the interaction picture is

$$i\hbar \frac{\partial}{\partial t} |\alpha, t\rangle_I = \mathcal{H}_1(t) |\alpha, t\rangle_I, \tag{7.6}$$

where

$$\mathcal{H}_1(t) = e^{iH_0 t/\hbar} H_1(t) e^{-iH_0 t/\hbar} \tag{7.7}$$

is the interaction Hamiltonian in the interaction picture. Note that other observables have the same connection via the unitary transformation with $\exp(-iH_0 t/\hbar)$.

Interaction picture kets at different times are related by a unitary transformation

$$|\alpha, t\rangle_I = U_I(t, t_0) |\alpha, t_0\rangle_I \tag{7.8}$$

where

$$i\hbar \frac{\partial U_I}{\partial t}(t, t_0) = \mathcal{H}_1(t) U_I(t, t_0). \tag{7.9}$$

This is the problem we solved in Section 2.14 with the Dyson expansion

$$U_I(t, t_0) = 1 + \sum_{n=1}^{\infty} \left(-\frac{i}{\hbar}\right)^n \int_{t_0}^{t} dt_1 \int_{t_0}^{t_1} dt_2 \int_{t_1}^{t_2} \cdots \int_{t_0}^{t_{n-1}} dt_n \, \mathcal{H}_1(t_1)\mathcal{H}_1(t_2) \cdots \mathcal{H}_1(t_n). \tag{7.10}$$

We recall that in the special circumstance that

$$[\mathcal{H}_1(t_a), \mathcal{H}_1(t_b)] = 0 \tag{7.11}$$

for arbitrary t_a, t_b, then

$$U_I(t, t_0) = \exp\left(-\frac{i}{\hbar} \int_{t_0}^{t} \mathcal{H}_1(t') dt'\right).$$ (7.12)

In general, we work with the Dyson series to a given order in $\mathcal{H}_1$.

Suppose that the system is in state $|m\rangle$ of H_0 for $t < t_1$. We can then ask what is the probability of the system being in the state $|n\rangle$ for $t > t_2$. Here the kets $|m\rangle$, $|n\rangle$ are Schrödinger eigenstates, but with the H_0 time dependence removed. The transition amplitude is evidently

$$a_{n,m}(t_2, t_1) = \langle n | U_I(t_2, t_1)| m\rangle.$$ (7.13)

To calculate this amplitude to first order, we expand

$$U_I^{(1)}(t_2, t_1) = 1 - \frac{i}{\hbar} \int_{t_1}^{t_2} \mathcal{H}(t') dt'.$$ (7.14)

Then

$$a_{n,m}^{(1)}(t_2, t_1) = \delta_{nm} - \frac{i}{\hbar} \int_{t_1}^{t_2} dt' \, \langle n | e^{iH_0 t'/\hbar} H_1(t') e^{-iH_0 t'/\hbar} | m\rangle$$

$$= \delta_{nm} - \frac{i}{\hbar} \int_{t_1}^{t_2} dt' \, e^{i(\omega_n - \omega_m)t'} \langle n | H_1(t') | m\rangle.$$ (7.15)

As an example, consider Problem 4.4, where $H_0 = -\mu_0 \boldsymbol{\sigma} \cdot \mathbf{B}_0 = -\mu_0 B_0 \sigma_z$ and $H_1 = -\mu_0 B_1 \sigma_x$ for $0 < t < T$. We want to find the probability for spin flip after time T. We have $\omega_\pm = \mp \mu_0 B_0/\hbar$ so that $\omega_n - \omega_m = 2\mu_0 B_0/\hbar = 2\omega_0$. Then

$$a_{-+}^{(1)}(T) = -\frac{i}{\hbar} \int_0^T dt \, e^{2i\omega_0 t} \langle -1/2 | -\mu_0 B_1 \sigma_x | +1/2\rangle$$

$$= \frac{i\mu_0 B_1}{\hbar} \int_0^T dt \, e^{2i\omega_0 t} = \frac{\mu_0 B_1}{2\hbar\omega_0} \left(e^{2i\omega_0 T} - 1\right)$$

$$= i\frac{B_1}{B_0} e^{i\omega_0 T} \sin(\omega_0 T).$$ (7.16)

The exact answer is

$$a_{-+}(T) = i\frac{B_1}{\sqrt{B_0^2 + B_1^2}} \sin\left[\frac{\sqrt{B_0^2 + B_1^2}}{B_0} \omega_0 T\right],$$ (7.17)

which agrees to lowest order in the perturbation.

As a second example, consider magnetic resonance, as described in Section 4.6. Here we have

$$H_0 = -\gamma \hbar B_0 J_z = -\hbar \omega_0 J_z$$

$$H_1 = -\hbar \Gamma (J_x \cos \omega t - J_y \sin \omega t),$$ (7.18)

where $\Gamma = \omega_0 B_1/B_0$. At $t = 0$, the system in the state with $J_z = m$. What is the probability of the system being in state m' at time t? We have $\omega_{m'} - \omega_m = (m - m')\omega_0$ and

$$H_1 = -\frac{\hbar \Gamma}{2} \left(J_+ e^{i\omega t} + J_- e^{-i\omega t}\right).$$ (7.19)

Thus we have

$$
\begin{aligned}
a_{m'm}^{(1)} &= +i\frac{\Gamma}{2}\int_0^t dt'\, e^{i(m-m')\omega_0 t}\left\langle m'\left|J_+ e^{i\omega t}+J_- e^{-i\omega t}\right|m\right\rangle \\
&= +i\frac{\Gamma}{2}\left\{\delta_{m',m+1}\sqrt{j(j+1)-m(m+1)}\int_0^t dt'\, e^{i[(m-m')\omega_0+\omega]t'}\right. \\
&\quad \left. +\,\delta_{m',m-1}\sqrt{j(j+1)-m(m-1)}\int_0^t dt'\, e^{i[(m-m')\omega_0-\omega]t'}\right\}.
\end{aligned}
\tag{7.20}
$$

So in first order, we find only $m' = m \pm 1$ with non-vanishing amplitudes and

$$
a_{m'm}^{(1)}(t) = i\Gamma\sqrt{j(j+1)-m(m\pm1)}\,\frac{e^{\pm\frac{i}{2}(\omega-\omega_0)t}}{\omega-\omega_0}\sin[\frac{(\omega-\omega_0)t}{2}].
\tag{7.21}
$$

The transition probability is therefore

$$
P_{m'm}^{(1)} = \frac{\Gamma^2}{(\omega_0-\omega)^2}\sin^2[\frac{(\omega-\omega_0)t}{2}]\cdot[j(j+1)-m(m\pm1)]
\tag{7.22}
$$

for $m' = m \pm 1$ and zero otherwise.

For $j = 1/2$ and $m = \pm 1/2$, $m' = \mp 1/2$ this becomes

$$
P_{\mp1/2,\pm1/2}^{(1)} = \frac{\Gamma^2}{(\omega-\omega_0)^2}\sin^2[\frac{(\omega-\omega_0)t}{2}].
\tag{7.23}
$$

This can be compared to the exact result obtained in Section 4.6

$$
P_{\mp1/2,\pm1/2}^{(1)} = \frac{\Gamma^2}{(\omega-\omega_0)^2+\Gamma^2}\sin^2\left[\frac{t\sqrt{(\omega-\omega_0)^2+\Gamma^2}}{2}\right].
\tag{7.24}
$$

We see that the overall factor Γ^2 in $P_{m'm}^{(1)}(t)$ is the result of the first order approximation in the amplitude; the rest of the amplitude must be zeroth order in Γ. Far from resonance ($\omega = \omega_0$), the first order result is a good approximation to the exact expression. This is not surprising, since at such frequencies the transition probability is small - it should be a good representation.

7.1.2 Adiabatic and Sudden Transitions

Consider a perturbation Hamiltonian $H_1(t; \mathbf{x}, \mathbf{p})$ that changes in time from zero to some final, time-independent form, $H_f(\mathbf{x}, \mathbf{p})$, at some finite time, T. We wish to know the transition amplitudes for various circumstances. Evidently, the Fourier components of the matrix element $\langle n|H_1(t; \mathbf{x}, \mathbf{p})|m\rangle$ at frequency $\omega = \omega_m - \omega_n$ govern the strength of the amplitude. For $m \neq n$, we integrate by parts in order to focus on the region where H_1 is changing in time:

$$
\begin{aligned}
a_{nm}^{(1)}(t) &= -\frac{i}{\hbar}\int_{-\infty}^{t>T} dt'\, e^{i(\omega_n-\omega_m)t'}\langle n|H_1(t;\mathbf{x},\mathbf{p})|m\rangle \\
&= -\frac{i}{\hbar}\frac{1}{i(\omega_n-\omega_m)}\langle n|H_1(t')|m\rangle e^{i(\omega_n-\omega_m)t'}\Big|_{t'=-\infty}^{t>T} \\
&\quad +\frac{1}{\hbar(\omega_n-\omega_m)}\int_{-\infty}^{\infty} dt\, e^{i(\omega_n-\omega_m)}\left\langle n\left|\frac{\partial H_1}{\partial t}(t;\mathbf{x},\mathbf{p})\right|m\right\rangle.
\end{aligned}
\tag{7.25}
$$

The state $|n\rangle$ is an eigenstate of the initial Hamiltonian. That state is not an eigenstate of the final Hamiltonian and to first order the corresponding final eigenstate is

$$|n\rangle_{\text{f}} = |n\rangle + \sum_{m \neq n} |m\rangle \frac{\langle m\,|H_{\text{f}}|\,n\rangle}{E_n - E_m}. \tag{7.26}$$

If we had used $|n_{\text{f}}\rangle$ instead of $|n\rangle$, we would have had an additional contribution to the transition matrix element

$$\frac{\langle n\,|H_{\text{f}}|\,m\rangle}{E_n - E_m} e^{i(\omega_n - \omega_m)t} \tag{7.27}$$

where we restored the time dependence of the initial eigenstates. This would precisely cancel the contribution to the end-point at $t > T$. That is to say, the end-point piece reflected the change in the eigenstate basis, not an actual transition. The final transition probability is thus

$$P_{nm}^{(1)} = \frac{1}{(E_n - E_m)^2} \left| \int_{-\infty}^{\infty} e^{i(\omega_n - \omega_m)t} \left\langle n \left| \frac{\partial H_1}{\partial t}(t; \mathbf{x}, \mathbf{p}) \right| m \right\rangle \right|^2. \tag{7.28}$$

If $H_1(t)$ builds up smoothly to its final value in a time τ, the derivative $\partial H_1/\partial t$ will involve frequency components up to $\omega_{\text{max}} = \mathcal{O}(1/\tau)$. Thus $\Delta\omega = 1/\tau$ set the scale for significant transitions. If $|\omega_n - \omega_m| \gg 1/\tau$, there will be negligible amplitudes for a transition as H_{f} turns on. Thus, for a tightly bound system (with large spacings between levels), the perturbation must vary rapidly in time in order to cause appreciable transitions. Alternatively, for a given level spacing, the slower the temporal variation of $H_1(t)$, the less likely there will be transitions. We speak of the limit $\tau \to \infty$ (really $(\omega_n - \omega_m)\tau \gg 1$) as the "adiabatic limit" in which no transitions occur. The initial state $|m\rangle_{\text{i}}$ evolves without change into $|m\rangle_{\text{f}}$ of the new Hamiltonian, $H_0 + H_{\text{f}}$.

There are physical situations in which the Hamiltonian changes abruptly from one form to another. For example, in nuclear beta decay, electron capture, or alpha emission, the atomic electrons experience a sudden change in the nuclear charge. In tritium decay, $Z_{\text{i}} = 1$, $Z_{\text{f}} = 2$, a dramatic change. We can consider the first-order effects of such sudden changes by writing

$$\frac{\partial H_1}{\partial t} = V\delta(t), \tag{7.29}$$

where V is the change in the Hamiltonian. Then we have

$$P_{nm}^{(1)} = \frac{1}{(E_n - E_m)^2} \left| \langle n\,|V|\,m\rangle \right|^2. \tag{7.30}$$

This is the first-order approximation to the exact result

$$P_{nm} = \left| \langle n, f | m, i \rangle \right|^2, \tag{7.31}$$

where the initial eigenkets $|m, i\rangle$ satisfy $H_0 |m, i\rangle = E_m^{(i)} |m, i\rangle$ and the final eigenkets $|n, f\rangle$ satisfy $(H_0 + V) |n, f\rangle = E_n^{(f)} |n, f\rangle$. This result follows merely by expanding the initial eigenket $|m, i\rangle$ in terms of the assumed complete orthonormal basis of the final Hamiltonian.

7.2 Fermi's Golden Rule

We can apply Eq. (7.15) for time-independent interactions. The meaningful quantity is then not the probability after a long time, but the time rate of increase of the probability, the transition rate, rather than the transition probability. Accordingly, we define

$$w_{nm} = \frac{dP_{nm}}{dt}. \tag{7.32}$$

An example is provided by a beam of particles impinging on a target. We seek the counting rate for a given incident flux at time $t \approx 0$. We want to represent a beam that has been running since long ago and we do that by writing

$$H_1(t) = e^{\eta t/\hbar} H'(\mathbf{x}, \mathbf{p}) \tag{7.33}$$

and take the limit $\eta \to 0^+$ at the end. For $m \neq n$ in Eq. (7.15),

$$
\begin{aligned}
a_{nm}^{(1)}(t) &= -\frac{i}{\hbar} \int_{-\infty}^{t} dt' \, e^{i(\omega_n - \omega_m)t'} e^{\eta t'/\hbar} \langle n | H' | m \rangle \\
&= -\frac{i}{\hbar} \langle n | H' | m \rangle \int_{-\infty}^{t} dt' \, e^{i(\omega_n - \omega_m - i\eta/\hbar)t'} \\
&= \frac{\langle n | H' | m \rangle}{E_m - E_n + i\eta} e^{\eta t/\hbar} e^{i(\omega_n - \omega_m)t}.
\end{aligned}
\tag{7.34}
$$

The probability as a function of time is

$$P_{nm}(t) = \frac{|\langle n | H' | m \rangle|^2}{(E_m - E_n)^2 + \eta^2} e^{2\eta t/\hbar} \tag{7.35}$$

and its time derivative, the transition rate, is

$$w_{nm} = \lim_{\eta \to 0^+} \frac{dP_{nm}}{dt}\bigg|_{t=0} = \lim_{\eta \to 0^+} \frac{2\pi}{\hbar} |\langle n | H' | m \rangle|^2 \frac{\eta/\pi}{(E_m - E_n)^2 + \eta^2} e^{2\eta t/\hbar}. \tag{7.36}$$

We recognize that

$$\int_{-\infty}^{\infty} dE_n \frac{\eta/\pi}{(E_m - E_n)^2 + \eta^2} = 1 \tag{7.37}$$

independent of η, so that as $\eta \to 0^+$ we obtain $\delta(E_n - E_m)$. Altogether, the transition rate is

$$w_{nm} = \frac{2\pi}{\hbar} |\langle n | H' | m \rangle|^2 \delta(E_n - E_m). \tag{7.38}$$

This is often called Fermi's Golden Rule No. 2, but it is due originally to Dirac: P. A. M. Dirac, *Proc. Roy. Soc. (London)* **A114**, 243 (1927), reprinted in J. S. Schwinger, *Selected Papers on Quantum Electrodynamics*, Dover reprint, 1958. See especially Section 5, Eq. (24).

This result is so useful, it is worth considering from different points of view. Beginning again from Eq. (7.15), for $n \neq m$

$$a_{n,m}^{(1)}(t) = \delta_{nm} - \frac{i}{\hbar} \int_{-\infty}^{t} dt' \, e^{i(\omega_n - \omega_m)t'} \langle n | H_1 | m \rangle,$$

so that

$$P_{nm}(t) = \frac{1}{\hbar^2} |\langle n | H_1 | m \rangle|^2 \left[\int_{-\infty}^{t} dt' e^{i(\omega_n - \omega_m)t'} \right] \times \left[\int_{-\infty}^{t} dt'' e^{-i(\omega_n - \omega_m)t''} \right]. \tag{7.39}$$

We then simply compute the transition rate

$$
\begin{aligned}
w_{nm} &= \frac{dP_{nm}}{dt} \\
&= \frac{1}{\hbar^2} |\langle n | H_1 | m \rangle|^2 \left\{ e^{i(\omega_n - \omega_m)t} \times \left[\int_{-\infty}^{t} dt' e^{-i(\omega_n - \omega_m)t'} \right] + \quad \text{complex conjugate} \right\} \\
&= \frac{1}{\hbar^2} |\langle n | H_1 | m \rangle|^2 \left\{ \int_{-\infty}^{t} dt' e^{i(\omega_n - \omega_m)(t-t')} + \quad \text{complex conjugate} \right\} \\
&= \frac{1}{\hbar^2} |\langle n | H_1 | m \rangle|^2 \left\{ \int_{-\infty}^{0} d\tau e^{i(\omega_n - \omega_m)\tau} + \int_{0}^{\infty} d\tau e^{i(\omega_n - \omega_m)\tau} \right\} \\
&= \frac{1}{\hbar^2} |\langle n | H_1 | m \rangle|^2 2\pi\delta(\omega_n - \omega_m) = \frac{2\pi}{\hbar} |\langle n | H_1 | m \rangle|^2 \delta(E_n - E_m). \tag{7.40}
\end{aligned}
$$

Just as in time-independent perturbation theory, we can consider transitions that are not possible through a single interaction by H', but require two such interactions. We have

$$U^{(2)}(t, t_0) = 1 - \frac{i}{\hbar} \int_{t_0}^{t} dt_1 \mathcal{H}_1(t_1) - \frac{1}{\hbar^2} \int_{t_0}^{t} dt_1 \mathcal{H}_1(t_1) \int_{t_0}^{t_1} dt_2 \mathcal{H}_1(t_2). \tag{7.41}$$

Then

$$
\begin{aligned}
a_{mn}^{(2)}(t, t_0) &= \delta_{nm} - \frac{i}{\hbar} \int_{t_0}^{t} dt_1 e^{i(\omega_n - \omega_m)t_1} \langle n | H_1(t_1) | m \rangle \\
&\quad - \frac{1}{\hbar^2} \sum_{p} \int_{t_0}^{t} dt_1 e^{i(\omega_n - \omega_p)t_1} \langle n | H_1(t_1) | p \rangle \int_{t_0}^{t_1} dt_2 \, e^{i(\omega_p - \omega_m)t_2} \langle p | H_1(t_2) | m \rangle. \tag{7.42}
\end{aligned}
$$

With $H_1(t) = e^{\eta t/\hbar} H'$, we get (with $t_0 = -\infty$)

$$
\begin{aligned}
a_{nm}^{(2)}(t) &= \delta_{nm} + \frac{\langle n | H' | m \rangle}{E_m - E_n + i\eta} e^{\eta t/\hbar} e^{i(\omega_n - \omega_m)t} \\
&\quad + \sum_{p} \frac{\langle n | H' | p \rangle \langle p | H' | m \rangle}{(E_m - E_n + 2i\eta)(E_m - E_p + i\eta)} e^{2\eta t/\hbar} e^{i(\omega_n - \omega_m)t}. \tag{7.43}
\end{aligned}
$$

Here we can usually assume that the sum excludes $p = n$ and $p = m$ because H' does not couple (strongly) to these states. If the first order matrix element vanishes, the transition rate is

$$w_{nm}^{(2)} = \frac{2\pi}{\hbar} \left| \sum_p \frac{\langle n|H'|p\rangle \langle p|H'|m\rangle}{E_m - E_p + i\eta} \right|^2 \delta(E_m - E_n).$$ (7.44)

Before applying the Golden Rule to scattering, we consider that topic in a more pedestrian fashion.

7.3 Scattering Amplitude

When a beam of particles of well-defined momentum is incident on a center of force, the particles interact with the force center and are deflected - scattered - emerging with different momentum. If the scattering center or the projectile has internal degrees of freedom, there can be internal excitation and the magnitude of the momentum of the scattered particle can be different from that of the incident particle. We begin with elastic scattering. In terms of Schrödinger wave functions, we have

$$\psi = \psi_{\text{incident}} + \psi_{\text{scattered}},$$ (7.45)

where $\psi_{inc} = \psi_0 = e^{i\mathbf{k}\cdot\mathbf{x}}$, and for large distances from the scatterer,

$$\psi_{\text{scattered}} \rightarrow f(\theta, \phi) \frac{e^{ikr}}{r}$$ (7.46)

corresponding to outgoing spherical waves at infinity.

It is customary to describe the scattering in terms of a *differential cross section*, defined as follows:

$$d\sigma = \frac{\text{Outgoing flux} \times dA}{\text{Incoming flux}},$$ (7.47)

where $dA = r^2 d\Omega$ is the element of area subtending the solid angle element $d\Omega$ at the origin at the angles θ, ϕ. See Figure 7.1.

$$\text{Incoming flux} = \frac{\hbar k}{m} \left| e^{i\mathbf{k}\cdot\mathbf{x}} \right|^2 = \frac{\hbar k}{m};$$

$$\text{Outgoing flux} = \frac{\hbar k}{m} \left| \frac{f(\theta, \phi)e^{ikr}}{r} \right|^2 = \frac{\hbar k}{m} \frac{|f(\theta, \phi)|^2}{r^2}.$$

(7.48)

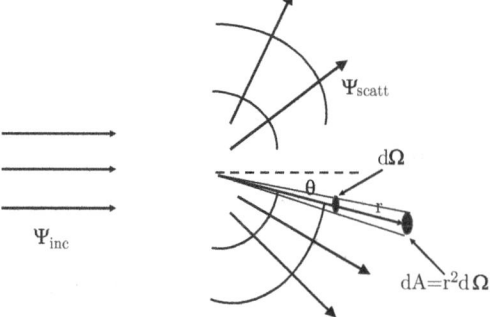

Figure 7.1 An incident beam of particles scatters through an angle θ into a solid angle element $d\Omega$. The differential cross section $d\sigma/d\Omega$ is the ratio of the outgoing flux per unit solid angle to the incident flux per unit area.

Thus the differential scattering cross section is

$$\frac{d\sigma}{d\Omega} = |f(\theta,\phi)|^2. \tag{7.49}$$

7.4 Born Approximation

We can turn the Schrödinger equation into an integral equation. Consider $H = H_0 + V$, where $H_0 = p^2/2m = -(\hbar^2/2m)\nabla^2$. The unperturbed Hamiltonian has solutions, that are plane waves, i.e.

$$\psi_0 = e^{-i\mathbf{k}\cdot\mathbf{x}}, \quad k = \sqrt{2mE/\hbar^2}. \tag{7.50}$$

The complete Schrödinger equation, $(H_0 + V)\psi = E\psi$, can be written

$$(\nabla^2 + k^2)\psi(\mathbf{x}) = \frac{2m}{\hbar^2}V(\mathbf{x})\psi(\mathbf{x}). \tag{7.51}$$

If we pretend that the right hand side is a source term, we can use the method of Green functions to convert the equation into an integral equation. Recall that the Helmholtz wave equation for its Green function

$$(\nabla^2 + k^2)G(\mathbf{x}, \mathbf{x}') = -\delta(\mathbf{x} - \mathbf{x}') \tag{7.52}$$

has solutions

$$\begin{aligned} G^{\pm}(\mathbf{x}, \mathbf{x}') &= \frac{1}{4\pi}\frac{e^{\pm ikR}}{R}, \quad R = |\mathbf{x} - \mathbf{x}'| \\ &= \frac{1}{8\pi^3}\int d^3q\, \frac{e^{-i\mathbf{q}\cdot(\mathbf{x}-\mathbf{x}')}}{q^2 - k^2 \mp i\epsilon}. \end{aligned} \tag{7.53}$$

Evidently, we have solutions of the Schrödinger equation of the form

$$\psi_{\mathbf{k}}^{(\pm)} = e^{i\mathbf{k}\cdot\mathbf{x}} - \int d^3x'\, G^{(\pm)}(\mathbf{x}, \mathbf{x}')W(\mathbf{x}')\psi_{\mathbf{k}}^{(\pm)}(\mathbf{x}'). \tag{7.54}$$

Taking $-i\epsilon$ in the denominator guarantees the outgoing wave, e^{+ikR}, and we henceforth drop the superscript $^+$. See Appendix B.2. If the initial wave is specified, as it is in our situation, we use the outgoing wave solution. Then the integral equation equivalent to the Schrödinger differential equation is

$$\psi(\mathbf{x}) = e^{i\mathbf{k}\cdot\mathbf{x}} - \frac{2m}{4\pi\hbar^2}\int d^3x'\, \frac{e^{ik|\mathbf{x}-\mathbf{x}'|}}{|\mathbf{x}-\mathbf{x}'|}V(\mathbf{x}')\psi(\mathbf{x}'). \tag{7.55}$$

We can find a formal expression for the scattering amplitude from this expression by examining the large-distance behavior of the Green function. We need $|\mathbf{x} - \mathbf{x}'| \approx |\mathbf{x}| - \hat{\mathbf{k}}'\cdot\mathbf{x}' = r - \hat{\mathbf{k}}'\cdot\mathbf{x}'$, where $\hat{\mathbf{k}}'$ is a unit vector in the direction of $\mathbf{x}$, which is the direction of observation, or the direction of the scattered momentum, $\mathbf{k}'$. Thus we have

$$\psi \rightarrow e^{i\mathbf{k}\cdot\mathbf{x}} - \frac{2m}{4\pi\hbar^2}\frac{e^{ikr}}{r}\int d^3x'\, e^{-i\mathbf{k}'\cdot\mathbf{x}'}V(\mathbf{x}')\psi(\mathbf{x}'). \qquad (7.56)$$

This means that the scattering amplitude is

$$f(\theta,\phi) = f(\mathbf{k}',\mathbf{k}) = -\frac{2m}{4\pi\hbar^2}\int d^3x\, e^{-i\mathbf{k}'\cdot\mathbf{x}}V(\mathbf{x})\psi(\mathbf{x}). \qquad (7.57)$$

This formal result shows how the scattering amplitude can be computed if only we knew $\psi(\mathbf{x})$.

If the potential is weak enough, we can treat the scattering in lowest order in the strength of $V(\mathbf{x})$. In the above expression, it is therefore permissible to replace $\psi(\mathbf{x})$ with its zeroth-order approximation, $\psi \rightarrow \psi_0 = e^{i\mathbf{k}\cdot\mathbf{x}}$. This gives the so-called Born approximation (or first Born approximation) to the scattering amplitude.

$$f^{(1)}(\mathbf{k}',\mathbf{k}) = -\frac{2m}{4\pi\hbar^2}\int d^3x\, e^{-i\mathbf{k}'\cdot\mathbf{x}}V(\mathbf{x})e^{i\mathbf{k}\cdot\mathbf{x}}. \qquad (7.58)$$

If $V(\mathbf{x})$ actually depended on operators involving $\mathbf{p}$, the ordering of the plane wave factors could be important. If it is an ordinary function of $\mathbf{x}$, the exponentials can be combined:

$$f^{(1)}(\mathbf{k}',\mathbf{k}) = -\frac{2m}{4\pi\hbar^2}\int d^3x\, e^{i\mathbf{q}\cdot\mathbf{x}}V(\mathbf{x}), \qquad (7.59)$$

where $\mathbf{q} = \mathbf{k} - \mathbf{k}'$ is the momentum transfer (to the scattering center). The Born approximation scattering amplitude is then just the Fourier transform of the scattering potential. For elastic scattering,

$$q^2 = 4k^2\sin^2(\theta/2). \qquad (7.60)$$

As an example, we consider a Gaussian potential,

$$V = V_0\, e^{-r^2/a^2}. \qquad (7.61)$$

Then the Born approximation amplitude is (changing variables)

$$f^{(1)}(\mathbf{q}) = -\frac{\lambda a}{4\pi}\int d^3y\, e^{i\mathbf{Q}\cdot\mathbf{y}}e^{-y^2}, \qquad (7.62)$$

where

$$\mathbf{y} = \mathbf{x}/a; \qquad \mathbf{Q} = \mathbf{q}a; \qquad \lambda = \frac{2mV_0 a^2}{\hbar^2}. \qquad (7.63)$$

Noting that

$$\int d^3y\, e^{-|\mathbf{y}|^2} = \pi^{3/2}, \qquad (7.64)$$

we compute

$$\int d^3y\, e^{i\mathbf{Q}\cdot\mathbf{y}}e^{-y^2} = \int d^3y\, e^{-(\mathbf{y}-i\mathbf{Q}/2)^2 - Q^2/4} = \pi^{3/2}e^{-Q^2/4}. \qquad (7.65)$$

Thus

$$f^{(1)}(\mathbf{q}) = -\frac{\lambda a}{4\pi}\pi^{3/2}e^{-Q^2/4} = -\frac{a\lambda\sqrt{\pi}}{4}e^{-k^2 a^2 \sin^2(\theta/2)}. \qquad (7.66)$$

Figure 7.2 The differential cross section in first Born approximation for a Gaussian potential $V = V_0 e^{-r^2/a^2}$ normalized to the value at $\cos\theta = 1$.

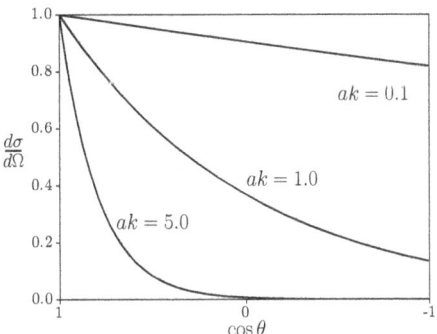

The differential cross section is

$$\frac{d\sigma^{(1)}}{d\Omega} = \frac{\pi\lambda^2 a^2}{16} e^{-k^2 a^2(1-\cos\theta)}. \tag{7.67}$$

The angular dependence several values of ak is shown in Figure 7.2.

The total scattering cross section is the integral over all angles:

$$\sigma_{scat} = \int_{-1}^{1} 2\pi d\cos\theta \frac{d\sigma}{d\Omega} = \frac{\pi^2\lambda^2 a^2}{8} \int_{-1}^{1} d\mu' e^{-k^2 a^2(1-\mu')}$$

$$= \frac{\pi^2\lambda^2}{8k^2}\left(1 - e^{-2k^2 a^2}\right). \tag{7.68}$$

Note that for $k \to 0$ (zero energy),

$$\frac{d\sigma}{d\Omega} \to \frac{\pi\lambda^2 a^2}{16}; \qquad \sigma_{scat} \to 4\pi\frac{d\sigma}{d\Omega} = \frac{\pi^2\lambda^2 a^2}{4}. \tag{7.69}$$

At high energies ($ka \gg 1$), $d\sigma/d\Omega$ is sharply peaked in the forward direction and $\sigma_{scatt} \approx \pi^2\lambda^2/(8k^2)$.

7.5 Scattering Theory from Fermi's Golden Rule

Fermi's Golden Rule provides the means to determine rates for quantum mechanical processes: decay, absorption, emission, and scattering. Here we apply it to scattering of particles confined to a box of volume V (not to be confused with a potential of the same name). We are interested in the transition rate for i $\to$ f due to a perturbation H' of a general nature. While the box normalization is not essential, it can be a useful check on the calculation and assuages any fears that unrenormalizable wave functions like $e^{i\mathbf{k}\cdot\mathbf{x}}$ are mathematically suspect.

$$dw_{fi} = \frac{2\pi}{\hbar}\left|\langle f|H'|i\rangle\right|^2 \rho_f(E_f), \tag{7.70}$$

where $\rho_f(E_f)$ is the density of final states of energy E_f into which the rate is measured. The usual counting tells us that in a cube of sides L, the condition $kL = n\pi$ fixes the number of state with wave numbers up to some value k to be

$$\frac{1}{8}\frac{4\pi}{3}\left(\frac{k^2 L^2}{\pi^2}\right)^{3/2} = \frac{\pi}{6}\frac{k^3}{\pi^3}V \tag{7.71}$$

and the number inside a shell at radius k of thickness dk is

$$\frac{\pi}{2}\frac{k^2 dk}{\pi^3} V.$$ (7.72)

If only a fraction $d\Omega/4\pi$ of the solid angle is included, the number of states is

$$\frac{\pi}{2}\frac{k^2 dk}{\pi^3}\frac{d\Omega}{4\pi}V = \frac{d^3 k}{(2\pi)^3}V.$$ (7.73)

Thus, the density of states in energy is

$$\rho(E) = \frac{V}{(2\pi)^3}k^2\frac{dk}{dE}d\Omega = \frac{V}{(2\pi\hbar)^3}p^2\frac{dp}{dE}d\Omega.$$ (7.74)

From $E^2 = p^2 c^2 + m^2 c^4$, we have in all instances

$$\frac{dp}{dE} = \frac{E}{pc^2}$$ (7.75)

so that

$$\rho(E) = \frac{V}{(2\pi\hbar)^3}\frac{pE}{c^2}d\Omega.$$ (7.76)

In the non-relativistic limit, this is

$$\rho(E) = \frac{V}{(2\pi\hbar)^3}m^2 v\, d\Omega,$$ (7.77)

where v is the velocity. In the extreme relativistic limit $E = pc$ and

$$\rho(E) = \frac{V}{(2\pi\hbar c)^3}E^2\, d\Omega.$$ (7.78)

For now, we consider the non-relativistic case and assume that the incoming and outgoing particles are the same, though allowing the possibility that the magnitude of the incoming and outgoing velocities, v and v', are different. The incoming flux is v/V. Since the rate is equal to the flux times the cross section, we have

$$dw_{\text{fi}} = \frac{v}{V}d\sigma = \frac{2\pi}{\hbar}\left|\langle f|H'|i\rangle\right|^2\frac{V}{(2\pi\hbar)^3}m^2 v'\, d\Omega'$$

$$\frac{d\sigma}{d\Omega'} = \left(\frac{V}{2\pi}\right)^2\frac{v'}{v}\frac{m^2}{\hbar^4}\left|\langle f|H'|i\rangle\right|^2.$$ (7.79)

7.5.1 Elastic and Inelastic Scattering in Born Approximation

Let us suppose that the collision is actually between two composite systems (though the compositeness could be just spin degrees of freedom) and that $m = m_1 m_2/(m_1 + m_2)$ is the reduced mass. Let $\mathbf{x}$ be the coordinate between the two centers of mass and $\mathbf{r}_1$ and $\mathbf{r}_2$ represent the internal coordinates in each system. Then the initial wave function is

$$\Psi_i = \frac{1}{\sqrt{V}} e^{i\mathbf{k}\cdot\mathbf{x}} \phi_1 \phi_2 \qquad (7.80)$$

and that of the final state is

$$\Psi_f = \frac{1}{\sqrt{V}} e^{i\mathbf{k}'\cdot\mathbf{x}} \phi_1' \phi_2', \qquad (7.81)$$

where the ϕs represent internal motion and may be different in the final state. The differential cross section is then

$$\frac{d\sigma}{d\Omega} = \frac{v'}{v} \left| \frac{2m}{4\pi\hbar^2} \int d^3x \, e^{-i\mathbf{k}'\cdot\mathbf{x}} \langle \phi_1' \phi_2' | \, H' \, |\phi_1 \phi_2 \rangle e^{i\mathbf{k}\cdot\mathbf{x}} \right|^2. \qquad (7.82)$$

Conservation of energy is expressed by

$$\frac{\hbar^2 k^2}{2m} + E_1 + E_2 = \frac{\hbar^2 k'^2}{2m} + E_1' + E_2', \qquad (7.83)$$

where the E_is are internal energies. This is a general formula that can cover a wide variety of situations.

7.5.2 Elastic Scattering by a Static Potential

If we have spinless, structureless elastic scattering by a static potential $V(\mathbf{x})$, this reduces to Eqs. (7.49) and (7.58). As a further example, consider a shielded Coulomb potential, $V = zZe^2 e^{-\mu r}/r$. Then

$$\tilde{V}(\mathbf{q}) = \int d^3x \, V(\mathbf{x}) e^{i\mathbf{q}\cdot\mathbf{x}} = zZe^2 \int d^3x \frac{e^{-\mu r}}{r} e^{i\mathbf{q}\cdot\mathbf{x}}$$

$$= \frac{4\pi zZe^2}{q} \int_0^\infty e^{-\mu r} \sin qr \, dr = \frac{4\pi zZe^2}{q^2 + \mu^2}. \qquad (7.84)$$

In general, even in the inelastic case,

$$q^2 = (\mathbf{k} - \mathbf{k}')^2 = k^2 + k'^2 - 2kk' \cos\theta = (k - k')^2 + 2kk'(1 - \cos\theta)$$

$$= (k - k')^2 + 4kk' \sin^2(\theta/2) \qquad (7.85)$$

and

$$d\Omega = \frac{1}{2kk'} dq^2 d\phi. \qquad (7.86)$$

Using this result, we have

$$\frac{d\sigma}{d\Omega} = \left(\frac{m}{2\pi\hbar^2} \frac{4\pi zZe^2}{q^2 + \mu^2} \right)^2 = \left(\frac{zZe^2}{2mv^2} \right)^2 \frac{1}{(\sin^2(\theta/2) + \mu^2/(4k^2))^2}. \qquad (7.87)$$

For $\mu \to 0$, this of course reduces to the Rutherford cross section.

7.5.3 Scattering by a Spin-Orbit Potential

Consider elastic scattering by a spin-orbit potential:

$$H' = \frac{1}{r}\frac{dV_{LS}(r)}{dr}\mathbf{L}\cdot\mathbf{S}, \tag{7.88}$$

where $\mathbf{S}$ may be the spin of a nucleon scattering from a nucleus or may be the sum of the spins of two particles that are scattering. Then the spin-space operator will be of the form

$$f^{(1)}(\mathbf{S},\mathbf{k}',\mathbf{k}) = \frac{m}{2\pi\hbar^2}\int d^3x\, e^{-i\mathbf{k}'\cdot\mathbf{x}}\frac{1}{r}\frac{dV_{LS}}{dr}\mathbf{S}\cdot\left(\frac{\mathbf{x}\times\mathbf{p}}{\hbar}\right)e^{i\mathbf{k}\cdot\mathbf{x}}$$

$$= -\frac{m}{2\pi\hbar^2}(\mathbf{S}\times\mathbf{k})\cdot\int d^3x\, e^{i\mathbf{q}\cdot\mathbf{x}}\frac{\mathbf{x}}{r}\frac{dV_{LS}}{dr}. \tag{7.89}$$

But we recognize

$$\frac{\mathbf{x}}{r}\frac{dV_{LS}}{dr} = \nabla V_{LS}, \tag{7.90}$$

and with an integration by parts we have

$$f^{(1)}(\mathbf{S},\mathbf{k}',\mathbf{k}) = \frac{im}{2\pi\hbar^2}(\mathbf{S}\times\mathbf{k})\cdot\mathbf{q}\int d^3x\, e^{i\mathbf{q}\cdot\mathbf{x}}V_{LS}(r)$$

$$= -\frac{im}{2\pi\hbar^2}\mathbf{S}\cdot(\mathbf{k}\times\mathbf{k}')\tilde{V}_{LS}(q^2). \tag{7.91}$$

In nuclear scattering, the central potential and the spin-orbit interaction are both attractive. The amplitude will thus have the form

$$f^{(1)} = \frac{m}{2\pi\hbar^2}\left[-\tilde{V}_{central}(q^2) - i\mathbf{S}\cdot(\mathbf{k}\times\mathbf{k}')\tilde{V}_{LS}(q^2)\right]. \tag{7.92}$$

The general form of the amplitude is

$$F = A(q^2) + i\mathbf{S}\cdot(\mathbf{k}\times\mathbf{k}')B(q^2). \tag{7.93}$$

Why should this be its form?

(a) With rotational invariance and parity conservation we must have a scalar quantity. We have at our disposal $\mathbf{S},\mathbf{k},\mathbf{k}'$. The pieces that are independent of $\mathbf{S}$ must be functions of $k^2, k'^2, \mathbf{k}\cdot\mathbf{k}'$. In elastic scattering $k^2 = k'^2$ and the possible variables are k^2 and $\cos\theta$. [In Born approximation, $\mathbf{q}$ is the only quantity and thus the variable is q^2.]

(b) Since $\mathbf{S}$ is an axial vector and we want an overall scalar, we need another axial vector and $\mathbf{k}\times\mathbf{k}'$ is the only thing available. Thus we get $\mathbf{S}\cdot(\mathbf{k}\times\mathbf{k}')$ times a function of k^2 and $\cos\theta$ (or just q^2 in Born approximation).

(c) The presence of i multiplying a real V_{SL} will result in no difference between scattering of up-polarized and down-polarized scattering. A more complete treatment would allow complex A and B. Then there is a different differential cross section for different polarizations and with unpolarized incident particles, the scattered particles are polarized. Proton-carbon scattering at 10s to 100s of MeV was an early source of secondary beams of polarized protons.

7.6 Inelastic Scattering

As example of the Born approximation, consider the inelastic scattering of a particle of mass M, charge ze, and initial speed v on a many-electron atom. The interaction is assumed to be electrostatic and spin is ignored. The initial state includes the incident particle with momentum $\hbar \mathbf{k}$ and the atom in its initial state, $|a\rangle$. The final state has the scattered particle of momentum $\hbar \mathbf{k}'$ and the atom in state $|b\rangle$:

$$|i\rangle = \frac{1}{\sqrt{V}} e^{i\mathbf{k}\cdot\mathbf{x}} |a\rangle; \qquad \langle f| = \frac{1}{\sqrt{V}} e^{-i\mathbf{k}'\cdot\mathbf{x}} \langle b|. \tag{7.94}$$

Our general formula is

$$\frac{d\sigma_{ba}}{d\Omega} = \left(\frac{V}{2\pi}\right)^2 \frac{v'}{v} |\langle f|H'|i\rangle|^2 \frac{M^2}{\hbar^4}, \tag{7.95}$$

where here the interaction Hamiltonian is

$$H' = \frac{zZe^2}{|\mathbf{x}|} - \sum_j \frac{ze^2}{|\mathbf{x} - \mathbf{r}_j|}. \tag{7.96}$$

The position of the incident particle is $\mathbf{x}$ while the electrons are at $\mathbf{r}_j$. The nucleus of charge Z is at the origin. The matrix element is thus

$$\langle f|H'|i\rangle = \frac{1}{V} \int d^3x \, e^{i\mathbf{q}\cdot\mathbf{x}} \langle b| \frac{zZe^2}{|\mathbf{x}|} - \sum_j \frac{ze^2}{|\mathbf{x} - \mathbf{r}_j|} |a\rangle. \tag{7.97}$$

The first term does not involve the electronic coordinates and vanishes because the distinct state $|a\rangle$ and $|b\rangle$ are orthogonal. In the second term, we write $\mathbf{x} = \mathbf{x}' + \mathbf{r}_j$ in each separate contribution j. Then we have

$$\langle f|H'|i\rangle = -\frac{ze^2}{V} \int d^3x' \frac{e^{i\mathbf{q}\cdot\mathbf{x}'}}{|\mathbf{x}'|} \langle b| \sum_j e^{i\mathbf{q}\cdot\mathbf{r}_j} |a\rangle$$

$$= -\frac{ze^2}{V} \frac{4\pi}{q^2} \langle b| \sum_j e^{i\mathbf{q}\cdot\mathbf{r}_j} |a\rangle \equiv -\frac{ze^2}{V} \frac{4\pi}{q^2} F_{ba}(\mathbf{q}). \tag{7.98}$$

We see that $F_{ba}(\mathbf{q})$ is the transition form factor, the Fourier transform of the product of the initial and final wave functions. This gives us a cross section

$$\frac{d\sigma_{ba}}{d\Omega'} = \left(\frac{2Mze^2}{\hbar^2}\right)^2 \frac{v'}{v} \frac{1}{(q^2)^2} |F_{ba}(\mathbf{q})|^2. \tag{7.99}$$

To convert this to $d\sigma/dq^2$, we use

$$q^2 = (\mathbf{k} - \mathbf{k}')^2 = k^2 + k'^2 - 2kk' \cos\theta'$$

$$dq^2 \, d\phi = 2kk' d\Omega', \tag{7.100}$$

so

$$\frac{v'}{v} d\Omega' = \frac{v'}{v} \frac{1}{2kk'} dq^2 d\phi = \frac{\hbar^2}{2M^2v^2} dq^2 d\phi. \tag{7.101}$$

Integrating over ϕ,

$$\frac{d\sigma}{dq^2} = \frac{\pi\hbar^2}{M^2v^2}\left(\frac{d\sigma}{d\Omega'}\frac{v}{v'}\right). \tag{7.102}$$

Applying this to our problem,

$$\frac{d\sigma}{dq^2} = 4\pi\left(\frac{ze^2}{\hbar vq^2}\right)^2|F_{ba}(\mathbf{q})|^2. \tag{7.103}$$

We examine briefly two situations:

(a) $1s \to 2p$ excitation in hydrogen - this is called an optically allowed inelastic collision.
(b) $1s \to 2s$ excitation in hydrogen - this is not optically allowed.

This language can be understood by considering the form factor evaluated at small q. Expanding the exponential in Eq. (7.98), we see that the first term vanishes because the states $|a\rangle$ and $|b\rangle$ are orthogonal while the next term, $\mathbf{q} \cdot \mathbf{x}$, gives the dipole matrix element for an E1 transition, which is allowed for $1s \to 2p$ but not for $1s \to 2s$. Because the dipole matrix element comes from $\mathbf{q} \cdot \mathbf{x}$, if we choose the coordinate system for the final state to have the z axis in the direction of $\mathbf{q}$, only the $m = 0$ magnetic substate of $2p$ will be generated. Introducing $Q = qa_0$, we can evaluate the form factor using atomic units. The relevant wave functions are

$$\psi_{1s} = \frac{2}{\sqrt{4\pi}}e^{-r}; \quad \psi_{2p,m=0} = \frac{1}{2\sqrt{6}}re^{-r/2}\sqrt{\frac{3}{4\pi}}\cos\theta; \quad \psi_{2s} = \frac{1}{\sqrt{2}}(1-r/2)e^{-r/2}\frac{1}{\sqrt{4\pi}}. \tag{7.104}$$

We calculate the form factors

$$\begin{aligned}
F_{2p,1s}(q^2) &= \frac{1}{\sqrt{2}}\frac{1}{4\pi}\int_0^\infty r^2dr\,e^{-3r/2}\int d\Omega\,r\cos\theta e^{iQr\cos\theta} \\
&= -\frac{i}{\sqrt{2}}\frac{d}{dQ}\int_0^\infty r^2dr\,e^{-3r/2}\frac{\sin Qr}{Qr} \\
&= i6\sqrt{2}\frac{Q}{(Q^2+9/4)^3} = i\frac{2^{15/2}}{3^5}\frac{qa_0}{(1+\frac{4}{9}q^2a_0^2)^3}.
\end{aligned} \tag{7.105}$$

Similarly, for $1s \to 2s$ we find, taking the isotropic piece of $e^{i\mathbf{q}\cdot\mathbf{x}} = \sin Qr/(Qr)$,

$$\begin{aligned}
F_{2s,1s}(q^2) &= \sqrt{2}\int_0^\infty r^2dr\,(1-r/2)e^{-3r/2}\frac{\sin Qr}{Qr} \\
&= \frac{4\sqrt{2}Q^2}{(Q^2+9/4)^3} = \frac{2^{17/2}}{3^6}\frac{q^2a_0^2}{(1+\frac{4}{9}q^2a_0^2)^3}.
\end{aligned} \tag{7.106}$$

Note the extra power of q in $F_{2s,1s}$ relative to $F_{2p,1s}$. The cross sections are

$$\frac{d\sigma}{dq^2}(1s \rightarrow 2p) = \frac{2^{17}}{3^{10}}\pi a_0^2 \left(\frac{e^2}{\hbar v}\right)^2 \frac{1}{q^2}\left(1 + \frac{4}{9}q^2 a_0^2\right)^{-6}$$

$$\frac{d\sigma}{dq^2}(1s \rightarrow 2s) = \frac{2^{19}}{3^{12}}\pi a_0^4 \left(\frac{e^2}{\hbar v}\right)^2 \left(1 + \frac{4}{9}q^2 a_0^2\right)^{-6}. \tag{7.107}$$

The optically allowed cross section grows as $1/q^2$ at low q^2, and falls significantly for $q^2 a_0^2 > 9/4$. There is a minimum q^2 required to excite the atom. For our example, the excitation energy needed is

$$\Delta E = \frac{3}{8}\frac{e^2}{a_0}. \tag{7.108}$$

Some minimum loss of the incident particle's momentum is required for the excitation, and thus there is a minimum value of q^2 for which this is possible.

$$\frac{\hbar^2 k'^2}{2m} = \frac{\hbar^2 k^2}{2m} - \Delta E. \tag{7.109}$$

Assuming the momentum lost is a small fraction of the total momentum $\hbar k = mv$, we have

$$k' \approx k - \frac{2m}{2\hbar^2 k}\frac{3}{8}\frac{e^2}{a_0} = k - \frac{3}{8a_0}\frac{e^2}{\hbar v}. \tag{7.110}$$

This sets the minimum $q^2 = (k - k')^2$ given by

$$\frac{4}{9}q_{min}^2 a_0^2 = \left(\frac{e^2}{4\hbar v}\right)^2 = \frac{\alpha^2}{16}(v/c)^{-2}. \tag{7.111}$$

Since the incident electron is moving much faster than a bound electron, this quantity is small.

The total cross sections are found by integrating over q^2 from q_{min}^2 to roughly $q_{max}^2 = (9/4)a_0^{-2}$. For the optically disallowed process, the total cross section is easily determined:

$$\sigma_{1s \rightarrow 2s} = \frac{2^{19}}{3^{12}}\pi a_0^4 \left(\frac{e^2}{\hbar v}\right)^2 \int_{q_{min}^2}^{q_{max}^2} dq^2 \left(1 + \frac{4}{9}q^2 a_0^2\right)^{-6} \tag{7.112}$$

$$\approx \frac{1}{5}\frac{2^{19}}{3^{12}}\pi a_0^2 \left(\frac{e^2}{\hbar v}\right)^2 \frac{9}{4}\int_0^\infty dx\,(1 + x)^{-6} \tag{7.113}$$

$$= \frac{1}{5}\frac{2^{17}}{3^{10}}\pi a_0^2 \left(\frac{e^2}{\hbar v}\right)^2. \tag{7.114}$$

Slightly more difficult is

$$\sigma_{1s \rightarrow 2p} \approx \frac{2^{17}}{3^{10}}\pi a_0^4 \left(\frac{e^2}{\hbar v}\right)^2 \int_{q_{min}^2}^{q_{max}^2} dq^2 \frac{1}{q^2 a_0^2}(1 + \frac{4}{9}q^2 a_0^2)^{-6}. \tag{7.115}$$

Let $(1 + \frac{4}{9}q^2 a_0^2) = 1/u$, then

$$a_0^2 q^2 = -\frac{9}{4}\left(\frac{1}{u} - 1\right); \qquad \frac{dq^2}{q^2} = -\frac{du}{u(1 - u)}. \tag{7.116}$$

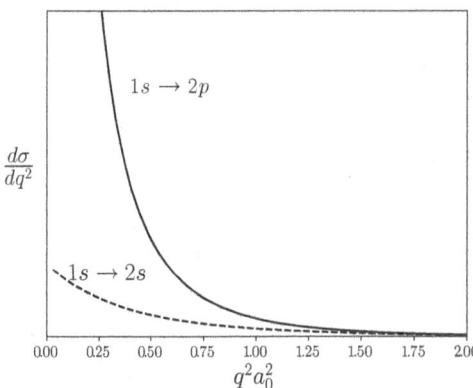

Figure 7.3 The differential cross sections for inelastic scattering of an electron on hydrogen in which the ground state $1s$ is excited to $2p$ or $2s$. The optically allowed transition $1s \rightarrow 2p$ has an extra factor of $1/q^2$, leading to a total cross section for the process that grows logarithmically with the incident energy.

Thus

$$\sigma_{1s \rightarrow 2p} \approx \frac{2^{17}}{3^{10}} \pi a_0^2 \left(\frac{e^2}{\hbar v} \right)^2 \int_{u_{min}}^{u_{max}} du \, \frac{u^5}{1 - u}. \tag{7.117}$$

Now $u_{max} = 1/(1 + (e^2/4\hbar v)^2)$ and $u_{min} \approx 0$, so we have

$$\sigma_{1s \rightarrow 2p} \approx \frac{2^{17}}{3^{10}} \pi a_0^2 \left(\frac{e^2}{\hbar v} \right)^2 \int_0^{1/(1+(e^2/4\hbar v)^2)} du \, \frac{u^5 - 1 + 1}{1 - u}$$

$$= \frac{2^{17}}{3^{10}} \pi a_0^2 \left(\frac{e^2}{\hbar v} \right)^2 \left[2 \ln(\frac{4\hbar v}{e^2}) - \frac{137}{60} \right]. \tag{7.118}$$

To translate these into more recognizable terms, we note that for an incident electron,

$$\left(\frac{e^2}{\hbar v} \right)^2 = \alpha^2 \frac{c^2}{v^2} = \frac{(1/2)\alpha^2 mc^2}{(1/2)mv^2} = \frac{13.6 \, \text{eV}}{\text{KE}}, \tag{7.119}$$

that is, the reciprocal of the incident electron energy in Rydberg units. For an incident kinetic energy of 400 eV, this is 0.034. The factor $2^{17}/3^{10}$ is near 2.22. We see that the $1s \rightarrow 2s$ excitation cross section at this energy is about 0.015 times the "nominal" size of the hydrogen atom, πa_0^2. The optically allowed transition has a total cross section larger by a factor of about

$$\frac{\sigma_{1s \rightarrow 2p}}{\sigma_{1s \rightarrow 2s}} \approx 5 \ln \left(\frac{4 \times \text{KE}}{13.6 \, \text{eV}} \right). \tag{7.120}$$

See Figure 7.3.

7.7 Optical Theorem

The optical theorem relates the total cross section to the imaginary part of the forward scattering amplitude. We rewrite Eq. (7.57) and the complex conjugate of Eq. (7.56) with $W = 2mV/\hbar^2$ for the special case of forward scattering, $\mathbf{k}' = \mathbf{k}$

$$f(\mathbf{k}, \mathbf{k}) = -\frac{1}{4\pi} \int d^3x \, e^{-i\mathbf{k}\cdot\mathbf{x}} W(\mathbf{x})\psi_{\mathbf{k}}(\mathbf{x})$$

$$e^{-i\mathbf{k}\cdot\mathbf{x}} = \psi_{\mathbf{k}}^*(\mathbf{x}) + \int d^3x' \, G^*(\mathbf{x}, \mathbf{x}')W(\mathbf{x}')\psi_{\mathbf{k}}^*(\mathbf{x}). \tag{7.121}$$

Substituting the latter into the former,

$$f(\mathbf{k}, \mathbf{k}) = -\frac{1}{4\pi} \int d^3x \, \psi_{\mathbf{k}}^*(\mathbf{x})W(\mathbf{x})\psi(\mathbf{x})$$

$$-\frac{1}{4\pi} \int d^3x \int d^3x' \, G^*(\mathbf{x}, \mathbf{x}')W(\mathbf{x}')\psi_{\mathbf{k}}^*(x')W(\mathbf{x})\psi_{\mathbf{k}}(\mathbf{x}).$$

The first term is real and our interest is in the imaginary part, so consider only the second and insert the momentum representation of $G^*(\mathbf{x}, \mathbf{x}')$,

$$\Im f(\mathbf{k}, \mathbf{k}) = -\frac{1}{4\pi}\Im \int \frac{d^3q}{(2\pi)^3(q^2 - k^2 + i\epsilon)} \int d^3x \, e^{-i\mathbf{q}\cdot\mathbf{x}}W(\mathbf{x})\psi_{\mathbf{k}}(\mathbf{x})$$

$$\times \int d^3x' \, e^{i\mathbf{q}\cdot x'} W(\mathbf{x}')\psi_{\mathbf{k}}^*(\mathbf{x}').$$

Comparing with Eq. (7.57), we see that this can be written as

$$\Im f(\mathbf{k}, \mathbf{k}) = -4\pi\Im \int \frac{d^3q}{(2\pi)^3(q^2 - k^2 + i\epsilon)}|f(\mathbf{q}, \mathbf{k})|^2. \tag{7.122}$$

Using the standard prescription,

$$\frac{1}{x + i\epsilon} \to P(1/x) - i\pi\delta(x) \tag{7.123}$$

and if $f(x_0) = 0$, then

$$\delta(f(x)) = \frac{\delta(x - x_0)}{|f'(x_0)|} \tag{7.124}$$

we have finally

$$\Im f(\mathbf{k}, \mathbf{k}) = 4\pi \int \frac{d^3q}{(2\pi)^3} \frac{\pi\delta(q - k)}{2q}|f(\mathbf{q}, \mathbf{k})|^2$$

$$= \frac{k}{4\pi} \int d\Omega_{\mathbf{k}}' |f(\mathbf{k}', \mathbf{k})|^2 = \frac{k}{4\pi}\sigma_{\text{tot}}. \tag{7.125}$$

The optical theorem is usually stated as

$$\sigma_{\text{tot}} = \frac{4\pi}{k}\Im f(0°). \tag{7.126}$$

The result is actually much more general than this spinless potential scattering situation. It depends only on conservation of probability and holds with inelastic channels, relativistic kinematics, etc.

7.8 Validity Criterion for the First Born Approximation

It is useful to have some criterion for the validity of the first Born approximation. Since the amplitude is given by Eq. (7.57),

$$f(\theta,\phi) = f(\mathbf{k}',\mathbf{k}) = -\frac{1}{4\pi}\int d^3x\, e^{-i\mathbf{k}'\cdot\mathbf{x}}W(\mathbf{x})\psi_{\mathbf{k}}(\mathbf{x}). \tag{7.127}$$

One criterion is that the differences between the amplitude in first Born approximation $\psi_{\mathbf{k}}^{(1)}(\mathbf{x})$ and $\psi_{\mathbf{k}}^{0}(\mathbf{x}) = e^{i\mathbf{k}\cdot\mathbf{x}}$

$$\Delta\psi_{\mathbf{k}}(\mathbf{x}) = \psi_{\mathbf{k}}^{(1)}(\mathbf{x}) - e^{i\mathbf{k}\cdot\mathbf{x}} = \int d^3x\, G(\mathbf{x},\mathbf{x}')W(\mathbf{x}')e^{i\mathbf{k}\cdot\mathbf{x}'} \tag{7.128}$$

be in some sense small, that is, $|\Delta\psi| \ll 1$ in the region where W is large. Since the potential, and thus W, is usually largest (though not always) at the origin, it is customary to consider $\Delta\psi(0)$ and thus use $G(0,\mathbf{x}')$

$$\Delta\psi(0) = -\frac{1}{4\pi}\int d^3x'\,\frac{e^{ikr'}}{r'}W(\mathbf{x}')e^{i\mathbf{k}\cdot\mathbf{x}'} \tag{7.129}$$

where $r' = |\mathbf{x}'|$. Thus, the criterion for the validity of the first Born approximation is

$$\left|\frac{1}{4\pi}\int d^3x'\,\frac{e^{ikr'}}{r'}W(\mathbf{x}')e^{i\mathbf{k}\cdot\mathbf{x}'}\right| \ll 1. \tag{7.130}$$

For a spherically symmetric potential, this is

$$\left|\frac{1}{2k}\int_0^\infty W(r)\left(e^{2ikr}-1\right)dr\right| \ll 1. \tag{7.131}$$

Two observations follow:

(a) At high energies, one eventually reaches a domain of validity provided $\int_0^\infty W(r)dr < \infty$. Define

$$2k_0 = \left|\int_0^\infty W(r)dr\right|. \tag{7.132}$$

Then for $k \gg k_0$ the criterion is satisfied.

(b) For $k \to 0$, the criterion reads

$$\left|\int_0^\infty rW(r)dr\right| \ll 1. \tag{7.133}$$

For attractive potentials, the potential must be weak enough that there are no bound states if the Born approximation is to hold down to $k = 0$. Check as an example, $V(r) = -V_0$ for $r < a$ and $V(r) = 0$ for $r > a$. Then the criterion is $2mV_0a^2/\hbar^2 \ll 2$, while for there to be a bound state $2mV_0a^2/\hbar^2 > \pi^2/4 = 2.467$.

7.9 Eikonal Approximation

At high energies, the wavelength of the particle being scattered is very short compared to the range of the forces. This means that the approximation of geometrical optics with diffractive corrections can be usefully applied.

Recall that we start with the Schrödinger equation in the form,

$$[\nabla^2 + K(\mathbf{x})^2]\psi(\mathbf{x}) = 0, \tag{7.134}$$

where

$$K(\mathbf{x})^2 = \frac{2m}{\hbar^2}(E - V(\mathbf{x})) = k^2 - \frac{2m}{\hbar^2}V(\mathbf{x}). \tag{7.135}$$

If V is finite everywhere, at high enough energies we can write

$$K(\mathbf{x}) \approx k - \frac{m}{\hbar^2 k}V(\mathbf{x}) = k - \frac{1}{\hbar v}V(\mathbf{x}), \tag{7.136}$$

where $\hbar k = p = mv$.

In the absence of the potential, $\psi_0 = e^{ikz}$. With the potential present, the eikonal approximation has

$$\ln \psi \approx i \int_{\text{path}} K(\mathbf{x})dz. \tag{7.137}$$

At each impact parameter, $\mathbf{b} = x\hat{\mathbf{x}} + y\hat{\mathbf{y}}$ the phase accumulates differently in z. The wave function is then

$$\psi(\mathbf{x}) = \psi(\mathbf{b}, z) = \exp\left(ikz - \frac{i}{\hbar v}\int_{z_0}^{z} V(\mathbf{b}, z)dz'\right). \tag{7.138}$$

The approximation is that the particle comes swiftly by the scattering center, is deflected only slightly by the interaction, but accumulates a phase given by the eikonal integral. See Figure 7.4. We shall discuss scattering in terms of phase shifts for a given angular momentum state, a concept to be more fully developed below. Since a definite impact parameter $|\mathbf{b}|$ implies a definite angular momentum $\ell\hbar$, we can make the semi-classical correspondence

$$|\mathbf{b}|\hbar k = (\ell + 1/2)\hbar. \tag{7.139}$$

Then

$$-\frac{1}{\hbar k}\int_{-\infty}^{\infty} V(\mathbf{b}, z')dz' = \phi(\mathbf{b}) \tag{7.140}$$

is related to the phase shift (actually $2\delta_\ell(k)$) for $\ell + 1/2 = |\mathbf{b}|k$. We chose $z_0 = -\infty$ to get a solution corresponding to $\psi_{\mathbf{k}}^+$. We now insert our eikonal wave function into the integral for the scattering amplitude

$$f(\mathbf{k}', \mathbf{k}) = -\frac{1}{4\pi}\frac{2m}{\hbar^2}\int d^3x\, e^{-i\mathbf{k}'\cdot\mathbf{x}}V(\mathbf{x})e^{ikz}e^{-i/(\hbar v)\int_{-\infty}^{z} V(x,y,z')dz'}, \tag{7.141}$$

where we have reverted to x, y in the phase to make clear the connection to the integration over d^3x.

At high energies, the scattering is dominantly at small angles. Thus we write

$$e^{ikz-i\mathbf{k}'\cdot\mathbf{x}} = e^{i(k-k_z')z}e^{i\mathbf{q}_\perp\cdot\mathbf{x}_\perp}, \tag{7.142}$$

where $\mathbf{q}_\perp$ is the component of $\mathbf{q} = \mathbf{k} - \mathbf{k}'$ that is perpendicular to $\mathbf{k}$. Now $k_z - k_z' = k(1 - \cos\theta) \simeq k\theta^2/2$, while $|\mathbf{q}_\perp| \simeq k\theta$. Thus $k - k_z'$ is second order in θ, while $\mathbf{q}_\perp$ is first order in θ. We therefore write

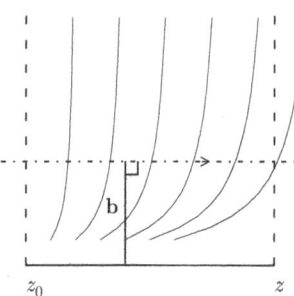

Figure 7.4 In the eikonal approximation, the scattered particle is treated as if it moves along a straight line at impact parameter b. It accumulates phase as it moves along the z direction. Lines of constant phase are shown as contours.

$$f(\mathbf{k}', \mathbf{k}) = -\frac{1}{4\pi}\frac{2m}{\hbar^2} \int d^2 x_\perp \, e^{i\mathbf{q}_\perp \cdot \mathbf{x}_\perp} \int_{-\infty}^{\infty} dz \, V(\mathbf{x}_\perp, z) e^{-i/(\hbar v)\int_{-\infty}^{z} V(\mathbf{x}_\perp, z')dz'}. \tag{7.143}$$

Now the integral over dz is easily seen to involve a perfect differential:

$$\int_{-\infty}^{\infty} dz \, V e^{-i\int_{-\infty}^{z} V/(\hbar v)} = i\hbar v \int_{-\infty}^{\infty} dz \frac{d}{dz} \exp\left(-\frac{i}{\hbar v}\int_{-\infty}^{z} V(x_\perp, z')dz'\right)$$

$$= i\hbar v\left[\exp\left(-\frac{i}{\hbar v}\int_{-\infty}^{\infty} V(x_\perp, z)dz\right) - 1\right]. \tag{7.144}$$

Combining the factors in front,

$$f(k, \cos\theta) = \frac{k}{2\pi i} \int d^2 x_\perp \, e^{i\mathbf{q}_\perp \cdot \mathbf{x}_\perp}\left[\exp\left(-\frac{i}{\hbar v}\int_{-\infty}^{\infty} V(x_\perp, z)\,dz\right) - 1\right]. \tag{7.145}$$

The result is called the eikonal or Glauber approximation for the scattering amplitude. It is useful at high energies with relatively weak potentials (which may, however, have long range and so give large phase shifts).

7.9.1 Spherically symmetric potential

For $V(\mathbf{x}) = V(r)$, we know that $f(\mathbf{k}', \mathbf{k}) = f(k, \cos\theta)$. This simplifies the Glauber approximation dependence upon $|\mathbf{q}_\perp| \simeq q \simeq k\theta$. We can rewrite the expression by integrating over the azimuthal angle and $d^2 x_\perp = b\,db\,d\phi$. Choosing $\mathbf{q}_\perp$ in the x-direction, we have $\mathbf{q}_\perp \cdot \mathbf{x}_\perp = kb\theta\cos\phi$.

But

$$\frac{1}{2\pi}\int_0^{2\pi} e^{iz\cos\phi}d\phi = J_0(z) \tag{7.146}$$

is the standard integral representation of the zeroth order Bessel function. Thus if we write

$$\delta(k, b) = -\frac{1}{2\hbar v}\int_{-\infty}^{\infty} V(\sqrt{b^2 + z^2})dz, \tag{7.147}$$

then we have the attractive form,

$$f(k, \cos\theta) = \frac{k}{i}\int_0^{\infty} b\,db\,J_0(kb\theta)[e^{2i\delta(k,b)} - 1]. \tag{7.148}$$

This is written in a form that parallels the partial wave expansion derived a bit below,

$$f(k, \cos\theta) = \sum_{\ell=0}^{\infty} (2\ell + 1) \left[\frac{e^{2i\delta_\ell(k)} - 1}{2ik} \right] P_\ell(\cos\theta). \tag{7.149}$$

We have already seen the correspondence $kb \leftrightarrow \ell + 1/2$. The connection is made more vivid by the approximation at large ℓ and small angle,

$$P_\ell(\cos\theta) = J_0((2\ell + 1)\sin(\theta/2)) + \mathcal{O}(\sin^2(\theta/2)). \tag{7.150}$$

(Magnus and Oberhettinger, *Formulas and Theorems for the Functions of Mathematical Physics* p. 72; Bateman and Erdélyi, *Higher Transcendental Functions*, Vol 2, pp. 56,56; Gradshteyn and Ryzhik, *Tables of Integrals, Series, and Products* Eq. 8.7822). The partial wave and eikonal approximation merge if there are many partial waves present and the scattering is dominantly in the forward direction.

7.9.2 Coulomb Scattering in the Eikonal Approximation

The infinite range of the Coulomb potential means that it involves many partial waves and has a sizable phase shift, even at large b. Use of the eikonal approximation is appropriate. With the potential being

$$V(r) = \frac{z_1 z_2 e^2}{r}, \tag{7.151}$$

we define dimensionless parameter

$$\eta = \frac{z_1 z_2 e^2}{\hbar v}. \tag{7.152}$$

Now consider the indefinite integral,

$$-\int^z \frac{1}{\hbar v} V(\sqrt{b^2 + z^2}) dz = -\eta \int^z \frac{dz}{\sqrt{b^2 + z^2}} = -\eta \ln(r + z); \qquad (r = \sqrt{b^2 + z^2})$$

$$= +\eta \ln(r - z) - \eta \ln b^2. \tag{7.153}$$

Using this, we have

$$\psi \simeq e^{ikz} e^{-i\eta \ln(kb^2)} e^{i\eta \ln[k(r-z)]}$$

$$= e^{-i\eta \ln(kb^2)} e^{i\phi(b,z)}, \tag{7.154}$$

where

$$\phi(b, z) = kz + \eta \ln[k(r - z)]. \tag{7.155}$$

If we take $\nabla\phi$ to represent the tangent to the path of a classical particle, this will define a hyperbola whose incoming direction forms one asymptote and whose outgoing direction forms the other asymptote at an angle $\theta = 2\eta/(kb)$, that is, the path of the classical scattering.

We see that we cannot directly take the limit of $\delta(k, z)$ as $z \to \infty$. For scattering from neutral atoms, the nuclear Coulomb potential is screened, so we expect a limiting process will give sensible results. Here are two models for this:

(a) Put

$$V = z_1 z_2 \frac{e^2}{r} e^{-\mu r} \tag{7.156}$$

and let $\mu \to 0^+$ at the end.

(b) Put

$$V = z_1 z_2 \frac{e^2}{r} \Theta(R - r). \tag{7.157}$$

We then have

(i)

$$2\delta(b) = -\frac{1}{\hbar v} \int V(b,z)dz = -\eta \int_{-\infty}^{\infty} \frac{e^{-\mu\sqrt{b^2+z^2}}}{\sqrt{b^2 + z^2}} dz = -2\eta K_0(\mu b)$$

$$\simeq -2\eta \ln\left(\frac{2e^{-\gamma}}{\mu b}\right) = -2\eta \ln\left(\frac{1.123}{\mu b}\right), \tag{7.158}$$

where we have used the limiting form of the modified Bessel function and $\gamma = 0.5772.$ is the Euler-Mascheroni constant.

(ii)

$$2\delta(b) = -2\eta \int_0^{\sqrt{R^2+b^2}} \frac{dz}{\sqrt{b^2 + z^2}} = -2\eta \ln\left(\frac{R + \sqrt{R^2 - b^2}}{b}\right) \simeq -2\eta \ln\left(\frac{2R}{b}\right) \tag{7.159}$$

for $R \gg b$.

In either case, we have $e^{2i\delta(b)} = e^{i\sigma} b^{2i\eta}$, where σ is independent of b: $(\sigma_a = -2\eta \ln(1.123/\mu), \sigma_b = -2\eta \ln(2R))$. The factor $e^{i\sigma}$ is therefore an inessential phase in the amplitude. We have then,

$$f(k, \cos\theta) = \frac{k}{i} \int_0^{\infty} b \, db \, J_0(qb) \left[e^{i\sigma} b^{2i\eta} - 1\right]. \tag{7.160}$$

The integral of the -1 term vanishes, except if $q = 0$. The full integral sends us to *Higher Transcendental Functions*, Vol.2, p. 49, Eq. (19), or Gradshteyn and Ryzhik, *Tables of Integrals, Series, and Products*, Eq. (6.561.14),

$$\lim_{\gamma \to 0^+} \int_0^{\infty} e^{-\gamma b} J_0(qb) b^{1+2i\eta} db = \frac{\eta}{2i} \left(\frac{2}{q}\right)^{2+2i\eta} \frac{\Gamma(1 + i\eta)}{\Gamma(1 - i\eta)}. \tag{7.161}$$

The ratio of gamma functions has modulus one. The eikonal approximation for Coulomb scattering is thus

$$f(k, \cos\theta) = -\frac{2\eta k}{q^2} \frac{\Gamma(1 + i\eta)}{\Gamma(1 - i\eta)} e^{i\sigma} e^{-i\eta \ln q^2}. \tag{7.162}$$

Despite the origin of this result and its need for $\theta \ll 1$, etc., we put $q^2 = 4k^2 \sin^2\theta/2$ and define $\beta = \sigma - \eta \ln(4k^2)$. Then we have

$$f(k, \cos\theta) = \left[-\frac{\eta}{2k} \frac{1}{\sin^2\theta/2}\right] \frac{\Gamma(1 + i\eta)}{\Gamma(1 - i\eta)} e^{-i\eta \sin^2\theta/2} e^{i\beta}. \tag{7.163}$$

The factor in brackets is the Born amplitude. This result now agrees completely with the full calculation. See, for example, Schiff, *Quantum Mechanics*, 3rd edition, p.140, Eq. (21.10). The phase β has no physical significance. However, the phases that depend on the scattering angle do have significance for scattering of identical particles, as seen in Figure 6.3. The eikonal approximation happens to give the exact answer, in part because many partial waves enter and the cross section is very peaked in the forward direction.

7.9.3 Complex Potential and the Optical Theorem in the Eikonal Approximation

A complex potential in the Schrödinger equation causes a violation of the conservation of probability. If the imaginary part is negative, one can see that it represents a leakage away of the amplitude, which can be interpreted as the disappearance of the particle into other channels. For example, for the collision of a muon with a hydrogen atom in its ground state, elastic scattering would leave the muon accompanied by ground-state hydrogen, but inelastic scattering would leave the hydrogen atom in an excited state. If we are concerned only with the elastic scattering, a complex potential can simulate the effects of the other channels. If we write $V = V_R - iV_I$ (note the $-i$) in the Schroödinger equation, it is easy to show that the continuity equation for probability becomes

$$\frac{\partial}{\partial t}|\psi|^2 = \nabla \cdot \left[\frac{\hbar}{2m}(\psi^* \nabla \psi - (\nabla \psi^*)\psi) \right] = -\frac{2}{\hbar}(\psi^* V_I \psi). \tag{7.164}$$

The idea of a complex potential is most fruitful in nuclear physics, where the strongly interacting, many-body targets have a great number of possible excitations and transformations.

Let us look at the optical theorem in the eikonal description with a complex potential. The phase shift is

$$\delta(k, \mathbf{b}) = -\frac{1}{2\hbar v} \int (V_R - iV_I)dz. \tag{7.165}$$

The scattering amplitude is

$$f(k, \cos\theta) = \frac{k}{2\pi i} \int d^2 x_\perp e^{i\mathbf{q}_\perp \cdot \mathbf{x}_\perp} \left[e^{-\frac{1}{\hbar v}\int V_I dz} e^{-\frac{i}{\hbar v}\int V_R dz} - 1 \right]. \tag{7.166}$$

Now let us look at the total elastic scattering cross section:

$$\sigma_{el} = \int d\Omega |f(k, \cos\theta)|^2 = \int d\phi \int d\cos\theta \, |f|^2 \simeq \frac{1}{k^2} \int d^2 q_\perp |f(k, \mathbf{q}_\perp)|^2$$

$$= \frac{1}{4\pi^2} \int d^2 q_\perp \int d^2 x_\perp \int d^2 x'_\perp e^{i\mathbf{q}_\perp \cdot (\mathbf{x}_\perp - \mathbf{x}'_\perp)} \left[e^{-\frac{1}{\hbar v}\int V_I(\mathbf{x}_\perp, z)dz} e^{-\frac{i}{\hbar v}\int V_R(\mathbf{x}_\perp, z)dz} - 1 \right]$$

$$\times \left[e^{-\frac{1}{\hbar v}\int V_I(\mathbf{x}'_\perp, z')dz'} e^{+\frac{i}{\hbar v}\int V_R(\mathbf{x}'_\perp, z')dz'} - 1 \right]. \tag{7.167}$$

But now do the $d^2 q_\perp$ integral first:

$$\int d^2 q_\perp e^{i\mathbf{q}_\perp \cdot (\mathbf{x}_\perp - \mathbf{x}'_\perp)} = (2\pi)^2 \delta^2(\mathbf{x}_\perp - \mathbf{x}'_\perp). \tag{7.168}$$

We can now rewrite the total elastic cross section:

$$\sigma_{el} = \int d^2 x_\perp \left| 1 - e^{-\frac{1}{\hbar v} \int V_I(x_\perp, z) dz} e^{-\frac{i}{\hbar v} \int V_R(x_\perp, z) dz} \right|^2$$

$$= \int d^2 x_\perp \left[1 + e^{-\frac{2}{\hbar v} \int V_I dz} - 2 e^{-\frac{1}{\hbar v} \int V_I dz} \cos \left(\frac{1}{\hbar v} \int V_R \, dz \right) \right]. \tag{7.169}$$

In the event that the potential is purely real, $V_I = 0$,

$$\sigma_{el} = 4 \int d^2 x_\perp \sin^2 \left(\frac{1}{\hbar v} \int V_R \, dz \right). \tag{7.170}$$

The absorptive cross section is obtained by considering the rate of loss of probability per unit volume integrated over all space:

$$\int d^3 x \, \frac{2}{\hbar} V_I(\mathbf{x}) |\psi(\mathbf{x})|^2 \tag{7.171}$$

dividing by the flux, v

$$\sigma_{abs} = \frac{2}{\hbar v} \int d^3 x \, V_I(\mathbf{x}) |\psi(\mathbf{x})|^2. \tag{7.172}$$

We have

$$\psi(\mathbf{x}) = e^{ikz - \frac{i}{\hbar v} \int (V_r - iV_I) dz'}, \tag{7.173}$$

so

$$|\psi(\mathbf{x})|^2 = e^{-\frac{2}{\hbar v} \int^z V_I dz'}. \tag{7.174}$$

Thus,

$$\sigma_{abs} = \frac{2}{\hbar v} \int d^2 x_\perp \int dz V_I(\mathbf{x}_\perp, z) e^{-\frac{2}{\hbar v} \int^z V_I dz'}. \tag{7.175}$$

Once again, we recognize a perfect differential. The result of the z integrations is

$$\sigma_{abs} = \int d^2 x_\perp \left[1 - e^{-\frac{2}{\hbar v} \int_{-\infty}^{\infty} V_I(x_\perp, z) dz} \right]. \tag{7.176}$$

Adding the elastic and absorptive pieces,

$$\sigma_{tot} = 2 \int d^2 x_\perp \left[1 - e^{-\frac{1}{\hbar v} \int V_I dz} \cos \left(\frac{1}{\hbar v} \int V_R dz \right) \right]. \tag{7.177}$$

On the other hand, we have

$$\frac{4\pi}{k} \Im m \, f(0^o) = 2 \int d^2 x_\perp \left[1 - e^{-\frac{1}{\hbar v} \int V_I dz} \cos \left(\frac{1}{\hbar v} \int V_R dz \right) \right].$$

$$= \sigma_{tot}, \tag{7.178}$$

confirming the optical theorem in the context of the eikonal approximation.

7.10   Method of Partial Waves

We consider scattering of a spinless particle of mass m by a central potential $V(r)$, whose range is such that beyond some radius $r = R$ it is sensibly zero. This excludes the Coulomb field. Angular momentum is conserved and the Schrödinger equation can be separated into one equation for each ℓ value. This means that the scattering amplitude can be written as a superposition of *partial wave amplitudes* $f_\ell(k)$ such that

$$f(k,\theta) = \sum_{\ell=0}^{\infty} f_\ell(k)(2\ell+1)P_\ell(\cos\theta); \quad f_\ell(k) = \frac{1}{2}\int_{-1}^{1} d\cos\theta f(k,\theta)P_\ell(\cos\theta), \quad (7.179)$$

where $\cos\theta = \hat{\mathbf{k}} \cdot \hat{\mathbf{k}}'$ is the cosine of the scattering angle.

Evidently we need to consider expansions of various wave functions in partial waves, i.e. amplitudes that are associated with a definite value of ℓ. The simplest example is

$$e^{ikz} = \sum_{\ell=0}^{\infty} i^\ell (2\ell+1)j_\ell(kr)P_\ell(\cos\theta), \quad (7.180)$$

where $j_\ell(z)$ is the spherical Bessel function of order ℓ. We could take this to be the definition of the spherical Bessel functions so

$$j_\ell(x) = \frac{1}{2i^\ell}\int_{-1}^{1} e^{ix\cos\theta}P_\ell(\cos\theta)d\cos\theta. \quad (7.181)$$

With a potential $V(r)$ and $W(r) = 2mV(r)/\hbar^2$, the Schrödinger equation

$$\left[\nabla^2 + k^2 - W(r)\right]\psi(\mathbf{x}) = 0 \quad (7.182)$$

can be solved using separation of variables by writing in analogy with the expansion of the plane wave

$$\psi_{\mathbf{k}}(\mathbf{x}) = \sum_{\ell=0}^{\infty} i^\ell (2\ell+1)\frac{u_\ell(r)}{kr}P_\ell(\cos\theta), \quad (7.183)$$

where it is assumed that $\hat{\mathbf{k}}$ is parallel to $\hat{\mathbf{z}}$. The radial wave function $u_\ell(r) = rR_\ell$ satisfies the equation

$$\frac{d^2u_\ell}{dr^2} + \left(k^2 - \frac{\ell(\ell+1)}{r^2} - W(r)\right)u_\ell = 0. \quad (7.184)$$

We assume that there is some finite range, R, for the potential so that for $r > R$, $W(r) = 0$. For $r > R$ the function $u_\ell(r)$ satisfies

$$\frac{d^2u_\ell}{dr^2} + \frac{2}{r}\frac{du_\ell}{dr} + \left(k^2 - \frac{\ell(\ell+1)}{r^2}\right)u_\ell = 0. \quad (7.185)$$

The assumption that

$$u = r^\alpha(1 + a_1r + a_12r^2 + \ldots), \quad (7.186)$$

the method of Frobenius, leads to an indicial equation for α

$$\alpha = -\frac{1}{2} \pm |\ell + \frac{1}{2}| \quad (7.187)$$

so α is either ℓ or $-\ell - 1$. The solution regular at the origin is j_ℓ, while the other solution is indicated by n_ℓ. With customary normalizations, the behavior of these solutions is for $z \to 0$

$$j_\ell(z) = \frac{z^\ell}{1 \cdot 3 \cdots (2\ell + 1)} + \mathcal{O}(z^{\ell+2})$$

$$n_\ell(z) = \frac{1 \cdot 3 \cdots (2\ell - 1)}{z^{\ell+1}} + \mathcal{O}(z^{-\ell+1}), \tag{7.188}$$

while for $z \to \infty$

$$j_\ell(z) \to \frac{1}{z} \sin(z - \frac{\ell\pi}{2})$$

$$n_\ell(z) \to -\frac{1}{z} \cos(z - \frac{\ell\pi}{2}). \tag{7.189}$$

For $r > R$ we must have

$$\frac{u_\ell(r)}{kr} = A j_\ell(kr) + B n_\ell(kr). \tag{7.190}$$

This means that

$$\lim_{r \gg R, \ell/k} u_\ell(r) = A \sin(kr - \frac{\ell\pi}{2}) - B \cos(kr - \frac{\ell\pi}{2}). \tag{7.191}$$

We define

$$-B/A = \tan \delta_\ell(k); \qquad C_\ell = \sqrt{A^2 + B^2}. \tag{7.192}$$

Then

$$\lim_{r \gg R, \ell/k} u_\ell(r) = C_\ell \sin(kr - \frac{\ell\pi}{2} + \delta_\ell). \tag{7.193}$$

This asymptotic behavior defines the phase shift, $\delta_\ell(k)$. It is determined by the radial Schrödinger equation for u_ℓ, which involves the potential $W(r)$ and the centrifugal barrier. This is displayed in Figure 7.5. The unshifted radial wave function is $j_\ell(kr)$.

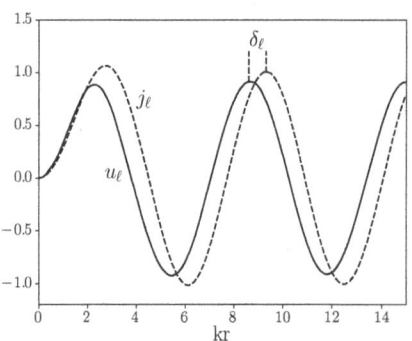

Figure 7.5 An attractive square-well potential "pulls" in the wave function leading to a positive phase shift δ_ℓ. Here the energy is $\hbar^2 k^2 / 2m$ and the $\ell = 1$ partial wave is shown for a square well ending at $kr = 3$. The depth of the square well is $-V_0 = (4/9)(\hbar^2 k^2 / 2m)$.

We want to find $f_\ell(k)$ in terms of the phase shift, δ_ℓ. To do that, we recall that $f(k, \cos\theta)$ is defined by the asymptotic behavior of $\psi_\mathbf{k}$:

$$\lim_{r\to\infty} \psi_\mathbf{k} = e^{ikz} + \frac{e^{ikr}}{r} f(k, \cos\theta). \tag{7.194}$$

All we need do is examine the asymptotic form of our wave functions and read off the expression for $f(k, \cos\theta)$ or $f_\ell(k)$. Thus we are led to compare

$$\lim_{r\to\infty} u_\ell(r) = i^\ell C_\ell \sin(kr - \frac{\ell\pi}{2} + \delta_\ell) = \frac{C_\ell}{2i}\left(i^\ell e^{i(kr - \frac{\ell\pi}{2} + \delta_\ell)} - i^\ell e^{-i(kr - \frac{\ell\pi}{2} + \delta_\ell)}\right) \tag{7.195}$$

with the ℓ component of the incident plane wave

$$\lim_{r\to\infty}[i^\ell kr j_\ell(kr)] = i^\ell \sin(kr - \frac{\ell\pi}{2}) = \frac{1}{2i}(e^{ikr} - e^{i\pi\ell}e^{-ikr}). \tag{7.196}$$

Asymptotically, the difference of the two must have only outgoing waves (e^{ikr}). This means that we should choose

$$i^\ell C_\ell e^{-i(kr - \frac{\ell\pi}{2} + \delta)} = e^{i\pi\ell}e^{-ikr}$$

$$C_\ell = e^{i\delta_\ell}. \tag{7.197}$$

Having adjusted the coefficients C_ℓ, which were at our disposal, to assure the boundary condition, we can read off the expression for f_ℓ from the difference:

$$f_\ell(k) = \frac{1}{2ik}\left(C_\ell e^{i\delta_\ell} - 1\right) = \frac{1}{2ik}\left(e^{2i\delta_\ell} - 1\right) = \frac{e^{i\delta_\ell}\sin\delta_\ell}{k}. \tag{7.198}$$

A convenient relation for elastic scattering is

$$\Im \frac{1}{kf_\ell(k)} = \Im \frac{e^{-i\delta_\ell}}{\sin\delta_\ell} = -i. \tag{7.199}$$

The full scattering amplitude is

$$f(k, \cos\theta) = \frac{1}{k}\sum_{\ell=0}^{\infty}(2\ell + 1)e^{i\delta_\ell}\sin\delta_\ell P_\ell(\cos\theta). \tag{7.200}$$

7.11 Behavior of the Cross Section and the Argand Diagram

The differential cross section is

$$\frac{d\sigma}{d\Omega} = \left|f(k, \cos\theta)\right|^2 = \sum_{\ell,\ell'} f_\ell(k)f_{\ell'}^*(k)(2\ell + 1)(2\ell' + 1)P_\ell(\cos\theta)P_{\ell'}(\cos\theta). \tag{7.201}$$

But we know from our work on rotations that

$$P_\ell(\cos\theta)P_{\ell'}(\cos\theta) = \sum_L |\langle\ell 0\ell'0|L0\rangle|^2 P_L(\cos\theta). \tag{7.202}$$

Therefore,

$$\frac{d\sigma}{d\Omega} = A_L P_L(\cos\theta), \quad \text{where} \quad A_L = \sum_{\ell,\ell'} f_\ell f_{\ell'}^* |\langle\ell 0\ell'0|L0\rangle|^2. \tag{7.203}$$

From this expression we can draw some general conclusions.

(a) If $f_\ell \neq 0$ only for $\ell \leq \ell_{max}$, the A_L are non-zero only for L up to $2\ell_{max}$.
(b) If $A_L \neq 0$ for L odd, at least two partial waves enter, with at least one pair having ℓ values differing by an odd number. This follows because $\langle \ell 0 \ell' 0 | L 0 \rangle$ vanishes unless $\ell + \ell' + L$ is even. For example, with $\ell_1 = 0, \ell_2 = 1$, we can have non-zero values for A_L with $L = 0, 1, 2$.

See Figure 7.6.

Integrating over the scattering angles gives the total elastic cross section

$$\sigma_{elastic} = 4\pi \sum_{\ell=0}^{\infty} (2\ell + 1)|f_\ell(k)|^2 = \frac{4\pi}{k^2} \sum_{\ell=0}^{\infty} (2\ell + 1) \sin^2 \delta_\ell. \tag{7.204}$$

From this we see that in any partial wave, the maximum cross section possible is

$$\sigma_{\ell,max} = \frac{4\pi}{k^2}(2\ell + 1). \tag{7.205}$$

The partial wave amplitude is elegantly displayed in an Argand diagram. We set

$$z = kf_\ell(k) = \frac{1}{2i}(e^{2i\delta_\ell} - 1) = \frac{i}{2} + \frac{1}{2}e^{i(2\delta_\ell - \pi/2)}. \tag{7.206}$$

Plotting z in the complex plane gives a circle whose center is at $(0, i/2)$ and whose radius is $1/2$. For elastic scattering, z lies on the circle at an angle $2\delta_\ell$ measured from the line joining the circle's center to the origin. See Figure 7.7.

For complex potentials ($V = V_R - iV_I$, where $V_I \geq 0$) simulating other reaction channels, the phase shift is complex, with $\Im m\, \delta_\ell > 0$. This means that the point z lies inside the circle. For partial waves with great absorption, $z \cong i/2$, independent of the real part of δ_ℓ.

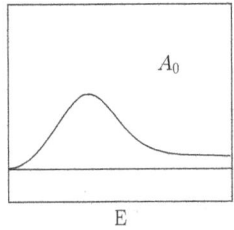
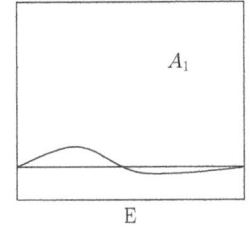
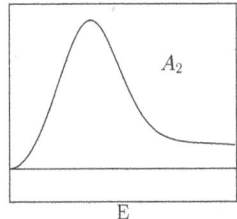

Figure 7.6 These graphs might represent cross section data for $\pi^+\pi^-$ elastic scattering in which a non-resonant $\ell = 0$ partial wave and a resonant $\ell = 1$ state are present.

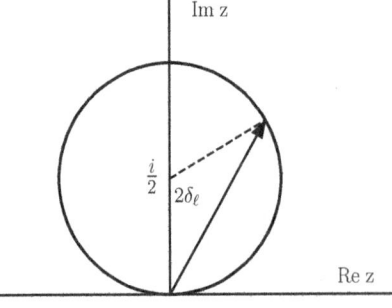

Figure 7.7 The Argand diagram. For elastic scattering, $z = kf_\ell(k)$ lies along the circle. For elastic scattering, the angle measured counterclockwise from the lower vertical radius to z is twice the phase shift. For inelastic scattering, z lies inside the circle.

The integral equation for the scattering amplitude,

$$f(k,\theta) = -\frac{1}{4\pi} \int d^3x \, e^{-i\mathbf{k}'\cdot\mathbf{x}} W(\mathbf{x})\psi_{\mathbf{k}}(\mathbf{x}), \tag{7.207}$$

can be turned into an integral expression for the partial wave amplitude by substituting expansions for both $e^{-i\mathbf{k}'\cdot\mathbf{x}}$ and the true wave function, $\psi_{\mathbf{k}}(\mathbf{x})$, with $\mathbf{k}$ taken along the z axis.

$$e^{-i\mathbf{k}'\cdot\mathbf{x}} = 4\pi \sum_{\ell'm'} (-i)^{\ell'} j_{\ell'}(k'r) Y_{\ell'm'}(\hat{\mathbf{k}}') Y^*_{\ell'm'}(\hat{\mathbf{x}})$$

$$\psi_{\mathbf{k}}(\mathbf{x}) = \sqrt{4\pi} \sum_{\ell} i^{\ell} (2\ell+1)^{1/2} \frac{u_{\ell}(r)}{kr} Y_{\ell 0}(\hat{\mathbf{x}}), \tag{7.208}$$

where we used the relation

$$P_{\ell}(\cos\theta) = \sqrt{\frac{4\pi}{2\ell+1}} Y_{\ell 0}(\theta,\phi). \tag{7.209}$$

We find

$$f_{\ell}(k) = \frac{e^{2i\delta_{\ell}} - 1}{2ik} = -\int_0^\infty r^2 dr \, j_{\ell}(kr) W(r) \frac{u_{\ell}(r)}{kr}. \tag{7.210}$$

The first Born approximation is obtained by setting $\frac{u_{\ell}(r)}{kr} \cong j_{\ell}(kr)$:

$$f_{\ell}(k) = -\int_0^\infty r^2 dr \, j_{\ell}(kr)^2 W(r). \tag{7.211}$$

This is manifestly real, so we should take $f_{\ell}(k) \to \delta_{\ell}/k$, and thus

$$\delta_{\ell} = -k \int_0^\infty r^2 dr \, j_{\ell}(kr)^2 W(r). \tag{7.212}$$

Since this integral can become large, some authors preserve unitarity by writing $\delta_{\ell} \to \tan\delta_{\ell}$, but this has no justification.

7.12 Hard Sphere Scattering

Scattering from a hard sphere is imposed by requiring that the radial wave function u_{ℓ} vanish at $r = R$. Outside R, we must have

$$\frac{u_{\ell}(r)}{kr} = A j_{\ell}(kr) + B n_{\ell}(kr), \tag{7.213}$$

and thus

$$\tan\delta_{\ell} = -\frac{B}{A} = \frac{j_{\ell}(kR)}{n_{\ell}(kR)}. \tag{7.214}$$

For $\ell = 0$, $j_0(x) = \sin x/x$, and $n_0(x) = -\cos x/x$ so

$$\frac{j_0(kR)}{n_0(kR)} = -\tan kR; \qquad \delta_0 = -kR; \qquad \sigma_0 = \frac{4\pi}{k^2} \sin^2(kR). \tag{7.215}$$

At low energies, $f_0(k) \cong -R$ and $\sigma_0 = 4\pi R^2$.

For $\ell \neq 0$, the result is simple only at low and high energies. For $kR \ll 1$

$$\frac{j_\ell(kR)}{n_\ell(kR)} \cong \delta_\ell \cong -\frac{(kR)^{2\ell+1}}{(2\ell+1)!!(2\ell-1)!!}, \tag{7.216}$$

while for $kR \gg 1$ and for $0 < \ell < \ell_{\max} \sim \mathcal{O}(kR)$

$$\delta_\ell = -kR + \frac{\ell\pi}{2}. \tag{7.217}$$

The presence of many partial waves means an angular distribution that is far from isotropic. As a matter of fact, the total cross section takes on a simple form, $\sigma_{\text{tot}} \cong 2\pi R^2$. Again, we have a geometrical aspect πR^2, this time with a factor 2, not 4. The explanation lies in the short wavelength limit (the same as quasi-geometrical optics).

We can write

$$f \cong \frac{1}{2ik} \sum_{\ell=0}^{\ell_{\max}} (2\ell+1)\left(e^{-2ikR}e^{i\pi\ell} - 1\right)P_\ell(\cos\theta). \tag{7.218}$$

We convert the sum to an integral with the substitution $\ell + 1/2 = kb$

$$f \cong \frac{k}{i} \int_0^R db\, b\left[-ie^{-2ikR}e^{i\pi kb} - 1\right]P_{kb-1/2}(\cos\theta). \tag{7.219}$$

At small angles, $P_\ell(\cos\theta) \cong J_0((2\ell+1)\sin(\theta/2)) \cong J_0(kb\theta)$. Then

$$f \cong \frac{1}{ik} \int_0^{kR} dx\, x\left(-ie^{-2ikR}e^{i\pi x} - 1\right)J_0(\theta x). \tag{7.220}$$

The first term oscillates rapidly in x, averaging to zero. Only the (-1) contributes significantly, and we have

$$f \cong \frac{i}{k} \int_0^{kR} dx\, xJ_0(\theta x) = \frac{i}{k\theta^2} \int_0^{\theta kR} dt\, tJ_0(t) = \frac{ikR^2}{2}\left[\frac{2J_1(kR\theta)}{kR\theta}\right]. \tag{7.221}$$

The term in square brackets goes to unity as $\theta \to 0$. The optical theorem tells us

$$\sigma_{\text{tot}} = \frac{4\pi}{k}\Im f(0°) = 2\pi R^2. \tag{7.222}$$

The portion of the cross section in the forward, diffractive region can be ascertained by integrating over the small θ region. We have

$$\frac{d\sigma}{d\Omega} = |f|^2 = k^2 R^4\left[\frac{J_1(kR\theta)}{kR\theta}\right]^2. \tag{7.223}$$

Since the definite integral is dominated by small values of θ, there is no harm in using this as the variable and extending the integration to infinity:

$$\sigma_{\text{tot}} = \int d\Omega\, \frac{d\sigma}{d\Omega} = 2\pi \int d\cos\theta\, \frac{d\sigma}{d\Omega} \cong 2\pi \int \theta d\theta\, \frac{d\sigma}{d\Omega}$$

$$= 2\pi \int_0^\infty \theta d\theta k^2 R^4\left[\frac{J_1(kR\theta)}{kR\theta}\right]^2 = 2\pi R^2 \int_0^\infty dx\, \frac{J_1(x)^2}{x}. \tag{7.224}$$

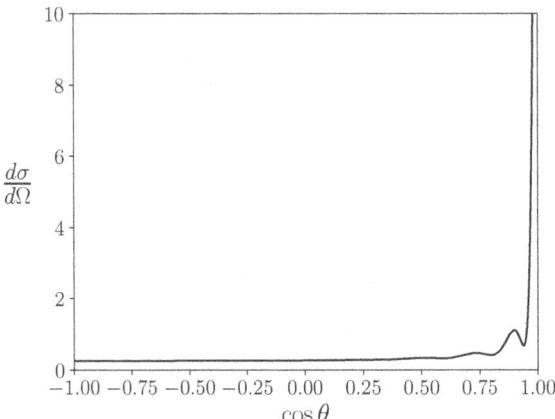

Figure 7.8 Differential cross section for scattering from a hard sphere of radius R for $kR = 10$. The differential cross section is given in units of R^2. The forward peak carries about half the total cross section, the remainder being roughly isotropic. In the limit of large kR, the total cross-section is $2\pi R^2$. For $kR = 10$, $\sigma_{\text{tot}} \simeq 1.16(2\pi R^2)$.

Using the definite integral

$$\int_0^\infty dx \frac{J_1(x)^2}{x} = \frac{1}{2},$$ (7.225)

we see that the diffractive peak, in the limit of high energy (and thus high k) contains a contribution to the total cross section of πR^2 inside an angle of order $\theta \sim 1/kR$. See Figure 7.8.

7.13 Strongly Attractive Potentials and Resonance

If the potential is strongly attractive, at most energies the behavior is just like that of a hard sphere. The point is that the wave faction must join "smoothly," i.e in value and slope, at the edge $r = R$, approached either from inside, R^- or outside, R^+. Thus,

$$u_i(R^-) = u_{\text{out}}(R^+); \qquad \frac{du_i}{dr}(R^-) = \frac{du_{\text{out}}}{dr}(R^+).$$ (7.226)

Combining these gives logarithmic derivatives:

$$\frac{1}{u_i} \frac{du_i}{dr}\bigg|_{r=R^-} = \frac{1}{u_{\text{out}}} \frac{du_{\text{out}}}{dr}\bigg|_{r=R^+}.$$ (7.227)

The wave vector inside the square well, K, is much larger that the wave vector k beyond $r = R$. At the edge, the logarithmic derivative for the wave inside is of order $\mathcal{O}(K)$, while for a typical solution outside the square well the logarithmic derivative there is of order $\mathcal{O}(k)$. To match at edge thus requires an anomalously small value of u_{out}, effectively zero. This makes the situation quite similar to that for a hard sphere.

Exceptions to general argument occur when $du_i/dr = 0$ at R. Then the amplitudes inside and out can be the same. (We are speaking here of $\ell = 0$; for higher partial waves the situation is qualitatively the same.)

For $\ell = 0$ inside the square well,

$$u_i(r) = \sin Kr; \qquad u_i' = K \cos Kr, \tag{7.228}$$

so $KR = (n + 1/2)\pi$. For the wave function outside the well,

$$u_{out}(r) = \sin(kr + \delta); \qquad u_{out}' = k \cos(kr + \delta). \tag{7.229}$$

To match slopes, we must have $kR + \delta = (n + 1/2)\pi$. These special energies define resonant states, but disappointingly do not produce rapid changes in the phase shift. That requires quasi-bound states.

To see resonant behavior, we consider a potential that is attractive at the interior but repulsive outside, and analyze it with the WKB method. We consider $\ell = 0$ for simplicity. The potential is sketched in Figure 7.9.

The WKB connection formulas

$$\frac{1}{\sqrt{\kappa}} e^{\xi'} \leftarrow -\frac{1}{\sqrt{k}} \sin(\xi - \frac{\pi}{4})$$

$$\frac{1}{2\sqrt{\kappa}} e^{-\xi'} \rightarrow \frac{1}{\sqrt{k}} \cos(\xi - \frac{\pi}{4}) \tag{7.230}$$

are used, starting with the solution in region I as

$$u_I(r) = \frac{2}{\sqrt{k}} \sin \int_0^r k(r')dr' = \frac{2}{\sqrt{k}} \sin(\Theta_I - \xi_1), \tag{7.231}$$

where

$$\Theta_I = \int_0^{r_1} k(r')dr'; \qquad \xi_1 = \int_r^{r_1} k(r')dr'. \tag{7.232}$$

We use the method of Fröman and Fröman, *JWKB Approximation: Contributions to the Theory*. The general solution in regions I and III are written as

$$\sqrt{ku_I} = Ae^{-i(\xi_1 + \pi/4)} + Be^{i(\xi_1 + \pi/4)}$$

$$= -(A + B)\sin(\xi_1 - \pi/4) + i(B - A)\cos(\xi_1 - \pi/4)$$

$$\sqrt{ku_{III}} = Ce^{i(\xi_2 - \pi/4)} + De^{-i(\xi_2 - \pi/4)}$$

$$= (C + D)\cos(\xi_2 - \pi/4) + i(C - D)\sin(\xi_2 - \pi/4). \tag{7.233}$$

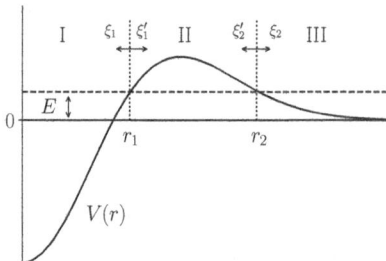

Figure 7.9 Potential that can produce resonances, with regions delimited by turning points. The various ξs are defined so that they are positive in the regions indicated and so that they are increasing as their positions move away from the turning points.

Using the connection formulae, we can express the solution in region II in two different ways, which we can then equate.

$$\sqrt{\kappa}u_{II} = (A+B)e^{\xi_1'} + i\frac{1}{2}(B-A)e^{-\xi_1'}$$

$$= (A+B)e^{\Theta_{II}-\xi_2'} + i\frac{1}{2}(B-A)e^{-(\Theta_{II}-\xi_2')}$$

$$= (C+D)\frac{1}{2}e^{-\xi_2'} - i(C-D)e^{\xi_2'} \qquad (7.234)$$

where

$$\Theta_{II} = \int_{r_1}^{r_2} dr\,\kappa(r); \qquad \kappa = \sqrt{\frac{2m}{\hbar^2}(V(r)-E)}. \qquad (7.235)$$

We can express the connection between $A, B, C,$ and D by

$$\begin{pmatrix} C \\ D \end{pmatrix} = \begin{pmatrix} e^{\Theta_{II}} + \frac{1}{4}e^{-\Theta_{II}} & e^{\Theta_{II}} - \frac{1}{4}e^{-\Theta_{II}} \\ e^{\Theta_{II}} - \frac{1}{4}e^{-\Theta_{II}} & e^{\Theta_{II}} + \frac{1}{4}e^{-\Theta_{II}} \end{pmatrix} \begin{pmatrix} A \\ B \end{pmatrix}. \qquad (7.236)$$

The more precise result is

$$\begin{pmatrix} C \\ D \end{pmatrix} = \begin{pmatrix} \sqrt{e^{2\Theta_{II}}+1} & e^{\Theta_{II}} \\ e^{\Theta_{II}} & \sqrt{e^{2\Theta_{II}}+1} \end{pmatrix} \begin{pmatrix} A \\ B \end{pmatrix};$$

$$\begin{pmatrix} A \\ B \end{pmatrix} = \begin{pmatrix} \sqrt{e^{2\Theta_{II}}+1} & -e^{\Theta_{II}} \\ -e^{\Theta_{II}} & \sqrt{e^{2\Theta_{II}}+1} \end{pmatrix} \begin{pmatrix} C \\ D \end{pmatrix}. \qquad (7.237)$$

If we consider

$$u_{III} = \frac{2}{\sqrt{k}}\sin(kr+\delta) \qquad (7.238)$$

as the desired form for the "outside" solution, the form of $u_I(r)$ results in $C = D^*$. Writing $C = |C|e^{i\delta'}$, we find

$$\delta = \delta' + \beta(k); \qquad \beta(k) = -kr_2 + \pi/4 - \int_{r_2}^{\infty}(k-k(r))dr. \qquad (7.239)$$

The effect of the potential inside r_2 is reflected in δ', which we compute below. We see that $\beta(k)$ is the analog of the phase shift from hard-sphere scattering. If the potential vanishes outside r_2, the energy dependence is the same for the two.

To use the Förman-Förman formula, we take

$$A = e^{i(\Theta_I - \pi/4)}, \quad B = e^{-i(\Theta_I - \pi/4)}. \qquad (7.240)$$

From the matrix, we find

$$C = D^* = \sqrt{e^{2\Theta_{II}}+1}\,e^{i(\Theta_I - \pi/4)} + e^{\Theta_{II}}e^{-i(\Theta_I - \pi/4)} \qquad (7.241)$$

and

$$\tan\delta' = \frac{\mathfrak{Im}\,C}{\mathfrak{Re}\,C} = \tan(\Theta_I - \pi/4)\left[\frac{\sqrt{e^{2\Theta_{II}}+1} - e^{\Theta_{II}}}{\sqrt{e^{2\Theta_{II}}+1} + e^{\Theta_{II}}}\right]. \qquad (7.242)$$

For a thick barrier, $e^{\Theta_{II}} \gg 1$, this reduces to

$$\tan \delta' \cong \frac{1}{4} e^{-2\Theta_{II}} \tan(\Theta_I - \pi/4). \tag{7.243}$$

Thus, we expect rather little action in the phase shift other than that due to $\beta(k)$, unless $\tan(\Theta_I - \pi/4) \to \infty$, i.e. unless $\Theta_I = (n + \frac{3}{4})\pi$. This is just the condition for an s-wave bound state in the WKB approximation, as if the barrier were infinitely thick.

The presence of the factor $e^{-2\Theta_{II}}/4$ multiplying $\tan(\Theta_I - \pi/4)$ means that it is only for energies close to the condition of "bound state" energy that $\tan \delta'$ is appreciably different from zero. The variation of $\Theta_I(E)$ is smooth on such a small energy interval and can be expanded in a Taylor series around $E = E_0$:

$$\Theta(E) - \frac{\pi}{4} = (n + \frac{1}{2})\pi + \left(\frac{d\Theta_I}{dE}\right)_{E_0}(E - E_0)) + \cdots \tag{7.244}$$

Then

$$\tan(\Theta_I - \frac{\pi}{4}) \cong \frac{1}{\left(\frac{d\Theta_I}{dE}\right)_{E_0}(E - E_0)}. \tag{7.245}$$

Define the 'width" of the resonance by

$$\Gamma = \frac{e^{-2\Theta_{II}}}{2\left(\frac{d\Theta_I}{dE}\right)_{E_0}}. \tag{7.246}$$

Then, with $\beta(E)$ evaluated at $E = E_0$, we have

$$\delta(E) \cong \tan^{-1}\left(\frac{\Gamma/2}{E_0 - E}\right) + \beta(E_0) \tag{7.247}$$

as the scattering phase shift within a few units of Γ on either side of $E = E_0$.

Examining Figure 7.10, we see that the amplitude starts off at (A), "far" below threshold for resonance. As the energy approaches E_0, the amplitude moves counterclockwise around the circle $(A \to A' \to B \to C \to C' \to D \to F)$. At A', the amplitude vanishes through destructive interference between the resonant and "potential/hard-sphere" scattering. At B, $E = E_0 - \Gamma/2$ and the phase shift is $\pi/4$ below its value at $E = E_0$. Resonance occurs in the region (C, C'). At D, $E = E_0 + \Gamma/2$ and the phase is $\pi/4$ past its value at $E = E_0$. Finally, the amplitude returns to near its non-resonant value at F. For a thick barrier, the width Γ is small and the resonance is narrow.

7.14 Levinson's Theorem

Levinson's theorem states that if the potential is such that

$$\int_0^\infty dr\, r|V(r)| < \infty \qquad \text{and} \qquad \int_0^\infty dr\, r^2|V(r)| < \infty, \tag{7.248}$$

then

$$\delta_\ell(k = 0) - \delta_\ell(k = \infty) = n_\ell \pi, \tag{7.249}$$

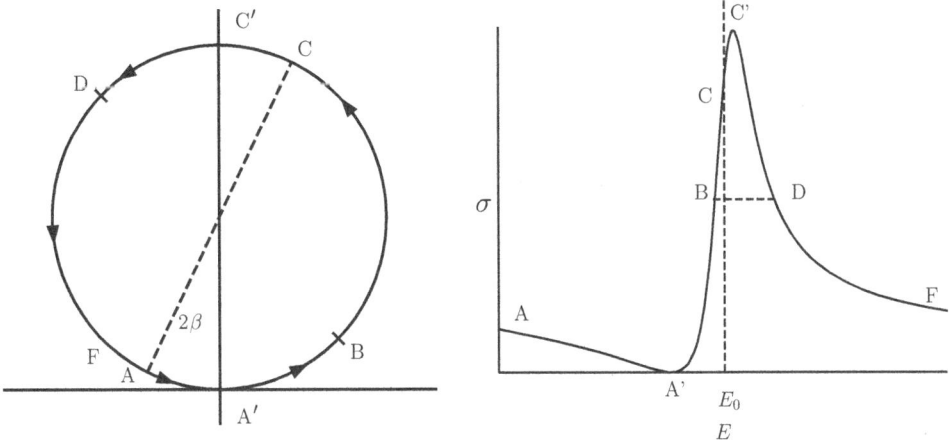

Figure 7.10 Argand diagram and cross section for resonance of width Γ. The asymmetrical shape arises from the interference between a resonant and non-resonant amplitude.

where n_ℓ is the number of bound states of angular momentum ℓ. Levinson's proof requires the use of analytic properties of the scattering amplitude, but we can give a heuristic demonstration. Consider $\ell = 0$ for simplicity. Then as $k \to 0$, according to Eq. (7.193),

$$e^{-i\delta} u_\ell(r) = \sin(kr + \delta) = \sin kr \cos \delta + \cos kr \sin \delta \propto 1 + kr \cot \delta(0). \qquad (7.250)$$

Indeed, we know that at $k = 0$ and outside the range of the potential, the solutions to the Schrödinger equation are linear functions. The slope is the limit of $k \cot \delta$ as $k \to 0$. Now if we write the potential as $V(r) = V_0 f(r)$, we can imagine increasing V_0 from zero to larger and larger values. As the strength of the potential increases, the wave function is "sucked in" and the phase shift increases. When the potential is just strong enough to make a bound state at zero energy, the wave function must be zero asymptotically, so $k \cot \delta = 0$ and $\delta = \pi/2$. Now increasing the energy again, when there is a second bound state at zero energy, we must have $k \cot \delta = 0$ and $\delta = 3\pi/2$.

A simple example of Levinson's theorem is shown in Figure 7.11. The potential is a three-dimensional square well:

$$V(r) = -\frac{\hbar^2}{2m} A^2 \quad r < R_0; \qquad V(r) = 0 \quad r > R_0. \qquad (7.251)$$

If the wave number outside the potential is k, then inside the wave number is $k' = \sqrt{k^2 + A^2}$. The solution inside the potential is $j_0(k'r)$, while outside it is a linear combination of $j_0(kr)$ and $n_0(kr)$. It is straightforward to calculate the phase shift as a function of kR_0. If we use R_0 as the unit of distance, the equation for bound states is

$$k' \cot k' = -\sqrt{A^2 - k^2} \equiv -\kappa. \qquad (7.252)$$

Figure 7.11 shows the s-wave phase shift as a function the wave number outside the potential and manifests Levinson's theorem for this particular potential.

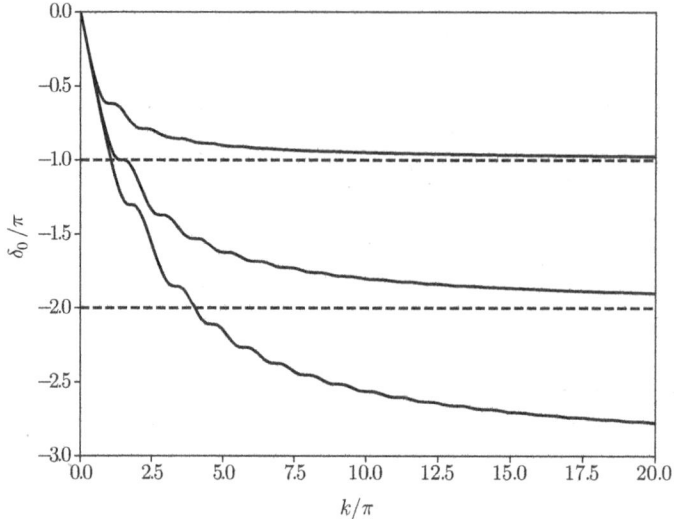

Figure 7.11 The phase shift for a three-dimensional attractive square well with radius R_0 set to unity and depth $\hbar^2 A^2/2m$, with A taken to be π, 2π, and 3π, giving one, two, and three bound s-wave states. In agreement with Levinson's theorem, we see that $\delta_0(k=0) - \delta_0(k=\infty) = n\pi$, where n is the number of s-wave bound states.

References and Suggested Reading

For the classical version of the optical theorem, see Jackson, *Classical Electrodynamics,*, 3rd edition, pp. 500-502.

For a careful presentation of scattering of wave packets see E. H. Wichmann, Am. J. Phys. **33**, 20 (1965).

Mott and Massey, *The Theory of Atomic Collisions*, 2nd edition, is a classic treatment of scattering in that context.

Fröman and Fröman, *JWKB Approximation: Contributions to the Theory*, is an advanced treatise on the method.

Problems

7.1 A system with unperturbed state vectors $|n\rangle$ with energy E_n is subjected to a time-dependent perturbation,

$$H'(t) = \frac{H_1}{\sqrt{\pi}\tau} e^{-t^2/\tau^2}, \tag{7.253}$$

where H_1 is a time-independent operator. At $t = -\infty$, the system is in its ground state $|0\rangle$.

(a) Using first-order time-dependent perturbation theory, show that the probability of a transition to the mth state at $t = +\infty$ is

$$P_{m0} = \frac{1}{\hbar^2} |\langle m\,|H_1|\,0\rangle|^2 \exp\left[-\frac{(E_m - E_0)^2\tau^2}{2\hbar^2}\right]. \qquad (7.254)$$

(b) The adiabatic limit occurs when

$$\tau \gg \frac{\hbar}{|E_1 - E_0|}. \qquad (7.255)$$

Why do all the transition probabilities go to zero in this limit? Discuss the behavior of the system as a function of time in this limit.

7.2 A three-dimensional harmonic oscillator with level spacing $\hbar\omega_0$ consists of a particle of electric charge e and mass m. A pulsed electric field E_0 is applied in the negative z direction for a time T. The oscillator is initially in its ground state (1s).

(a) What is the first-order probability that after the pulse the oscillator is in the first excited state (2p)?

(b) Suppose the time T becomes very small and the electric field E_0 very large in such a way that the electric field delivers an impulse $p = eE_0T$ to the oscillator. What is the probability of excitation to the first excited state in terms of p?

7.3 A particle of mass m and kinetic energy E is scattered by a fixed field of force described by the potential $V(r) = -V_0 e^{-r/a}$.

(a) Calculate the differential elastic scattering cross section in the first Born approximation. Sketch the differential cross section versus the cosine of the scattering angle for low energies ($ka \ll 1$) and high energies ($ka \gg 1$). What is the total cross section as a function of energy?

(b) If $a = 0.7 \times 10^{-13}$ cm, what value of V_0 in MeV will give agreement with the total neutron-proton cross section at momenta in the range 0.1 - 0.5 GeV/c? Make a fit with $\sigma = A[P_{lab}(GeV/c)]^{-2}$ to the n-p total cross section as shown in Figure 7.12, data taken from the compilation of the Particle Data Group, UCRL-20000NN (1970). Remember to convert cms wave numbers to laboratory wave numbers in this calculation.

(c) Determine the criterion for the validity of the first Born approximation and check whether the V_0 value determined in (b) satisfies this criterion at all energies or any energies.

(d) Use the WKB quantization rule to estimate V_0 such that there is just one bound $\ell = 0$ state. Compare this with the value found in part (b).

7.4 Consider the elastic scattering of a pion of charge e, mass M, and incident speed v by a hydrogen atom in the first Born approximation. Treat the pion non-relativistically, the proton as infinitely massive, and neglect spin.

(a) Show that the differential scattering cross section is given by the Rutherford formula for point charges, times $|1 - F(q^2)|^2$, where $\hbar\mathbf{q} = \mathbf{p} - \mathbf{p}'$ is the momentum transfer and $F(q^2)$ is the form factor of the ground state charge distribution

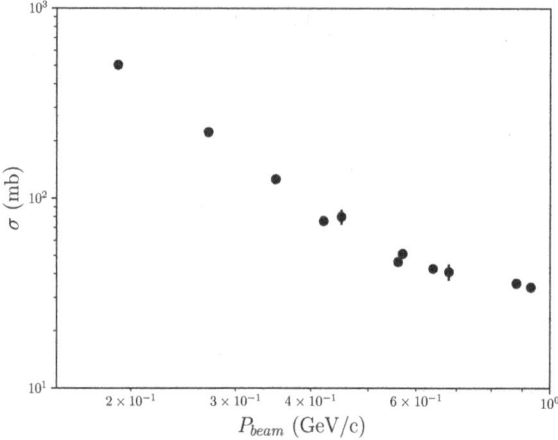

Figure 7.12 Data for pp and np total cross sections from UCRL 20000NN (1970).

$$F(q^2) = \int d^3x \, e^{i\mathbf{q}\cdot\mathbf{x}} |\Psi_0(\mathbf{x})|^2. \tag{7.256}$$

(b) The form factor can be expanded in powers of q^2. Find the general expressions for $F(0)$ and $dF/dq^2(0)$ in terms of $\langle r^2 \rangle$, where the average is in the ground state. What is the expression for the limiting value of the differential scattering cross section in the forward direction?

(c) Calculate $F(q^2)$ for the 1s state of hydrogen and an expression for the differential cross section, both as $d\sigma/d\Omega$ and $d\sigma/dq^2$. Sketch $d\sigma/dq^2$ as a function of q^2 and calculate the total elastic scattering cross section. What is the value of $d\sigma/d\Omega$ at zero degrees in units of a_0?

7.5 A swiftly moving, but non-relativistic, charged particle of charge ze, mass M, and initial speed v interacts with the electrons in a many-electron atom with an infinitely massive nucleus (electrons of charge $-e$ and mass m). The atom is initially in its ground state, $|a\rangle$.

The Born approximation for the differential cross section for excitation to the state $|b\rangle$ in a collision with wave number transfer q^2 is Eq. (7.99)

$$\frac{d\sigma}{dq^2} = 4\pi \left(\frac{ze^2}{\hbar v}\right)^2 \frac{1}{(q^2)^2} |F_{ba}(\mathbf{q})|^2, \tag{7.257}$$

where

$$F_{ba}(\mathbf{q}) = \langle b| \sum_j e^{i\mathbf{q}\cdot\mathbf{x}_j} |a\rangle \tag{7.258}$$

is a transitional form factor.

(a) Specialize the atom to one electron, bound in an isotropic harmonic oscillator potential with energy spacing $\hbar\omega_0$. Show that the differential cross section for excitation of the first excited state is

$$\frac{d\sigma_{10}}{dq^2} = \frac{2\pi z^2 e^4}{\hbar m \omega_0 v^2 q^2} e^{-\hbar q^2/2m\omega_0}.\tag{7.259}$$

(b) Use conservation of energy to determine the minimum and maximum values of q^2. Show that the total cross section for this inelastic process at energies such that $Mv^2/2 \gg \hbar\omega_0$ is

$$\sigma_{10} = \frac{2\pi z^2 e^4}{mv^2 \hbar\omega_0}[\ln(2mv^2/\hbar\omega_0) - 0.5772\ldots].\tag{7.260}$$

(c) If there are N atoms per unit volume, what is the result for the energy loss dE/dx by the massive particle from these excitations? Compare with Eq. 13.6 in Jackson's *Classical Electrodynamics*, 3rd edition.

7.6 **(a)** Define the commutator $C_n(A)$ by the relation $C_n(A) = [H, C_{n-1}(A)]$, $C_0 = A$, where H is the Hamiltonian. Show that the completeness relation for the eigenstates of H allows one to write a general sum rule

$$\sum_b (E_b - E_a)^n |\langle b| A |a\rangle|^2$$

$$= \frac{1}{2}[\langle a| A^\dagger C_n(A) |a\rangle + (-1)^n \langle a| C_n(A^\dagger) A |a\rangle].\tag{7.261}$$

(b) Specialize to a non-relativistic atom with Z electrons and ignore spin. Choosing $A = \sum_j z_j$, where z_j is the z co-ordinate of the jth electron, prove the many-electron Thomas-Reiche-Kuhn sum rule for oscillator strengths,

$$\sum_b (E_b - E_a)|\langle b| \sum_j z_j |a\rangle|^2 = Z\hbar^2/2m.\tag{7.262}$$

(c) By choosing $A = \sum_j e^{i\mathbf{q}\cdot\mathbf{x}_j}$, show that there is a sum rule for generalized oscillator strengths

$$\sum_b (E_b - E_a)|\langle b| \sum_j e^{i\mathbf{q}\cdot\mathbf{x}_j} |a\rangle|^2 = Zq^2\hbar^2/2m.\tag{7.263}$$

This sum rule permits one to compute energy loss, summing over all excitations. See Bethe and Jackiw, *Intermediate Quantum Mechanics*, 3rd edition, p. 303-308. For small q^2, this reproduces the Thomas-Reiche-Kuhn sum rule.

7.7 **(a)** The differential scattering cross section for a particle of mass m by a very weak, real, attractive potential is $d\sigma/d\Omega = A\exp(-q^2a^2/2)$, where q is the momentum transfer (divided by $\hbar$). What is the scattering potential $V(r)$? Assume spherical symmetry.

(b) The strength of the potential is increased greatly. Calculate the eikonal phase shift $\delta(k, b)$ and sketch $\sin^2 \delta(k, b)$ versus b/a for $|\delta(k, 0)| \gg 1$. Estimate the total elastic scattering cross section in this limit in terms of $\delta(k, 0)$ and a.

(c) Suppose the potential is purely imaginary instead of real, with strength such that $|\delta(k, 0)| \gg 1$. Sketch the analog of $\sin^2 \delta(k, b)$ versus b/a and estimate the total cross section.

(d) Hadronic scattering cross sections are observed to have roughly the angular dependence given in part (a) at relativistic energies, but with an energy-dependent coefficient such that the integrated scattering cross sections and total cross section are constant or only varying slowly with energy. What dependence on wave number must the strength of the potential have if the integrated scattering cross section of part (a) is independent of energy? Assuming that the scattering can be described by such a potential, what energy dependence occurs for the total cross sections (elastic or absorption) of parts (b) and (c)? Do these dependences have any relation to actual data? (See the Particle Data Group website for graphs of hadronic cross sections versus momentum.)

Note that in part (d), relativistic effects are important. For the eikonal phase shifts and amplitude it can be shown that $v = c^2 p/E$ is the speed of the particle and k is the center-of-mass wave number.

7.8 **(a)** From the expressions for σ_{tot} and $d\sigma/dt$ in terms of the scattering amplitude $f(k, \theta)$, show that

$$\sigma_{tot}^2 = \frac{16\pi}{1 + \rho^2} \frac{d\sigma}{dt}\bigg|_{t=0}. \tag{7.264}$$

Here, $t = (p - p')^2$ is the four-vector equivalent of $-q^2$ and ρ is the ratio of the real to the imaginary part of the forward elastic scattering amplitude, i.e. $\rho = \Re f(t = 0)/\Im f(t = 0)$.

(b) In the very forward direction, the hadronic scattering amplitude interferes with the Coulomb amplitude. From part (a), in the absence of the Coulomb scattering, we can write

$$\frac{d\sigma}{dt} = \pi \left| (i + \rho) \frac{\sigma_{tot}}{4\pi} e^{Bt/2} \right|^2. \tag{7.265}$$

If we include the Coulomb interaction between the protons, treating them as point (spinless) particles, we have

$$\frac{d\sigma}{dt} = \pi \left| -\frac{2\alpha}{|t|} + (i + \rho) \frac{\sigma_{tot}}{4\pi} e^{Bt/2} \right|^2. \tag{7.266}$$

The ATLAS Collaboration measured $\sigma_{tot} = 100$ mb at $\sqrt{s} = 8$ TeV. At what value of $|t|$ are the Coulomb and hadronic amplitudes approximately equal? Ignore ρ, whose value is roughly 0.1. At what scattering angle does this occur if the center-of-mass energy is 8 TeV?

(c) Sketch the elastic differential cross section as a function of $-t$ in the region where interference between the hadronic and Coulomb amplitudes for the values $\rho = 0, \rho = 0.15$. Note that scattering at very low $|t|$ can be used to calibrate the data and determine ρ. In fact, it is necessary to include a correction to the above formula to account for the combined effect of the nuclear and Coulombic interactions. The result is a phase $e^{i\alpha\phi(t)}$ multiplying the Coulomb, first calculated by Bethe.

7.9* A model for the Ramsauer-Townsend effect (very long mean free paths for scattering of electrons by noble gases at low energies) is the low-energy scattering of a particle of mass m by a spherically symmetric, attractive, square-well potential of radius a and depth V_0 such that at least one bound s-state exists. It is useful to define $\chi_0^2 = 2mV_0 a^2/\hbar^2, ka = x, \chi^2 = \chi_0^2 + x^2$.

(a) For zero angular momentum, solve the Schrödinger equation for positive energy and find expressions for the s-wave scattering cross section at zero energy and at a finite positive energy in terms of $\chi_0 \cot \chi_0$, $\chi \cot \chi$, $x \cot x$, and x.

(b) Plot the total s-wave scattering cross section at zero energy in units of $4\pi a^2$ as a function of χ_0 for $0 \leq \chi_0 \leq 10$. Note on the abscissa the values of χ_0 where the successive bound states come in.

(c) For $\chi_0^2 = 20, 25$, plot a semilog graph (with four or five decades for ordinate) of the s-wave cross section in units of $4\pi a^2$ versus x for $0 \leq x \leq 10$. How many bound states occur in the two cases. With Levinson's theorem in mind, make a sketch or plot of the s-wave phase shift versus x for the two cases. On your graph of the cross sections, show the unitary bound for s-wave scattering.

Two remarks: (1) it is useful in doing part (a) to write the cross section as $4\pi a^2 (\sin x / x)^2$ times a function of $\chi \cot \chi$, etc.; (2) do the calculation in such a way that you can exhibit the phase shift as a function of x. This is useful for the next problem.

7.10 (a) For an attractive square well potential strength of V_0 and radius a, calculate the first Born approximation for the s-wave $\sin \delta$ (or $\tan \delta$, or δ). In the notation of the Problem 7.9, plot the phase shift as a function of $x = ka$ for $\chi_0^2 = 1.25$ and 2.50.

(b) Calculate the exact phase shifts for these two values of potential strength and plot on the same graph as the Born approximation. How well does the Born approximation do for the two strengths? Correlate with the criterion for the validity of the Born approximation.

7.11 Scattering of neutrons from silicon (Z=14, A=28, J=0) at thermal and epithermal energies shows a nearly constant cross section of 2.27 barns (10^{-24} cm^2). Assuming this is described by hard sphere scattering, what is the radius of the silicon nucleus? Data from the "Barn Book", BNL-325 Vol. II (1976) for neutron silicon scattering at higher energies are used to create Table 7.1 and Figure 7.13 at higher energies.

The prominent feature is an s-wave resonance with a pronounced interference dip on the low side of the resonance. The dip would be even more pronounced, but for the

Table 7.1 Representative cross sections based on data from the "Barn Book", BNL-325 Vol. II (1976) for neutron-silicon scattering.

E(keV)	σ (barns)	E(keV)	σ (barns)
60	1.35	184	8.2
80	1.16	192	12.3
100	0.81	200	13.0
120	0.44	208	11.5
140	0.06	216	10.4
160	0.31	224	8.9
168	1.35	240	7.0
176	3.85	260	5.7

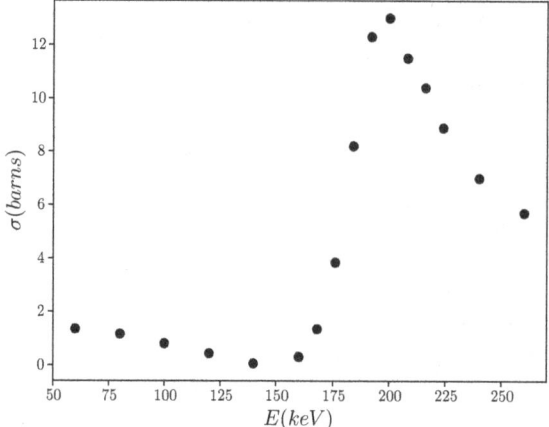

Figure 7.13 Total cross section for neutrons on silicon in the range 60 - 260 keV kinetic energy based on data taken from BNL-325 (1976).

energy spread of about 10% at 150 keV. From these data and the radius inferred from the very low energy cross section, deduce the energy and width of the resonance. Write the phase shift as the sum of a resonant phase and a non-resonant hard-scattering phase. Then deduce the resonant phase from the data and plot it as a function of energy. The resonant energy is where that phase is $\pi/2$ and the width is found from its $3\pi/4$ and $\pi/4$ points (or you may parameterize the resonant phase as $\tan^{-1}(\Gamma/2(E_0 - E)))$. On an Argand diagram, indicate the locations at the various energies.

8

Semi-Classical and Quantum Electromagnetic Field

Electromagnetism was a forerunner of quantum mechanics. Interference of light beams was demonstrated already in 1801 by Young. Planck's solution in 1900 to the ultra-violet catastrophe in black-body radiation with the introduction of the now eponymous constant gave birth to quantum mechanics. Five years later, Einstein showed how quanta of light explained the photoelectric effect. In some sense then, the quantum mechanics of electromagnetism was established before even the Bohr atom of 1912. In place of the Schrödinger equation, there were the Maxwell equations. All that was needed was a translation of electromagnetism into the language of quantum theory. The theory of coherent states developed by Glauber completed the merging of classical electrodynamics with quantum mechanics.

8.1   Electromagnetic Hamiltonian and Gauge Invariance

We have considered from time to time the interaction of a charged particle with other charges or with external electromagnetic fields. We turn now to a systematic formal discussion. Let the particles have mass m_i and charge e_i and canonically conjugate momenta and coordinates $\mathbf{p}_i$ and $\mathbf{x}_i$. Let the mutual interactions, whether electromagnetic in origin or not, be collectively denoted by U and the external electromagnetic field by a scalar and vector potential $\Phi(\mathbf{x}, t)$ and $\mathbf{A}(\mathbf{x}, t)$. Then the Hamiltonian is

$$H = \sum_j \frac{1}{2m_j} \left(\mathbf{p}_j - \frac{e_j}{c} \mathbf{A}(\mathbf{x}_j, t) \right)^2 + \sum_j e_j \Phi(\mathbf{x}_j, t) + U. \tag{8.1}$$

Note that $m_j \dot{\mathbf{x}}_j = \mathbf{p}_j - (e_j/c)\mathbf{A}(\mathbf{x}_j, t)$ is the kinetic momentum. This quantum mechanical equation follows from

$$\dot{\mathbf{x}}_j = \frac{i}{\hbar}[H, \mathbf{x}_j], \quad \text{with } [\mathbf{x}_j, \mathbf{p}_k] = i\hbar \delta_{jk}. \tag{8.2}$$

The time-dependent Schrödinger equation is

$$i\hbar \frac{\partial \Psi}{\partial t} = H\Psi; \quad [\Psi = \Psi(\mathbf{x}_1, \dots, \mathbf{x}_n; t)]. \tag{8.3}$$

We recall that the physically meaningful quantities are not $\mathbf{A}$ and Φ, but $\mathbf{E}$ and $\mathbf{B}$. This is implemented by requiring that "a change in gauge does not affect the physics." A gauge transformation of the potential is defined as follows:

John David Jackson: A Course in Quantum Mechanics, First Edition. Robert N. Cahn.
© 2024 John Wiley & Sons, Inc. Published 2024 by John Wiley & Sons, Inc.
Companion website: www.wiley.com/go/Jackson/QuantumMechanics

$$\mathbf{A} \to \mathbf{A}' = \mathbf{A} + \nabla\chi; \quad \Phi \to \Phi' = \Phi - \frac{1}{c}\frac{\partial\chi}{\partial t}, \tag{8.4}$$

which leave $\mathbf{E}$ and $\mathbf{B}$ unchanged. However, this transformation does modify the Hamiltonian, $H' \neq H$. For the physics to remain unchanged, we require that there exist a new wave amplitude, Ψ' such that

$$i\hbar\frac{\partial\Psi'}{\partial t} = H'\Psi', \tag{8.5}$$

with Ψ' differing from Ψ by at most a phase factor of modulus unity. This will guarantee that expectation values of coordinates and functions of them are unchanged. We will look at momenta and functions of them later.

To see whether there is such a phase factor and what it might be, let us look at the extra time derivative terms in the left- and right-hand sides of the new Schrödinger equation. Put $\Psi' = e^{i\alpha}\Psi$. Then comparing the new equation with the old, multiplied by $e^{i\alpha}$

$$\Delta\left(i\hbar\frac{\partial\Psi'}{\partial t}\right) = -\hbar\frac{\partial\alpha}{\partial t}\left(e^{i\alpha}\Psi\right). \tag{8.6}$$

On the right-hand side, looking just at the scalar potential part

$$\Delta(H'\Psi') = \sum_j e_j \left(\Phi'(\mathbf{x}_j, t) - \Phi(\mathbf{x}_j, t)\right) e^{i\alpha}\Psi = -\sum_j \frac{e_j}{c}\frac{\partial\chi}{\partial t}(\mathbf{x}_j, t)\left(e^{i\alpha}\Psi\right). \tag{8.7}$$

The comparison suggests

$$\alpha = \frac{1}{\hbar c}\sum_j e_j \chi(\mathbf{x}_j, t). \tag{8.8}$$

Now we look at the kinetic energy part to see that this works. Again, we want to compare $H'\Psi'$ with $e^{i\alpha}H\Psi$, but now looking only at the terms of the form $(\mathbf{p} - (e/c)\mathbf{A})^2\Psi$. Consider first just

$$\left(\mathbf{p} - \frac{e}{c}\mathbf{A}'\right)\Psi' = \left(\mathbf{p} - \frac{e}{c}\mathbf{A} - \frac{e}{c}\nabla\chi\right)e^{ie\chi/\hbar c}\Psi$$

$$= e^{ie\chi/\hbar c}\left(\mathbf{p} - \frac{e}{c}\mathbf{A}\right)\Psi. \tag{8.9}$$

It follows that

$$\left(\mathbf{p} - \frac{e}{c}\mathbf{A}'\right)^2\Psi' = e^{ie\chi/\hbar c}\left(\mathbf{p} - \frac{e}{c}\mathbf{A}\right)^2\Psi. \tag{8.10}$$

Now using this for all the $\mathbf{p}_j$, we establish that indeed that if Ψ is a solution in the original gauge, then

$$\Psi' = \exp\left(\frac{i}{\hbar c}\sum_j e_j \chi(\mathbf{x}_j, t)\right)\Psi \tag{8.11}$$

is a solution in the new gauge.

8.2 Aharonov-Bohm Effect

There is a peculiar and unexpected (until the Aharonov-Bohm paper, Phys. Rev. **115**, 485(1959); see also W. Furry and N. Ramsey, Phys. Rev. **118**, 623 (1960)) result associated with the gauge transformations business. Suppose we have a single charged particle in the presence of static electromagnetic fields described by $\mathbf{A}$ and Φ and that $\mathbf{B} = \nabla \times \mathbf{A}$ is confined to some region. We wish to discuss the motion of the particle in the regions where $\mathbf{B} = 0$, for example a region so remote from the locale where $\mathbf{B} \neq 0$ that the particle's wave function does not overlap the region where $\mathbf{B}$ is non-zero.

We can solve the Schrödinger equation

$$i\hbar \frac{\partial \Psi}{\partial t} = H\Psi; \quad [\Psi = \Psi(\mathbf{x}_1, \ldots, \mathbf{x}_n; t)] \tag{8.12}$$

by choosing a gauge in which $\mathbf{A}$ vanishes with the gauge transformation

$$\mathbf{A} \rightarrow \mathbf{A}' = \mathbf{A} + \nabla \chi; \qquad \Phi \rightarrow \Phi' = \Phi \tag{8.13}$$

by taking

$$\nabla \chi = -\mathbf{A}(\mathbf{x}). \tag{8.14}$$

This is achieved if we choose for

$$\chi = -\int_{\text{path}}^{\mathbf{x}} \mathbf{A}(\mathbf{x}') \cdot d\ell, \tag{8.15}$$

where "path" means any path that ends at the point $\mathbf{x}$. Any path that lies far from the region where $\mathbf{B}$ is non-vanishing will work because the difference between the two paths is simply a path that returns to the start and

$$\Delta \chi = \oint \mathbf{A} \cdot d\ell = \int_S (\nabla \times \mathbf{A}) \cdot d\mathbf{S} = 0. \tag{8.16}$$

Thus we can take as our solution in this new gauge

$$\psi^{(0)} = e^{i\frac{e}{\hbar c}\chi}\psi, \tag{8.17}$$

where ψ^0 satisfies the equation

$$i\hbar \frac{\partial \psi^{(0)}}{\partial t} = \left[\frac{p^2}{2m} + e\Phi(\mathbf{x}) \right] \psi^{(0)}. \tag{8.18}$$

This all looks innocuous enough, but present is a startling interference effect, the Aharonov-Bohm effect. Consider an electron diffraction experiment of a peculiar sort, with a solenoid between the slits, as shown in Figure 8.1. The paths of the wave packets through the two slits never go near the region of $\mathbf{B} \neq 0$. But with $\mathbf{A} \neq 0$, the waves interfering at the point P have an additional phase difference,

$$\Delta \theta = \frac{e}{\hbar c} \int_{\text{path 1}}^{P} \mathbf{A} \cdot d\ell - \frac{e}{\hbar c} \int_{\text{path 2}}^{P} \mathbf{A} \cdot d\ell = \frac{e}{\hbar c} \oint \mathbf{A} \cdot d\ell = \frac{e}{\hbar c} \int_{\text{solenoid}} \mathbf{B} \cdot d\mathbf{S} \tag{8.19}$$

so the phase difference is

$$\Delta \theta = -\frac{e}{\hbar c}(\text{Flux of } \mathbf{B}). \tag{8.20}$$

If the flux through the solenoid is changed, the fringe pattern will shift, even though the electrons never pass through the magnetic field! Experiments have been done: see R. G. Chambers, *Phys. Rev. Letters* **5**, 3 (1960).

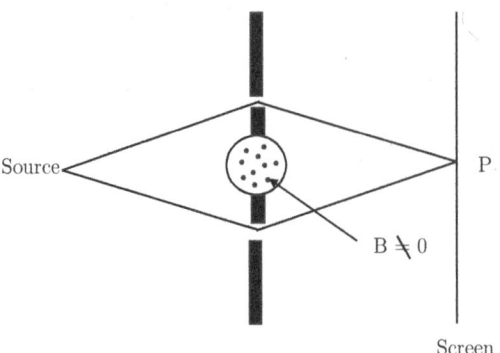

Figure 8.1 The Aharonov-Bohm effect results in an interference pattern between paths that never encounter a magnetic field, but which pattern depends on the magnetic field present in a region not overlapped by the particle wave function.

Sometimes the Aharonov-Bohm effect is cited as evidence for the observability of the vector potential, not at the classical level, but because of quantum mechanics. What is actually observable is the magnetic flux of the line integral of $\mathbf{A}$ around a closed path. Thus it is permitted to continue to say that only the fields are observable. Note in particular that the Aharonov-Bohm effect does not depend on a particular choice of gauge, a thing that you might expect if the vector potential were actually observable.

8.3 Semi-Classical Radiation Theory

Before quantizing the radiation field, let us discuss the ideas of semi-classical radiation theory as applied to the absorption of light by a quantum system of charged particles. The light beam can be described by a vector potential $\mathbf{A}_{\mathrm{rad}}(\mathbf{x}, t)$ alone. We choose the "radiation gauge," where $\nabla \cdot \mathbf{A} = 0$ and the scalar potential Φ_{rad} satisfies the instantaneous Coulomb equation. (See Jackson, *Classical Electrodynamics*, 3rd edition, p. 241.) The radiation potential, $\mathbf{A}_{\mathrm{rad}}$, is associated with no charge, so we can take $\Phi = 0$ and then the Hamiltonian reads

$$H = H_0 - \sum_j \frac{e}{2m_j c}\left(\mathbf{p}_j \cdot \mathbf{A}_{\mathrm{rad}} + \mathbf{A}_{\mathrm{rad}} \cdot \mathbf{p}_j\right) + \sum_j \frac{e_j^2}{2m_j c^2}\mathbf{A}_{\mathrm{rad}} \cdot \mathbf{A}_{\mathrm{rad}}. \tag{8.21}$$

We neglect the second order term at present and do perturbation theory in the term linear in $\mathbf{A}_{\mathrm{rad}}$:

$$H'_{\mathrm{rad}} = -\sum_j \frac{e}{2m_j c}\left(\mathbf{p}_j \cdot \mathbf{A}_{\mathrm{rad}} + \mathbf{A}_{\mathrm{rad}} \cdot \mathbf{p}_j\right). \tag{8.22}$$

Because $\nabla \cdot \mathbf{A}_{\mathrm{rad}} = 0$, we need not worry about the order of $\mathbf{p}$ and $\mathbf{A}_{\mathrm{rad}}$, so we can write

$$H'_{\mathrm{rad}} = -\sum_j \frac{e}{m_j c}\mathbf{A}_{\mathrm{rad}}(\mathbf{x}, t) \cdot \mathbf{p}_j. \tag{8.23}$$

8.3.1 Semi-classical Normalization of $\mathbf{A}_{\mathrm{rad}}$

If $\mathbf{A}_{\mathrm{rad}}$ describes a beam with n photons with polarization λ, wave vector $\mathbf{k}$, and frequency $\omega = ck$ in a box of volume V, we will have

$$\mathbf{A}_{\mathrm{rad}}(\mathbf{x}, t) = A_0 \boldsymbol{\epsilon}_{\mathbf{k}}^{(\lambda)} \cos(\mathbf{k} \cdot \mathbf{x} - \omega t), \tag{8.24}$$

where we still need to determine A_0. In the radiation gauge, we have $\epsilon_{\mathbf{k}}^{(\lambda)} \cdot \mathbf{k} = 0$, and $\lambda = 1, 2$ specifies two independent polarizations. The corresponding physical fields are

$$\mathbf{E} = -\frac{1}{c}\frac{\partial \mathbf{A}}{\partial t} = -kA_0\epsilon_{\mathbf{k}}^{(\lambda)}\sin(\mathbf{k}\cdot\mathbf{x} - \omega t)$$

$$\mathbf{B} = \nabla \times \mathbf{A} = -\mathbf{k} \times \epsilon_{\mathbf{k}}^{(\lambda)}A_o\sin(\mathbf{k}\cdot\mathbf{x} - \omega t). \tag{8.25}$$

Now the energy density in the fields is

$$w = \frac{1}{8\pi}\left(E^2 + B^2\right) = \frac{1}{4\pi}k^2 A_0^2 \sin^2(\mathbf{k}\cdot\mathbf{x} - \omega t). \tag{8.26}$$

Equating the time-average energy density times the volume to the photon total energy gives

$$\frac{1}{8\pi}k^2 A_0^2 V = n_{\mathbf{k},\lambda}\hbar\omega = n_{\mathbf{k},\lambda}\hbar ck; \qquad A_0 = \sqrt{\frac{8\pi\hbar c}{Vk}}\sqrt{n_{\mathbf{k},\lambda}}. \tag{8.27}$$

So

$$\mathbf{A}_{\text{rad}}(\mathbf{x}, t) = \sqrt{\frac{2\pi\hbar c}{Vk}}\sqrt{n_{\mathbf{k},\lambda}}\left(\epsilon_{\mathbf{k}}^{(\lambda)}e^{i\mathbf{k}\cdot\mathbf{x}-i\omega t} + \epsilon_{\mathbf{k}}^{(\lambda)*}e^{-i\mathbf{k}\cdot\mathbf{x}+i\omega t}\right) \tag{8.28}$$

and the interaction Hamiltonian thus becomes

$$H'_{\text{rad}} = -\sqrt{\frac{2\pi\hbar c}{Vk}}\sqrt{n_{\mathbf{k},\lambda}}\sum_j \frac{e_j}{m_j c}\left[e^{i\mathbf{k}\cdot\mathbf{x}-i\omega t}\epsilon_{\mathbf{k}}^{(\lambda)}\cdot\mathbf{p}_j + e^{-i\mathbf{k}\cdot\mathbf{x}+i\omega t}\epsilon_{\mathbf{k}}^{(\lambda)*}\cdot\mathbf{p}_j\right]. \tag{8.29}$$

8.3.2 Transition Amplitudes

We showed above that for a transition from a state α to a state β, the first order amplitude, for $\alpha \neq \beta$,

$$a_{\beta\alpha}^{(1)} = -\frac{i}{\hbar}\int_{-\infty}^{t} dt'\, e^{i(\omega_\beta - \omega_\alpha)t'}\langle\beta|H'(t')|\alpha\rangle. \tag{8.30}$$

We now have harmonic time dependence so that the time integral becomes

$$\int_{-\infty}^{t} dt' e^{i(\omega_\beta - \omega_\alpha \pm \omega)t'} \to 2\pi\delta(\omega_\beta - \omega_\alpha \pm \omega). \tag{8.31}$$

We thus will have transitions obeying the Einstein-Bohr frequency condition, $\omega_\beta = \omega_\alpha \pm \omega$. Clearly this must represent absorption and emission of photons of energy $\hbar\omega$.

The effective vector potentials are

Absorption: $\qquad \mathbf{A}_{\text{absorption}} = c\sqrt{\frac{2\pi\hbar}{V\omega}}\sqrt{n_{\mathbf{k},\lambda}}e^{i\mathbf{k}\cdot\mathbf{x}-i\omega t}\epsilon_{\mathbf{k}}^{(\lambda)}$

Emission: $\qquad \mathbf{A}_{\text{emission}} = c\sqrt{\frac{2\pi\hbar}{V\omega}}\sqrt{n_{\mathbf{k},\lambda}}e^{-i\mathbf{k}\cdot\mathbf{x}+i\omega t}\epsilon_{\mathbf{k}}^{(\lambda)*}. \tag{8.32}$

For absorption, $n_{\mathbf{k},\lambda}$ is the number of such photons in the initial state, while for emission, it is the number of photons in the final state. Where this interpretation goes beyond classical arguments, it benefits from knowing the correct answer.

8.4   Scalar Field Quantization

A brief review of classical mechanics for discrete and continuous media is in order before moving to the quantization of the electromagnetic field in the radiation gauge: $\nabla \cdot \mathbf{A} = 0$, $\Phi = 0$. The analogy between discrete point mechanics and quantum mechanics begins with the Lagrangian $L = L(q_j, \dot{q}_j, t)$. The canonical momenta are

$$p_j = \frac{\partial L}{\partial \dot{q}_j} \tag{8.33}$$

and the Hamiltonian is

$$H = \sum_j \dot{q}_j p_j - L. \tag{8.34}$$

In the Lagrangian, the variables are $q_j, \dot{q}_j$, while in the Hamiltonian they are q_j, p_j, where $j = 1, 2, \dots N$ labels the particles.

In quantum mechanics, we assume for the operators q_j, p_k

$$[q_j, p_k] = i\hbar \delta_{jk} \tag{8.35}$$

and the time-dependent Schrödinger equation is

$$H\Psi = i\hbar \frac{\partial \Psi}{\partial t}. \tag{8.36}$$

Returning to classical dynamics, we consider continuous media, $\phi_k(\mathbf{x}, t)$ with $k = 1, 2, \dots N$, with the correspondences between the discrete indices j and the space-time points $(\mathbf{x}, t)$ and

$$q_j(t) \rightarrow \phi_k(\mathbf{x}, t); \tag{8.37}$$

$$L(q_j, \dot{q}_j, t) \rightarrow \mathcal{L}(\phi_k, \nabla\phi_k, \partial\phi_k/\partial t, t) = \mathcal{L}(\phi_k, \partial^\mu \phi_k, t), \tag{8.38}$$

and for the canonical momenta

$$p_j(t) \rightarrow \pi_k(\mathbf{x}, t) = \frac{\partial \mathcal{L}}{\partial(\partial\phi_k/\partial t)}. \tag{8.39}$$

The Euler-Lagrange equations for discrete dynamics

$$\frac{d}{dt}\left(\frac{\partial L}{\partial \dot{q}_j}\right) - \frac{\partial L}{\partial q_j} = 0 \tag{8.40}$$

become

$$\sum_\mu \partial^\mu \left(\frac{\partial \mathcal{L}}{\partial(\partial^\mu \phi_k)}\right) - \frac{\partial \mathcal{L}}{\partial \phi_k} = 0. \tag{8.41}$$

To form the Hamiltonian we need,

$$\pi_k = \frac{\partial \mathcal{L}}{\partial \dot{\phi}_k}. \tag{8.42}$$

The Hamiltonian for discrete dynamics has its counterpart

$$H = \sum_j p_j \dot{q}_j - L \rightarrow \mathcal{H} = \sum_k \pi_k \dot{\phi}_k - \mathcal{L}. \tag{8.43}$$

The quantization condition has the correspondence

$$[q_i, p_j] = i\hbar\delta_{ij} \rightarrow [\phi_k(\mathbf{x}, t), \pi_{k'}(\mathbf{x}', t)] = i\hbar\delta_{kk'}\delta(\mathbf{x} - \mathbf{x}'). \tag{8.44}$$

8.5 Quantization of the Radiation Field

In the radiation gauge we have only the vector potential $\mathbf{A}(\mathbf{x}, t)$, with

$$\mathbf{E} = -\frac{1}{c}\frac{\partial \mathbf{A}}{\partial t}, \qquad \mathbf{B} = \nabla \times \mathbf{A}. \tag{8.45}$$

We know the equations of motion are

$$\left(\frac{1}{c^2}\frac{\partial^2}{\partial t^2} - \nabla^2\right)\mathbf{A} = 0; \qquad \nabla \cdot \mathbf{A} = 0. \tag{8.46}$$

The Lagrangian density, $\mathcal{L}$, is a Lorentz scalar of necessity. Because of its connection to $\mathcal{H}$ and energy, it needs to be quadratic in the fields. The only quadratic Lorentz scalar (not pseudoscalar) is $E^2 - B^2$. (See Jackson, *Classical Electrodynamics*, 3rd edition, pp. 598-600.) In fact,

$$\mathcal{L} = \frac{1}{8\pi}\left(E^2 - B^2\right) = \frac{1}{8\pi}\left[\dot{\mathbf{A}}^2 - (\nabla \times \mathbf{A})^2\right], \tag{8.47}$$

where we will put $c = 1$ and, as well, $\hbar = 1$ most of the time. Let us check the Euler-Lagrange equations:

$$\frac{\partial \mathcal{L}}{\partial \dot{A}_k} = \frac{1}{4\pi}\dot{A}_k \equiv \pi_k. \tag{8.48}$$

We need

$$\frac{\partial \mathcal{L}}{\partial(\partial A_k/\partial x_j)} = -\frac{\partial \mathcal{L}}{\partial(\partial^j A^k)} = -\frac{1}{4\pi}\left(\frac{\partial A_k}{\partial x_j} - \frac{\partial A_j}{\partial x_k}\right), \tag{8.49}$$

where we have used conventional relativistic notation. We also observe that

$$\frac{\partial \mathcal{L}}{\partial A_j} = 0. \tag{8.50}$$

The Euler-Lagrange equation is

$$0 = \sum_{\mu}\partial^{\mu}\left(\frac{\partial \mathcal{L}}{\partial(\partial^{\mu}A_k)}\right) - \frac{\partial \mathcal{L}}{\partial A_k} = \frac{1}{4\pi}\left\{\ddot{A}_k - \sum_j \frac{\partial}{\partial x_j}\left(\frac{\partial A_k}{\partial x_j} - \frac{\partial A_j}{\partial x_k}\right)\right\}$$

$$0 = \ddot{A}_k - \nabla^2 A_k + \frac{\partial}{\partial x_k}\nabla \cdot \mathbf{A}. \tag{8.51}$$

This is the expected wave equation, provided $\nabla \cdot \mathbf{A} = 0$.

To find the Hamiltonian density, we follow canonical procedures.

$$\mathcal{H} = \pi \cdot \dot{\mathbf{A}} - \mathcal{L} = \frac{1}{4\pi}\dot{\mathbf{A}}^2 - \frac{1}{8\pi}\left[\dot{\mathbf{A}}^2 - (\nabla \times \mathbf{A})^2\right]$$

$$= \frac{1}{8\pi}\left[\dot{\mathbf{A}}^2 + (\nabla \times \mathbf{A})^2\right] = \frac{1}{8\pi}\left(E^2 + B^2\right). \tag{8.52}$$

So everything is consistent. We are tempted to impose quantization (setting $\hbar = 1$)

$$[A_j(\mathbf{x}, t), \dot{A}_{j'}(\mathbf{x}', t)] = 4\pi i \delta_{jj'} \delta(\mathbf{x} - \mathbf{x}'). \tag{8.53}$$

The astute reader will detect a problem here. We have imposed the gauge condition $\nabla \cdot \mathbf{A} = 0$, removing a degree of freedom from $\mathbf{A}$. Can we still impose this condition on all three Cartesian directions indicated by j and j'? This will become clearer once we move to Fourier space.

The quantities $\mathbf{A}$ and $\dot{\mathbf{A}}$ are operators. To obtain the spectral (Fourier) decomposition of the fields and Hamiltonian, we expand them in an orthonormal set of basis functions with the coefficients being operators. The basis states can be (a) plane waves, (b) spherical waves, or (c) electromagnetic eigen-modes in a cavity. The "photons" so described are different in each case. We choose here to work with plane waves. To emphasize that the fields are real, we expand in real functions.

$$\mathbf{A}(\mathbf{x}, t) = \sqrt{\frac{4\pi}{V}} \sum_{\mathbf{k}} \left[\mathbf{b_k} \sin(\mathbf{k} \cdot \mathbf{x} - \omega_k t) + \mathbf{d_k} \cos(\mathbf{k} \cdot \mathbf{x} - \omega_k t)\right]. \tag{8.54}$$

We are working in a box of volume V, so the $\mathbf{k}$ take on discrete values. The wave equation is satisfied if $\omega_k = ck$. The reality of the classical vector potential translates into $\mathbf{A}$, $\mathbf{b_k}$, and $\mathbf{d_k}$ being Hermitian operators. That $\nabla \cdot \mathbf{A} = 0$ implies $\mathbf{b_k} \cdot \mathbf{k} = \mathbf{d_k} \cdot \mathbf{k} = 0$. It is convenient to exhibit this transversality explicitly with polarization vectors $\epsilon_{\mathbf{k}\lambda}, (\lambda = 1, 2)$ such that

$$\epsilon_{\mathbf{k}\lambda} \cdot \mathbf{k} = 0; \quad \epsilon_{\mathbf{k}1} \times \epsilon_{\mathbf{k}2} = \frac{1}{k}\mathbf{k} = \hat{\mathbf{k}}; \quad \epsilon_{\mathbf{k}\lambda} \cdot \epsilon_{\mathbf{k}\lambda'} = \delta_{\lambda,\lambda'};$$

$$\hat{\mathbf{k}} \times \epsilon_{\mathbf{k}1} = \epsilon_{\mathbf{k}2}; \quad \hat{\mathbf{k}} \times \epsilon_{\mathbf{k}2} = -\epsilon_{\mathbf{k}1}. \tag{8.55}$$

There is one subtlety here. If we had three independent polarizations, we would have imposed

$$\sum_{\lambda}(\epsilon_{\mathbf{k}\lambda})_j(\epsilon_{\mathbf{k}\lambda})_{j'} = \delta_{jj'}, \tag{8.56}$$

but this is inconsistent with the previous equation, which says that ϵ never points along $\hat{\mathbf{k}}$. Thus, the true statement is

$$\sum_{\lambda}(\epsilon_{\mathbf{k}\lambda})_j(\epsilon_{\mathbf{k}\lambda})_{j'} = \delta_{jj'} - \hat{k}_j\hat{k}_{j'}. \tag{8.57}$$

We have taken the $\epsilon_{\mathbf{k}\lambda}$ to be real vectors so that $\mathbf{A}$ is manifestly real/Hermitian. In practice, we often want to consider circular polarization with

$$\mathbf{e_{k,\pm}} = \frac{1}{\sqrt{2}}(\epsilon_{\mathbf{k}1} \pm i\epsilon_{\mathbf{k}2}); \quad \lambda = (+, -) \tag{8.58}$$

so that

$$\mathbf{e}_{\mathbf{k},\lambda} \cdot \hat{\mathbf{k}} = 0; \quad \mathbf{e}_{\mathbf{k},+} \times \mathbf{e}_{\mathbf{k},-} = -i\hat{\mathbf{k}}; \quad \mathbf{e}_{\mathbf{k},\lambda}^* \cdot \mathbf{e}_{\mathbf{k},\lambda'} = \delta_{\lambda,\lambda'};$$

$$\sum_{\lambda=+,-} (\mathbf{e}_{\mathbf{k},\lambda})_j (\mathbf{e}_{\mathbf{k},\lambda}^*)_{j'} = \delta_{jj'} - \hat{\mathbf{k}}_j \hat{\mathbf{k}}_{j'}. \tag{8.59}$$

We will proceed in Fourier space, where this is simple, and ultimately see what commutation relation we can impose on **A** in real space.

Now writing

$$\mathbf{b}_{\mathbf{k}} = \sum_{\lambda=1}^{2} \boldsymbol{\epsilon}_{\mathbf{k}\lambda} Q_{\mathbf{k}\lambda}; \quad \mathbf{d}_{\mathbf{k}} = \sum_{\lambda=1}^{2} \boldsymbol{\epsilon}_{\mathbf{k}\lambda} \frac{P_{\mathbf{k}\lambda}}{\omega_k} \tag{8.60}$$

will guarantee $\nabla \cdot \mathbf{A} = 0$. We can rewrite the vector potential as

$$\mathbf{A}(\mathbf{x},t) = \sqrt{\frac{4\pi}{V}} \sum_{\mathbf{k},\lambda} \boldsymbol{\epsilon}_{\mathbf{k}\lambda} \left[Q_{\mathbf{k}\lambda} \sin(\mathbf{k} \cdot \mathbf{x} - \omega_k t) + \frac{P_{\mathbf{k}\lambda}}{\omega_k} \cos(\mathbf{k} \cdot \mathbf{x} - \omega_k t) \right]. \tag{8.61}$$

The question now is what properties must the operators P and Q have to assure that something like the canonical commutation relations, Eq. (8.53), are satisfied. As the notation itself suggests, we speculate that

$$[Q_{\mathbf{k},\lambda}, Q_{\mathbf{k}',\lambda'}] = 0; \quad \text{all } (\mathbf{k},\lambda), (\mathbf{k}',\lambda') \tag{8.62}$$

$$[P_{\mathbf{k},\lambda}, P_{\mathbf{k}',\lambda'}] = 0; \quad \text{all } (\mathbf{k},\lambda), (\mathbf{k}',\lambda') \tag{8.63}$$

$$[Q_{\mathbf{k},\lambda}, P_{\mathbf{k}',\lambda'}] = i\delta_{\lambda,\lambda'} \delta_{\mathbf{k},\mathbf{k}'} \tag{8.64}$$

are sufficient to satisfy the field commutation relations and proceed.

Now we calculate the commutator symbolically

$$[A_j(\mathbf{x},t), \dot{A}_{j'}(\mathbf{x}',t)] = \frac{4\pi}{V} \sum_{\mathbf{k},\lambda} \sum_{\mathbf{k}',\lambda'} (\boldsymbol{\epsilon}_{\mathbf{k}\lambda})_j (\boldsymbol{\epsilon}_{\mathbf{k}\lambda})_{j'} \{-\omega_{k'}[Q,Q'] \sin() \cos(')$$

$$- [P/\omega_k, Q'] \omega_{k'} \cos() \cos(') + [Q,P'] \sin() \sin(')$$

$$+ (1/\omega_k)[P,P'] \cos() \sin(') \}. \tag{8.65}$$

Using the postulated commutation relations,

$$[A_j(\mathbf{x},t), \dot{A}_{j'}(\mathbf{x}',t)] = \frac{4\pi i}{V} \sum_{\mathbf{k},\lambda} (\boldsymbol{\epsilon}_{\mathbf{k}\lambda})_j (\boldsymbol{\epsilon}_{\mathbf{k}\lambda})_{j'} \{\sin(\mathbf{k} \cdot \mathbf{x} - \omega_k t) \sin(\mathbf{k} \cdot \mathbf{x}' - \omega_k t)$$

$$+ \cos(\mathbf{k} \cdot \mathbf{x} - \omega_k t) \cos(\mathbf{k} \cdot \mathbf{x}' - \omega_k t)\}$$

$$= \frac{4\pi i}{V} \sum_{\mathbf{k}} (\delta_{jj'} - \hat{\mathbf{k}}_j \hat{\mathbf{k}}_{j'}) \cos[\mathbf{k} \cdot (\mathbf{x} - \mathbf{x}')]. \tag{8.66}$$

As anticipated, Eq. (8.53) cannot really be correct. Instead, we find a form that is consistent in Fourier space with our gauge condition, $\nabla \cdot \mathbf{A} = 0$. To recover the proper commutation relation in real space, we need to perform a Fourier transform.

Now, an examination of Fourier series on a finite interval shows that for a large volume

$$\frac{1}{V}\sum_{\mathbf{k}}\cos[(\mathbf{k}\cdot(\mathbf{x}-\mathbf{x}'))] \to \delta(\mathbf{x}-\mathbf{x}'). \tag{8.67}$$

However, the presence of $\hat{\mathbf{k}}_j\hat{\mathbf{k}}_{j'}$ in Eq. (8.66) prevents us from obtaining our originally hypothesized commutation relation. Using the correspondence

$$\frac{1}{V}\sum_{\mathbf{k}} \to \int\frac{d^3k}{(2\pi)^3}, \tag{8.68}$$

we see that

$$\frac{1}{V}\sum_{k}(\delta_{jj'} - \hat{\mathbf{k}}_j\hat{\mathbf{k}}_{j'})\cos\mathbf{k}\cdot(\mathbf{x}-\mathbf{x}')$$

$$= \frac{1}{2}\int\frac{d^3k}{(2\pi)^3}(\delta_{jj'} - \hat{\mathbf{k}}_j\hat{\mathbf{k}}_{j'})(e^{i\mathbf{k}\cdot(\mathbf{x}-\mathbf{x}')} + e^{-i\mathbf{k}\cdot(\mathbf{x}-\mathbf{x}')})$$

$$= \int\frac{d^3k}{(2\pi)^3}\left(\delta_{jj'} - \frac{1}{k^2}\frac{\partial}{\partial x_j}\frac{\partial}{\partial x'_{j'}}\right)\left(e^{i\mathbf{k}\cdot(\mathbf{x}-\mathbf{x}')} + \text{complex conjugate.}\right). \tag{8.69}$$

Now we can do the d^3k integral with $\mathbf{R} = \mathbf{x} - \mathbf{x}'$

$$\int d^3k\frac{e^{i\mathbf{k}\cdot\mathbf{R}}}{k^2} = 4\pi\int_0^\infty dk\frac{\sin kR}{kR} = \frac{2\pi^2}{R}, \tag{8.70}$$

so that

$$\int\frac{d^3k}{(2\pi)^3}(\delta_{jj'} - \hat{\mathbf{k}}_j\hat{\mathbf{k}}_{j'})\cos\mathbf{k}\cdot(\mathbf{x}-\mathbf{x}') = \delta_{jj'}\delta(\mathbf{x}-\mathbf{x}') - \frac{1}{4\pi}\frac{\partial}{\partial x_j}\frac{\partial}{\partial x'_{j'}}\frac{1}{|\mathbf{x}-\mathbf{x}'|}. \tag{8.71}$$

This then shows that in Eq. (8.53) there is an additional piece to $\delta_{jj'}\delta(\mathbf{x}-\mathbf{x}')$, viz.

$$-\frac{1}{4\pi}\frac{\partial}{\partial x_j}\frac{\partial}{\partial x'_{j'}}\frac{1}{|\mathbf{x}-\mathbf{x}'|}. \tag{8.72}$$

so that

$$[A_j(\mathbf{x},t),\dot{A}_{j'}(\mathbf{x}',t)] \quad = 4\pi i\left[\delta_{jj'}\delta(\mathbf{x}-\mathbf{x}') - \frac{1}{4\pi}\frac{\partial}{\partial x_j}\frac{\partial}{\partial x'_{j'}}\frac{1}{|\mathbf{x}-\mathbf{x}'|}\right]. \tag{8.73}$$

See Schiff, *Quantum Mechanics*, 3rd edition, p. 515. From this, we can compute the commutation relation between the physical fields $\mathbf{E}$ and $\mathbf{B}$:

$$[E_j(\mathbf{x},t),B_k(\mathbf{x}',t)] = [-\dot{A}_j(\mathbf{x}.t),\epsilon_{k\ell m}\partial'_\ell A_m(\mathbf{x}',t)]$$

$$= \epsilon_{k\ell m}\partial'_\ell[A_m(\mathbf{x}',t),\dot{A}_j(\mathbf{x},t)]$$

$$= 4\pi i\epsilon_{k\ell m}\partial'_\ell\left[\delta_{jm}\delta(\mathbf{x}-\mathbf{x}') - \frac{1}{4\pi}\frac{\partial}{\partial x_j}\frac{\partial}{\partial x'_m}\frac{1}{|\mathbf{x}-\mathbf{x}'|}\right]$$

$$= 4\pi i\epsilon_{jk\ell}\partial'_\ell\delta(\mathbf{x}-\mathbf{x}'). \tag{8.74}$$

With the quantization of the electromagnetic field in hand, we turn to the Hamiltonian.

$$H = \int \mathcal{H} d^3x = \frac{1}{8\pi} \int_V (E^2 + B^2) d^3x = \frac{1}{8\pi} \int_V [\dot{\mathbf{A}}^2 + (\nabla \times \mathbf{A})^2] d^3x. \quad (8.75)$$

Now we insert the expansion for **A**, Eq. (8.61).

$$\dot{\mathbf{A}}(\mathbf{x}, t) = \sqrt{\frac{4\pi}{V}} \sum_{\mathbf{k},\lambda} \epsilon_{\mathbf{k}\lambda} \omega_k \left[-Q_{\mathbf{k}\lambda} \cos(\mathbf{k} \cdot \mathbf{x} - \omega_k t) + \frac{P_{\mathbf{k}\lambda}}{\omega_k} \sin(\mathbf{k} \cdot \mathbf{x} - \omega_k t) \right]$$

$$\nabla \times \mathbf{A}(\mathbf{x}, t) = \sqrt{\frac{4\pi}{V}} \sum_{\mathbf{k},\lambda} \hat{\mathbf{k}} \times \epsilon_{\mathbf{k}\lambda} \omega_k \left[Q_{\mathbf{k}\lambda} \cos(\mathbf{k} \cdot \mathbf{x} - \omega_k t) - \frac{P_{\mathbf{k}\lambda}}{\omega_k} \sin(\mathbf{k} \cdot \mathbf{x} - \omega_k t) \right].$$

$$(8.76)$$

In computing H, we will need to evaluate integrals

$$\int d^3x \, (\sin(\mathbf{k} \cdot \mathbf{x}), \cos(\mathbf{k} \cdot \mathbf{x})) \times (\sin(\mathbf{k}' \cdot \mathbf{x}), \cos(\mathbf{k}' \cdot \mathbf{x})), \quad (8.77)$$

where **k** and **k**′ meet the boundary conditions at the surfaces of our large rectangular solid volume. The only non-vanishing pieces are

$$\int d^3x \, \sin(\mathbf{k} \cdot \mathbf{x}) \sin(\mathbf{k}' \cdot \mathbf{x}) = \int d^3x \, \cos(\mathbf{k} \cdot \mathbf{x}) \cos(\mathbf{k}' \cdot \mathbf{x}) = \frac{1}{2} V \delta_{\mathbf{k}\mathbf{k}'}. \quad (8.78)$$

Armed with these integrals, we evaluate H using our linear polarization vectors

$$H = \frac{1}{8\pi} \frac{4\pi}{V} \frac{V}{2} \sum_{\mathbf{k},\lambda\lambda'} \left\{ \epsilon_{\mathbf{k},\lambda} \cdot \epsilon_{\mathbf{k},\lambda'} + (\hat{\mathbf{k}} \times \epsilon_{\mathbf{k},\lambda}) \cdot (\hat{\mathbf{k}} \times \epsilon_{\mathbf{k},\lambda'}) \right\} \left[P_{\mathbf{k},\lambda} P_{\mathbf{k},\lambda'} + \omega_k^2 Q_{\mathbf{k},\lambda} Q_{\mathbf{k},\lambda'} \right].$$

$$(8.79)$$

The orthogonality of the two independent polarization vectors ϵ to each other and to $\hat{\mathbf{k}}$ gives, finally,

$$H = \frac{1}{2} \sum_{\mathbf{k},\lambda} [P_{\mathbf{k},\lambda}^2 + \omega_k^2 Q_{\mathbf{k},\lambda}^2]. \quad (8.80)$$

This form, together with the commutation relation $[Q_{\mathbf{k},\lambda}, P_{\mathbf{k}',\lambda'}] = i\delta_{\lambda,\lambda'}\delta_{\mathbf{k},\mathbf{k}'}$, describe a denumerably infinite set of non-interacting harmonic oscillators, two for each wave vector **k**. Naturally, we treat this with raising and lowering operators

$$a_{\mathbf{k},\lambda} = \frac{1}{\sqrt{2\omega_k}} \left(P_{\mathbf{k},\lambda} - i\omega_k Q_{\mathbf{k},\lambda} \right); a_{\mathbf{k},\lambda}^\dagger = \frac{1}{\sqrt{2\omega_k}} \left(P_{\mathbf{k},\lambda} + i\omega_k Q_{\mathbf{k},\lambda} \right). \quad (8.81)$$

Correspondingly,

$$Q_{\mathbf{k},\lambda} = \frac{i}{\sqrt{2\omega_k}}(a_{\mathbf{k},\lambda} - a_{\mathbf{k},\lambda}^\dagger); \quad P_{\mathbf{k},\lambda} = \sqrt{\frac{\omega_k}{2}}(a_{\mathbf{k},\lambda} + a_{\mathbf{k},\lambda}^\dagger). \quad (8.82)$$

The only non-vanishing commutator is

$$[a_{\mathbf{k},\lambda}, a_{\mathbf{k}',\lambda'}^\dagger] = \delta_{\lambda\lambda'}\delta_{\mathbf{k},\mathbf{k}'}. \quad (8.83)$$

In terms of the creation and annihilation operators,

$$\mathbf{A}(\mathbf{x}, t) = \sqrt{\frac{4\pi}{V}} \sum_{\mathbf{k},\lambda} \frac{1}{\sqrt{2\omega_k}} \left[\epsilon_{\mathbf{k},\lambda} e^{i\mathbf{k}\cdot\mathbf{x}-i\omega_k t} a_{\mathbf{k},\lambda} + \epsilon_{\mathbf{k},\lambda}^* e^{-i\mathbf{k}\cdot\mathbf{x}+i\omega_k t} a_{\mathbf{k},\lambda}^\dagger \right], \tag{8.84}$$

where we have allowed for the possibility of circular polarization with complex polarization vectors. The fields themselves follow:

$$\mathbf{E}(\mathbf{x}, t) = i\sqrt{\frac{4\pi}{V}} \sum_{\mathbf{k},\lambda} \sqrt{\frac{\omega_k}{2}} \left[\epsilon_{\mathbf{k},\lambda} e^{i\mathbf{k}\cdot\mathbf{x}-i\omega_k t} a_{\mathbf{k},\lambda} - \epsilon_{\mathbf{k},\lambda}^* e^{-i\mathbf{k}\cdot\mathbf{x}+i\omega_k t} a_{\mathbf{k},\lambda}^\dagger \right]. \tag{8.85}$$

The magnetic field has the same form as $\mathbf{E}$ with the replacement $\epsilon \to \mathbf{k} \times \epsilon$. Comparison with the preceding classical description shows that the factor $\sqrt{n_{\mathbf{k},\lambda}}$ has been replaced by the operators $a_{\mathbf{k},\lambda}$ and $a_{\mathbf{k},\lambda}^\dagger$ in the expression for $\mathbf{A}(\mathbf{x}, t)$.

8.6 States of the Electromagnetic Field

We have reduced the representation of photons to the form of the simple harmonic oscillator (again with $\hbar = 1$):

$$H = \frac{1}{2}(P^2 + \omega^2 Q^2) = \omega(a^\dagger a + \frac{1}{2}); \quad H |n\rangle = E_n |n\rangle, \tag{8.86}$$

with $E_n = (n + \frac{1}{2})\omega$. The states at various levels of excitation are represented as

$$a |n\rangle = \sqrt{n} |n-1\rangle; \quad a^\dagger |n\rangle = \sqrt{n+1} |n+1\rangle. \tag{8.87}$$

The vacuum or ground state $|0\rangle$ is defined by

$$a |0\rangle = 0, \tag{8.88}$$

and thus

$$|n\rangle = \frac{(a^\dagger)^n}{\sqrt{n!}}. \tag{8.89}$$

For our set of decoupled oscillators, we have a general state specified by the diagonal operators $a_{\mathbf{k},\lambda}^\dagger a_{\mathbf{k},\lambda}$, i.e. the occupation numbers $n_{\mathbf{k},\lambda}$:

$$|n_{\mathbf{k}_1\lambda_1}, n_{\mathbf{k}_2\lambda_2}, \cdots, n_{\mathbf{k}_i\lambda_i}\rangle = |n_{\mathbf{k}_1\lambda_1}\rangle |n_{\mathbf{k}_2\lambda_2}\rangle \cdots |n_{\mathbf{k}_i\lambda_i}\rangle. \tag{8.90}$$

The vacuum state is

$$|0\rangle = |0_{\mathbf{k}_1\lambda_1}\rangle |0_{\mathbf{k}_2\lambda_2}\rangle \cdots |0_{\mathbf{k}_i\lambda_i}\rangle. \tag{8.91}$$

A single photon state is $a_{\mathbf{k},\lambda}^\dagger$ and a state with two identical photons is

$$\frac{1}{\sqrt{2!}}(a_{\mathbf{k},\lambda}^\dagger)^2 |0\rangle. \tag{8.92}$$

Evidently, the general state is

$$|n_{\mathbf{k}_1\lambda_1}, n_{\mathbf{k}_2\lambda_2}, \cdots, n_{\mathbf{k}_i\lambda_i}\rangle = \prod_i \frac{(a_{\mathbf{k}_i\lambda_i}^\dagger)^{n_{\mathbf{k}_i\lambda_i}}}{\sqrt{n_{\mathbf{k}_i\lambda_i}!}}. \tag{8.93}$$

This is just the product of the individual n-photon states for each separate $(\mathbf{k}, \lambda)$ oscillator. Note that because the $a^\dagger$s all commute with each other, the state is totally symmetric in the exchange of any two photons, respecting Bose-Einstein statistics.

8.7 Vacuum Expectation Values of E, E · E over Finite Volume

We define a spatially averaged field

$$\overline{\mathbf{E}}(\mathbf{x}, t) = \int d^3 x' f(\mathbf{x}') \mathbf{E}(\mathbf{x} - \mathbf{x}', t), \tag{8.94}$$

where $f(\mathbf{x})$ is a normalized test function, e.g. a Gaussian. We want first to determine the vacuum expectation value of operator $\overline{\mathbf{E}}(\mathbf{x}, t)$. From Eq. (8.85) we have

$$\overline{\mathbf{E}}(\mathbf{x}, t) = i \sqrt{\frac{4\pi}{V}} \sum_{\mathbf{k}, \lambda} \sqrt{\frac{\omega_k}{2}} \left[\boldsymbol{\epsilon}_{\mathbf{k}, \lambda} e^{i\mathbf{k}\cdot\mathbf{x} - i\omega_k t} a_{\mathbf{k}, \lambda} F(\mathbf{k}) - \boldsymbol{\epsilon}_{\mathbf{k}, \lambda}^* e^{-i\mathbf{k}\cdot\mathbf{x} + i\omega_k t} a_{\mathbf{k}, \lambda}^\dagger F^*(\mathbf{k}) \right], \tag{8.95}$$

where

$$F(\mathbf{k}) = \int e^{-i\mathbf{k}\cdot\mathbf{x}'} f(\mathbf{x}') d^3 x' \tag{8.96}$$

is the Fourier transform of $f(\mathbf{x})$.

Since this operator is still linear in a and $a^\dagger$, it has no diagonal matrix elements

$$\langle 0 | \overline{\mathbf{E}}(\mathbf{x}, t) | 0 \rangle = 0. \tag{8.97}$$

But we now examine the vacuum expectation value of $\overline{\mathbf{E}} \cdot \overline{\mathbf{E}}$. Since

$$\langle 0 | aa | 0 \rangle = \langle 0 | a^\dagger a^\dagger | 0 \rangle = 0, \tag{8.98}$$

we need consider only the terms

$$\left\langle 0 \left| \overline{\mathbf{E}} \cdot \overline{\mathbf{E}} \right| 0 \right\rangle = \frac{4\pi}{V} \sum_{\mathbf{k}, \lambda} \sum_{\mathbf{k}'\lambda'} \frac{\sqrt{\omega_k \omega_{k'}}}{2} \left\langle \boldsymbol{\epsilon}_{\mathbf{k}, \lambda} \cdot \boldsymbol{\epsilon}_{\mathbf{k}'\lambda'}^* a_{\mathbf{k}, \lambda} a_{\mathbf{k}'\lambda'}^\dagger e^{i(\mathbf{k}-\mathbf{k}')\cdot\mathbf{x} - i(\omega_k - \omega_{k'})t} F(\mathbf{k}) F^*(\mathbf{k}') \right.$$

$$\left. + \boldsymbol{\epsilon}_{\mathbf{k}, \lambda}^* \cdot \boldsymbol{\epsilon}_{\mathbf{k}'\lambda'} a_{\mathbf{k}, \lambda}^\dagger a_{\mathbf{k}'\lambda'} e^{-i(\mathbf{k}-\mathbf{k}')\cdot\mathbf{x} + i(\omega_k - \omega_{k'})t} F^*(\mathbf{k}) F(\mathbf{k}') \right\rangle. \tag{8.99}$$

Using

$$\left\langle 0 | a_{\mathbf{k}} a_{\mathbf{k}'\lambda'}^\dagger | 0 \right\rangle = \delta_{\mathbf{k}\mathbf{k}'} \delta_{\lambda\lambda'}; \qquad \boldsymbol{\epsilon}_{\mathbf{k}, \lambda} \cdot \boldsymbol{\epsilon}_{\mathbf{k}, \lambda}^* = 1, \tag{8.100}$$

we find

$$\langle 0 | \overline{\mathbf{E}} \cdot \overline{\mathbf{E}} | 0 \rangle = \frac{2\pi}{V} \sum_{\mathbf{k}, \lambda} \omega_{\mathbf{k}} |F(\mathbf{k})|^2 = \frac{4\pi}{V} \sum_{\mathbf{k}} \omega_{\mathbf{k}} |F(\mathbf{k})|^2. \tag{8.101}$$

First we note that the result is independent of $\mathbf{x}$ and t, as befits a vacuum expectation value. To examine it further, convert the sum over $\mathbf{k}$ to an integral

$$\frac{1}{V} \sum_{\mathbf{k}} \to \frac{1}{(2\pi)^3} \int d^3 k. \tag{8.102}$$

Then

$$\langle 0|\overline{\mathbf{E}} \cdot \overline{\mathbf{E}}|0\rangle = \frac{1}{2\pi^2} \int d^3k \, k|F(\mathbf{k})|^2 = \frac{2}{\pi} \int_0^\infty dk \, k^3|F(k)|^2, \tag{8.103}$$

where the last expression assumes $F(\mathbf{k})$ is spherically symmetric. To see what this means, choose

$$f(\mathbf{x}) = \frac{1}{(2\pi a^2)^{3/2}} e^{-r^2/2a^2}; \qquad F(k) = e^{-a^2k^2/2} \tag{8.104}$$

and

$$\langle 0|\overline{\mathbf{E}} \cdot \overline{\mathbf{E}}|0\rangle = \frac{2}{\pi} \int_0^\infty dk \, k^3 e^{-a^2k^2} = \frac{1}{\pi a^4}. \tag{8.105}$$

Now to restore the factors of $\hbar$ and c, we note that $\overline{\mathbf{E}} \cdot \overline{\mathbf{E}}$ is an energy density. Now $\hbar c$ has dimensions of energy times distance, so this is just what is needed.

$$\langle 0|\overline{\mathbf{E}} \cdot \overline{\mathbf{E}}|0\rangle = \frac{\hbar c}{\pi a^4}. \tag{8.106}$$

Averaging over a volume with dimensions of the order of a, we find the r.m.s. field fluctuations is $\mathcal{O}(\sqrt{\hbar c}/a^2)$. The distribution in k peaks at $k_{\text{dom}} \approx \sqrt{3/2}/a$. The total energy in the region captured by our Gaussian is

$$\frac{1}{4\pi} \langle 0|\overline{\mathbf{E}} \cdot \overline{\mathbf{E}}|0\rangle \frac{4\pi}{3} a^3 \sim \frac{\hbar c}{3\pi a} \sim \frac{\hbar c k_{\text{dom}}}{3\pi} \sim \frac{\hbar \omega_{\text{dom}}}{3\pi}. \tag{8.107}$$

We can picture this as roughly one typical photon of wavelength $\lambda \sim a$ within the volume a^3.

8.8 Classical vs. Quantum Radiation

To decide whether or not classical considerations are adequate, we can ask if the beam of radiation under consideration has many or few quanta of frequency ω in a box of volume k^{-3}. If the radiation flux is I, which we might measure in the units watts/cm^2, we can compare this to the "natural" quantum expression which has the proper dimensions:

$$I_0 = \hbar c^2 k^4. \tag{8.108}$$

This represents an intensity equivalent to one photon in a volume roughly one wavelength on a side. Substituting the values of $\hbar$ and c and converting k to $2\pi/\lambda$, we find that

$$I_0(\text{watts/cm}^2) \approx \frac{1.5 \times 10^6}{[\lambda(\mu\text{m})]^4}. \tag{8.109}$$

Representing the intensity of some radiation as $I = nI_0$, we can determine whether the situation is classical, $n \gg 1$, or not, $n \sim 1$. As a very rough and simplistic estimate, suppose an FM station broadcasting at 100 MHz emits 10 kilowatts of power isotropically. At 10 km from the station, the flux is about 10^{-9} watts/cm^2, while $I_0(\lambda = 3\text{m}) = 1.8 \times 10^{-20}$ watt/cm^2, so $n \approx 10^{11}$, very classical, indeed.

At the other extreme of wavelength is a laser with an intensity of 4×10^{13} watts/cm^2 at a wavelength of 1.060 μm (see P. Agostini et al., *Phys. Rev. Letters*, **42**,1127 (1979)). Using these parameters, we can estimate the density of photons in a volume one wavelength on a side:

$$n \approx \frac{4 \cdot 10^{13}}{1.5 \cdot 10^6} \approx 2.7 \cdot 10^7, \tag{8.110}$$

so again, this is very classical.

8.9 Quasi-Classical Fields and Coherent States

The development of the theory of coherent states is due in significant measure to R. J. Glauber (*Phys. Rev.* **131**, 2766 (1963)). We can compare the quantum operator

$$\mathbf{A}(\mathbf{x}, t) = \sqrt{\frac{4\pi}{V}} \sum_{\mathbf{k},\lambda} \frac{1}{\sqrt{2\omega_k}} \left[\epsilon_{\mathbf{k},\lambda} e^{i\mathbf{k}\cdot\mathbf{x} - i\omega_k t} a_{\mathbf{k},\lambda} + \epsilon^*_{\mathbf{k},\lambda} e^{-i\mathbf{k}\cdot\mathbf{x} + i\omega_k t} a^\dagger_{\mathbf{k},\lambda} \right] \tag{8.111}$$

to an expansion of the classical vector potential

$$\mathbf{A}_{\mathrm{cl}}(\mathbf{x}, t) = \sqrt{\frac{4\pi}{V}} \sum_{\mathbf{k},\lambda} \frac{1}{\sqrt{2\omega_k}} \left[\epsilon_{\mathbf{k},\lambda} e^{i\mathbf{k}\cdot\mathbf{x} - i\omega_k t} \alpha_{\mathbf{k},\lambda} + \epsilon^*_{\mathbf{k},\lambda} e^{-i\mathbf{k}\cdot\mathbf{x} + i\omega_k t} \alpha^*_{\mathbf{k},\lambda} \right], \tag{8.112}$$

where the $\alpha_{\mathbf{k},\lambda}$ are classical (c-number) coefficients of the Fourier expansion of the classical field, $A_{\mathrm{cl}}(\mathbf{x}, t)$. This shows that there is somehow a correspondence between $\alpha_{\mathbf{k},\lambda}$ and $a_{\mathbf{k},\lambda}$. To make this precise, we ask for a quantum state $|\alpha\rangle$ such that

$$\langle \alpha | \, \mathbf{A}_{\mathrm{op}}(\mathbf{x}, t) \, |\alpha\rangle = \mathbf{A}_{\mathrm{cl}}(\mathbf{x}, t) \tag{8.113}$$

and

$$\langle \alpha | \, \mathcal{H} \, |\alpha\rangle = \mathcal{H}_{\mathrm{cl}} = \textit{classical energy density}. \tag{8.114}$$

The unique solution is (see Kroll, 1964 Les Houches Lectures, *Quantum Optics and Electronics*)

$$a_{\mathbf{k},\lambda} \, |\alpha\rangle = \alpha_{\mathbf{k},\lambda} \, |\alpha\rangle. \tag{8.115}$$

Consider this one mode at a time. We demand

$$a \, |\alpha\rangle = \alpha \, |\alpha\rangle. \tag{8.116}$$

Reiterating, a is a destruction operator, while α is a c-number, which we use to label the state $|\alpha\rangle$ that actually satisfies this condition. Any state for this single mode must be a superposition of states of occupation number n, that is of n photons, so write

$$|\alpha\rangle = \sum_n |n\rangle \langle n \, |\alpha\rangle \tag{8.117}$$

To find $\langle n|\alpha\rangle$, we look at $\langle n - 1| \, na \, |\alpha\rangle$. But $\langle n - 1| \, a = \sqrt{n} \langle n|$, so

$$\langle n|\alpha\rangle = \frac{\alpha}{\sqrt{n}} \langle n - 1|\alpha\rangle. \tag{8.118}$$

This recursion relation has the solution

$$\langle n|\alpha\rangle = \frac{\alpha^n}{\sqrt{n!}}\,\langle 0|\alpha\rangle. \tag{8.119}$$

The coherent superposition state (called a "coherent state") is thus

$$|\alpha\rangle = \sum_n \frac{\alpha^n}{\sqrt{n!}}\,|n\rangle\,\langle 0|\alpha\rangle. \tag{8.120}$$

The overall coefficient, $\langle 0|\alpha\rangle$, can be determined by normalizing the state

$$1 = \langle \alpha|\alpha\rangle = \sum_{n,n'} \frac{\alpha^n}{\sqrt{n!}}\frac{\alpha^{*n'}}{\sqrt{n'!}}\,\langle n'|n\rangle\,|\langle 0|\alpha\rangle|^2$$

$$= \sum_n \frac{|\alpha|^{2n}}{n!}|\langle 0|\alpha\rangle|^2 = e^{|\alpha|^2}|\langle 0|\alpha\rangle|^2. \tag{8.121}$$

So we see that

$$\langle 0|\alpha\rangle = e^{-|\alpha|^2/2} \tag{8.122}$$

up to a possible overall phase.

Because $|n\rangle = (a^\dagger)^n\,|0\rangle/\sqrt{n!}$, we can write the coherent state for each $(\mathbf{k}, \lambda)$ as

$$|\alpha\rangle = e^{-|\alpha|^2/2}\sum_n \frac{(\alpha a^\dagger)^n}{n!}\,|0\rangle = e^{-|\alpha|^2/2}\sum_n \frac{\alpha^n}{\sqrt{n!}}\,|n\rangle. \tag{8.123}$$

The coherent state has a distribution of the populations

$$|\langle n|\alpha\rangle|^2 = \frac{|\alpha|^{2n}}{n!}e^{-|\alpha|^2}. \tag{8.124}$$

This is a Poisson distribution with $\overline{n} = |\alpha|^2$ and r.m.s. spread $\Delta n = \sqrt{\overline{n}} = |\alpha|$.

We can now see the connection between the semi-classical arguments above and the correspondence $\alpha_{\mathbf{k},\lambda} \to \sqrt{n_{\mathbf{k},\lambda}}$. To make the classical field, it is not sufficient to populate just the state with $\overline{n}_{\mathbf{k},\lambda}$ quanta. We must have a coherent superposition of different numbers of photons of each variety. The state for $\mathbf{A}_{cl}(\mathbf{x}, t)$ is

$$|\alpha\rangle = \exp^{-\frac{1}{2}\Sigma_{\mathbf{k},\lambda}|\alpha_{\mathbf{k},\lambda}|^2}\prod_{\mathbf{k},\lambda}\left[\sum_n \frac{(\alpha_{\mathbf{k},\lambda}a^\dagger_{\mathbf{k},\lambda})^n}{n!}\right]|0\rangle. \tag{8.125}$$

It can be shown that the states $|\alpha\rangle$ are linearly independent and can be used as states for expanding fields, even though they are not orthogonal.

References and Supplementary Reading

Our treatment of the fluctuations of the electromagnetic field follows that of Sakurai, *Advanced Quantum Mechanics*, pp. 32-36.

For coherent states:

Glauber, Vol. 42 of the "Enrico Fermi" Varenna series, *Quantum Optics*, New York : Academic Press, 1969.

Klauder and Sudarshan *Introduction to Quantum Optics,* W. A Benjamin (1968), especially Chapter 7 on coherent states.

Problems

8.1 A particle with charge e and mass m is bound around a fixed center of force and has a number of discrete bound states. A semi-monochromatic incoherent beam of light of plane polarization ϵ and intensity $J(\omega)$ (energy per unit area per unit time per unit interval of circular frequency) in the direction $\hat{\mathbf{n}}$ is incident on the system and causes transitions from the ground state $|i\rangle$ to some final state $|f\rangle$.

Treat the absorption of radiation semi-classically, using the interaction Hamiltonian

$$H_{\text{int}} = -\frac{e}{mc}\mathbf{A}\cdot\mathbf{p}; \qquad \mathbf{A} = 2A_0\epsilon\cos(k\mathbf{n}\cdot\mathbf{r}-\omega t) \tag{8.126}$$

for each of the incoherent components of the beam. Find the relation between A_0 and $J(\omega)$. Assuming the intensity distribution of the beam is broad in frequency compared with the natural line widths of the "discrete" states (which are not actually infinitely narrow), show that the transition rate from state $|i\rangle$ to state $|f\rangle$ is

$$w_{\text{fi}} = \frac{4\pi^2 e^2}{m^2\omega^2 c\hbar^2}J(\omega_{\text{fi}})|\langle f|e^{ik_{\text{fi}}\mathbf{n}\cdot\mathbf{r}}\epsilon\cdot\mathbf{p}|i\rangle|^2. \tag{8.127}$$

8.2 Starting from the Fourier expansion for the vector potential of the radiation field, show that the electromagnetic field momentum operator,

$$\mathbf{P} = \frac{1}{4\pi c}\int \mathbf{E}\times\mathbf{B}\,d^3x, \tag{8.128}$$

is diagonal in photon occupation-number space and is equal to

$$\mathbf{P} = \sum_{\mathbf{k},\lambda}\hbar\mathbf{k}\,a^\dagger_{\mathbf{k},\lambda}a_{\mathbf{k},\lambda}, \tag{8.129}$$

where $a^\dagger_{\mathbf{k},\lambda}a_{\mathbf{k},\lambda}$ is the number operator for photons of momentum $\hbar\mathbf{k}$ and polarization λ.

8.3* This concerns the angular momentum of a photon and is the quantum version of Problem 7.27 on p. 350 of Jackson, *Classical Electrodynamics,* 3rd edition. The angular momentum of an electromagnetic field in vacuum is

$$\mathbf{L} = \frac{1}{4\pi c}\int_V \mathbf{r}\times(\mathbf{E}\times\mathbf{B})d^3x. \tag{8.130}$$

In quantum mechanics, this becomes an operator.

(a) Show that if the magnetic field is eliminated in favor of the vector potential, the angular momentum operator takes the form

$$\mathbf{L} = \frac{1}{4\pi c}\int_V\left[\mathbf{E}\times\mathbf{A}+\sum_{j=1}^{3}E_j(\mathbf{r}\times\nabla)A_j\right]d^3x, \tag{8.131}$$

where repeated indices are summed from 1 to 3. The presence of the angular momentum operator in the second term suggests identification of the first term with the intrinsic "spin" of the photons.

(b) Starting with the same Fourier expansion as in Problem 8.1, but being sure to use a circular polarization basis with

$$\epsilon_{\mathbf{k},\pm} = \frac{1}{\sqrt{2}}(\epsilon_{\mathbf{k},1} \pm i\epsilon_{\mathbf{k},2}) \tag{8.132}$$

and the corresponding creation and destruction operators, show that the spin part of **L** is

$$\mathbf{L}_{spin} = \sum_{\mathbf{k}} \hat{\mathbf{k}}(a^{\dagger}_{\mathbf{k},+} a_{\mathbf{k},+} - a^{\dagger}_{\mathbf{k},-} a_{\mathbf{k},-}). \tag{8.133}$$

Interpret.

8.4 A complex amplitude function $g_{\lambda}(\mathbf{k})$ is such that

$$\sum_{\lambda} \int d^3k \, |g_{\lambda}(\mathbf{k})|^2 = 1. \tag{8.134}$$

A quantum state of radiation representing a wave packet is defined by

$$|G\rangle = \sum_{\lambda} \int d^3k \, g_{\lambda}(\mathbf{k}) a^{\dagger}_{\mathbf{k},\lambda} \, |0\rangle . \tag{8.135}$$

(a) Show that $|G\rangle$ is normalized to unity if the vacuum state is so normalized.

(b) The number operator N for the radiation field is defined by

$$N = \sum_{\lambda} \int d^3k \, a^{\dagger}_{\mathbf{k},\lambda} a_{\mathbf{k},\lambda}. \tag{8.136}$$

Is $|G\rangle$ an eigenket of N? What is the eigenvalue, if there is one? Interpret.

9

Emission and Absorption of Radiation

Detailed spectra obtained already in the nineteenth century were a challenge to quantum mechanics. Previous chapters showed how quantum mechanics met the challenge of these spectra, not just in hydrogen, but in many-electron atoms. The full explanation of emission and absorption requires the determination not just of photon energies, but of the rates and angular distributions for these processes. The quantized electromagnetic field shifts the energies of stationary states and gives them widths. The former breaks the degeneracy of the $2s_{1/2}$ and $2p_{1/2}$ in hydrogen, an effect known as the Lamb shift. Modifying the phase space available to virtual intermediate states can change both the energy and width of ordinary electromagnetic transitions.

9.1 Matrix Elements and Rates

We have considered particles interacting by static forces, e.g. Coulomb interactions, and the radiation field by itself. Now we consider the combination, treating the coupling between them as a perturbation. The Hamiltonian is

$$H = H_{\text{particles}} + H_{\text{radiation}} + H_{\text{int}}, \tag{9.1}$$

where

$$H_{\text{particles}} = \sum_j \frac{p_j^2}{2m_j} + U(\mathbf{x}_1, \mathbf{x}_2, \dots \mathbf{x}_n)$$

$$H_{\text{radiation}} = \sum_{\mathbf{k},\lambda} \hbar\omega_k a_{\mathbf{k},\lambda}^\dagger a_{\mathbf{k},\lambda}$$

$$H_{\text{int}} = -\sum_j \frac{e_j}{m_j c} \mathbf{A}_{\text{rad}}(\mathbf{x}_j, t) \cdot \mathbf{p}_j + \sum_j \frac{e_j^2}{2m_j c^2} \left(\mathbf{A}_{\text{rad}}(\mathbf{x}, t)\right)^2 \tag{9.2}$$

with $\mathbf{A}_{\text{rad}}$ having the expansion derived above, Eq. (8.84). If we include spin interactions for the particles described by Pauli spinors, there is an additional magnetic moment coupling

$$H_{\text{int}}^{\text{spin}} = -\sum_j \mu_j \frac{e\hbar}{2m_j c} \boldsymbol{\sigma}_j \cdot \mathbf{B}_{\text{rad}}(\mathbf{x}_j, t), \tag{9.3}$$

John David Jackson: A Course in Quantum Mechanics, First Edition. Robert N. Cahn.
© 2024 John Wiley & Sons, Inc. Published 2024 by John Wiley & Sons, Inc.
Companion website: www.wiley.com/go/Jackson/QuantumMechanics

where e is the charge of the proton and μ_j is the magnetic moment in the units of the appropriate magneton ("Bohr" for electrons, "nuclear" for nucleons: $\mu_e = -1, \mu_p = 2.7928\ldots, \mu_n = -1.9130\ldots$).

The unperturbed matter states we indicate by $|a\rangle$ with $H_{\text{particles}} |a\rangle = E_a |a\rangle$. These are energy eigenstates of the atomic or nuclear system. The radiation field states we indicate by $|n_\alpha\rangle = |n_{\mathbf{k}_1\lambda_1}, n_{\mathbf{k}_2\lambda_2}, \ldots n_{\mathbf{k}_n\lambda_n}\rangle$.

The first-order transitions arise from terms linear in e_j or μ_j. These are linear in $\mathbf{A}$ or $\mathbf{B}$ and so are linear in $a_{\mathbf{k},\lambda}$ or $a_{\mathbf{k},\lambda}^\dagger$. As a consequence, the photon occupation number changes only by ± 1.

Suppose only photons of one kind are present. Then a first order transition involving absorption of a photon has a transition amplitude

$$\langle b, n_{\mathbf{k},\lambda} - 1| H_{\text{int}} |a, n_{\mathbf{k},\lambda}\rangle$$

$$= -\sqrt{\frac{4\pi n_{\mathbf{k},\lambda}}{2V\omega_k}} \times \langle b| \sum_j e^{i\mathbf{k}\cdot\mathbf{x}_j}\boldsymbol{\epsilon}_{\mathbf{k},\lambda} \cdot \left(\frac{e_j}{m_j}\mathbf{p}_j - i\frac{\mu_j e}{2m_j}(\mathbf{k}\times\boldsymbol{\sigma}_j)\right) |a\rangle e^{-i\omega_k t}. \quad (9.4)$$

Similarly for emission, we have

$$\langle b, n_{\mathbf{k},\lambda} + 1| H_{\text{int}} |a, n_{\mathbf{k},\lambda}\rangle$$

$$= -\sqrt{\frac{4\pi(n_{\mathbf{k},\lambda} + 1)}{2V\omega_k}} \times \langle b| \sum_j e^{-i\mathbf{k}\cdot\mathbf{x}_j}\boldsymbol{\epsilon}_{\mathbf{k},\lambda}^* \cdot \left(\frac{e_j}{m_j}\mathbf{p}_j + i\frac{\mu_j e}{2m_j}(\mathbf{k}\times\boldsymbol{\sigma}_j)\right) |a\rangle e^{+i\omega_k t}. \quad (9.5)$$

Our time-dependent perturbation theory result for harmonic matrix elements gives us for the spontaneous emission rate, through Fermi's Golden Rule,

$$dw_{\mathbf{k},\lambda} = 2\pi| \langle b, 1_{\mathbf{k},\lambda}| H_{\text{int}} |a, 0_{\mathbf{k},\lambda}\rangle |^2 \rho_f(E)\delta(E_b - E_a + \hbar\omega_k). \quad (9.6)$$

For an emitted photon there is a continuum of available states and

$$\rho_f\delta(E_b - E_a + \hbar\omega) = \frac{V}{(2\pi)^3}k^2 d\Omega dk\,\delta(E_b - E_a + \hbar\omega) = \frac{V}{(2\pi)^3}k^2 d\Omega, \quad (9.7)$$

where $\hbar$ and c will be restored at the end. Inserting this, we find

$$\frac{dw_{\mathbf{k},\lambda}}{d\Omega} = \frac{2\pi V}{(2\pi)^3} \times \frac{4\pi k}{2V\omega_k} \left|\langle b| \sum_j e^{-i\mathbf{k}\cdot\mathbf{x}_j}\boldsymbol{\epsilon}_{\mathbf{k},\lambda}^* \cdot \left(\frac{e_j}{m_j}\mathbf{p}_j + i\frac{\mu_j e}{2m_j}(\mathbf{k}\times\boldsymbol{\sigma}_j)\right) |a\rangle\right|^2$$

$$= \frac{k}{2\pi} \left|\langle b| \sum_j e^{-i\mathbf{k}\cdot\mathbf{x}_j}\boldsymbol{\epsilon}_{\mathbf{k},\lambda}^* \cdot \left(\frac{e_j}{m_j}\mathbf{p}_j + i\frac{\mu_j e}{2m_j}(\mathbf{k}\times\boldsymbol{\sigma}_j)\right) |a\rangle\right|^2. \quad (9.8)$$

To restore $\hbar$ and c, divide by $\hbar c^2$.

The power radiated per unit solid angle is

$$\frac{dP_{\mathbf{k},\lambda}}{d\Omega} = \hbar\omega_k\frac{dw_{\mathbf{k},\lambda}}{d\Omega} = \frac{k^2}{2\pi c} \left|\langle b| \sum_j e^{-i\mathbf{k}\cdot\mathbf{x}_j}\boldsymbol{\epsilon}_{\mathbf{k},\lambda}^* \cdot \left(\frac{e_j}{m_j}\mathbf{p}_j + i\frac{\mu_j e}{2m_j}(\mathbf{k}\times\boldsymbol{\sigma}_j)\right) |a\rangle\right|^2. \quad (9.9)$$

While it is $\mathbf{p}$ that naturally occurs in the matrix element for radiation, we might expect from the classical correspondence $\mathbf{p}/m = \mathbf{v} = \dot{\mathbf{x}} = \pm i\omega\mathbf{x}$, which obtains for $\mathbf{x} \propto e^{\pm i\omega t}$, that it is possible to use matrix elements of $\mathbf{x}$ instead. Indeed, if we have static interactions $U(\mathbf{x}_1, \mathbf{x}_2, \dots \mathbf{x}_n)$, we have

$$\dot{\mathbf{x}}_j = \frac{i}{\hbar}[H,\mathbf{x}_j] = \frac{i}{\hbar}\frac{1}{2m_j}[p_j^2,\mathbf{x}_j] = \mathbf{p}_j/m_j. \tag{9.10}$$

So we can evaluate the matrix element of $\mathbf{p}_j/m_j$ as

$$\langle b|\, e^{\pm i\mathbf{k}\cdot\mathbf{x}_j}\boldsymbol{\epsilon}\cdot\mathbf{p}_j/m_j\,|a\rangle = \frac{i}{\hbar}\langle b|\, e^{\pm i\mathbf{k}\cdot\mathbf{x}_j}[H,\boldsymbol{\epsilon}\cdot\mathbf{x}_j]\,|a\rangle. \tag{9.11}$$

Now

$$He^{\pm i\mathbf{k}\cdot\mathbf{x}_j}\,|b\rangle = \left(-\frac{\hbar^2}{2m_j}\nabla_j^2 e^{\pm i\mathbf{k}\cdot\mathbf{x}_j}\right)|b\rangle + e^{\pm i\mathbf{k}\cdot\mathbf{x}_j}H\,|b\rangle$$

$$= \left(E_b + \frac{\hbar^2 k^2}{2m_j}\right)e^{\pm i\mathbf{k}\cdot\mathbf{x}_j}\,|b\rangle. \tag{9.12}$$

Thus,

$$\langle b|\, e^{\pm i\mathbf{k}\cdot\mathbf{x}_j}\boldsymbol{\epsilon}\cdot\mathbf{p}_j/m_j\,|a\rangle = \frac{i}{\hbar}\left(E_b - E_a + \frac{\hbar^2 k^2}{2m_j}\right)\langle b|\, e^{\pm i\mathbf{k}\cdot\mathbf{x}_j}\boldsymbol{\epsilon}\cdot\mathbf{x}_j\,|a\rangle$$

$$= i\left(\omega_{ba} + \frac{\hbar k^2}{2m_j}\right)\langle b|\, e^{\pm i\mathbf{k}\cdot\mathbf{x}_j}\boldsymbol{\epsilon}\cdot\mathbf{x}_j\,|a\rangle. \tag{9.13}$$

In the long wave-length limit, $k \to 0$, this becomes the electric dipole amplitude

$$\langle b|\, e^{\pm i\mathbf{k}\cdot\mathbf{x}_j}\boldsymbol{\epsilon}\cdot\mathbf{p}_j\,|a\rangle \approx i\omega_{ba}\langle b|\sum_j e_j\mathbf{x}_j\cdot\boldsymbol{\epsilon}\,|a\rangle \equiv i\omega_{ba}\boldsymbol{\epsilon}\cdot\langle b|\,\mathbf{d}\,|a\rangle, \tag{9.14}$$

where

$$\mathbf{d} = \sum_j e_j\mathbf{x}_j \tag{9.15}$$

is the system's electric dipole operator.

9.2 Dipole Transitions

The electric dipole transition rate is thus

$$\frac{dw_{\mathbf{k},\lambda}}{d\Omega} = \frac{\omega_{ba}^3}{2\pi\hbar c^3}\left|\langle b|\,\boldsymbol{\epsilon}_\lambda^*\cdot\mathbf{d}\,|a\rangle\right|^2. \tag{9.16}$$

The corresponding magnetic dipole transition rate is

$$\frac{dw_{\mathbf{k},\lambda}}{d\Omega} = \frac{\omega_{ba}^3}{2\pi\hbar c^3}\left|\langle b|\,\boldsymbol{\epsilon}_\lambda^*\cdot\boldsymbol{\mu}\,|a\rangle\right|^2, \tag{9.17}$$

where

$$\boldsymbol{\mu} = \sum_j\left[\frac{\mu_j e\hbar}{2m_j c}\boldsymbol{\sigma}_j + \frac{e_j\hbar}{2m_j c}\mathbf{L}_j\right]. \tag{9.18}$$

The second piece is the orbital magnetic moment contribution. See Jackson, *Classical Electrodynamics*, 3rd edition, Section 9.3, p. 413. The total transition rate, summed over polarizations and integrated over the full solid angle, is computed using

$$\sum_\lambda \epsilon^*_{\lambda,i}\epsilon_{\lambda,j} = \delta_{ij} - \hat{k}_i\hat{k}_j,\tag{9.19}$$

with the result for electric dipole

$$\Gamma_{ba} = \frac{4}{3\hbar}\left(\frac{\omega_{ba}}{c}\right)^3 |\langle b|\ \mathbf{d}\ |a\rangle|^2.\tag{9.20}$$

To check that the dimensions of the result are correct, note that $d \sim ex$, so $d^2 \sim e^2x^2$. But the dimensions of e^2 are the same as those of $\hbar c$, so that the dimensions of the right-hand side are those of $c(\omega/c)^3x^2$, which are indeed $[T^{-1}]$.

Analogously, for the magnetic dipole transitions,

$$\Gamma_{ba} = \frac{4}{3\hbar}\left(\frac{\omega_{ba}}{c}\right)^3 |\langle b|\ \boldsymbol{\mu}\ |a\rangle|^2.\tag{9.21}$$

Selection rules for dipole transitions follow directly from the matrix elements

$$E1 \propto \langle b\,|\mathbf{d}|\,a\rangle$$
$$M1 \propto \langle b\,|\boldsymbol{\mu}|\,a\rangle.\tag{9.22}$$

The operators $\mathbf{d}$ and $\boldsymbol{\mu}$ are vectors, so the initial and final angular momenta can differ by no more than one. Clearly the matrix elements would vanish if both the initial and final states had zero angular momentum. Moreover, the parity of $\mathbf{d}$ is negative, while that for $\boldsymbol{\mu}$ must be like that of the operator $\mathbf{r}\times\mathbf{p}$, which is positive. Thus for an E1 transition, the initial and final states must have opposite parity, while for an M1 transition they must have the same parity.

9.3 General Selection Rules

We have seen that, for dipole radiation, E1 transitions require a change of parity between the initial and final states, while for M1 transitions there is no change of parity. This generalizes to higher multipoles. In Problem 9.3, we see how this works for quadrupole transitions. In general, the matrix element for the process necessarily is a product of a tensor depending on the atoms and a tensor from the emitted photon. The simplest form for the first tensor is

$$T^E_{i_1 i_2 \dots i_\ell} = \langle b\,|x_{i_1}\,x_{i_2}\,\dots\,x_{i_\ell}|\,a\rangle.\tag{9.23}$$

This is a rank ℓ tensor and clearly has parity $(-1)^\ell$. If we insert a factor p_{i_0}, we will have a tensor of rank $\ell+1$, and parity $(-1)^{\ell+1}$, but if we multiply by $\epsilon_{\alpha i_0 i_1}$, we again have a tensor of rank ℓ since there are ℓ indices. However, the parity is still $(-1)^{\ell+1}$.

$$T^M_{\alpha i_2 \dots i_\ell} = \langle b\,|\epsilon_{i_1 i_0 \alpha}\,p_{i_0}\,x_{i_1}\,x_{i_2}\,\dots\,x_{i_\ell}|\,a\rangle = \langle b\,|L_\alpha\,x_{i_1}\,x_{i_2}\,\dots\,x_{i_\ell}|\,a\rangle.\tag{9.24}$$

Put another way, the tensor giving the atomic factor can be made from polar vectors like $\mathbf{x}$ with parity -1 and from axial vectors like $\mathbf{L}$ or $\boldsymbol{\sigma}$ with parity $+1$. Combinations that have parity $(-1)^\ell$ are called Eℓ and those that have parity $(-1)^{\ell+1}$ are called Mℓ. More properly, we should use the irreducible component of the rank-ℓ tensor for Eℓ or Mℓ transitions. It is these tensors that are appropriate for use in the Wigner-Eckart theorem. If the initial state has angular momentum j and the angular momentum of the final state is j', then the allowed Eℓ or Mℓ transitions must satisfy the triangular inequalities $|j - j'| \le \ell \le j + j'$. Whether the transition is Eℓ or Mℓ is then determined by the parities of the initial and final states.

The tensor associated with the outgoing photon must be constructed from a single copy of the polarization vector complex-conjugated ϵ_λ^* and some number of copies of the wave vector $\mathbf{k}$. The product of these two tensors must be a scalar, so both must be of the form T_m^ℓ. Moreover, since the electromagnetic interaction conserves parity, the two tensors must have the same parity. This is manifest in Problem 9.3.

9.4 Charged Particle in a Central Field

For a charged particle in a central force field, e.g. an atom or a two-body system that reduces to the equivalent, a quite detailed analysis is possible. If we ignore spin, the wave function is necessarily of the form

$$\psi_{n\ell m}(\mathbf{x}) = R_{n\ell} Y_{\ell,m}(\theta, \phi) \tag{9.25}$$

with energy $E_{n\ell}$. The rate for an electric dipole transition between an initial state (n, ℓ, m) and a final state (n', ℓ', m') is

$$\Gamma(n', \ell', m'; n, \ell, m) = \frac{4}{3\hbar} e^2 k^3 \left| \langle n', \ell', m' | \mathbf{r} | n, \ell, m \rangle \right|^2 , \tag{9.26}$$

where we have written $\mathbf{d} = e\mathbf{r}$ [$\mathbf{r} = \mathbf{r}_1 - \mathbf{r}_2$ in a two-body problem with charges e and $-e$]. The reader may prefer to see αc, with $\alpha \approx 1/137$, the fine-structure constant, in place of $e^2/\hbar$.

In the absence of external fields, the various m and m' states are degenerate. Thus, the quantity of interest is the above partial rate summed over the final states, m', and averaged over the initial states, m. We thus define the transition rate between the energy levels as

$$\Gamma(n', \ell'; n, \ell) = \frac{1}{2\ell + 1} \sum_{m,m'} \Gamma(n', \ell', m'; n, \ell, m)$$

$$= \frac{1}{2\ell + 1} \cdot \frac{4}{3\hbar} e^2 k^3 \sum_{m,m'} \left| \langle n', \ell', m' | \mathbf{r} | n, \ell, m \rangle \right|^2 . \tag{9.27}$$

With the sums over m and m', all reference to any particular direction is lost. Thus, each Cartesian component contributes equally and we can select the z-direction and then multiply the result by three.

$$\sum_{m,m'} | \langle n', \ell', m | \mathbf{r} | n, \ell, m \rangle |^2 = 3 \sum_{m,m'} | \langle n', \ell', m | z | n, \ell, m \rangle |^2 . \tag{9.28}$$

With $z = \sqrt{4\pi/3} r Y_{10}(\hat{\mathbf{x}})$, we get

$$\Gamma(n', \ell'; n, \ell) = \frac{16\pi}{3(2\ell + 1)} \frac{e^2 k^3}{\hbar} \sum_{m,m'} | \langle n', \ell', m' | r Y_{10} | n, \ell, m \rangle |^2$$

$$= \frac{16\pi}{3(2\ell + 1)} \frac{e^2 k^3}{\hbar} | \langle n', \ell' | r | n, \ell \rangle |^2 \sum_{m,m'} \left| \langle \ell', m' | Y_{10} | \ell, m \rangle \right|^2 . \tag{9.29}$$

The angular integral can be addressed using Eqs. (4.172) and (4.177). Thus our matrix element is

$$\int Y_{\ell',m'}^{*} Y_{\ell,m} Y_{10} \, d\Omega = (-1)^{m'} \int Y_{\ell',-m'} Y_{\ell,m} Y_{10} \, d\Omega$$

$$= (-1)^{m'} \sqrt{\frac{3(2\ell'+1)(2\ell+1)}{4\pi}} \begin{pmatrix} \ell' & \ell & 1 \\ -m' & m & 0 \end{pmatrix} \begin{pmatrix} \ell' & \ell & 1 \\ 0 & 0 & 0 \end{pmatrix}. \quad (9.30)$$

From the identity

$$\sum_{m_1 m_2} \begin{pmatrix} \ell_1 & \ell_2 & \ell_3 \\ m_1 & m_2 & m_3 \end{pmatrix}^2 = \frac{1}{2\ell_3 + 1}, \quad (9.31)$$

provided the triangular inequality $|\ell_1 - \ell_2| \le \ell_3 \le \ell_1 + \ell_2$ is satisfied, we find

$$\sum_{m,m'} |\langle \ell', m' | Y_{10} | \ell, m \rangle|^2 = \frac{(2\ell'+1)(2\ell+1)}{4\pi} \begin{pmatrix} \ell' & \ell & 1 \\ 0 & 0 & 0 \end{pmatrix}^2. \quad (9.32)$$

Using this explicit result,

$$\begin{pmatrix} \ell' & \ell & 1 \\ 0 & 0 & 0 \end{pmatrix}^2 = \frac{\ell_>}{(2\ell'+1)(2\ell+1)}, \quad (9.33)$$

where $\ell_>$ is the greater of ℓ' and ℓ and where $\ell' = \ell \pm 1$, altogether

$$\Gamma(n', \ell'; n, \ell) = \frac{4}{3} \frac{\ell_>}{2\ell+1} \frac{e^2 k^3}{\hbar} \left| \int_0^{\infty} r^2 dr \, R_{n'\ell'}^{*}(r) r R_{n\ell}(r) \right|^2. \quad (9.34)$$

9.5 Decay Rates with *LS* Coupling

We consider the situation in which the Russell-Saunders approximation applies. The states are described by kets $|\alpha L S J M\rangle$, but the fine structure is small enough that the spatial wave functions are essentially the same for all the different J states in a given multiplet specified by L and S, with $|L-S| \le J \le L+S$. Also, the energy differences between the fine-structure states are small compared to the transition energy. We define the transition rate between states of definite $J \to J'$ but with fixed M and M':

$$\Gamma(\alpha', L'S'J'M'; \alpha L, S, J, M) = \frac{4}{3} k^3 \left| \langle \alpha', L'S'J'M' | \, \mathbf{d} \, | \alpha L, S, J, M \rangle \right|^2, \quad (9.35)$$

where $\mathbf{d} = \sum_j e_j \mathbf{r}_j$.

Because the electric dipole transitions do not affect the spin, the spins S and S' of the initial and final states must be the same. If we have a well-defined initial state, $|\alpha L, S, J, M\rangle$, but observe with such low resolution that we do not separate the fine structure separating the different values of J', we can imagine treating all the final states as truly degenerate. In that case, we could work in the basis $|L'M_L' S M_S\rangle$. Each of these transition rates would be calculated using only the $L M_L$ quantum numbers. Because we sum over the final values of M_L', the rates cannot depend on the initial M, which is determined by orientation relative to an arbitrarily chosen direction.

We define S to be the sum of the rates for a fixed initial state J, M to all the final states J', M':

$$S = \sum_{J'M'} \Gamma(\alpha', L'S'J'M'; \alpha L, S, J, M).$$ (9.36)

The argument above shows then that

$$S = \sum_{M_L'M_S'} \Gamma(\alpha', L'M_L'S'M_S'; \alpha L, S, J, M)$$

$$\propto \sum_{M_L'M_S'M_LM_S} \left|\langle \alpha'L'M_L'S'M_S'| \mathbf{d} |\alpha L M_L S M_S\rangle\right|^2 \left|\langle L M_L S M_S| JM \rangle\right|^2.$$ (9.37)

Now the first factor cannot depend on M_L or M_S because of the sum over M_L' and M_S', nor does it depend on S since the electric dipole operator works only on the spatial degrees of freedom. So, in fact, we can write

$$\sum_{M_L'M_S'} \left|\langle \alpha'L'M_L'S'M_S'| \mathbf{d} |\alpha L M_L S M_S\rangle\right|^2 = \delta_{S'S} \sum_{M_L'} \left|\langle \alpha'L'M_L'| \mathbf{d} |\alpha L M_L\rangle\right|^2.$$ (9.38)

This quantity, which is summed over M_L', cannot depend then on M_L. But then we can separately do the sum over M_L and M_S

$$\sum_{M_L M_S} \left|\langle L M_L S M_S| JM \rangle\right|^2 = 1.$$ (9.39)

We conclude that S does not depend on J and we can write

$$S = \frac{4}{3} k^3 \delta_{SS'} \sum_{M_L'} \left|\langle \alpha'L'M_L'| \mathbf{d} |\alpha L M_L\rangle\right|^2$$

$$\equiv \delta_{SS'} \Gamma(\alpha', L'; \alpha, L).$$ (9.40)

This is the fine-structure sum rule:

$$\sum_{J'M'} \Gamma(\alpha'L'S'J'M'; \alpha LSJM) = \delta_{SS'} \Gamma(\alpha'L'; \alpha L).$$ (9.41)

The conclusion is that to find the decay rate of any member of the multiplet defined by J, L, S, M to the entirety of the multiplet defined by J', L', S, M', just use the spinless result for the transition from any particular L, M to the totality of the decays to $L'M'$.

On the other hand, if the resolution is good enough to separate the fine structure, that is, to distinguish each of the distinct transitions $J \rightarrow J'$, the intensity of each line will depend on the detailed coupling of L and S for each. This can be analyzed with the useful adage "average over initial states, sum over final." In this context, we average over the initial values of M associated with the initial state $|\alpha L, S, J, M\rangle$, and sum over final values of M':

$$\Gamma(\alpha'J'L'; \alpha JL) = \frac{1}{2J+1} \sum_{MM'} \Gamma(\alpha', L'S'J'M'; \alpha, LSJM).$$ (9.42)

This means that the reverse process is related to this one by

$$(2J+1)\Gamma(\alpha'J'L'; \alpha JL) = (2J'+1)\Gamma(\alpha JL; \alpha'J'L').$$ (9.43)

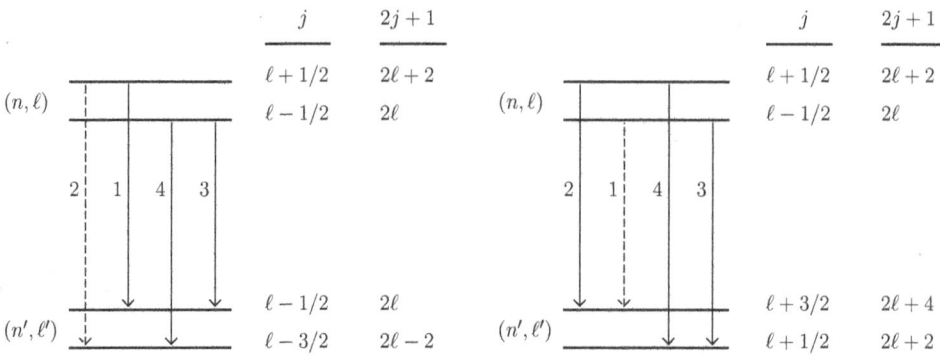

Figure 9.1 Line intensities for the case of doublets, i.e. $S = 1/2$. Left $\ell' = \ell - 1$. Right $\ell' = \ell + 1$. The relative rates can be determined from the relation $(2J + 1)\Gamma(\alpha'J';\alpha J) = (2J' + 1)\Gamma(\alpha J;\alpha'J')$.

To see how these relations enable us to compute the line intensities, consider the case of doublets, where we write conventionally: $S = s = 1/2, L = \ell$. This is shown in Figure 9.1. Let Γ_0 indicate the rate for $(n, \ell) \rightarrow (n', \ell')$ without spin. For the spinless result, we can pick any of the J values for the initial state and sum over all the allowed values for J'. Then considering decays from the $j = \ell + 1/2$ state in Figure 9.1, we have $\Gamma_0 = \Gamma_1$ because Γ_2 vanishes since it requires $\Delta J = 2$, which is forbidden in an electric dipole (E1) transition. From transitions starting at $j = \ell - 1/2$, we have $\Gamma_0 = \Gamma_3 + \Gamma_4$. Next we consider relations that begin in the (n', ℓ') and end at (n, ℓ). These are not real emissions, but the relations for the Γs still apply and the matrix elements for $(n', \ell') \rightarrow (n, \ell)$ would apply for absorption. Considering the two possible values for ℓ', we see that $\Gamma_0' = \Gamma_4' = \Gamma_1' + \Gamma_3'$.

From Eq. (9.43), we see that the statistical weight of the initial state times the decay rate is the same for the decay and its reverse. Thus, for the rates where we do not distinguish between the fine-structure lines in the final state, we derive

$$(2\ell + 1)\Gamma_0 = (2\ell' + 1)\Gamma_0'. \tag{9.44}$$

Similarly, again for $\ell' = \ell - 1$, using the statistical weights $2j + 1$ as shown in Figure 9.1,

$$\Gamma_1' = \frac{2\ell + 2}{2\ell}\Gamma_1 \tag{9.45}$$

$$\Gamma_2' = 0 \tag{9.46}$$

$$\Gamma_3' = \frac{2\ell}{2\ell}\Gamma_3 \tag{9.47}$$

$$\Gamma_4' = \frac{2\ell}{2\ell - 2}\Gamma_4. \tag{9.48}$$

Solving these simultaneous equations, we find

$$\Gamma_1 = \Gamma_0; \quad \Gamma_3 = \frac{1}{\ell(2\ell - 1)}\Gamma_0; \quad \Gamma_4 = \frac{(\ell - 1)(2\ell + 1)}{\ell(2\ell - 1)}\Gamma_0. \tag{9.49}$$

The analogous calculation for $\ell' = \ell + 1$ gives

$$\Gamma_1 = \frac{(\ell + 2)(2\ell + 1)}{(\ell + 1)(2\ell + 3)}\Gamma_0; \quad \Gamma_3 = \frac{1}{(\ell + 1)(2\ell + 3)}\Gamma_0; \quad \Gamma_4 = \Gamma_0. \tag{9.50}$$

For the specific case $3d \rightarrow 2p$, we find $\Gamma(3d_{5/2} \rightarrow 2p_{3/2}) = \Gamma_0$; $\Gamma(3d_{3/2} \rightarrow 2p_{3/2}) = \frac{1}{6}\Gamma_0$; $\Gamma(3d_{3/2} \rightarrow 2p_{1/2}) = \frac{5}{6}\Gamma_0$.

The relative intensities of spectral lines may differ from these relative values depending upon the mechanism of excitation. Most commonly, the excitation is independent of J values. The levels are populated uniformly in M and so $E(\alpha, J)$ has a population proportional to $2J+1$. In the example of $3d_j \rightarrow 2p_{j'}$ transitions, the $d_{5/2}$ intensity will be enhanced relative to the $d_{3/2}$ by a factor of $6/4 = 3/2$. Then the above values of $6 : 1 : 5$ become intensity ratios of $9 : 1 : 5$.

9.6 Line Breadth and Level Shift

We now consider, in a more systematic way, the time development of the states involved in a radiative transition. We consider the interaction to order e^2. As before, we take as our basis states the combined matter and radiation states $|b\rangle |n_{k,\lambda}\rangle = |b, n_{k,\lambda}\rangle$. These are such that

$$H_{\text{matter}} |b, n_{k,\lambda}\rangle = E_b |b, n_{k,\lambda}\rangle \tag{9.51}$$

$$H_{\text{rad}} |b, n_{k,\lambda}\rangle = \sum_{k,\lambda} n_{k,\lambda} \hbar\omega |b, n_{k,\lambda}\rangle . \tag{9.52}$$

We consider a time-dependent state of the form

$$|t\rangle = \sum_{\alpha = b, n_{k,\lambda}} c_\alpha(t) |\alpha\rangle e^{-iE_\alpha t}. \tag{9.53}$$

We shall assume that at time $t = 0$, the system is in the state $|a, 0\rangle$, with energy $E = E_a$. Since there is an interaction Hamiltonian, the state will evolve in time, that is, $c_\alpha(t)$ will change in time. It is useful to write

$$H_{\text{int}} = \sum_{k,\lambda} \left(H_{k,\lambda} e^{-i\omega_k t} + H_{k,\lambda}^\dagger e^{+i\omega_k t} \right), \tag{9.54}$$

where

$$H_{k,\lambda} = -\sqrt{\frac{4\pi}{2V\omega_k}} \epsilon_{k,\lambda} \cdot \sum_j e^{ik \cdot x_j} \left(\frac{e_j}{m_j} p_j + ... \right). \tag{9.55}$$

For states $|b, n_{k,\lambda}\rangle$ other than $|a\rangle$, the equation of motion for the coefficient $c_\alpha(t)$ is

$$i\dot{c}_{b,k,\lambda} = \langle b| H_{k,\lambda}^\dagger |a\rangle e^{i\omega_k t} e^{i(E_b - E_a)t} c_{a,0}(t), \tag{9.56}$$

while for the state $|\alpha\rangle$, we have

$$i\dot{c}_{a,0} = \sum_{b',k,\lambda'} \langle a| H_{k',\lambda'} |b'\rangle e^{-i\omega_{k'} t} e^{i(E_a - E_{b'})t} c_{b',k',\lambda'}(t). \tag{9.57}$$

To solve these equations, Weisskopf and Wigner made an ansatz (effectively using Laplace transforms):

$$c_{a,0}(t) = e^{-i\Delta E_a t}, \tag{9.58}$$

where ΔE_a is a complex energy. The motivation is that for a damped classical electromagnetic system (e.g. resonant cavities), one has a shift in the energy and exponential decay . If $\Im m \, \Delta E_a < 0$, we get $e^{-\Im m \, \Delta E_a t}$ in the amplitude.

Inserting the ansatz into the equation for $c_{a,0}(t)$ and integrating from zero to t,

$$c_{b,\mathbf{k},\lambda}(t) = \langle b| H^{\dagger}_{\mathbf{k},\lambda} |a\rangle \frac{1 - e^{i(\omega_k + E_b - E_a - \Delta E_a)t}}{\omega_k + E_b - E_a - \Delta E_a}. \tag{9.59}$$

Taking the time derivative of Eq. (9.58) and inserting Eq. (9.59), we obtain

$$\Delta E_a = \sum_{b,\mathbf{k},\lambda} \left|\langle b| H^{\dagger}_{\mathbf{k},\lambda} |a\rangle\right|^2 \frac{1 - e^{-i(\omega_k + E_b - E_a - \Delta E_a)t}}{E_a - E_b - \omega_k + \Delta E_a}. \tag{9.60}$$

Since we are working to lowest order in ΔE_a, we can set $\Delta E_a = 0$ on the right-hand side. The right-hand side is a function of t, while the left-hand side is not. We take the limit at $t \to \infty$ to get rid of transient behavior and identify the dominant ΔE_a (the simple exponential for $c_{a,0}(t)$). We use a device to define the limit of the exponential

$$\lim_{t \to \infty} f(t) = \lim_{\epsilon \to 0} \epsilon \int_0^{\infty} e^{-\epsilon t} f(t) dt. \tag{9.61}$$

Note that, assuming $f(t)$ is finite everywhere, the value of $f(t)$ in any finite range cannot contribute to the limit on the right-hand side, whereas if $f(t)$ does has a limit as $t \to \infty$, then this expression captures it.

Now set $\nu = E_a - E_b - \omega_k$. Then we want

$$\lim_{t \to \infty} \left(\frac{1 - e^{i\nu t}}{\nu}\right) = \frac{1}{\nu} - \lim_{\epsilon \to 0} \frac{\epsilon}{\nu} \int_0^{\infty} e^{-\epsilon t} e^{i\nu t} dt = \frac{1}{\nu} - \lim_{\epsilon \to 0} \frac{\epsilon}{\nu} \frac{1}{\epsilon - i\nu} \tag{9.62}$$

$$= \frac{1}{\nu} \lim_{\epsilon \to 0} \left(1 - \frac{\epsilon}{\epsilon - i\nu}\right) = \frac{1}{\nu + i\epsilon}. \tag{9.63}$$

Thus we have

$$\Delta E_a = \sum_{\mathbf{k},\lambda,b} \frac{|\langle b| H^{\dagger}_{\mathbf{k},\lambda} |a\rangle|^2}{E_a - E_b - \omega_k + i\epsilon}. \tag{9.64}$$

As always, we understand the prescription

$$\lim_{\epsilon \to 0} \frac{1}{z + i\epsilon} = \lim_{\epsilon \to 0} \frac{z - i\epsilon}{z^2 + \epsilon^2} = P(1/z) - i\pi\delta(z), \tag{9.65}$$

where P stands for principal part. Thus we have

$$\Im m \, \Delta E_a = -\pi \sum_{b,\mathbf{k},\lambda} \left|\langle b| H^{\dagger}_{\mathbf{k},\lambda} |a\rangle\right|^2 \delta(E_a - E_b - \omega_k). \tag{9.66}$$

This is just $-1/2$ times Fermi's Golden Rule expression for the total transition rate for a state a to all energetically allowed states b, i.e.

$$\Im m \, \Delta E_a = -\frac{1}{2} \Gamma_{total}(a). \tag{9.67}$$

Figure 9.2 The Breit-Wigner line shape showing the radiative shift of the center of the resonance.

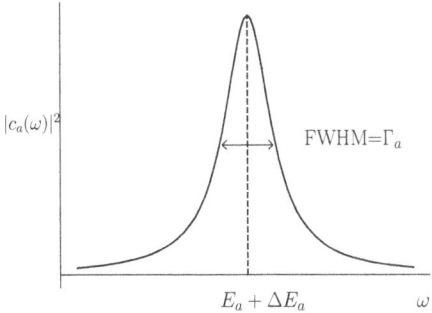

We thus have $c_{a,0}(t) = e^{-i\Re \Delta E_a t} e^{-\Gamma_a t/2}$. If we include the regular energy dependence and compute the one-sided Fourier transform, we find

$$c_a(\omega) = \int_0^\infty dt\, e^{i\omega t} e^{-i(\Delta E_a + E_a)t - \Gamma_a t/2} \tag{9.68}$$

$$|c_a(\omega)|^2 \propto \frac{1}{(E_a + \Delta E_a - \omega)^2 + \Gamma_a^2/4}. \tag{9.69}$$

This is a Breit-Wigner line shape with width Γ_a. See Figure 9.2. Weak but broad spectral lines were a puzzle until the Weisskopf-Wigner paper. (V. Weisskopf and E. Wigner, Zeit. f. Phys., 63, 54-73 (1930)). The total width is the sum of all the partial widths

$$\Gamma_a = \Gamma_{b_1 a} + \Gamma_{b_2 a} + \Gamma_{b_3 a} + \dots \tag{9.70}$$

If the sum is large, the initial state is broad, but individual terms in the sum can be small.

The real part of ΔE_a is a shift in the energy of the state. We make explicit the density of photon states

$$\sum_{\mathbf{k},\lambda} = \frac{V}{(2\pi)^3} \sum_\lambda \int d^3 k = \frac{V}{(2\pi)^3} \sum_\lambda \int k^2 dk\, d\Omega_k. \tag{9.71}$$

Then we have

$$\Re(\Delta E_a) = \frac{V}{(2\pi)^3} \sum_{b,\lambda} P \int k^2 dk\, d\Omega_k \frac{\left|\langle b| H^\dagger_{\mathbf{k},\lambda} |a\rangle\right|^2}{E_a - E_b - \omega_k}. \tag{9.72}$$

We now define a generalized (off-the-energy-shell transition rate)

$$\Gamma_{ba}(k) = 2\pi \frac{V}{(2\pi)^3} \sum_\lambda k^2 \int d\Omega_k \left|\langle b| H^\dagger_{\mathbf{k},\lambda} |a\rangle\right|^2 \tag{9.73}$$

in which $k \neq E_a - E_b$. Then we have the level shift

$$\Re(\Delta E_a) = \frac{1}{2\pi} P \int_0^\infty dk \sum_b \frac{\Gamma_{ba}(k)}{E_a - E_b - k}. \tag{9.74}$$

In the dipole approximation, we have

$$\Gamma_{ba} = \frac{4}{3} \frac{e^2 k}{m^2} |\langle b| \mathbf{p} |a\rangle|^2. \tag{9.75}$$

Then we have

$$\Re(\Delta E_a) = -\frac{2}{3\pi}\frac{e^2}{m^2}\sum_b |\langle b|\ \mathbf{p}\ |a\rangle|^2\ P \int_0^\infty dk \left[1 + \frac{E_a - E_b}{k - (E_a - E_b)}\right]. \tag{9.76}$$

For the first term in the integral, we can use the completeness of the states $|b\rangle$ to rewrite the expression as

$$\Re(\Delta E_a) = -\frac{2}{3\pi} \langle a|\ (\mathbf{p})^2\ |a\rangle \int_0^\infty dk[1]$$

$$+ \frac{2}{3\pi}\frac{e^2}{m^2}\sum_b (E_b - E_a)\,|\langle b|\ \mathbf{p}\ |a\rangle|^2\ P \int_0^\infty dk\ \frac{1}{k - (E_a - E_b)}. \tag{9.77}$$

We note that the first term depends only on the state a. It can be interpreted as a radiative correction to the mass of the electron: suppose the "base" mass is m_0 and the actual mass is $m = m_0 + \delta$, then the kinetic energy will be

$$T = \frac{p^2}{2m} = \frac{p^2}{2(m_0 + \delta m)} \approx \frac{p^2}{2m_0}\left(1 - \frac{\delta m}{m_0}\right). \tag{9.78}$$

Comparison shows

$$-\frac{p^2\delta m}{2m^2} = -\frac{2}{3\pi}\frac{e^2}{m^2}p^2 \int_0^\infty dk\ [1] \tag{9.79}$$

or

$$\delta m = \frac{4}{3\pi}\left(\frac{e^2}{\hbar c}\right)\int_0^\infty dk[1]. \tag{9.80}$$

If we agree to use the experimental mass in our formulas, we can omit this mass renormalization term in the expression for $\Re \Delta E_a$. Thus, the observable energy shift is

$$Re(\Delta E_a) = \frac{2}{3\pi}\frac{e^2}{m^2}\sum_b (E_b - E_a)\,|\langle b|\ \mathbf{p}\ |a\rangle|^2\ P \int_0^\infty dk\ \frac{1}{k - (E_a - E_b)}. \tag{9.81}$$

The integral is logarithmically divergent. We follow the original work of Bethe, *Phys. Rev.* **72**, 339 (1947), in which he gave the first explanation of the Lamb shift.

While the Dirac equation predicts that the $2s_{1/2}$ and $2p_{1/2}$ states of hydrogen are absolutely degenerate, Lamb and Retherford measured a splitting of about 1000 Mhz. The direct calculation of the shift from the radiative correction we study here led to the divergent integral above. Bethe argued that the relativistic corrections set in for momenta of order $k_{\max} = m(c/\hbar)$ and truncated the integral there:

$$P \int_0^m \frac{dk}{k - (E_a - E_b)} = \ln\left(\frac{m - (E_a - E_b)}{|E_a - E_b|}\right) \approx \ln\left(\frac{m}{|E_a - E_b|}\right). \tag{9.82}$$

With the substitution of $\mathbf{p} = m\dot{\mathbf{x}} = im[H, \mathbf{x}]$ in the matrix element, we get

$$Re(\Delta E_a) = \frac{2}{3\pi}e^2 \sum_b (E_b - E_a)^3\,|\langle b|\ \mathbf{x}\ |a\rangle|^2 \ln\left(\frac{m}{|E_a - E_b|}\right). \tag{9.83}$$

Because of the weak dependence on $|E_b - E_a|$ aside from the factor $(E_b - E_a)^3$, it is useful to define the "Bethe logarithm"

$$\ln(K_a) \equiv \frac{\sum_b (E_b - E_a)^3 |\langle b| \mathbf{x} |a\rangle|^2 \ln |E_a - E_b|}{\sum_b (E_b - E_a)^3 |\langle b| \mathbf{x} |a\rangle|^2}. \tag{9.84}$$

Then we can write

$$Re(\Delta E_a) = \frac{2}{3\pi} e^2 \left\{ \sum_b (E_b - E_a)^3 |\langle b| \mathbf{x} |a\rangle|^2 \right\} \ln\left(\frac{m}{K_a}\right). \tag{9.85}$$

The piece in brackets satisfies a sum rule:

$$\sum_b (E_b - E_a)^3 |\langle b| \mathbf{x} |a\rangle|^2 = \sum_b (E_b - E_a) |\langle b| \frac{\mathbf{p}}{m} |a\rangle|^2$$

$$= \frac{1}{2m^2} \sum_b \{\langle a| \mathbf{p} |b\rangle \cdot \langle b| [H, \mathbf{p}] |a\rangle - \langle a| [H, \mathbf{p}] |b\rangle \cdot \langle b| \mathbf{p} |a\rangle\}$$

$$= \frac{1}{2m^2} \langle a| [\mathbf{p}, [H, \mathbf{p}]] |a\rangle.$$

Now $[H, \mathbf{p}] = i\nabla V$, where $V(\mathbf{x})$ is the potential energy and thus $[\mathbf{p}, [H, \mathbf{p}]] = \nabla^2 V$. Thus,

$$\sum_b (E_b - E_a)^3 |\langle b| \mathbf{x} |a\rangle|^2 = \frac{1}{2m^2} \langle a| \nabla^2 V |a\rangle \tag{9.86}$$

and the energy shift is

$$Re(\Delta E_a) = \frac{1}{3\pi} \frac{e^2}{m^2} \langle a| \nabla^2 V |a\rangle \ln\left(\frac{m}{K_a}\right). \tag{9.87}$$

For a Coulomb field around a nucleus of charge Ze, we have

$$\nabla^2 V = 4\pi Ze^2 \delta(\mathbf{x}). \tag{9.88}$$

As a result, this perturbation changes the energy only of s-wave states. This is recognizable as the source of the Lamb shift. For these, we have

$$|\psi_n(0)|^2 = \frac{Z^3}{\pi a_0^3} \frac{1}{n^3}. \tag{9.89}$$

Altogether, the shift in the s-wave energy is

$$Re(\Delta E_a) = \frac{2}{3\pi} e^2 \frac{1}{2m^2} 4\pi Ze^2 \frac{Z^3}{\pi a_0^3} \frac{1}{n^3} \ln\left(\frac{m}{K_a}\right) \tag{9.90}$$

$$= \frac{4}{3\pi} \frac{Z^4 \alpha^5 m}{n^3} \ln \frac{m}{K_a}. \tag{9.91}$$

As a frequency, with the units restored,

$$\nu = Re(\Delta E_a)/h = \frac{2}{3\pi^2} \frac{Z^4 \alpha^5 mc^2}{\hbar n^3} \ln \frac{m}{K_a} \tag{9.92}$$

$$= \frac{2Z^4 \alpha^5}{3\pi^2 n^3} \frac{0.511 \text{ MeV}}{6.52 \times 10^{-22} \text{ MeV s}} \ln \frac{m}{K_a} \tag{9.93}$$

$$= \frac{Z^4}{n^3} \ln \frac{m}{K_a} \times 1.085 \times 10^3 \text{MHz.} \tag{9.94}$$

For $n = 2$, Bethe and Salpeter, *Quantum Mechanics of One- and Two-Electron Atoms*, gives for the 2s state $K = (16.6 \, \alpha^2 m/2)$ so the logarithm is $\ln(1/8.3\alpha^2) = 7.72$ and we find

$$\nu = 1047 \text{ MHz.} \tag{9.95}$$

Current theoretical and experimental results for the splitting between the $2s_{1/2}$ and $2p_{1/2}$ states in hydrogen are 1057.833(4)MHz (theory) and 1057.845(3)MHz (experiment). (M. I. Eides, H. Grotch, and V. A. Shelyuto, V. A., "Theory of light hydrogen-like atoms", Physics Reports, 342, 63-261(2001).) The large value of K_0 comes from the detailed calculation in (H. Bethe, L. Brown, and J. Stehn, Phys. Rev. 77, 370 (1950)), J. M. Harriman, Phys. Rev. 101, 594 (1956). See Bethe and Salpeter, *Quantum Mechanics of One- and Two-Electron Atoms*, pp. 318–320, especially the table on p. 319. The calculations are described in Bethe and Salpeter, *Quantum Mechanics of One- and Two-Electron Atoms*, pp. 97–107.

9.7 Alteration of Spontaneous Emission from Changed Density of States

The shift of the resonant energy can be made manifest by modifying the density of states available to the virtual photons responsible for the shift. To see how the complex energy shift depends on the photon density of states, we write out explicitly, with $\hbar = c = 1$,

$$\Delta E_a = \sum_{\mathbf{k}\lambda} \sum_b \frac{\left| \left\langle b \left| H_{\mathbf{k}\lambda}^\dagger \right| a \right\rangle \right|^2}{E_a - E_b - \omega_{\mathbf{k}} + i\epsilon} \tag{9.96}$$

with

$$\sum_{\mathbf{k}} = \frac{V}{(2\pi)^3} \int d^3 k \frac{\rho(\mathbf{k})}{\rho_{\text{free}}}, \tag{9.97}$$

where $\rho(\mathbf{k})$ is the density of states in which the decaying atom finds itself, e.g. inside a resonant cavity. If we then define a generalized decay distribution

$$\frac{d\Gamma_{ba}(k)}{d\Omega_k} = 2\pi \frac{V}{(2\pi)^3} \sum_\lambda k^2 \left| \left\langle b \left| H_{\mathbf{k}\lambda}^\dagger \right| a \right\rangle \right|^2 \tag{9.98}$$

in terms of which we can write

$$\Delta E_a = \frac{1}{2\pi} \int_0^\infty dk \int d\Omega_k \frac{d\Gamma_{ba}(k)}{d\Omega_k} \frac{1}{(E_a - E_b - k + i\epsilon)} \cdot \left[\frac{\rho(\mathbf{k})}{\rho_{\text{free}}} \right]. \tag{9.99}$$

This shows explicitly how ΔE_a depends on the density of photon states.

With the identification

$$\frac{1}{x + i\varepsilon} \rightarrow P\frac{1}{x} - i\pi\delta(x) \tag{9.100}$$

and

$$\Delta E_a = \delta\omega - \frac{i}{2}\delta\Gamma, \tag{9.101}$$

we see that the shifts in the resonant energy and width are

$$\delta\omega = \frac{1}{2\pi}P\int_0^\infty dk \int d\Omega_k \frac{d\Gamma_{ba}(k)}{d\Omega_k}\frac{1}{(E_a - E_b - k)}\cdot\left[\frac{\rho(\mathbf{k})}{\rho_{\text{free}}}\right]$$

$$\delta\Gamma = \int d\Omega_k \frac{d\Gamma_{ba}(k_{\text{res}})}{d\Omega_k}\left[\frac{\rho(\mathbf{k}_{\text{res}})}{\rho_{\text{free}}}\right], \tag{9.102}$$

where we have set $\hbar$ and c to unity so $|\mathbf{k}_{\text{res}}| = \omega_{\text{res}}$. The observable shifts are obtained by subtracting from these quantities, the same expressions, but with the bracketed quantity set to unity.

We discuss the experiment reported by Heinzen and Feld, *Phys. Rev. Lett.* **59**, 2623 (1987). It involves a barium transition at 553 nm ($^1S_0 \rightarrow {}^1P_1$). The $^1P_1$ state has a radiative width in free space of 19 MHz. The barium atoms were placed at the center of a concentric optical resonator formed from two spherical mirrors whose radius of curvature was half the separation of their midpoints. The resonator altered the density of states over the part of the solid angle subtended by the mirrors. The resonant frequency could be tuned by adjusting slightly the position of one mirror. With that fixed, the polarized laser light was scanned through the region of the barium line.

The response of the resonator determines the density of states. One measures the frequency response by examining the time evolution of a sharp pulse sent out to the right and being reflected successively, with reflectivity R (ratio of the reflected to incident intensity).

Thus the response to the initial delta function is approximately a series of delta functions of decreasing magnitude and equal spacing. We analyze the time spectrum with

$$\sum_{n=0}^\infty R^n\delta\left(t - \frac{2nL}{c}\right) = \frac{1}{\sqrt{2\pi}}\int_{-\infty}^\infty d\omega A(\omega)e^{-i\omega t}. \tag{9.103}$$

So we find

$$A(\omega) = \frac{1}{\sqrt{2\pi}}\int_{-\infty}^\infty dt e^{i\omega t}\sum_{n=0}^\infty R^n\delta\left(t - \frac{2nL}{c}\right)$$

$$= \frac{1}{\sqrt{2\pi}}\sum_{n=0}^\infty R^n e^{2ni\omega L/c} = \frac{1}{\sqrt{2\pi}}\left(1 - Re^{2i\omega L/c}\right)^{-1} \tag{9.104}$$

and

$$|A(\omega)|^2 = \frac{1}{2\pi}\left(1 + R^2 - 2R\cos(2\omega L/c)\right)^{-1}, \tag{9.105}$$

where we have assumed R (actually $\sqrt{R}$) to be real. See Figure 9.3. The variation as a function of frequency is expressed through

$$\mathcal{L}(\omega) = \frac{\sqrt{1+F}}{1 + F\sin^2(\omega L/c)}; \qquad F = \frac{4R}{(1-R)^2}, \tag{9.106}$$

which is normalized as

$$\int_0^{\pi c/L} d\omega\,\mathcal{L}(\omega) = \frac{\pi c}{L} = \int_0^{\pi c/L} d\omega, \tag{9.107}$$

so that the value of $\mathcal{L}$ is unity, averaged over one or many resonant frequencies. The effect of the resonator is to redistribute the states from a uniform distribution to one peaked at the resonant frequencies. In optics, $\mathcal{L}(\omega)$ is called the Airy function (not to be confused with the Airy function that appears in the WKB method!). The modified density of states is represented in Figure 9.4.

The peaks occur at multiples of the resonant frequency, $\omega_{res} = \pi c/L$. The full-width at half-max of each peak is $2\omega_{res}/\pi\sqrt{F}$. The Q of the system, the ratio of the resonant frequency to the half-width, is also called the "Finesse."

$$\mathcal{F} = \frac{\omega_{res}}{\text{FWHM}} = \frac{\pi\sqrt{F}}{2}. \tag{9.108}$$

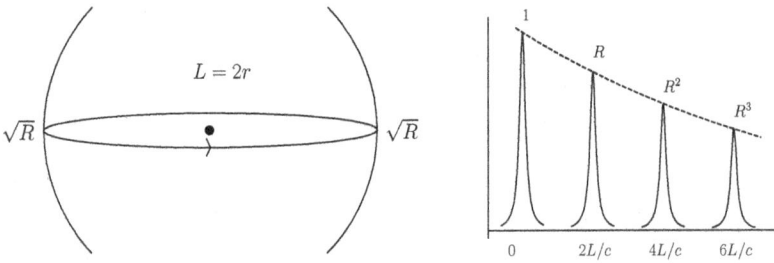

Figure 9.3 Left: A resonator of diameter L reflects a fraction $\sqrt{R}$ of the intensity at each surface, leading to a reduction by R in a time interval $2L/c$. Right: The response in time to a pulse at zero measured at the source for a resonator.

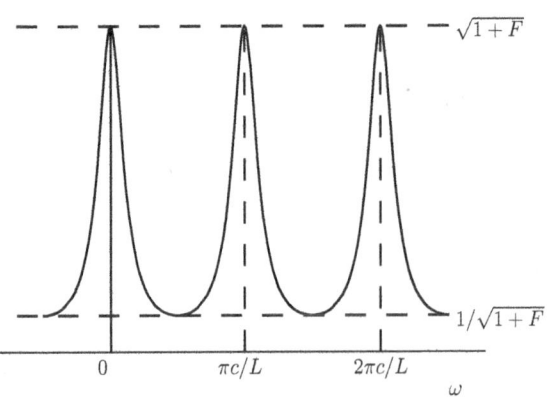

Figure 9.4 The function $\mathcal{L}(\omega)$ describing the response of the resonant cavity of diameter L and reflectivity R. Here $F = 4R/(1-R)^2$ is taken to be 8.

With the resonator covering only a portion of the solid angle, the actual density of states is a function of $\mathbf{k}$:

$$\frac{\rho(\mathbf{k})}{\rho_{\text{free}}} = \begin{cases} \mathcal{L}(\omega) \text{ for } \mathbf{k} \text{ inside solid angle of resonator} \\ 1 \text{ for } \mathbf{k} \text{ outside.} \end{cases} \tag{9.109}$$

The experimental setup is shown in Figure 9.5.

The consequences of the difference in phase space densities is observed in three distinct ways. The intensity of decays from the excited atoms in directions away from the resonator was compared to the intensity within the solid angle subtended by the mirrors. The results are shown in the top two plots in Figure 9.6.

At each setting of the cavity tuning, the laser frequency was scanned through the resonance. This made it possible to track both the resonant frequency and the resonance width. These were both modified by the phase-space correction $\mathcal{L}(\omega)$.

To predict the effect of the modification of phase space, first imagine that the effect is isotropic. Later we can correct for the finite coverage of the resonator. Then we can write Eq. (9.102) with $\hbar = c = 1$, $E_a - E_b = k_{\text{res}}$ as

$$\delta\omega = \frac{1}{2\pi} P \int_0^\infty dk \frac{1}{k_{\text{res}} - k} \mathcal{L}(k) \cdot \int d\Omega_k \frac{d\Gamma_{ba}(k)}{d\Omega_k}$$

$$\delta\Gamma = \int d\Omega_k \frac{d\Gamma_{ba}(k_{\text{res}})}{d\Omega_k} \mathcal{L}(k_{\text{res}}). \tag{9.110}$$

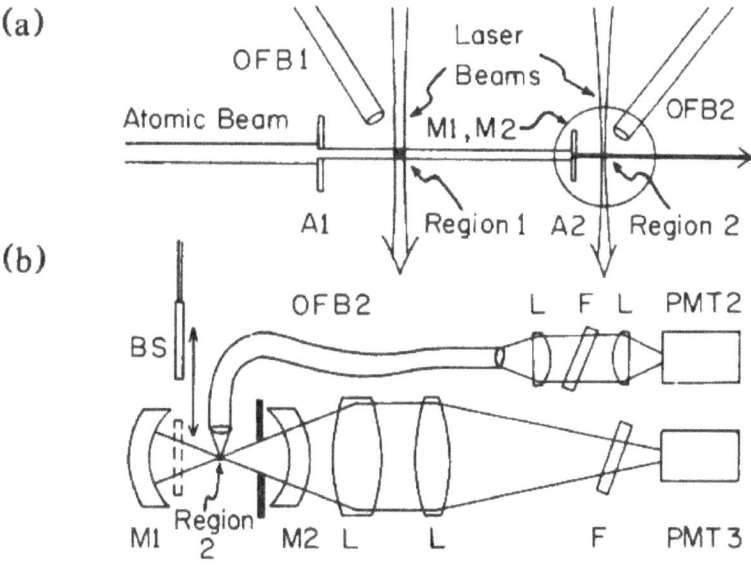

Figure 9.5 The experimental apparatus used in Heinzen and Feld 1987 / American Physical Society. The atomic-beam excitation geometry is shown in (a). The view from the side is shown in (b). The laser beam excited the barium atoms to the $^1P_1$ state and their decays were observed ultimately in the photomultipliers, PMT2 and PMT3. The rate of those observed in PMT2 were governed by the free phase-space density, while those observed in PMT3 determined the modified density, $\rho(\mathbf{k})$. The laser beam at M1 was used to generate excitations that could be studied in the absence of the optical resonator.

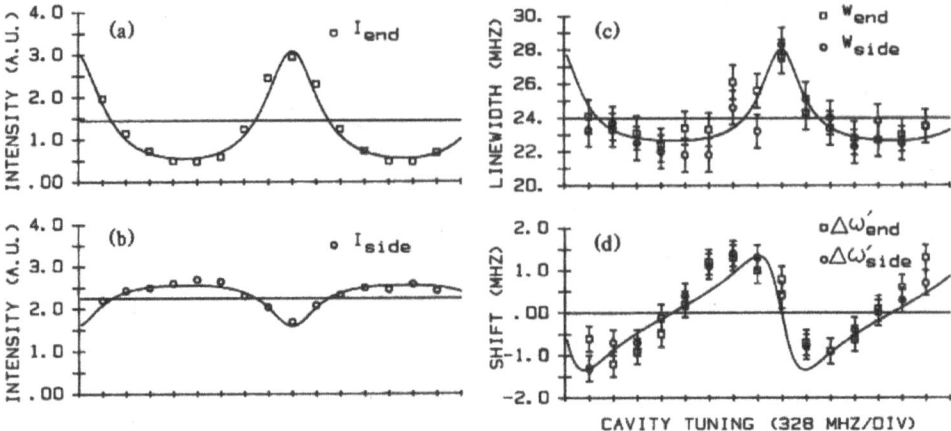

Figure 9.6   Results obtained by Heinzen and Feld 1987 / American Physical Society. The top two plots show the intensity of decays within the solid angle of the resonator and outside it. The slight depression at the center of the latter is due to the reduction in excited states resulting from the increase shown in the plot above it. The behavior of the resonance energy shift and the linewidth are well described by the theory. In all four plots, the horizontal axis is the cavity tuning, which was obtained by moving the position of one of the mirrors.

The shift in the resonant energy $\delta\omega$ diverges, but we need to subtract the same quantity with $\mathcal{L}(k)$ replaced by unity. The phase space factor is periodic, but in only one period is the resonance condition satisfied. Elsewhere, we can use the average value of $\mathcal{L}$, which is itself unity, so that the observable quantity is convergent. A very non-trivial calculation shows that

$$P \int_0^\infty dk \frac{1}{k_{\text{res}} - k}(\mathcal{L}(k) - 1) = \frac{\pi}{2} \frac{F \sin(2k_{\text{res}}L)}{1 + F \sin^2(k_{\text{res}}L)}. \tag{9.111}$$

Consequently, if the resonator covered the full solid angle, the resonance energy shift would be

$$\delta\omega = \frac{\Gamma_{\text{free}}}{4} \frac{F \sin(2k_{\text{res}}L)}{1 + F \sin^2(k_{\text{res}}L)}. \tag{9.112}$$

In the experiment, the resonator covered only a fraction of the solid angle and the angular distribution of the radiation needed to be accounted for. In the end, the effective coverage was 11% of 4π, taking into account the non-isotropic nature of the radiation. While the nominal value of F was 21.2, various factors resulted in an effective value of $F = 8.0$. Whether the shift of the resonant energy was positive or negative was determined by whether the density of states was greater above or below the resonance. At the peak and minimum of the density of states, these balanced and the energy shift vanished, as seen in Figure 9.6. This behavior is reflected in Eq. (9.112).

References and Suggested Reading

For spontaneous emission and the photoeffect:
Bethe and Salpeter, *Quantum Mechanics of One- and Two-Electron Atoms*, p. 248 ff.

A complete treatment of multipole radiation in the classical context, which is nearly identical to the quantum context, is given in:

Jackson, *Classical Electrodynamics*, 3rd edition, pp. 429–444.

For fine structure and relative line intensities:

Condon and Shortley, *The Theory of Atomic Spectra*, Chapter IX.

For Weisskopf-Wigner and its application to the Lamb shift by Bethe, we follow a path similar to that in:

Sakurai, J., *Advanced Quantum Mechanics*, pp. 64–72.

For alteration of phase space:

E.M. Purcell, Phys. Rev. **69**, 681(A),1946.

D.J. Heinzen et al., Phys. Rev. Lett. **58**, 1320 (1987).

D.J. Heinzen and M.S. Feld, Phys. Rev. Lett. **59**, 2623 (1987).

De Martini et al., Phys. Rev. Lett. **59**, 2955 (1987).

Problems

9.1* In the non-relativistic Pauli limit for the wave functions and in the electric dipole approximation, but including in the kinematics such effects as the fine-structure splittings and the Lamb shift (i.e. using the experimental energy differences), calculate the (partial) lifetimes in seconds (or days or years, if appropriate) for the following transitions in hydrogen: (a) $2p_{1/2} \rightarrow 1s_{1/2}$, (b) $2p_{3/2} \rightarrow 1s_{1/2}$, (c) $2s_{1/2} \rightarrow 2p_{1/2}$, (d) $2s_{1/2} \rightarrow 1s_{1/2}$. Sum over the final magnetic substates and average over initial, as well as integrating over the direction of the emitted photon.

Compare the calculated line widths $\Delta \nu \ (= (2\pi\tau)^{-1})$ in MHz with the fine-structure splitting between the $2p_{3/2}$ and $2p_{1/2}$ states. Determine the oscillator strength $f_{n'n}$ and determine the value for the sum of the reversed transitions $1s_{1/2} \rightarrow 2p_{1/2}$ and $1s_{1/2} \rightarrow 2p_{3/2}$.

9.2 Bound states of a charmed quark and a charmed antiquark form a fairly non-relativistic system, called charmonium, in the mass range of 3 - 4 GeV. The rest energies and spin-parities are shown (J^{PC}) in Figure 9.7. The postulated orbital and spin assignments in the Pauli approximation are given, together with the radial quantum number. The parities are opposite what one might naively expect because an extra -1 comes from the particle-antiparticle feature for spin-1/2 particles. The state at 3770 MeV is above threshold for the strong decay into charmed mesons ($D\bar{D}$ pairs). The states below the dashed line are relatively stable; they can be viewed as almost stationary states. The $\psi(3686)$ is found to decay some fraction of the time by radiative transitions to the three intermediate states (^3P_J), denoted χ_{cJ}. These χ states then decay some of the time by radiative transitions to the $\psi(3097)$, although mostly they decay into hadrons.

(a) Noting that the level pattern crudely resembles that of an isotropic harmonic oscillator, determine a reasonable value of the oscillator scale parameter ξ (defined below) from the level spacing between the ground and first radially excited $^3S_1$ states and the assumption that the system is a weakly bound state of two quarks.

```
                                         1⁻⁻(3.770)
   1⁻⁻(3.686)      ------------------------------
   0⁻⁺(3.637)
                        2⁺⁺(3.556)
                        1⁺⁺(3.511)
                        0⁺⁺(3.415)


   1⁻⁻(3.097)
   0⁻⁺(2.984)


      S              P              D
```

Figure 9.7 A portion of the charmonium spectrum with the states in columns labeled by orbital angular momentum: $S(L = 0)$; $P(L = 1)$; $D(L = 2)$. For a fermion - anti-fermion bound state the value of parity is $P=(-1)^{L+1}$, while the charge conjugate value is $C=(-1)^{L+S}$. The states at 3.637 GeV and 3.686 GeV are radial excitations: $n_r = 1$. All the other states have $n_r = 0$. The standard particle names and the spectroscopic classifications are $[2.984 : \eta(1S), {}^1S_0]$; $[3.097 : J/\psi, {}^3S_1]$; $[3.415 : \chi_{c0}, {}^3P_0]$; $[3.511 : \chi_{c1}, {}^3P_1]$.

(b) Compute the electric dipole radiative widths in keV for the six transitions $\psi(3686) \rightarrow \chi_J + \gamma$, $\chi_J \rightarrow \psi(3097) + \gamma$ for $J = 0, 1, 2$. Use the harmonic oscillator wave functions given below to evaluate the dipole matrix elements, but use the experimental photon energies in the formulas for the transition rates. Assume that the charge of the charm quark is $(2/3)e$. Compare your results with the experimental data as reported by the Particle Data Group.

(c) What type of radiative transition can occur from the state at 3686 MeV to the one at 3097 MeV? Explain. What type from the state at 3097 MeV to the state at 2981 MeV? From 3686 MeV to 2981? MeV? What about transitions from the χ states to the state at 2981 MeV? Are they seen? Explain. The isotropic harmonic oscillator in three dimension has normalized wave functions

$$\Psi_{L,M,n_r}(r, \theta\phi) = \left[\frac{2^{L+2}}{\sqrt{\pi}(2L + 1)!!}\right]^{1/2} \xi^{3/2} x^L e^{-x^2/2} w_{L,n_r}(x) Y_{LM}(\theta, \phi), \qquad (9.113)$$

where $x = \xi r$, $\xi^2 = m\omega/\hbar$, where m is the reduced mass of the two-particle system and ω is the osciallator frequency. The function w_{L,n_r} for the first three n_r values are

$$w_{L,0}(x) = 1$$

$$w_{L,1}(x) = [(2L + 3)/2]^{1/2}\left[1 - \frac{2x^2}{2L + 3}\right]$$

$$w_{L,2}(x) = [(2L + 3)(2L + 5)/8]^{1/2}\left[1 - \frac{4x^2}{2L + 3} + \frac{4x^4}{(2L + 3)(2L + 5)}\right]. \qquad (9.114)$$

9.3* The matrix element for the spontaneous emission of radiation by a group of spin-1/2 charged particles is

$$M_{ba} = \langle b| \sum_j e^{-i\mathbf{k}\cdot\mathbf{r}_j} \boldsymbol{\epsilon}_{k,\lambda}^* \cdot \left(\frac{e_j\mathbf{p}_j}{m_j} + i\frac{\mu_j e}{2m_j}\mathbf{k} \times \boldsymbol{\sigma}_j\right) |a\rangle. \qquad (9.115)$$

(a) By expanding the exponential, show that M_{ba} has a long-wavelength multipole expansion

$$M_{ba} = M_{ba}(E1) + M_{ba}(M1) + M_{ba}(E2) + \dots, \qquad (9.116)$$

where

$$M_{ba}(E1) = \langle b| \sum_j i\omega_{ba}\boldsymbol{\epsilon}^*_{\mathbf{k},\lambda} \cdot \left(e_j\mathbf{r}_j + i\frac{\omega_{ba}\mu_j e}{4m_j}\mathbf{r}_j \times \boldsymbol{\sigma}_j\right)|a\rangle$$

$$M_{ba}(M1) = \langle b| \sum_j i\boldsymbol{\epsilon}^*_{\mathbf{k},\lambda} \cdot \mathbf{k} \times \sum_j \left(\frac{\mu_j e\boldsymbol{\sigma}_j + e_j\mathbf{L}_j}{2m_j}\right)|a\rangle$$

$$M_{ba}(E2) = -\langle b| \frac{\omega_{ba}}{2} \sum_j e_j\mathbf{k} \cdot \mathbf{r}_j\boldsymbol{\epsilon}^*_{\mathbf{k},\lambda} \cdot \mathbf{r}_j |a\rangle. \tag{9.117}$$

There is a neglected term, symmetric in $\mathbf{r}_j$ and $\boldsymbol{\sigma}_j$, which contributes to the magnetic quadrupole matrix element. Note that in these formulas $\hbar$ and c have been suppressed.

(b) The second term in $M_{ba}(E1)$ is called the induced electric dipole contribution. Discuss its size relative to the dominant $E1$ term for a typical atomic transition. Will it contribute to $\Delta S = 0$ transitions? To $\Delta S = 1$ transitions? There is a resonant line in mercury at 2537 Å that is a $^3P_1 \rightarrow{}^1 S_0$ transition with parity change. Could the induced $E1$ amplitude explain this transition? Why or why not?

(c) Show that the electric quadrupole matrix element can be expressed in terms of the traceless symmetric quadrupole tensor, familiar from electrostatics.

9.4 Calculate the magnetic dipole transition rate for the decay of the charmonium $^3S_1$ state at 3097 MeV to its $^1S_0$ partner at 2984 MeV. Assume the spin-dependent forces responsible for the splitting between the triplet and singlet states do not appreciably change the radial wave functions. Use the charge of the charm quark, $+2/3$, and a mass of 1.6 GeV and assume they have normal magnetic moments ($g = 2$, but remember the antiquark has the opposite charge and magnetic moment to the particle). Express your answer in keV. Compare with experiment using tables published by the Particle Data Group.

9.5 Calculate the transition rate for the $3d \rightarrow 1s$ transition in hydrogen. This is an electric quadrupole transition.

(a) Show that the transition rate is

$$\Gamma(3d \rightarrow 1s) = \frac{e^2k^5}{75\hbar} \left|R_{1s}(r)R_{3d}(r)r^4dr\right|^2. \tag{9.118}$$

(b) Evaluate the radial matrix element and find the mean lifetime in seconds. A useful unit of lifetime is $(a_0/\alpha^4 c) = 0.6225 \times 10^{-10}$s. Compare with the result of Bethe and Salpeter, *Quantum Mechanics of One- and Two-Electron Atoms*, p. 281.

10

Relativistic Quantum Mechanics

From the outset, it was clear that quantum mechanics needed to conform to special relativity. The obvious relativistic equation, the Klein-Gordon equation, failed to explain the hydrogen spectrum, and Schrödinger and others retreated to the non-relativistic domain. Paul Dirac, having already established much of what became canonical non-relativistic quantum mechanics, wrote down with astonishing conciseness a first-order differential equation to describe the electron and its interaction with electromagnetism. In particular, it explained the spectrum of hydrogen in great detail. It did much more. It predicted the existence of the positron and ultimately described the fundamental fermions that are responsible for matter of all known kinds. The Foldy-Wouthuysen transformation provides a methodical procedure for expanding the Dirac equation in powers of $1/m$ and can be used with advantage as well to analyze neutron-electron scattering. With Dirac's equation, we can calculate the relativistic scattering of electrons and photons. We show briefly how traditional perturbation theory can be reorganized into a Lorentz-invariant prescription and presented in the form of a Feynman diagram.

10.1 Klein-Gordon Equation

Let us recall the basic features of non-relativistic quantum mechanics with an eye towards their generalization for relativistic quantum mechanics. The classical position and momentum are promoted to operators, which in the spatial representation are

$$\mathbf{x} \to \mathbf{x}_{\text{op}} = \mathbf{x}; \qquad \mathbf{p} \to \mathbf{p}_{\text{op}} = \frac{\hbar}{i}\nabla$$

$$E = T + V;$$

$$T \to T_{\text{op}} = -\frac{\hbar^2}{2m}\nabla^2$$

$$V \to V_{\text{op}} = V(\mathbf{x})$$

$$E \to E_{\text{op}} = i\hbar\frac{\partial}{\partial t}. \tag{10.1}$$

From these identifications we are led to the Schrödinger equation:

$$i\hbar\frac{\partial\psi(\mathbf{x},t)}{\partial t} = -\frac{\hbar^2}{2m}\nabla^2\psi + V(\mathbf{x})\psi. \tag{10.2}$$

John David Jackson: A Course in Quantum Mechanics, First Edition. Robert N. Cahn.
© 2024 John Wiley & Sons, Inc. Published 2024 by John Wiley & Sons, Inc.
Companion website: www.wiley.com/go/Jackson/QuantumMechanics

The probability interpretation of the wave function is expressed by

$$\rho(\mathbf{x}) = \psi^*(\mathbf{x})\psi(\mathbf{x}) \tag{10.3}$$

and the corresponding conservation of probability as

$$\frac{\partial \rho}{\partial t} + \nabla \cdot \mathbf{j} = 0; \quad \mathbf{j} = \frac{\hbar}{2mi}(\psi^* \nabla \psi - \psi \nabla \psi^*). \tag{10.4}$$

Associated with each physical observable is an operator $\mathcal{O}(p,q)$, which is Hermitian with matrix elements

$$\langle a \,|\mathcal{O}|\, b \rangle = \int d^3 x \, \psi_a^* \mathcal{O}(p,q) \psi_b. \tag{10.5}$$

Spin can be represented by $2s + 1$-dimensional vector if the spin is s.

The idea is to make the minimal changes incorporating relativity (here we take $c = \hbar = 1$). We must always have

$$E^2 - p^2 = m^2. \tag{10.6}$$

We write

$$(E, \mathbf{p}) = p^\alpha \rightarrow p_{\mathrm{op}}^\alpha = i\partial^\alpha = (i\frac{\partial}{\partial t}, -i\nabla). \tag{10.7}$$

We introduce a scalar wave function $\psi(x^\beta)$ and have the equation

$$-\partial_\alpha \partial^\alpha \psi = m^2 \psi, \tag{10.8}$$

which is called the Klein-Gordon equation. It is natural to define a probability current analogous to the familiar Schrödinger current density

$$j^\alpha = \frac{i}{2m}(\psi^* \partial^\alpha \psi - \psi \partial^\alpha \psi^*) = \frac{i}{2m}(\psi^* \overset{\leftrightarrow}{\partial}^\alpha \psi). \tag{10.9}$$

It follows easily that $\partial_\alpha j^\alpha = 0$. The spatial part of the four-vector current is the familiar Schrödinger current density, but what about the time component?

$$j_0 = \frac{i}{2m}(\psi^* \frac{\partial \psi}{\partial t} - \psi \frac{\partial \psi^*}{\partial t}). \tag{10.10}$$

First consider $\psi \propto e^{-iEt}$, so $j_0 = \frac{i}{2m}(-2iE)|\psi|^2 = \frac{E}{m}|\psi|^2$. In the non-relativistic limit, $E \rightarrow m$ and $j_0 \rightarrow |\psi|^2$. That is reassuring. Secondly, however, consider $E = \pm\sqrt{p^2 + m^2}$. The Klein-Gordon equation is indifferent to the sign of E. What if $E < 0$ is possible? Then $j_0 < 0$.

More generally, $j_0 = \frac{1}{m}\Re(i\psi^* \frac{\partial \psi}{\partial t})$ is the real part of a complex quantity and therefore does not have to be semi-positive-definite. This failure of the time-component of the current four-vector to satisfy the basic requirement of a probability density led physicists in the 1920s to discard the Klein-Gordon equation as a generalization of non-relativistic quantum mechanics. (Actually, Schrödinger tried it first in 1926, with $E \rightarrow E - e\Phi$, $\mathbf{p} \rightarrow \mathbf{p} - e\mathbf{A}/c$, but discarded it because it gave the wrong fine-structure splitting by a factor $4n/(2n-1)$ for the level n of the Coulomb problem. See Schiff, *Quantum Mechanics*, 3rd edition, pp. 467-471, 486.)

The Klein-Gordon equation was restored to quantum mechanics in "second-quantized" form by Pauli and Weisskopf in 1933-34. They showed that the complex $\psi(\mathbf{x}, t)$ could represent a non-Hermitian quantized field

$$\psi(\mathbf{x}, t) = \sum_{\mathbf{k}} \frac{1}{\sqrt{2\omega_k}} \left(a_{\mathbf{k}} e^{i(\mathbf{k}\cdot\mathbf{x}-\omega_k t)} + b_{\mathbf{k}}^\dagger e^{-i(\mathbf{k}\cdot\mathbf{x}-\omega_k t)} \right), \tag{10.11}$$

where $\omega_k - \sqrt{k^2 + m^2}$ is the energy of a particle of momentum $\mathbf{k}$ and mass m.

Since $\psi^\dagger \neq \psi$, the operators $a_{\mathbf{k}}, a_{\mathbf{k}}^\dagger$ destroy and create one kind of particle and $b_{\mathbf{k}}, b_{\mathbf{k}}^\dagger$ destroy and create another kind. Now ej^α becomes the charge-current four-vector density. The two kinds of quanta have the same mass, opposite charges, and zero spin. The particles are thus charged, spinless bosons. The Pauli-Weisskopf formalism describes $\pi^\pm$, $K^\pm$, etc. Note that the positive and negative energy (frequency) components occur on an equal footing in the second-quantized formulation, where ψ is an operator, not a wave function. This closely parallels the treatment of the electromagnetic field.

10.2 Dirac Equation

In 1928, Paul Adrien Maurice Dirac made one of the great discoveries of all time - a miracle, almost! Fresh from the success of his transformation theory, the program of replacing classical mechanics (described by the Hamilton equation) with quantum mechanics by substituting commutators for Poisson brackets (work that he personally viewed as his most important paper), Dirac turned to the problem of relativistic quantum mechanics. He wanted a relativistic equation of the form

$$i\hbar \frac{\partial \psi}{dt} = H\psi \tag{10.12}$$

and, at the same time, a non-negative probability density.

He fooled around with $\boldsymbol{\sigma} \cdot \mathbf{p}$ as a warm-up and observed that $\boldsymbol{\sigma} \cdot \mathbf{p}\,\boldsymbol{\sigma} \cdot \mathbf{p} = p^2$ could give the kinetic energy. Notice that $\boldsymbol{\sigma} \cdot \mathbf{p}$ is sort of the square root of the Hamiltonian for a massless particle. Then Dirac turned to four-dimensional space-time and decided to write an equation that was linear in $\mathbf{p}_{\text{op}} = (\hbar/i)\nabla$, as well as in $E_{\text{op}} = i\hbar(\partial/\partial t)$. Knowing that spin-1/2 needed $\boldsymbol{\sigma}$ as 2 x 2 matrices and spins,

$$\begin{pmatrix} \psi_1 \\ \psi_2 \end{pmatrix}, \tag{10.13}$$

Dirac proposed the equation

$$i\hbar \frac{\partial \psi}{\partial t} = c\boldsymbol{\alpha} \cdot \mathbf{p}\psi + \beta mc^2 \psi$$
$$= \frac{\hbar c}{i} \sum_j \alpha_j \frac{\partial \psi}{\partial x_j} + \beta mc^2 \psi, \tag{10.14}$$

where $\alpha_1, \alpha_2, \alpha_3$, and β were $N \times N$ constant matrices and

$$\psi = \begin{pmatrix} \psi_1 \\ \psi_2 \\ \cdot \\ \cdot \\ \cdot \\ \psi_N \end{pmatrix} \tag{10.15}$$

is an N-component column vector. Henceforth, we set $c = 1$.

The matrices are constrained in several ways: the equations must have free-particle solutions with energy E and momentum p satisfying $E^2 - p^2 = m^2$. To see the consequences of this constraint, "square" the Dirac equation:

$$(\alpha \cdot \mathbf{p} + \beta m)i\frac{\partial \psi}{\partial t} = (\alpha \cdot \mathbf{p} + \beta m)^2 \psi$$

$$-\frac{\partial^2 \psi}{\partial t^2} = (\alpha \cdot \mathbf{p} + \beta m)^2 \psi. \tag{10.16}$$

This should turn into $E^2\psi = (p^2 + m^2)\psi$. The right-hand side is

$$\sum_{i,j} \frac{1}{2}(\alpha_i\alpha_j + \alpha_j\alpha_i)p_ip_j + \sum_j (\beta\alpha_j + \alpha_j\beta)mp_j + \beta^2 m^2. \tag{10.17}$$

To obtain the required form, Dirac needed

$$\alpha_i\alpha_j + \alpha_j\alpha_i = 2\delta_{ij}I$$
$$\beta\alpha_j + \alpha_j\beta = 0$$
$$\beta^2 = I. \tag{10.18}$$

The first line demands $\alpha_i^2 = I$. Altogether, α_i and β must anticommute with each other and have squares equal to I. Since their squares are I, their eigenvalues must be ± 1. We also want the matrices to be Hermitian, so our Hamiltonian will be Hermitian. It follows from the anticommutation relations that the α_i and β are traceless:

$$0 = \alpha_i\beta + \beta\alpha_i \rightarrow \alpha_i = -\beta\alpha_i\beta$$
$$\mathrm{Tr}\,\alpha_i = -\mathrm{Tr}\,\beta\alpha_i\beta = -\mathrm{Tr}\,\beta^2\alpha_i = -\mathrm{Tr}\,\alpha_i \tag{10.19}$$

and analogously for β. Since their eigenvalues are ± 1 and they are traceless, N must be even. We cannot have $N = 2$ because there are only three anti-commuting 2 x 2 matrices, the sigma matrices. Thus $N = 4$ is the smallest possibility. The Pauli representation of the Dirac matrices is

$$\beta = \begin{pmatrix} I & 0 \\ 0 & -I \end{pmatrix}; \quad \alpha = \begin{pmatrix} 0 & \sigma \\ \sigma & 0 \end{pmatrix}. \tag{10.20}$$

Because the Dirac equation is first-order in time, it is possible that the probability density will evade the problems associated with the Klein-Gordon equation. We try

$$\rho = \psi^\dagger\psi = \sum_{\alpha=1}^{4} \psi_\alpha^*\psi_\alpha. \tag{10.21}$$

The continuity equation can be written

$$\nabla \cdot \mathbf{j} = -\frac{\partial \rho}{\partial t} = -\psi^\dagger\frac{\partial \psi}{\partial t} - \frac{\partial \psi^\dagger}{\partial t}\psi$$

$$= \frac{1}{\hbar}\left(\psi^\dagger H\psi - (H\psi)^\dagger\psi\right)$$

$$= \frac{i}{\hbar}\left[\frac{\hbar c}{i}\psi^\dagger\alpha \cdot \nabla\psi + mc^2\psi^\dagger\beta\psi + \frac{\hbar c}{i}(\nabla\psi^\dagger) \cdot \alpha^\dagger\psi - mc^2\psi^\dagger\beta^\dagger\psi\right]. \tag{10.22}$$

But $\alpha^\dagger = \alpha$ and $\beta^\dagger = \beta$, so

$$\nabla \cdot \mathbf{j} = c[\psi^\dagger \alpha \cdot \nabla\psi + (\nabla\psi)^\dagger \cdot \alpha\psi] = c\nabla(\psi^\dagger \alpha\psi). \tag{10.23}$$

We can thus identify the Dirac current as

$$\mathbf{j} = c(\psi^\dagger \alpha\psi), \tag{10.24}$$

with the usual proviso that we could add a piece with vanishing divergence. Evidently $c\alpha$ is the velocity operator. Note that in presence of external electromagnetic fields and the substitution $p^\mu \to p^\mu - eA^\mu$ in the Hamiltonian, the current still retains the same form. This is different from Schrödinger non-relativistic quantum mechanics, where $\Delta\mathbf{j} = -\frac{e}{mc}\mathbf{A}\psi^*\psi$. But see Problem 10.1.

10.3 Angular Momentum in Dirac Equation

A great triumph of the Dirac equation is that it "explains" electron spin, which was put in by hand in quantum mechanical descriptions prior to Dirac's work. We can ask whether orbital angular momentum is conserved by computing

$$[H, L_z] = [(c\alpha \cdot \mathbf{p} + \beta mc^2, (xp_y - yp_x)]$$
$$= c\alpha_x[p_x, x]p_y - c\alpha_y[p_y, y]p_x = -i\hbar(\alpha_x p_y - \alpha_y p_x) = -i\hbar c(\alpha \times \mathbf{p})_z. \tag{10.25}$$

So definitively, **L** is not conserved. But spin comes to the rescue. Let us generalize σ to a 4 x 4 matrix:

$$\sigma_D = \begin{pmatrix} \sigma & 0 \\ 0 & \sigma \end{pmatrix}. \tag{10.26}$$

We compute $[H, \mathbf{b} \cdot \sigma_D]$, where $\mathbf{b}$ is some constant vector, which will simplify the algebra.

$$\left[\begin{pmatrix} mc^2 I & c\alpha \cdot \mathbf{p} \\ c\alpha \cdot \mathbf{p} & -mc^2 I \end{pmatrix}, \begin{pmatrix} \mathbf{b}\cdot\sigma & 0 \\ 0 & \mathbf{b}\cdot\sigma \end{pmatrix}\right] = \begin{pmatrix} 0 & c[\sigma\cdot\mathbf{p}, \mathbf{b}\cdot\sigma] \\ c[\sigma\cdot\mathbf{p}, \mathbf{b}\cdot\sigma] & 0 \end{pmatrix}$$
$$= \begin{pmatrix} 0 & 2ic\mathbf{b}\cdot(\sigma\times\mathbf{p}) \\ 2ic\mathbf{b}\cdot(\sigma\times\mathbf{p}) & 0 \end{pmatrix} = 2ic\mathbf{b}\cdot(\alpha\times\mathbf{p}), \tag{10.27}$$

so that

$$[H, \tfrac{1}{2}\sigma_D] = ic\alpha \times \mathbf{p}$$
$$[H, \mathbf{L} + \tfrac{1}{2}\sigma_D] \equiv [H, \mathbf{J}] = 0, \tag{10.28}$$

where we have dropped the $\hbar$. Because $[H, \mathbf{L}] \neq 0$, we cannot speak of the orbital angular momentum, except non-relativistically. The helicity of a free spin-one particle, the projection of its spin along the direction of its momentum, is conserved as we see from

$$[H, \sigma \cdot \mathbf{p}] = 2ic(\alpha \times \mathbf{p}) \cdot \mathbf{p} = 0. \tag{10.29}$$

10.4 Two-Component Equation and Plane-Wave Solutions

Consider a Dirac electron of mass m, charge e interacting with an external field $A^\mu = (A^0 = \Phi, \mathbf{A})$. The equation of motion is

$$i\hbar\frac{\partial\psi}{\partial t} = \left[c\alpha \cdot (\mathbf{p} - \frac{e}{c}\mathbf{A}) + \beta mc^2 + e\Phi\right]\psi. \tag{10.30}$$

We introduce two two-component spinors, ϕ and χ, through

$$\psi = \begin{pmatrix} \phi_1 \\ \phi_2 \\ \chi_1 \\ \chi_2 \end{pmatrix} \equiv \begin{pmatrix} \phi \\ \chi \end{pmatrix}. \tag{10.31}$$

Then we get two coupled equations for ϕ and χ:

$$i\hbar\frac{\partial\phi}{\partial t} = c\sigma \cdot (\mathbf{p} - \frac{e}{c}\mathbf{A})\chi + mc^2\phi + e\Phi\phi$$

$$i\hbar\frac{\partial\chi}{\partial t} = c\sigma \cdot (\mathbf{p} - \frac{e}{c}\mathbf{A})\phi - mc^2\chi + e\Phi\chi. \tag{10.32}$$

Before determining the consequences of the Dirac equation in the presence of an electromagnetic field, we first seek plane-wave solutions in the absence of the electromagnetic field for a particle with definite momentum $\mathbf{p}$ and definite energy E, setting $\hbar = c = 1$.

$$\psi_j = A_j e^{i(\mathbf{p}\cdot\mathbf{x} - Et)} \tag{10.33}$$

so that

$$E\phi = \sigma \cdot \mathbf{p}\,\chi + m\phi$$

$$E\chi = \sigma \cdot \mathbf{p}\,\phi - m\chi. \tag{10.34}$$

It follows that

$$\chi = \frac{\sigma \cdot \mathbf{p}}{E + m}\phi$$

$$E\phi = \frac{\sigma \cdot \mathbf{p}\sigma \cdot \mathbf{p}}{E + m}\phi + m\phi$$

$$(E + m)(E - m)\phi = p^2\phi. \tag{10.35}$$

So

$$E = \pm E(p), \qquad E(p) = +\sqrt{p^2 + m^2}. \tag{10.36}$$

Just as with the Klein-Gordon equation, we find two signs for the energy.

(a) Consider first the positive energy solutions $E > 0$. We have $\phi \propto \begin{pmatrix} 1 \\ 0 \end{pmatrix}$ or $\phi \propto \begin{pmatrix} 0 \\ 1 \end{pmatrix}$, corresponding to spin up or spin down in the particle rest frame. There are different choices for normalization:

    (i) $|\phi|^2 + |\chi|^2 = 1$, the conventional normailzation so $(\psi^\dagger\psi) \propto j_0$.

    (ii) $|\phi|^2 - |\chi|^2 = 1$, invariant normalization so $(\psi^\dagger\beta\psi)$ is a Lorentz scalar.

    (iii) $|\phi|^2 - |\chi|^2 = 2m$, invariant normalization so $(\psi^\dagger\beta\psi)$ is a Lorentz scalar.

We see that in any event,

$$\chi^2 = \frac{p^2}{(E+m)^2}\phi^2 \tag{10.37}$$

so for choice (i)

$$|\phi|^2\left(1 + \frac{p^2}{(E+m)^2}\right) = 1; \qquad |\phi|^2 = \frac{E+m}{2E}, \tag{10.38}$$

while for choice (ii)

$$|\phi|^2\left(1 - \frac{p^2}{(E+m)^2}\right) = 1; \qquad |\phi|^2 = \frac{E+m}{2m}. \tag{10.39}$$

The positive energy solutions are therefore

$$\psi_+(\mathbf{x},t) = N_i\sqrt{E(p)+m}\begin{pmatrix}\phi \\ \frac{\sigma\cdot\mathbf{p}}{E(p)+m}\phi\end{pmatrix}e^{i(\mathbf{p}\cdot\mathbf{x}-E(p)t)}, \tag{10.40}$$

where $N_1 = 1/\sqrt{2E(p)}$, $N_2 = 1/\sqrt{2m}$, $N_3 = 1$.

(b) For the negative energy solution, $E = -E(p)$, we have

$$-E(p)\phi = \sigma\cdot\mathbf{p}\chi + m\phi$$

$$\phi = -\frac{\sigma\cdot\mathbf{p}}{E(p)+m}\chi. \tag{10.41}$$

Then

$$\psi_-(\mathbf{x},t) = N_i\sqrt{E(p)+m}\begin{pmatrix}\frac{-\sigma\cdot\mathbf{p}}{E(p)+m}\chi \\ \chi\end{pmatrix}e^{i(\mathbf{p}\cdot\mathbf{x}+E(p)t)}. \tag{10.42}$$

We conventionally define the functions $w_j(\mathbf{p})$, $j = 1, 2, 3, 4$ as the spinors (without the phase waves) normalized according to type (1):

$$w_j^\dagger(\mathbf{p})w_{j'}(\mathbf{p}) = \delta_{jj'}, \tag{10.43}$$

where

$j = 1$ is $E > 0$, "spin up" with $\phi = \begin{pmatrix}1 \\ 0\end{pmatrix}$

$j = 2$ is $E > 0$, " spin down" with $\phi = \begin{pmatrix}0 \\ 1\end{pmatrix}$

$j = 3$ is $E < 0$, " spin up" with $\chi = \begin{pmatrix}1 \\ 0\end{pmatrix}$

$j = 4$ is $E < 0$, " spin down" with $\chi = \begin{pmatrix}0 \\ 1\end{pmatrix}$. \qquad (10.44)

That is,

$$w_1(\mathbf{p}) = \sqrt{\frac{E(p)+m}{2E(p)}} \begin{pmatrix} \begin{pmatrix} 1 \\ 0 \end{pmatrix} \\ \dfrac{\boldsymbol{\sigma}\cdot\mathbf{p}}{E(p)+m}\begin{pmatrix} 1 \\ 0 \end{pmatrix} \end{pmatrix}$$

$$w_2(\mathbf{p}) = \sqrt{\frac{E(p)+m}{2E(p)}} \begin{pmatrix} \begin{pmatrix} 0 \\ 1 \end{pmatrix} \\ \dfrac{\boldsymbol{\sigma}\cdot\mathbf{p}}{E(p)+m}\begin{pmatrix} 0 \\ 1 \end{pmatrix} \end{pmatrix}$$

$$w_3(\mathbf{p}) = \sqrt{\frac{E(p)+m}{2E(p)}} \begin{pmatrix} -\dfrac{\boldsymbol{\sigma}\cdot\mathbf{p}}{E(p)+m}\begin{pmatrix} 1 \\ 0 \end{pmatrix} \\ \begin{pmatrix} 1 \\ 0 \end{pmatrix} \end{pmatrix}$$

$$w_4(\mathbf{p}) = \sqrt{\frac{E(p)+m}{2E(p)}} \begin{pmatrix} -\dfrac{\boldsymbol{\sigma}\cdot\mathbf{p}}{E(p)+m}\begin{pmatrix} 0 \\ 1 \end{pmatrix} \\ \begin{pmatrix} 0 \\ 1 \end{pmatrix} \end{pmatrix}.$$

$$(10.45)$$

It is conventional to define separately spinors

$$u_1(\mathbf{p}) = (E(p)/m)^{1/2}w_1(\mathbf{p}), \qquad\qquad u_2(\mathbf{p}) = (E(p)/m)^{1/2}w_2(\mathbf{p})$$
$$v_1(\mathbf{p}) = (E(p)/m)^{1/2}w_4(-\mathbf{p}), \qquad\qquad v_2(\mathbf{p}) = (E(p)/m)^{1/2}w_3(-\mathbf{p}), \qquad (10.46)$$

as is done in Bjorken and Drell, *Relativistic Quantum Mechanics*.

10.5   Dirac's Treatment of Negative Energy States

To explain negative energy states, Dirac postulated that the vacuum consisted of all the negative energy states being filled, according to the Pauli principle. The departures from the vacuum were observed as positrons (though Dirac originally hoped these would be protons). A photon hitting an electron in the negative-energy sea could liberate it, and leave behind a hole in the sea, which would appear as a positive particle, thus resulting in creation of an electron-positron pair. See Figure 10.1. Nowadays, we do not speak of "hole theory," but rather

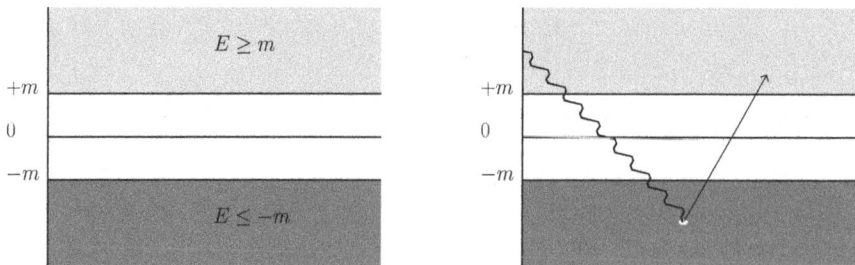

Figure 10.1   Left: Dirac's picture of negative energy states, $E < -mc^2$, fully occupied in the vacuum. Right: A photon of energy $E > 2mc^2$ kicks an electron out of the negative-energy sea, leaving behind a hole equivalent to a positron, together with the liberated electron.

of the negative frequency components of the wave function as related in a second-quantized formalism with the creation of positrons, while the positive energy components are involved in the destruction of electrons. There is a one-to-one correspondence between Dirac's hole theory and the present-day description.

10.6 Heisenberg Operators and Equations of Motion

Matrix elements like

$$\langle f | \alpha_x | i \rangle = \int d^3x \psi_f^\dagger(\mathbf{x}, t) \alpha_x \psi_i(\mathbf{x}, t) \tag{10.47}$$

can be interpreted in the Heisenberg picture by writing

$$\langle f | \alpha_x | i \rangle = \int d^3x \, \psi_f^\dagger(\mathbf{x}, 0) \alpha_x(t) \psi_i(\mathbf{x}, 0), \tag{10.48}$$

where

$$\alpha_x(t) = e^{+iHt} \alpha e^{-iHt} \tag{10.49}$$

is the Heisenberg representation of α_x, which coincides with the Schrödinger picture constant α_x at $t = 0$.

We have already considered the time development of some operators:

$$\frac{d\mathbf{L}}{dt} = \frac{i}{\hbar}[H, \mathbf{L}]; \qquad \frac{d\boldsymbol{\sigma}_D}{dt} = \frac{i}{\hbar}[H, \boldsymbol{\sigma}_D]. \tag{10.50}$$

Quite generally, the development of operators in the Heisenberg picture is

$$\frac{d\mathcal{O}}{dt} = \frac{i}{\hbar}[H, \mathcal{O}] + \frac{\partial \mathcal{O}}{\partial t}. \tag{10.51}$$

We apply this to $\boldsymbol{\pi} = \mathbf{p} - e\mathbf{A}$ with $H = \boldsymbol{\alpha} \cdot (\mathbf{p} - e\mathbf{A}) + e\Phi$. Here $\mathbf{p}$ is the canonical momentum, represented by $(\hbar/i)\nabla$, and $\boldsymbol{\pi}$ is the ordinary physical momentum. We calculate

$$\dot{\pi}_i = \frac{i}{\hbar}[\boldsymbol{\alpha} \cdot (\mathbf{p} - e\mathbf{A}) + \beta m + e\Phi, p_i - eA_i] - e\frac{\partial A_i}{\partial t}. \tag{10.52}$$

Contributions arise from the gradient in $\mathbf{p}$.

$$\dot{\pi}_i = \frac{i}{\hbar} \sum_j \alpha_j \frac{\hbar}{i} \left[\frac{\partial}{\partial x_j}(-eA_i) + e\frac{\partial}{\partial x_i}A_j \right] - e\frac{\partial}{\partial x_i}\Phi - e\frac{\partial A_i}{dt}$$

$$= e \sum_j \alpha_j \epsilon_{ijk} \mathbf{B}_k + e\mathbf{E}_i = e(\boldsymbol{\alpha} \times \mathbf{B})_i + e\mathbf{E}_i. \tag{10.53}$$

With our identification of α with $\mathbf{v}$, this is just the Lorentz force.

Consider $\boldsymbol{\sigma}_D \cdot \boldsymbol{\pi}$, the product of the spin and physical momentum operators. Writing out the matrices

$$H = \begin{bmatrix} (m + e\Phi)I & \boldsymbol{\sigma} \cdot \boldsymbol{\pi} \\ \boldsymbol{\sigma} \cdot \boldsymbol{\pi} & (-m + e\Phi)I \end{bmatrix}; \qquad \boldsymbol{\sigma}_D \cdot \boldsymbol{\pi} = \begin{bmatrix} \boldsymbol{\sigma} \cdot \boldsymbol{\pi} & 0 \\ 0 & \boldsymbol{\sigma} \cdot \boldsymbol{\pi} \end{bmatrix}, \tag{10.54}$$

where we have chosen for simplicity to calculate at time $t = 0$ so that we can use the values of the Heisenberg operators at that time, when they have their simple forms in the Schrödinger picture. We find directly

$$[H, \sigma_D \cdot \boldsymbol{\pi}] = \frac{e\hbar}{i} \begin{bmatrix} -\boldsymbol{\sigma} \cdot \nabla\Phi & 0 \\ 0 & -\boldsymbol{\sigma} \cdot \nabla\Phi \end{bmatrix} = \frac{e\hbar}{i} \begin{bmatrix} \boldsymbol{\sigma} \cdot \mathbf{E} & 0 \\ 0 & \boldsymbol{\sigma} \cdot \mathbf{E} \end{bmatrix} = \frac{e\hbar}{i} \sigma_D \cdot \mathbf{E}$$

$$\frac{d}{dt}(\sigma_D \cdot \boldsymbol{\pi}) = e\sigma_D \cdot \mathbf{E}. \tag{10.55}$$

In a purely magnetic field ($\mathbf{E} = 0$), the helicity is preserved. This is a consequence of $g = 2$ and is not exactly true for electrons since $g/2 = 1 + a_e$, with $a_e \approx \alpha/2\pi$. See Jackson, *Classical Electrodynamics*, 3rd edition, p. 565. Note that the time derivative of $\sigma_D \cdot \boldsymbol{\pi}$ is not simply $\sigma_D \cdot \dot{\boldsymbol{\pi}}$. While α, σ_D, β are matrices in the Schrödinger representation, in the Heisenberg representation they are operators that evolve in time, like any other operator. However, commutation and anticommutation relations that have constants on the right-hand side will still hold in the Heisenberg representation, since the constants are invariant under conjugation with the time-development operator.

10.7 Hydrogen in the Dirac Equation

The central role of hydrogen in the development of quantum mechanics required that the Dirac equation be tested in this essential application. The complete solution was given by C. G. Darwin and by W. Gordon in 1928. (C. G. Darwin, *Proc. Roy. Soc. Lond.* Ser. A 118, 654; Z. Gordon Phys. 48, 11). However, we obtain more insight working in a more pedestrian fashion. If we consider the two-component form of the Dirac equation

$$i\hbar\frac{\partial\phi}{dt} = c\boldsymbol{\sigma} \cdot (\mathbf{p} - \frac{e\mathbf{A}}{c})\chi + mc^2\phi + e\Phi\phi$$

$$i\hbar\frac{\partial\chi}{dt} = c\boldsymbol{\sigma} \cdot (\mathbf{p} - \frac{e\mathbf{A}}{c})\phi - mc^2\chi + e\Phi\chi \tag{10.56}$$

and write

$$\phi = \phi_{\text{NR}}e^{-imc^2t/\hbar}; \qquad \chi = \chi_{\text{NR}}e^{-imc^2t/\hbar}; \tag{10.57}$$

to remove the rest-energy contribution to the energy, we find

$$i\hbar\frac{\partial\phi_{\text{NR}}}{\partial t} = c\boldsymbol{\sigma} \cdot (\mathbf{p} - \frac{e\mathbf{A}}{c})\chi_{\text{NR}} + e\Phi\phi_{\text{NR}}$$

$$(2mc^2 - e\Phi + i\hbar\frac{\partial}{\partial t})\chi_{\text{NR}} = c\boldsymbol{\sigma} \cdot (\mathbf{p} - \frac{e\mathbf{A}}{c})\phi_{\text{NR}}. \tag{10.58}$$

In lowest approximation for χ, we neglect all but $2mc^2$ on the left-hand side and write

$$\chi_{\text{NR}} \approx \frac{\boldsymbol{\sigma} \cdot (\mathbf{p} - e\mathbf{A}/c)}{2mc}\phi_{\text{NR}}. \tag{10.59}$$

Substituting in the equation for ϕ_{NR},

$$i\hbar\frac{\partial\phi_{\text{NR}}}{\partial t} \approx \frac{1}{2m}\boldsymbol{\sigma} \cdot (\mathbf{p} - e\mathbf{A}/c)\boldsymbol{\sigma} \cdot (\mathbf{p} - e\mathbf{A}/c)\phi_{\text{NR}} + e\Phi\phi_{\text{NR}}$$

$$= \frac{1}{2m}(\mathbf{p} - e\mathbf{A}/c)^2\phi_{\text{NR}} + e\Phi\phi_{\text{NR}} + \frac{i}{2m}\boldsymbol{\sigma} \cdot (\mathbf{p} - e\mathbf{A}/c) \times (\mathbf{p} - e\mathbf{A}/c)\phi_{\text{NR}}. \tag{10.60}$$

In the last term, only the $\mathbf{p}$ operator acting on $\mathbf{A}$ survives:

$$\frac{i}{2m}\sigma\cdot(\mathbf{p}-e\mathbf{A}/c)\times(\mathbf{p}-e\mathbf{A}/c) = -\frac{e\hbar}{2mc}\sigma\cdot(\nabla\times\mathbf{A}) = -\frac{e\hbar}{2mc}\sigma\cdot\mathbf{B}. \tag{10.61}$$

Thus

$$i\hbar\frac{\partial\phi_{\mathrm{NR}}}{\partial t} = \frac{1}{2m}(\mathbf{p}-e\mathbf{A}/c)^2\phi_{\mathrm{NR}} + e\Phi\phi_{\mathrm{NR}} - \frac{e\hbar}{2mc}\sigma\cdot\mathbf{B}\phi_{\mathrm{NR}}. \tag{10.62}$$

The magnetic moment $\mu = \frac{e\hbar}{2mc}\sigma = 2\frac{e\hbar}{2mc}\mathbf{s}$ correponds to $g=2$. This emerges naturally from the Dirac equation.

10.8 Foldy-Wouthuysen Transformation

If we wish to go beyond this approximation (which is, in a sense, an expansion in powers of $1/m$, to order $1/m$, inclusive), it is best to use a systematic procedure. The Foldy-Wouthuysen transformation (L. Foldy and S. Wouthuysen, *Phys. Rev.* **78**, 29, 1950) is one such scheme. We define even and odd Dirac operators

$$\mathcal{O}_{\mathrm{even}} = \begin{pmatrix} \mathcal{O}_1 & 0 \\ 0 & \mathcal{O}_2 \end{pmatrix}; \qquad \mathcal{O}_{\mathrm{odd}} = \begin{pmatrix} 0 & \mathcal{O}_3 \\ \mathcal{O}_4 & 0 \end{pmatrix}. \tag{10.63}$$

Even (odd) operators commute (anticommute) with β. Any operator can be written as a sum of even and odd components:

$$\mathcal{O} = \mathcal{O}_{\mathrm{even}} + \mathcal{O}_{\mathrm{odd}}$$

$$\mathcal{O}_{\mathrm{even}} = \frac{1}{2}(\mathcal{O}+\beta\mathcal{O}\beta)$$

$$\mathcal{O}_{\mathrm{odd}} = \frac{1}{2}(\mathcal{O}-\beta\mathcal{O}\beta). \tag{10.64}$$

Now the coupled two-component equations arise because the Hamiltonian has both odd ($\alpha\cdot\mathbf{p}$) and even ($\beta mc^2, e\Phi$) operators in it. The Foldy-Wouthuysen transformation is a unitary transformation that removes all odd operators from the Hamiltonian, in practice to some order in a parameter, e.g. $(1/m)^n$.

Write $H = \beta m + \xi + \mathcal{O}$, where ξ is an even operator and $\mathcal{O}$ is an odd operator. We postulate a unitary transformation $\psi' = e^{iS}\psi$. Then

$$i\hbar\frac{\partial\psi'}{\partial t} = H'\psi', \tag{10.65}$$

where

$$H' = e^{iS}\left(H - i\hbar\frac{\partial}{\partial t}\right)e^{-iS}. \tag{10.66}$$

Now we have the expansions (see Problem 2.6)

$$e^{iS}He^{-iS} = H + i[S,H] + \frac{i^2}{2!}[S,[S,H]] + \cdots$$

$$e^{iS}\frac{\partial}{\partial t}e^{-iS} = \dot{S} + \frac{i}{2!}[S,\dot{S}] + \frac{i^2}{3!}[S,[S,\dot{S}]] + \cdots \tag{10.67}$$

This means that H' is given by, setting $\hbar = 1$,

$$H' = H + i[S, H] - \frac{1}{2}[S, [S, H]] - i\frac{1}{6}[S, [S, [S, H]]] + \frac{1}{24}[S, [S, [S, [S, H]]]] + \cdots$$
$$- \dot{S} - \frac{i}{2}[S, \dot{S}] + \frac{1}{6}[S, [S, \dot{S}]] + \cdots \quad (10.68)$$

It will turn out that $S = \mathcal{O}(1/m)$. Then to get H' even to order m^0, we need only consider the leading term in H in the commutator, e.g. $H \approx \beta m$. Then we would have

$$H' = \beta m + \xi + \mathcal{O} + i[S, \beta m]. \quad (10.69)$$

We wish to choose S so that the $\mathcal{O}$ term is cancelled. Since $\mathcal{O}$ anticommutes with β, so must S: $[S, \beta m] = -2\beta mS$, so we demand

$$-2i\beta mS = -\mathcal{O}; \qquad S = -i\frac{\beta\mathcal{O}}{2m}. \quad (10.70)$$

This leads to a tedious calculation, which depends on a few basic relations:

$$[\mathcal{O}, \beta] = -2\beta\mathcal{O}; \qquad [\beta\mathcal{O}, \beta] = -2\mathcal{O}; \qquad [\beta\mathcal{O}, \mathcal{O}] = 2\beta\mathcal{O}^2$$
$$[\beta\mathcal{O}, \xi] = \beta[\mathcal{O}, \xi]; \qquad [\beta\mathcal{O}, \mathcal{O}^n] = 2\beta\mathcal{O}^{n+1}; \qquad [\beta\mathcal{O}, \beta\mathcal{O}^n] = -2\mathcal{O}^{n+1}. \quad (10.71)$$

We use these to compute to order $1/m^2$:

$$H' = \beta m + \xi + \mathcal{O} + \frac{1}{2m}[\beta\mathcal{O}, (\beta m + \xi + \mathcal{O})] + \frac{1}{8m^2}[\beta\mathcal{O}, [\beta\mathcal{O}, (\beta m + \xi + \mathcal{O})]]$$
$$+ \frac{1}{48m^3}[\beta\mathcal{O}[\beta\mathcal{O}, [\beta\mathcal{O}, \beta m]]] + \cdots + \frac{i}{2m}\beta\dot{\mathcal{O}} + \frac{i}{8m^2}[\beta\mathcal{O}, \beta\dot{\mathcal{O}}] + \cdots, \quad (10.72)$$

substituting

$$[\beta\mathcal{O}, \beta m + \xi + \mathcal{O}] = -2m\mathcal{O} + 2\beta\mathcal{O}^2 + \beta[\mathcal{O}, \xi]$$
$$[\beta\mathcal{O}, [\beta\mathcal{O}, \beta m + \xi + \mathcal{O}]] = -4m\beta\mathcal{O}^2 - 4\mathcal{O}^3 - [\mathcal{O}, [\mathcal{O}, \xi]]$$
$$[\beta\mathcal{O}, [\beta\mathcal{O}, [\beta\mathcal{O}, \beta]] = 8\mathcal{O}^3, \quad (10.73)$$

we find

$$H' = \beta\left(m + \frac{\mathcal{O}^2}{2m}\right) + \xi - \frac{1}{8m^2}[\mathcal{O}, [\mathcal{O}, \xi]] - \frac{i}{8m^2}[\mathcal{O}, \dot{\mathcal{O}}]$$
$$+ i\frac{\beta\dot{\mathcal{O}}}{2m} - \frac{\mathcal{O}^3}{3m^2} + \frac{\beta}{2m}[\mathcal{O}, \xi]. \quad (10.74)$$

The last three terms are still odd, but now with leading order $1/m$. Now we repeat the process with $H = \beta m + \xi' + \mathcal{O}'$ and

$$S' = -i\frac{\beta\mathcal{O}'}{2m} = -i\frac{\beta}{2m}\left(\frac{\beta}{2m}[\mathcal{O}, \xi] - \frac{\mathcal{O}^3}{3m^2} + i\frac{\beta\dot{\mathcal{O}}}{2m}\right). \quad (10.75)$$

To obtain the new Hamiltonian, H'', we have

$$H'' = H' + i[S', H'] - \frac{1}{2}[S', [S', H'] + \cdots - \dot{S}' - \frac{i}{2}[S', \dot{S}'] + \cdots \quad (10.76)$$

We find that while the previous odd terms have been eliminated, there are new odd terms generated, but now of order $1/m^2$. These will be cancelled by the next iteration, which will have an S'' of order $\mathcal{O}(1/m^3)$. As a consequence, the Hamiltonian, good to order $\mathcal{O}(1/m^2)$, is

$$H'' = \beta\left(m + \frac{\mathcal{O}^2}{2m}\right) + \xi - \frac{1}{8m^2}[\mathcal{O},[\mathcal{O},\xi]] - \frac{i}{8m^2}[\mathcal{O},\dot{\mathcal{O}}] + \cdots \tag{10.77}$$

Let us apply this to the special circumstance of an external electromagnetic field. Then

$$\mathcal{O} = \boldsymbol{\alpha} \cdot (\mathbf{p} - e\mathbf{A}); \qquad \xi = e\Phi$$

$$\mathcal{O}^2 = \boldsymbol{\alpha} \cdot (\mathbf{p} - e\mathbf{A})\boldsymbol{\alpha} \cdot (\mathbf{p} - e\mathbf{A}) = (\mathbf{p} - e\mathbf{A})^2 - e\boldsymbol{\sigma}_D \cdot \mathbf{B}$$

$$[\mathcal{O},\xi] = [\boldsymbol{\alpha} \cdot (\mathbf{p} - e\mathbf{A}), e\Phi] = e(\boldsymbol{\alpha} \cdot \mathbf{p}\Phi - \Phi\boldsymbol{\alpha} \cdot \mathbf{p}) = -ie\boldsymbol{\alpha} \cdot \nabla\Phi. \tag{10.78}$$

We need

$$[\mathcal{O},\xi] + i\dot{\mathcal{O}} = -ie\boldsymbol{\alpha} \cdot \nabla\Phi - ie\boldsymbol{\alpha} \cdot \dot{\mathbf{A}} = ie\boldsymbol{\alpha} \cdot \mathbf{E}. \tag{10.79}$$

Then

$$[\mathcal{O},[\mathcal{O},\xi]] + i[\mathcal{O},\dot{\mathcal{O}}] = [\boldsymbol{\alpha} \cdot (\mathbf{p} - e\mathbf{A}), ie\boldsymbol{\alpha} \cdot \mathbf{E}]$$

$$= ie(\mathbf{p} - e\mathbf{A}) \cdot \mathbf{E} - ie\mathbf{E} \cdot (\mathbf{p} - e\mathbf{A})$$

$$\quad - e\boldsymbol{\sigma}_D \cdot [(\mathbf{p} - e\mathbf{A}) \times \mathbf{E} - \mathbf{E} \times (\mathbf{p} - e\mathbf{A})]$$

$$= e\nabla \cdot \mathbf{E} + ie\boldsymbol{\sigma}_D \cdot (\nabla \times \mathbf{E}) + 2e\boldsymbol{\sigma}_D \cdot (\mathbf{E} \times (\mathbf{p} - e\mathbf{A})). \tag{10.80}$$

The Hamiltonian, to order $1/m^2$ inclusive and to first order in the electromagnetic fields and potentials, is

$$H''' = \beta\left(m + \frac{(\mathbf{p} - e\mathbf{A})^2}{2m}\right) + e\Phi - \frac{e}{2m}\beta\boldsymbol{\sigma}_D \cdot \mathbf{B} - \frac{e\nabla \cdot \mathbf{E}}{8m^2}$$

$$\quad - \frac{e}{4m^2}\boldsymbol{\sigma}_D \cdot (\mathbf{E} \times \mathbf{p}) - i\frac{e}{8m^2}\boldsymbol{\sigma}_D \cdot (\nabla \times \mathbf{E}). \tag{10.81}$$

The first two terms are the conventional result, which would be there for a spinless particle. (Actually, we have dropped the purely relativistic term, $-p^4/8m^3$, which we found directly before.) The next term is the magnetic moment for $g = 2$. This is followed by the Darwin term and then the spin-orbit interaction with the Thomas factor of $(g-1)/g$. The final term is a higher order magnetic interaction ($\nabla \times E = -\frac{1}{c}\frac{\partial \mathbf{B}}{\partial t}$).

Of particular interest is the Darwin term.

$$H'_{Darwin} = -\frac{e}{8m^2}\nabla \cdot \mathbf{E} = -\frac{e\pi}{2m^2}\rho(\mathbf{x}) = \frac{Ze^2\pi}{2m^2}\delta(\mathbf{x}), \tag{10.82}$$

evaluated for a point nucleus. In perturbation theory, we have an energy level shift in first order as

$$\Delta E_{Darwin} = \frac{Ze^2\pi}{2m^2}|\psi_{n\ell m}(0)|^2 = \frac{Ze^2\pi}{2m^2}\frac{Z^3}{\pi n^3 a_0^3}\delta_{\ell,0} \tag{10.83}$$

$$= \frac{Z^4\alpha^2}{n^3} \cdot \left(\frac{e^2}{2a_0}\right)\delta_{\ell,0}. \tag{10.84}$$

This is the replacement for the fine-structure splitting for $\ell \neq 0$. We found for $\ell \neq 0$

$$\Delta E_{L \cdot S} = \frac{Z^4 \alpha^2}{n^3} \frac{e^2}{2a_0} \begin{cases} \frac{1}{(\ell+1)(2\ell+1)} & \text{for } j = \ell + 1/2; \\ -\frac{1}{\ell(2\ell+1)} & \text{for } j = \ell - 1/2 \end{cases}. \tag{10.85}$$

We see that the $\langle 1/r^3 \rangle \langle \mathbf{L} \cdot \mathbf{S} \rangle$ term for $\ell \neq 0$ states extrapolates to $\ell = 0$.

10.9 Lorentz Covariance

So far we have talked about relativity and non-relativistic reductions, but we have stayed in one reference frame. We must prove that the Dirac equation satisfies the requirements of special relativity.

Consider the free particle Dirac equation first. It reads

$$i \frac{\partial \psi}{\partial t} = (\boldsymbol{\alpha} \cdot \mathbf{p} + \beta m)\psi = (-i\boldsymbol{\alpha} \cdot \nabla + \beta m)\psi. \tag{10.86}$$

With $\partial_\mu = (\partial_0, \nabla)$, we can define $\gamma^\mu = (\beta, \beta\boldsymbol{\alpha})$ and rewrite the Dirac equation as

$$(i\gamma^\mu \partial_\mu - m)\psi = 0. \tag{10.87}$$

This has the appearance of a covariant equation, but things are a bit subtle. Let a homogeneous proper Lorentz transformation be specified by

$$x'^\mu = a^\mu_\nu x^\nu. \tag{10.88}$$

(See Jackson, *Classical Electrodynamics*, 3rd edition, pp. 543–548.) Covariance of the Dirac equation means that, under such a boost to the primed inertial frame, it reads

$$(i\tilde{\gamma}^\mu \partial'_\mu - m)\psi' = 0, \qquad \text{where } \psi' = \psi'(\mathbf{x}', t'), \tag{10.89}$$

where the $\tilde{\gamma}^\mu$ must obey the same anticommutation rules as the original γ^μ in order to preserve $E^2 - p^2 = m^2$ and the Klein-Gordon equation for the square of the Dirac equation. It is plausible (and can be proven, see R. H. Good, Rev. Mod. Phys, **27**, 187 (1953), Sec. III, p. 1909) that all sets of four such 4×4 matrices are unitarily equivalent. Without loss of generality, we can choose them to be the same. Thus the γ^μ are not components of a four-vector, they are constant, independent of reference frame.

$$\gamma_0 = \gamma^0 = \begin{pmatrix} I & 0 \\ 0 & -I \end{pmatrix}; \qquad \boldsymbol{\gamma} = \begin{pmatrix} 0 & \boldsymbol{\sigma} \\ -\boldsymbol{\sigma} & 0 \end{pmatrix}. \tag{10.90}$$

Thus we have

$$(i\gamma^\mu \partial'_\mu - m)\psi'(\mathbf{x}', t') = 0. \tag{10.91}$$

Now $\partial'_\mu = a_\mu^\nu \partial_\nu$. We write $\psi' = S\psi$, where S is a non-singular 4×4 linear operator that depends on a_μ^ν:

$$i\gamma^\mu a_\mu^\nu \partial_\nu S\psi - mS\psi = 0. \tag{10.92}$$

Operating from the left with S^{-1}, we have

$$S^{-1} i\gamma^\mu a_\mu^\nu \partial_\nu S\psi - m\psi = 0, \tag{10.93}$$

so we must have $S^{-1}\gamma'^\mu a_{\mu'}^\nu S = \gamma^\nu$ Multiplying by a^μ_ν and summing over ν, we have

$$S^{-1}\gamma^{\mu'}S(a^\kappa_\nu a_{\mu'}^\nu) = a^\mu_\nu \gamma^\nu. \tag{10.94}$$

But

$$a^\mu_\nu a^\nu_\lambda = \delta_{\mu\lambda} \quad \text{or} \quad a^\mu_\nu a^\lambda_\mu = \delta_{\nu\lambda}. \tag{10.95}$$

Thus

$$S^{-1}\gamma^\mu S = a^\mu_\nu \gamma^\nu. \tag{10.96}$$

If we write

$$\gamma = \begin{pmatrix} \gamma_0 \\ \gamma_1 \\ \gamma_2 \\ \gamma_3 \end{pmatrix}, \tag{10.97}$$

then $S^{-1}\gamma S = A\gamma$, where for a boost in the x direction (see *Jackson, Classical Electrodynamics*, 3rd edition, Sec. 11.7),

$$A = \begin{pmatrix} \cosh\xi & -\sinh\xi & 0 & 0 \\ -\sinh\xi & \cosh\xi & 0 & 0 \\ 0 & 0 & 1 & 0 \\ 0 & 0 & 0 & 1 \end{pmatrix} \tag{10.98}$$

and

$$\cosh\xi = \gamma; \quad \sinh\xi = \beta\gamma. \tag{10.99}$$

We see that for this choice of direction,

$$S^{-1}\gamma^2 S = \gamma^2; \quad S^{-1}\gamma^3 S = \gamma^3, \tag{10.100}$$

while

$$S^{-1}\gamma^0 S = \cosh\xi\gamma^0 - \sinh\xi\gamma^1$$
$$S^{-1}\gamma^1 S = -\sinh\xi\gamma^0 + \cosh\xi\gamma^1. \tag{10.101}$$

Operate from the left with $\gamma^0 S$ on the first equation and similarly with $\gamma^1 S$ on the second

$$S = \cosh\xi\gamma^0 S\gamma^0 - \sinh\xi\gamma^0 S\gamma^1$$
$$S = \sinh\xi\gamma^1 S\gamma^0 - \cosh\xi\gamma^1 S\gamma^1. \tag{10.102}$$

Comparison of the two forms for S shows that we must have

$$\gamma^0 S\gamma^0 = -\gamma^1 S\gamma^1$$
$$\gamma^1 S\gamma^0 = -\gamma^0 S\gamma^1. \tag{10.103}$$

The second equation shows that $[S, \gamma^0\gamma^1] = 0$ and also that S cannot have terms linear in γ^0 or γ^1. Thus,

$$A = a(\xi) + b(\xi)\gamma^0\gamma^1. \tag{10.104}$$

We now easily find

$$\frac{b}{a} = -\frac{\sinh\xi}{1+\cosh\xi} = -\tanh(\xi/2).$$

(10.105)

Because the inverse Lorentz transformation has $A^{-1}(\xi) = A(-\xi)$, we have

$$S^{-1} = a(\xi) - b(\xi)\gamma^0\gamma^1.$$

(10.106)

Since $S^{-1}S = I = (a - b\gamma^0\gamma^1)(a + b\gamma^0\gamma^1) = a^2 - b^2$, we find

$$a(\xi) = \cosh(\xi/2); \qquad b(\xi) = -\sinh(\xi/2).$$

(10.107)

Thus, the Dirac wave function transformation operator for a boost in the x direction with $\beta = \tanh\xi$ is

$$S(\xi) = I\cosh(\xi/2) - \gamma^0\gamma^1\sinh(\xi/2).$$

(10.108)

The presence of "half angles" is a property of the spin-1/2 nature of the particle described by the Dirac equation.

Rotations are more easily described. A rotation through an angle ω about the x axis is clearly

$$A = \begin{pmatrix} 1 & 0 & 0 & 0 \\ 0 & 1 & 0 & 0 \\ 0 & 0 & \cos\omega & \sin\omega \\ 0 & 0 & -\sin\omega & \cos\omega \end{pmatrix}.$$

(10.109)

A calculation analogous to the one above shows that

$$S = I\sin(\omega/2) - \gamma^2\gamma^3\sin(\omega/2).$$

(10.110)

Note that we are doing "passive" rotations, i.e. the coordinate axes are being rotated, not the states. We can compare the result here with the non-relativistic expression for rotation of a spin-1/2 state, $U^\dagger = \exp(i\omega\sigma_x/2)$. The analogy is $i\sigma_x \leftrightarrow -\gamma^2\gamma^3 = \alpha_2\alpha_3 = i(\boldsymbol{\sigma}_D)_1$. This gives us the appropriate generalization of Pauli spinors.

Now we can obtain a covariant expression for boosts and rotations. We define the second rank tensor

$$\sigma^{\mu\nu} = \frac{i}{2}[\gamma^\mu,\gamma^\nu],$$

(10.111)

which has six independent elements. When μ and ν are both space indices we have

$$\sigma^{jk} = \epsilon_{jk\ell}(\boldsymbol{\sigma}_D)_\ell = i\gamma^j\gamma^k.$$

(10.112)

For $\mu = 0$, and ν a space index, we have $\sigma^{0j} = i\gamma^0\gamma^j$. Thus $-\gamma^0\gamma^1 = i\sigma^{01}$, $-\gamma^2\gamma^3 = i\sigma^{23}$. We can write boosts in the x direction and rotations about the x axis as

$$S_{\text{boost}}(\xi) = I\cosh(\xi/2) + i\sigma^{01}\sinh(\xi/2)$$

$$S_{\text{rotation}}(\omega) = I\cos(\omega/2) + i\sigma^{23}\sin(\omega/2).$$

(10.113)

We can use these results to recover our standard spinors. Suppose we have an electron at rest in the frame K' described by the spinor

$$\psi' = \begin{pmatrix} \chi_j \\ 0 \end{pmatrix}.$$

(10.114)

If we make a transformation to frame K moving with speed βc in the negative x direction, the electron will have energy E and momentum $\mathbf{p} = p\hat{x}$, related to the boost by

$$E = m \cosh \xi; \quad p = m \sinh \xi. \tag{10.115}$$

The wave function in K will be $\psi = S^{-1}\psi'$, where

$$S^{-1} = I \cosh(\xi/2) + \gamma^0\gamma^1 \sinh(\xi/2) = I \cosh(\xi/2) + \alpha_x \sinh(\xi/2). \tag{10.116}$$

With $1 + \cosh \xi = 2 \cosh^2(\xi/2)$, we have $\cosh(\xi/2) = \sqrt{(E + m)/(2m)}$, so

$$S^{-1} = \sqrt{\frac{E+m}{2m}} \begin{pmatrix} I & \frac{\sigma_x p}{E+m} \\ \frac{\sigma_x p}{E+m} & I \end{pmatrix}. \tag{10.117}$$

The wave function in K, describing an electron with momentum $\mathbf{p}$ and energy $E = \sqrt{p^2 + m^2}$, is evidently

$$\psi_{\mathbf{p}} = \sqrt{\frac{E+m}{2m}} \begin{pmatrix} \chi_j \\ \frac{\sigma \cdot \mathbf{p}}{E+m}\chi_j \end{pmatrix}, \tag{10.118}$$

where χ_j is the Pauli spinor in the rest frame of the electron. The normalization is the invariant one, with $\psi^{\dagger}\gamma^0\psi$ invariant. This follows from $S^{-1} = \gamma^0 S^{\dagger}\gamma^0$.

10.10 Discrete Symmetries

Lorentz invariance is the most sacred of the symmetries we expect to be respected by our physical theories and, more important, by experiment. Lorentz invariance, together with translation and rotation invariance, are continuous symmetries: There are arbitrarily small boosts, translations, and rotations. The parity operation, reversing all spatial coordinates, is a discrete action. Two other discrete operations are of particular interest: Charge conjugation and time reversal. Time reversal invariance has a classical realization. It is conventional to say that classical physics respects time reversal invariance. A video of someone diving into a swimming pool, run backwards, shows a possible event, though a very improbable one. The assertion is based on the assumption that the interactions governing classical mechanics are symmetric under time reversal, as is generally true. Charge conjugation invariance is the assertion that if the charges of all particles are reversed, the motion will still be the same. Implicit is that for every particle there is a corresponding antiparticle, which for neutral particles may or may not be the particle itself.

Whether parity, charge conjugation, and time reversal are indeed good symmetries are experimental questions. Parity is a symmetry of the electromagnetic interaction and parity symmetry was revealed through the patterns of radiative transitions. The existence of the positron, the charge conjugate of the electron, was predicted as a consequence of the negative energy solutions to Dirac's equation. Five years after Dirac's inspiration, Carl Anderson discovered the positron in cosmic rays. These three operations, P, C, and T, can be represented as quantum mechanical operations on states or wave-functions, whether or not the symmetries are respected by nature. For electromagnetism, where the equations are known with confidence, the answer is that these are indeed symmetries. In 1956, T. D. Lee and C. N. Yang noted that there was no evidence that parity was a good symmetry for weak interactions and proposed tests to decide the matter. Early in 1957, results from Co^{60} decay in an experiment led by Chien-Shiung Wu and from muon decay showed that parity was indeed very much violated in weak interactions. The latter experiment showed that both P and C were violated.

While all theoretical assertions are ultimately subject to experimental scrutiny, some predictions seem particularly invulnerable. The spin and statistics theorem, which requires that, under identical particle interchange, bosons are symmetric and fermions are antisymmetric, is such a prediction. At nearly the same level of certainty is the demonstration, initially due to Pauli and Luders in 1954, that the combination CPT is always a good symmetry.

The discovery that both P and C are violated in weak interactions left open the possibility that their combination, CP, is a good symmetry. In 1964, an experiment with neutral kaons led by Val Fitch and James Cronin showed that CP was violated, though at a small level. The importance of this result increased when Andrei Sakharov pointed out in 1967 that CP violation is a necessary ingredient in any explanation of the excess of baryons over antibaryons in the universe.

10.10.1 Parity in the Dirac Equation

Because of the four-component nature of Dirac spinors, parity is slightly more complicated than for Pauli-Schrödinger states. We can see the problem by looking at the plane-wave states.

$$\text{NR}: \quad \psi_{\mathbf{p}}(\mathbf{x}) = e^{i\mathbf{p}\cdot\mathbf{x}}$$

$$P\psi_{\mathbf{p}}(\mathbf{x}) = \psi_{\mathbf{p}}(-\mathbf{x}) = e^{-i\mathbf{p}\cdot\mathbf{x}} = \psi_{-\mathbf{p}}(\mathbf{x})$$

$$\text{Dirac}: \quad \psi_{\mathbf{p},s}(\mathbf{x}) = N\begin{pmatrix} \phi_s \\ \frac{\sigma\cdot\mathbf{p}}{E+m}\phi_s \end{pmatrix} e^{i\mathbf{p}\cdot\mathbf{x}}$$

$$\psi_{\mathbf{p},s}(-\mathbf{x}) = N\begin{pmatrix} \phi_s \\ \frac{\sigma\cdot\mathbf{p}}{E+m}\phi_s \end{pmatrix} e^{-i\mathbf{p}\cdot\mathbf{x}} \neq \psi_{-\mathbf{p},s}(\mathbf{x}). \tag{10.119}$$

The "small" components don't change sign, as is needed to get $\psi_{-\mathbf{p},s}(\mathbf{x})$. In fact, the presence of $\sigma\cdot\mathbf{p}$ (or $-i\sigma\cdot\nabla$, in general) is a "flag." This is a pseudoscalar operator. Thus, whatever parity the "large" components have, the small components have the opposite parity.

The operator β will change the sign of the small components. We thus make the hypothesis that parity, P, acts as follows:

$$\psi_P(\mathbf{x}) = P\psi(\mathbf{x}) = \beta\psi(-\mathbf{x}). \tag{10.120}$$

To see whether this hypothesis make general sense, consider the Dirac equation

$$i\frac{\partial\psi}{\partial t} = (\alpha\cdot\mathbf{p} + \beta m + H_{\text{int}})\psi. \tag{10.121}$$

Then

$$i\frac{\partial\psi_P}{\partial t} = P(\frac{1}{i}\alpha\cdot\nabla + \beta m + H_{\text{int}})P^{-1}\psi_P. \tag{10.122}$$

Since $P\nabla P^{-1} = -\nabla$, the β is welcome to anticommute with α. Thus,

$$P(\alpha\cdot\frac{1}{i}\nabla + \beta m)P^{-1} = (\alpha\cdot\frac{1}{i}\nabla + \beta m). \tag{10.123}$$

Now

$$PH_{\text{int}}(\mathbf{x},t)P^{-1} = \beta H_{\text{int}}(-\mathbf{x},t)\beta. \tag{10.124}$$

Hence, if the interaction is spherically symmetric and does not involve odd powers of $\boldsymbol{\alpha}$, e.g. the Coulomb potential, the energy eigenstates are, or can be chosen to be, eigenstates of parity. Also, with a uniform magnetic field, where $\mathbf{A}(\mathbf{x}) = \frac{1}{2}(\mathbf{B} \times \mathbf{r})$, we have

$$PH_{\text{int}}(\mathbf{x}, t)P^{-1} = H_{\text{int}}(-\mathbf{x}, t) \tag{10.125}$$

and parity can be a valid concept, even though there is an odd power of $\boldsymbol{\alpha}$.

10.10.2 Charge Conjugation Invariance

Dirac hole theory is a language to deal with what is really a many-particle theory with creation and destruction of particles. If an energy greater than $2mc^2$ is available, a one-electron initial state can be transformed into two electrons and a positron. This process is called trident production for its appearance in a photographic emulsion. The process is $e^- + (Z, A) \rightarrow e^- + (Z, A) + e^+e^-$. The negative energy sea, the holes in which act as positrons, is a clumsy way of dealing with the negative energy states. A better way is to exploit the negative energy solution - positive energy positron correspondence by means of a formalism called charge conjugation.

We have for Dirac particles

$$(i\gamma^\mu \partial_\mu - e\gamma^\mu A_\mu - m)\psi = 0. \tag{10.126}$$

We want the negative energy solutions, ψ_-, of this equation to correspond to positive energy solutions for a particle of opposite charge. We denote this wave function as ψ_+^c, where the superscript "c" stands for "charge conjugate," that is, we want ψ_+^c to satisfy

$$(i\gamma^\mu \partial_\mu + e\gamma^\mu A_\mu - m)\psi_+^c = 0. \tag{10.127}$$

To find the connection between ψ^c and ψ, we first note that the sign of the interaction term can be changed relative to that of the kinetic energy by complex conjugation. Thus, we postulate

$$\psi^c = CK\psi \equiv C\psi^*, \tag{10.128}$$

where K takes the complex conjugate of everything to its right. Then we have

$$CK(i\gamma^\mu \partial_\mu - e\gamma^\mu A_\mu - m)C^{-1}KCK\psi = 0$$
$$C(-i\gamma^{\mu*} \partial_\mu - e\gamma^{\mu*} A_\mu - m)C^{-1}\psi^c = 0. \tag{10.129}$$

We thus want

$$C\gamma^{\mu*}C^{-1} = -\gamma^\mu. \tag{10.130}$$

Then we will have

$$(i\gamma^\mu \partial_\mu + e\gamma^\mu A_\mu - m)\psi^c = 0, \tag{10.131}$$

as befits a positron.

We recall that with our choice of γs, Eq. (10.90), all the γs are real except γ^2, which is purely imaginary. Thus we have

$$\gamma^{\mu*} = \gamma^\mu \qquad \mu = 0, 1, 3$$
$$\gamma^{\mu*} = -\gamma^\mu \qquad \mu = 2. \tag{10.132}$$

Evidently, we want

$$C\gamma^\mu C^{-1} = -\gamma^\mu \qquad \mu = 0, 1, 3$$
$$C\gamma^\mu C^{-1} = \gamma^\mu \qquad \mu = 2. \tag{10.133}$$

Clearly $C = \gamma^2$ fills the bill. There is the possibility of an arbitrary phase factor. Since γ^2 is purely imaginary, it is customary to define

$$C = i\gamma^2 = \begin{pmatrix} 0 & 0 & 0 & 1 \\ 0 & 0 & -1 & 0 \\ 0 & -1 & 0 & 0 \\ 1 & 0 & 0 & 0 \end{pmatrix}. \tag{10.134}$$

This antidiagonal form is just the right thing to turn negative energy states into positive and vice versa. For example,

$$e^{i\mathbf{p}\cdot\mathbf{x}-iEt}u_1(\mathbf{p}) = e^{i\mathbf{p}\cdot\mathbf{x}-iEt}\sqrt{E+m}\begin{pmatrix} \begin{pmatrix} 1 \\ 0 \end{pmatrix} \\ \dfrac{\boldsymbol{\sigma}\cdot\mathbf{p}}{E+m}\begin{pmatrix} 1 \\ 0 \end{pmatrix} \end{pmatrix} = e^{i\mathbf{p}\cdot\mathbf{x}-Et}\sqrt{E+m}\begin{pmatrix} 1 \\ 0 \\ \dfrac{p_z}{E+m} \\ \dfrac{p_x+ip_y}{E+m} \end{pmatrix}. \tag{10.135}$$

Now we calculate the corresponding spinor under charge conjugation:

$$Cu_1^*(\mathbf{p}) = \sqrt{E+m}\begin{pmatrix} \dfrac{p_x-ip_y}{E+m} \\ -\dfrac{p_z}{E+m} \\ 0 \\ 1 \end{pmatrix}$$

$$= \sqrt{E+m}\begin{pmatrix} \dfrac{\boldsymbol{\sigma}\cdot\mathbf{p}}{E+m}\begin{pmatrix} 0 \\ 1 \end{pmatrix} \\ \begin{pmatrix} 0 \\ 1 \end{pmatrix} \end{pmatrix} = v_1(\mathbf{p}) = \sqrt{2E}w_4(-\mathbf{p}). \tag{10.136}$$

Similarly, we find

$$Cu_2^*(\mathbf{p}) = -\sqrt{E+m}\begin{pmatrix} \dfrac{\boldsymbol{\sigma}\cdot\mathbf{p}}{E+m}\begin{pmatrix} 0 \\ 1 \end{pmatrix} \\ \begin{pmatrix} 0 \\ 1 \end{pmatrix} \end{pmatrix} = -v_2(\mathbf{p}) = -\sqrt{2E}w_3(-\mathbf{p}). \tag{10.137}$$

It is straightforward to show that

$$CKCK\psi = CK\psi^c = +\psi, \tag{10.138}$$

that is,

$$i\gamma^2 K i\gamma^2 K = (i\gamma^2)^2 = -\gamma^2\gamma^2 = +1. \tag{10.139}$$

10.10.3 Time Reversal

The equations of motion in classical and quantum mechanics permit time-reversed solutions in the absence of damping forces (no complex potential in the Schrödinger equation). Coordinates are not reversed if the sign of time is changed, but momenta and angular momenta are. As can easily be seen by reversing time ($t \to -t$) in the Schrödinger equation, time reversal must involve complex conjugation of the wave function, among other things. This is made quite explicit by considering the rotation operator for spatial degrees of freedom, $e^{-i\theta \cdot \mathbf{L}}$. If T represents time reversal, then $Te^{-i\theta \cdot \mathbf{L}} T = e^{-i\theta \cdot \mathbf{L}}$ because time reversal has no effect on rotations. On the other hand, $T\mathbf{L}T = -\mathbf{L}$ since time reversal inverts $\mathbf{p}$ but not $\mathbf{r}$. The conclusion is that $TiT = -i$. The transformation $i \to -i$ is clearly not a linear transformation. More generally, T must be realized as an antilinear transformation, as first discovered by Wigner in 1932. It is conventional to write $T = UK$, where again K stands for complex conjugation. For a spinless state, we have $\Psi_T(\mathbf{r}, t) = T\Psi = \Psi^*(\mathbf{r}, -t)$. Thus $e^{i\mathbf{p}\cdot\mathbf{r}-iEt} \to e^{-i\mathbf{p}\cdot\mathbf{r}-iEt}$ and a plane-wave of moment $\mathbf{p}$ becomes a plane-wave of momentum $-\mathbf{p}$.

Now include spin, but in a non-relativistic context. We need to pay attention to the representation of the spin. For spin-1/2, the convention is to use Pauli matrices, with σ_1 and σ_3 real, but σ_2 pure imaginary. So we can take $T = K\sigma_2$. Now

$$T^2 = \sigma_2 K\sigma_2 K = \sigma_2(-\sigma_2) = -1, \tag{10.140}$$

so that $T^{-1} = -T$. We can confirm that this choice of T works.

$$T\sigma_1 T^{-1} = K\sigma_2\sigma_1[-K\sigma_2] = [-\sigma_2\sigma_1][-K^2\sigma_2] = \sigma_2\sigma_1\sigma_2 = -\sigma_1$$

$$T\sigma_3 T^{-1} = K\sigma_2\sigma_3[-K\sigma_2] = [-\sigma_2\sigma_3][-K^2\sigma_2] = \sigma_2\sigma_3\sigma_2 = -\sigma_3$$

$$T\sigma_2 T^{-1} = K\sigma_2\sigma_2[-K\sigma_2] = [\sigma_2\sigma_2][-K^2\sigma_2] = -\sigma_2\sigma_2\sigma_2 = -\sigma_2. \tag{10.141}$$

So, indeed, $T\sigma T^{-1} = -\sigma$ as required by analogy to $T\mathbf{L}T = -\mathbf{L}$. We are free to multiply T by a phase, and the choice $-i\sigma_2 K$ makes the matrix factor real.

Since we can build up any spin, integral or half-integral, from spin-1/2, it follows that $T^2 = (-1)^{2J}$. For the Dirac equation, the analogous argument leads to $T = -i\sigma_{D_2} K = \alpha_1\alpha_3 K = \gamma^3\gamma^1 K$, though again the overall phase can be chosen freely.

10.11 Bilinear Covariants

The general 4×4 matrix has 16 elements. There are thus 16 independent matrices. We have examined some of them:

$$I; \gamma^0, \gamma^1, \gamma^2, \gamma^3; \sigma^{\mu\nu}(\text{six of these}). \tag{10.142}$$

To these 11, five more must be added. Traditionally, the choices are $\gamma^5 = i\gamma^0\gamma^1\gamma^2\gamma^3$ and $\gamma^5\gamma^\mu$. These can be organized into groups:

Name	Matrices	Number in group
Scalar	I	1
Vector	γ^μ	4
Tensor	$\sigma^{\mu\nu}$	6
Axial vector	$\gamma^5\gamma^\mu$	4
Pseudoscalar	γ^5	1

The name and numbers are suggestive of the Lorentz transformation properties of the terms that can be constructed by sandwiching them between Dirac spinors.

(a) Scalar: We define $\psi^\dagger \beta = \overline{\psi} = (\beta\psi)^\dagger$. Consider a Lorentz transformation corresponding to S, with $\psi' = S\psi$:

$$\overline{\psi}' \, \psi' = (\psi')^\dagger \beta S \psi = (S\psi)^\dagger \beta S \psi = \psi^\dagger S^\dagger \beta S \psi = \overline{\psi}(\beta S^\dagger \beta S)\psi. \tag{10.143}$$

Examination of the explicit forms for S for boosts and rotations, and recognizing that $\gamma^{0\dagger} = \gamma^0; \gamma^{i\dagger} = -\gamma^i$, shows that

$$\beta S^\dagger \beta = S^{-1}, \tag{10.144}$$

so that

$$\overline{\psi}' \, \psi' = \overline{\psi}\psi \tag{10.145}$$

and I gives a scalar when sandwiched between $\overline{\psi}$ and ψ.

(b) Consider

$$\overline{\psi}' \, \gamma^\mu \psi' = (S\psi)^\dagger \beta \gamma^\mu S \psi = \overline{\psi}(S^{-1}\gamma^\mu S)\psi. \tag{10.146}$$

But S was defined to have the property

$$S^{-1}\gamma^\mu S = a^\mu_{\ \nu}\gamma^\nu \tag{10.147}$$

so that

$$\overline{\psi}' \, \gamma^\mu \psi' = a^\mu_{\ \nu}(\overline{\psi}\gamma^\nu \psi), \tag{10.148}$$

so this bilinear indeed transforms as a four-vector.

(c) Consider next the tensor bilinear

$$\overline{\psi}' \, \sigma^{\mu\nu}\psi' = \frac{i}{2}\overline{\psi}' \, [\gamma^\mu, \gamma^\nu]\psi' = \frac{i}{2}\overline{\psi}S^{-1}[\gamma^\mu, \gamma^\nu]S\psi$$

$$= \frac{i}{2}\overline{\psi}[S^{-1}\gamma^\mu S, S^{-1}\gamma^\nu S]\psi = \frac{i}{2}a^\mu_{\ \lambda}a^\nu_{\ \lambda'}\overline{\psi}[\gamma^\lambda, \gamma^{\lambda'}]\psi$$

$$= a^\mu_{\ \lambda}a^\nu_{\ \lambda'}\overline{\psi}\sigma^{\lambda\lambda'}\psi. \tag{10.149}$$

This confirms its character as a second-rank tensor.

(d,e) The axial vector and pseudoscalar bilinears are the same as (b) and (a) above, because γ^5 commutes with S, except for the improper Lorentz transformations, i.e. parity, we get opposite behavior because there is an extra anticommutation of β with γ^5.

10.12 Applications to Electromagnetic Form Factors

One can probe the electromagnetic structure of the proton by elastic scattering of electrons at GeV energies. See Figure 10.2. We are interested in the matrix elements of the electromagnetic current operator between proton states, i.e. $\langle \mathbf{p}' | j^\mu_{\text{op}} | \mathbf{p} \rangle$.

Figure 10.2   The electromagnetic form factors of the proton can be measured by elastic scattering of electrons on the proton. Here, an electron of momentum k is scattered by a proton of momentum p. The outgoing momenta are k' and p'.

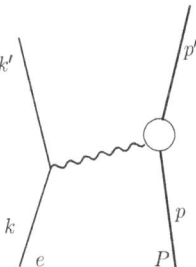

We can write the matrix element in terms of Dirac spinors

$$\langle \mathbf{p}' \left| j^\mu_{\text{op}} \right| \mathbf{p} \rangle = \overline{w}(\mathbf{p}') J^\mu w(\mathbf{p}). \tag{10.150}$$

We proceed to write out all the combinations of γs and momenta that can leave us with the single index μ, excluding parity violating terms with γ^5:

$$J^\mu = a_1(p + p')^\mu + a_2(p' - p)^\mu + a_3\gamma^\mu + a_4\sigma^{\mu\nu}(p' + p)_\nu + a_5\sigma^{\mu\nu}(p' - p)_\nu, \tag{10.151}$$

where the a_i are Lorentz invariant functions of the kinematic variables, i.e. $p^2, p'^2, p \cdot p'$ or alternatively p^2, p'^2, q^2 where $q = (p' - p)$.

We now use the equations of motion and current conservation to see what terms are actually permitted. From

$$(i\gamma^\mu \partial_\mu - m)\psi = 0; \quad w(\mathbf{p}) \propto e^{-i\mathbf{p} \cdot \mathbf{x}}, \tag{10.152}$$

we see that, with $\gamma^\mu p_\mu = \not{p}$,

$$(\not{p} - m)w = 0 \qquad \text{for } E > 0, \tag{10.153}$$

and recalling that $\gamma^{\mu\dagger} = \gamma^0 \gamma^\mu \gamma^0$,

$$\overline{w}(\not{p} - m) = 0. \tag{10.154}$$

Current conservation $\partial_\mu J^\mu_{\text{op}} = 0$ means that in our momentum description, $q_\mu J^\mu = 0$. We use this to constrain the possible form for J^μ:

$$\begin{aligned}
q_\mu J^\mu &= a_1(p' - p) \cdot (p' + p) + a_2(p' - p) \cdot (p' - p) + a_3(\not{p}' - \not{p}) \\
&\quad + a_4\sigma^{\mu\nu}(p' - p)_\mu(p' + p)_\nu + a_5\sigma^{\mu\nu}(p' - p)_\mu(p' + p)_\nu \\
&= a_1(m^2 - m^2) + a_2 q^2 + a_3(m - m) + a_4\sigma^{\mu\nu}('p' - p)_\mu(p' + p)_\nu + a_5\sigma^{\mu\nu}q_\mu q_\nu.
\end{aligned} \tag{10.155}$$

We see that the a_1, a_3, and a_5 terms satisfy current conservation while the others do not. We are left with

$$J^\mu = a_1(p' + p)^\mu + a_3\gamma^\mu + a_5\sigma^{\mu\nu}(p' - p)_\nu. \tag{10.156}$$

Even these are not three separate forms, as shown in Problem 10.1, where the Dirac current is written in terms of a Schrödinger-like current

$$j^\mu = \overline{\psi}\gamma^\mu\psi = j^\mu_{\text{Schr}} + j^\mu_{\text{spin}}, \tag{10.157}$$

where

$$j^\mu_{\text{Schr}} = \frac{i}{2m}[\bar\psi\partial^\mu\psi - (\partial^\mu\bar\psi)\psi]; \qquad j^\mu_{\text{spin}} = \frac{e}{2m}\partial_\nu(\bar\psi\sigma^{\mu\nu}\psi). \tag{10.158}$$

If we write $\psi = e^{-ip\cdot x}w(\mathbf{p}) = e^{i\mathbf{p}\cdot\mathbf{x}-iEt}w(\mathbf{p})$ and $\bar\psi = e^{ip'\cdot x}\bar w(\mathbf{p}')$, we can express the linear relationship between three terms as

$$\bar w(\mathbf{p}')\gamma^\mu w(\mathbf{p}) = \frac{1}{2m}\bar w(\mathbf{p}')(p' + p)^\mu w(\mathbf{p}) + \frac{i}{2m}\bar w(\mathbf{p}')\sigma^{\mu\nu}(p' - p)_\nu w(\mathbf{p}), \tag{10.159}$$

an expression known as the Gordon decomposition. It follows that the most general expression for the electromagnetic current for any spin-1/2 particle is

$$\bar w(\mathbf{p}')J^\mu w(\mathbf{p}) = e\bar w(\mathbf{p}')\left[F_1(q^2)\gamma^\mu + \frac{i}{2m}\sigma^{\mu\nu}(p' - p)_\nu F_2(q^2)\right]w(\mathbf{p}), \tag{10.160}$$

where F_1 and F_2 are called the Dirac form factors. The particle charge Q and magnetic moment μ in units of e and $e\hbar/(2mc)$, respectively, are in the $q^2 = 0$ limit

$$Q = F_1(0); \qquad \mu = F_1(0) + F_2(0). \tag{10.161}$$

Away from $q^2 = 0$, the form factors will vary unless the particle is truly elementary, that is, has no internal structure. Often, linear combinations of F_1 and F_2 are used instead, because they have a more physical interpretation. We define

$$G_E(q^2) = F_1(q^2) + \frac{q^2}{4m^2}F_2(q^2)$$

$$G_M(q^2) = F_1(q^2) + F_2(q^2). \tag{10.162}$$

The subscripts E and M reflect the conditions $G_E(0) = 1$, $G_M(0) = \mu$ and G_E and G_M are referred to as the charge and magnetic form factors.

The meaning of the charge form factor is apparent in the "brick-wall" reference frame, where the incident nucleon has three-momentum $\mathbf{p} = \mathbf{P}$ and the electromagnetic current carries momentum $\mathbf{q} = -2\mathbf{P}$ so $\mathbf{p}' = -\mathbf{P}$. See Figure 10.3. Using the Gordon decomposition, we can rewrite the current as

$$\bar w(\mathbf{p}')J^\mu w(\mathbf{p}) = e\bar w(\mathbf{p}')\left[F_1\frac{(p + p')^\mu}{2m} + i\frac{F_1 + F_2}{2m}\sigma^{\mu\nu}q_\nu\right]w(\mathbf{p}) \tag{10.163}$$

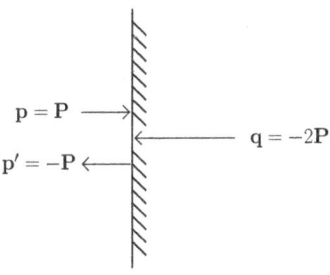

Figure 10.3 In the brickwall frame, the three-momentum of the incident particle is exactly reversed by the collision, as if it hit brickwall. These specific kinematics can allow for a clear physical interpretation. The incident momentum **P** is reversed by the collision, with momentum transfer **q** = −2**P**.

and consider the charge density J^0:

$$\overline{w}(\mathbf{p}')J^0 w(\mathbf{p}) = e\overline{w}(\mathbf{p}')\left[\frac{2E}{2m}F_1 + i\frac{F_1 + F_2}{2m}\sigma^{0j}(2p_j)\right]w(\mathbf{p})$$

$$= \frac{eE}{m}\overline{w}(\mathbf{p}')\left[F_1 - \frac{F_1 + F_2}{E}\boldsymbol{\alpha}\cdot\mathbf{p}\right]w(\mathbf{p})$$

$$= \frac{eE}{m}\overline{w}(\mathbf{p}')\left[F_1 - (F_1 + F_2)\left(1 - \frac{m}{E}\beta\right)\right]w(\mathbf{p}). \tag{10.164}$$

In the brick-wall frame, we have the special relations

$$(\boldsymbol{\alpha}\cdot\mathbf{p} + \beta m)w(\mathbf{p}) = Ew(\mathbf{p})$$

$$(-\boldsymbol{\alpha}\cdot\mathbf{p} + \beta m)w(\mathbf{p}') = Ew(\mathbf{p})$$

$$w(\mathbf{p}')^\dagger(-\boldsymbol{\alpha}\cdot\mathbf{p} + \beta m)w(\mathbf{p}) = Ew(\mathbf{p}')^\dagger w(\mathbf{p})$$

$$w(\mathbf{p}')^\dagger(-E + 2\beta m)w(\mathbf{p}) = Ew(\mathbf{p}')^\dagger w(\mathbf{p})$$

$$w(\mathbf{p}')^\dagger\beta m w(\mathbf{p}) = Ew(\mathbf{p}')^\dagger w(\mathbf{p})$$

$$\overline{w}(\mathbf{p}')m w(\mathbf{p}) = E\overline{w}(\mathbf{p}')\beta w(\mathbf{p}). \tag{10.165}$$

In the brick-wall frame, we have

$$q^2 = -4|\mathbf{p}|^2; \qquad \left(1 - \frac{m^2}{E^2}\right) = -\frac{q^2/4m^2}{1 - q^2/4m^2}. \tag{10.166}$$

So

$$\overline{w}(\mathbf{p}')J^0 w(\mathbf{p}) = \frac{eE}{m}\frac{m^2}{E^2}\overline{w}(\mathbf{p}')\left[F_1\left(1 - \frac{q^2}{4m^2}\right) + \frac{q^2}{4m^2}(F_1 + F_2)\right]w(\mathbf{p})$$

$$= \frac{em}{E}\overline{w}(\mathbf{p}')\left[F_1 + \frac{q^2}{4m^2}F_2\right]w(\mathbf{p}) = \frac{em}{E}\overline{w}(\mathbf{p}')G_E(q^2)w(\mathbf{p}). \tag{10.167}$$

Thus, we can say that the electric form factor, $G_E(q^2)$, is the Fourier transform of the charge density in the brick-wall frame.

10.13 Potential Scattering of a Dirac Particle

As an interlude to show that the Dirac equation, as well as being beautiful, has practical consequences, we consider the scattering of a Dirac particle by a potential operator $V(\mathbf{x}; \boldsymbol{\alpha}, \beta)$. For a fixed energy E, we have

$$E\psi = (\frac{1}{i}\boldsymbol{\alpha}\cdot\nabla + \beta m + V)\psi$$

$$(E + i\boldsymbol{\alpha}\cdot\nabla - \beta m) = V\psi. \tag{10.168}$$

Operating from the left with $(E - i\boldsymbol{\alpha}\cdot\nabla + \beta m)$, we have

$$[(E^2 - m^2) + \nabla^2]\psi = (p^2 + \nabla^2)\psi = (E - i\boldsymbol{\alpha}\cdot\nabla + \beta m)V\psi. \tag{10.169}$$

This is just like the Schrödinger equation, Eq. (7.51), but with a more complicated potential operator. The scattering amplitude for $(j, \mathbf{p}) \rightarrow (j', \mathbf{p}')$, where j, j' represent the spins, is

$$\langle j'\mathbf{p}' \,|f|\, j\mathbf{p} \rangle = -\frac{1}{4\pi} w_{j'}(\mathbf{p}')^\dagger \int d^3x \, e^{-i\mathbf{p}'\cdot\mathbf{x}}(E - i\boldsymbol{\alpha}\cdot\nabla + \beta m)V\psi_{j\mathbf{p}}^{(+)}(\mathbf{x}), \qquad (10.170)$$

where

$$\psi_{j\mathbf{p}}^{(+)}(\mathbf{x}) \rightarrow w_j(\mathbf{p})e^{i\mathbf{p}\cdot\mathbf{x}} + \frac{e^{ikr}}{r}f(...) \qquad (10.171)$$

at large distances. This is the analog of Eq. (7.57).

The operator $(E - i\boldsymbol{\alpha}\cdot\nabla + \beta m)$ can act to the left on the plane wave, giving a factor of $2E$ for elastic scattering and a positive energy spinor. Thus we have,

$$\langle j'\mathbf{p}' \,|f|\, j\mathbf{p} \rangle = -\frac{2E}{4\pi} w_{j'}(\mathbf{p}')^\dagger \int d^3x \, e^{-i\mathbf{p}'\cdot\mathbf{x}}V\psi_{j,\mathbf{p}}^{(+)}(\mathbf{x}). \qquad (10.172)$$

In Born approximation, $\psi_{j,\mathbf{p}}^{(+)}(\mathbf{x}) \rightarrow w_j(\mathbf{p})e^{i\mathbf{p}\cdot\mathbf{x}}$ and we get

$$f_{\mathrm{op}}^{(1)} = -\frac{2E}{4\pi} \int d^3x \, e^{-i\mathbf{p}'\cdot\mathbf{x}}V_{\mathrm{op}}e^{i\mathbf{p}\cdot\mathbf{x}}, \qquad (10.173)$$

where

$$\langle j'\mathbf{p}' \,\big|f^{(1)}\big|\, j\mathbf{p} \rangle = w_{j'}^\dagger(\mathbf{p}')f_{\mathrm{op}}^{(1)}w_j(\mathbf{p}). \qquad (10.174)$$

Now V is a matrix in the space of Dirac spinors, allowing for a variety of potential forms. If we want the scattering cross section for an unpolarized beam, and if the outgoing spins are not observed, we have

$$\frac{d\sigma}{d\Omega} = \frac{1}{2}\sum_{j=1}^{2}\sum_{j'=1}^{2} |\langle j'\mathbf{p}' \,|f|\, j\mathbf{p} \rangle|^2 = \frac{1}{2}\sum_{j=1}^{2}\sum_{j'=1}^{2} \left\langle j\mathbf{p} \,\big|f_{\mathrm{op}}^\dagger\big|\, j'\mathbf{p}' \right\rangle\left\langle j'\mathbf{p}' \,\big|f_{\mathrm{op}}\big|\, j\mathbf{p} \right\rangle. \qquad (10.175)$$

The spin sums are over the two states with momentum $\mathbf{p}$ or $\mathbf{p}'$ and positive energy. The projection operator for positive energy, obtained in Problem 10.2, is

$$\Lambda_+(\mathbf{p}) = \frac{1}{2}\left[1 + \frac{\boldsymbol{\alpha}\cdot\mathbf{p} + \beta m}{E(\mathbf{p})}\right]. \qquad (10.176)$$

This leaves us a trace over four-dimensional matrices:

$$\frac{d\sigma}{d\Omega} = \frac{1}{2}\mathrm{Tr}\left[f_{\mathrm{op}}^\dagger\Lambda_+(\mathbf{p}')f_{\mathrm{op}}\Lambda_+(\mathbf{p})\right]. \qquad (10.177)$$

Suppose that $V = e\Phi$, where Φ is an external scalar potential. Then the Dirac trace above is

$$\frac{1}{2}\mathrm{Tr}\left[\Lambda_+(\mathbf{p}')\Lambda_+(\mathbf{p})\right] = \frac{1}{8}\mathrm{Tr}\left[(1 + \frac{\boldsymbol{\alpha}\cdot\mathbf{p}' + \beta m}{E'})(1 + \frac{\boldsymbol{\alpha}\cdot\mathbf{p} + \beta m}{E})\right]$$

$$= \frac{1}{8EE'}\mathrm{Tr}\left[EE' + \boldsymbol{\alpha}\cdot\mathbf{p}\boldsymbol{\alpha}\cdot\mathbf{p}' + \beta\beta m^2\right], \qquad (10.178)$$

where we observed that the trace of a single α or β vanishes, as does the trace of $\beta\alpha_j$, while $\mathrm{Tr}\,\alpha_i\alpha_j = 4\delta_{ij}$ and $\mathrm{Tr}\,\beta\beta = 4$. Thus,

$$\frac{1}{2}\mathrm{Tr}\left[\Lambda_+(\mathbf{p}')\Lambda_+(\mathbf{p})\right] = \frac{1}{2EE'}[EE' + \mathbf{p}\cdot\mathbf{p}' + m^2]. \qquad (10.179)$$

For elastic scattering then,

$$\frac{1}{2}\text{Tr}\,[\Lambda_+(\mathbf{p}')\Lambda_+(\mathbf{p})] = \frac{1}{2}(1 + 1 - \beta^2 + \beta^2\cos\theta) = \frac{2 - \beta^2(1 - \cos\theta)}{2} = 1 - \beta^2\sin^2\frac{\theta}{2} \tag{10.180}$$

Comparing Eq. (10.172) with the analogous expression in non-relativistic scattering, we see that we have an extra factor of $(E/m)w(\mathbf{p}')^\dagger w(\mathbf{p})$ here. We thus conclude that

$$\left(\frac{d\sigma}{d\Omega}\right)^{\text{Born}}_{\text{Dirac}} = \left(1 - \beta^2\sin^2\frac{\theta}{2}\right)\left(\frac{E}{m}\right)^2\left(\frac{d\sigma}{d\Omega}\right)^{\text{Born}}_{\text{NR}}, \tag{10.181}$$

where the non-relativistic cross section here is to be expressed in terms of $\mathbf{p}$ and $\mathbf{q} = \mathbf{p}' - \mathbf{p}$.

Of particular interest is Coulomb scattering. With $V = z_1 z_2 e^2/q^2$, we find

$$V(q^2) = \int d^3x\, e^{i\mathbf{q}\cdot\mathbf{x}} V(r) = \frac{4\pi z_1 z_2 e^2}{q^2}. \tag{10.182}$$

The relativistic Rutherford cross section, where the particles are spinless, is obtained by including the factor E/m from Eq. (10.172), but not the trace factor:

$$\left(\frac{E}{m}\right)^2\left(\frac{d\sigma}{d\Omega}\right)^{\text{Born}}_{\text{NR}} = \left|-\frac{E}{2\pi}V(q^2)\right|^2 = \frac{(z_1 z_2 e^2 E)^2}{4p^4\sin^4\frac{\theta}{2}}$$

$$= \left(\frac{z_1 z_2 e^2}{2p\upsilon}\right)^2\frac{1}{\sin^4\frac{\theta}{2}}. \tag{10.183}$$

Thus we have,

$$\left(\frac{d\sigma}{d\Omega}\right)^{\text{Born}}_{\text{Dirac}} = \left(\frac{d\sigma}{d\Omega}\right)^{\text{Born}}_{\text{Relativistic Rutherford}}\left(1 - \beta^2\sin^2\frac{\theta}{2}\right). \tag{10.184}$$

The exact Dirac Coulomb scattering cannot be written down in closed form - only as a partial-wave expansion. The second Born approximation can be determined with the result [McKinley, W. A. and Feshbach, H., Phys. Rev. 74, 1759 (1948)]

$$\left(\frac{d\sigma}{d\Omega}\right)^{\text{Born}}_{\text{Dirac}} = \left(\frac{d\sigma}{d\Omega}\right)^{\text{Born}}_{\text{Relativistic Rutherford}}\left[1 - \beta^2\sin^2\frac{\theta}{2} - \pi z_1 z_2 \alpha\beta\sin\frac{\theta}{2}\left(1 - \sin\frac{\theta}{2}\right)\right]. \tag{10.185}$$

The electromagnetic scalar potential Φ is the fourth component of a four-vector. If V is instead a Lorentz scalar, the corresponding calculation must use βV in place of V, just as it is βm that appears in the Dirac equation. The trace now is

$$\frac{1}{2}\text{Tr}\,[\beta\Lambda_+(\mathbf{p}')\beta\Lambda_+(\mathbf{p})] = \frac{1}{2}\text{Tr}\,[\Lambda_+(-\mathbf{p}')\Lambda_+(\mathbf{p})] = 1 - \beta^2\cos^2\frac{\theta}{2}. \tag{10.186}$$

This leads to a dramatically different behavior at high energy.

10.14 Neutron-Electron Scattering

Even before the measurement of the neutron's magnetic moment in 1940 by Luis Alvarez and Felix Bloch, there were attempts to calculate the interaction between a low-energy neutron and the electrons in an atom. Fermi and Marshall tried to measure the effect by scattering thermal neutrons from xenon. In 1951, Foldy pointed out [*Phys. Rev.* **83**, 688 (1951)] that if a Dirac-like equation is written for the neutron, including the neutron's magnetic moment, this alone will produce a neutron interaction with a static electric field, in addition to the obvious magnetic moment interaction with the static magnetic field. We suppose the neutron to be structureless with mass m_n and magnetic moment $\mu = ge\hbar/(2m_p c)$, with $g = -1.913$. The proton mass appears because the neutron's magnetic moment is measured in nuclear magnetons, which use m_p, not m_n. The relativistic spin-1/2 equation is then

$$\left[-i\gamma^\alpha \partial_\alpha + m_n - \frac{ge}{4m_p}\sigma^{\alpha\beta}F_{\beta\alpha} \right]\Psi = 0, \tag{10.187}$$

where $\sigma^{\alpha\beta} = i[\gamma^\alpha, \gamma^\beta]/2$ and $F_{\beta\alpha} = \partial_\beta A_\alpha - \partial_\alpha A_\beta$ is the electromagnetic field tensor. Let us be explicit and let Greek letters run 0 to 3 and Latin letters run 1 to 3. For cartesian components a, we always use the a component of the ordinary vector, A^a. We write ∇_a for the ordinary gradient, so $\nabla_a A^a = \nabla \cdot \mathbf{A}$.

$$\partial^\alpha = (\frac{\partial}{\partial t}, -\nabla), \qquad A^\alpha = (A^0, \mathbf{A})$$

$$F^{ba} = -\nabla_b A^a + \nabla_a A^b = \epsilon_{abc}B^c$$

$$\sigma^{ab} = \epsilon_{abc}\sigma_D^c$$

$$F^{0a} = \frac{\partial A^a}{\partial t} + \nabla_a A^0 = -E^a$$

$$\sigma^{0a} = i\alpha_a. \tag{10.188}$$

Combining these,

$$\sum_{\mu > \nu} \sigma_{\mu\nu}F^{\mu\nu} = \sigma_D \cdot \mathbf{B} - i\alpha \cdot \mathbf{E}. \tag{10.189}$$

Then

$$\left[-i(\gamma^0 \partial_t + \gamma \cdot \nabla) + m_n - \frac{ge}{4m_p}2(\sigma_D \cdot \mathbf{B} - i\alpha \cdot \mathbf{E}) \right]\Psi = 0$$

$$i\beta\partial_t \Psi = \left[-i\beta\alpha \cdot \nabla + m_n - \frac{ge}{2m_p}(\sigma_D \cdot \mathbf{B} - i\alpha \cdot \mathbf{E}) \right]\Psi$$

$$H\Psi = \left[\alpha \cdot \mathbf{p} + \beta m_n - \frac{ge}{2m_p}(\beta\sigma_D \cdot \mathbf{B} - i\beta\alpha \cdot \mathbf{E}) \right]. \tag{10.190}$$

Following the Foldy-Wouthuysen prescription, we write

$$H = \beta m_n + \xi + \mathcal{O}, \tag{10.191}$$

where ξ is even and $\mathcal{O}$ is odd:

$$\xi = -\frac{ge}{2m_p}\beta\sigma_D \cdot \mathbf{B}$$

$$\mathcal{O} = \alpha \cdot \mathbf{p} + i\frac{ge}{2m_p}\beta\alpha \cdot \mathbf{E}. \tag{10.192}$$

We want the Foldy-Wouthuysen expansion correct to order $1/m_n^2$ and linear in g. Because ξ is already of order $1/m_n$, we do not need to compute $[\mathcal{O}, [\mathcal{O}, \xi]]$. The only piece needed is

$$\frac{\mathcal{O}^2}{2m_n} = \frac{1}{2m_n}(\boldsymbol{\alpha} \cdot \mathbf{p} \mid i\frac{ge}{2m_p}\beta\boldsymbol{\alpha} \cdot \mathbf{E})(\boldsymbol{\alpha} \cdot \mathbf{p} + i\frac{ge}{2m_p}\beta\boldsymbol{\alpha} \cdot \mathbf{E})$$

$$= \frac{p^2}{2m_n} - i\frac{ge}{4m_p^2}\beta\,(\boldsymbol{\alpha} \cdot \mathbf{p}\,\boldsymbol{\alpha} \cdot \mathbf{E} - \boldsymbol{\alpha} \cdot \mathbf{E}\,\boldsymbol{\alpha} \cdot \mathbf{p}) + \mathcal{O}(1/m_n^3), \tag{10.193}$$

where we ignored the difference between m_n and m_p in the second term. Now

$$\alpha_i\alpha_j = \delta_{ij} + i\epsilon_{ijk}\sigma_{Dk}, \tag{10.194}$$

so that, paying attention to the action of $\mathbf{p}$ to the right,

$$(\boldsymbol{\alpha} \cdot \mathbf{p}\,\boldsymbol{\alpha} \cdot \mathbf{E} - \boldsymbol{\alpha} \cdot \mathbf{E}\,\boldsymbol{\alpha} \cdot \mathbf{p}) = -i\boldsymbol{\nabla} \cdot \mathbf{E} + \boldsymbol{\sigma}_D \cdot (\boldsymbol{\nabla} \times \mathbf{E}) - 2i\boldsymbol{\sigma}_D \cdot (\mathbf{E} \times \mathbf{p}). \tag{10.195}$$

The F-W transformed Hamiltonian is then

$$H = \beta\left(m_n + \frac{p^2}{2m_n} + \cdots\right) - \frac{ge}{2m_p}\beta\boldsymbol{\sigma}_D \cdot \mathbf{B} - \frac{ge}{4m_p^2}\boldsymbol{\nabla} \cdot \mathbf{E}$$

$$- \frac{ge}{2m_p^2}\boldsymbol{\sigma}_D \cdot (\mathbf{E} \times \mathbf{p}) - i\frac{ge}{4m_p^2}\boldsymbol{\sigma}_D \cdot (\boldsymbol{\nabla} \times \mathbf{E}). \tag{10.196}$$

For static fields, $\boldsymbol{\nabla} \times \mathbf{E} = 0$, so our interaction Hamiltonian is

$$H_{\text{int}} = -\frac{ge}{2m_p}\beta\boldsymbol{\sigma}_D \cdot \mathbf{B} - \frac{ge}{4m_p^2}\boldsymbol{\nabla} \cdot \mathbf{E} - \frac{ge}{2m_p^2}\boldsymbol{\sigma}_D \cdot (\mathbf{E} \times \mathbf{p}). \tag{10.197}$$

Let us compare this with the analogous expression derived from the Dirac equation for a charged particle:

$$H_{\text{Dirac}} = -\frac{e}{2m}\beta\boldsymbol{\sigma}_D \cdot \mathbf{B} - \frac{e\boldsymbol{\nabla} \cdot \mathbf{E}}{8m^2} - \frac{e}{4m^2}\boldsymbol{\sigma}_D \cdot (\mathbf{E} \times \mathbf{p}). \tag{10.198}$$

The first term, the magnetic coupling, is the same if the magnetic moment of the neutral particle is the same as that of an electron (up to sign). However, the second (Darwin) term and the third (spin-orbit) term are each reduced by a factor of two in the charged Dirac case. The reason is that while the Thomas-precession factor $g \to (g-1)$ enters for the charged particle, that does not occur for the neutral particle because it is not accelerating; it is electrically neutral. See Jackson, *Classical Electrodynamics*, 3rd edition, p. 552, Eq. 11.121).

The electron-neutron interaction is of the magnetic type between the neutron's magnetic moment and the magnetic field of the electron, analogous to the hyperfine interaction explored in Problem 10.8. The spin-orbit coupling here is known as the Schwinger interaction. There is also the Darwin term, called the Foldy term in this context. The electron's electric field has a divergence,

$$\boldsymbol{\nabla} \cdot \mathbf{E} = -4\pi e\delta(\mathbf{r}_n - \mathbf{r}_e), \tag{10.199}$$

giving a Hamiltonian

$$H_{\text{int}}^F = -\frac{ge}{4m_p^2}\boldsymbol{\nabla} \cdot \mathbf{E} = \pi\frac{ge^2\hbar^2}{m_p^2c^2}\delta^3(\mathbf{r}_n - \mathbf{r}_e). \tag{10.200}$$

It is traditional to express H_{int}^F as if it were a square-well potential of strength V_0 inside a distance of the classical electron radius $e^2/(m_e c^2) = 2.8179$ fm. Thus we have,

$$V_0 \frac{4\pi}{3}\left(\frac{e^2}{m_e c^2}\right)^3 = \pi g e^2 \left(\frac{\hbar}{m_p c}\right)^2. \tag{10.201}$$

We use the known values

$$\frac{\hbar}{m_p c} = \frac{197.33 \text{ MeV fm}}{938.272 \text{ MeV}} = 0.2103 \text{ fm}$$

$$g_n = -1.91304 \tag{10.202}$$

to compute

$$V_0 = \frac{3}{4} g(m_e c^2)\frac{(\hbar/m_p c)^2}{(e^2/m_e c^2)^2} = -\frac{3}{4}(1.91304)\left(\frac{0.2103}{2.8179}\right)^2 \times 511.00343 \text{ keV}$$

$$= -4.084 \text{ keV}. \tag{10.203}$$

We study these interactions with low-energy neutrons scattering off atoms with all electrons paired so that the effect is not overwhelmed by the interaction with magnetic moments of the electrons. At very low energy, cross-sections can be studied in Born approximation

$$f(k) = -\frac{m}{2\pi\hbar^2}V(q^2 = 0) = -b, \tag{10.204}$$

where b is called the scattering length. For example, in hard-sphere s-wave scattering, if $kR \ll 1$, then $b = -f = R$, where R is the radius of the hard sphere. See Eq. (7.215). The coherent scattering of the neutron by the nuclear force can be represented by a scattering length b_{N}, with a total cross section $4\pi b_{\text{N}}^2$. The smaller scattering amplitude that includes the neutron's interaction with the electrons (and with the electric charge of the nucleus) interferes with this dominant process. At extremely low momentum transfer, the scattering is coherent off all the electrons and the nuclear electric charge. As the momentum transfer is increased, the electron amplitudes come in with a suppression factor equal to the Fourier transfer (at the measured q^2) of the electron distribution. From the variation in the differential cross section as a function of q^2, the total cross section acquires an energy dependence. Thus, the neutron-electron interaction can be studied through energy variation of total cross-section measurements.

We can construct a crude model of the neutron with two parameters. The first is the anomalous magnetic moment with $g = -1.913$. The second comes from a charge density, $\rho(\mathbf{x})$, whose integral is, of course, zero. A single parameter suffices to determine its contribution to low-energy scattering from an electron. Consider,

$$V(q^2) = -\frac{4\pi e^2}{q^2}G_E(q^2), \tag{10.205}$$

where at low q^2

$$G_E(q^2) = \int d^3x \rho(\mathbf{x})e^{i\mathbf{q}\cdot\mathbf{x}} \simeq -\frac{\langle r_E^2\rangle}{6}q^2 \tag{10.206}$$

is given in terms of the charge-radius-squared of the neutron. For the scattering length, this result, together with the formula for the Born scattering amplitude (Eq. (7.59)), gives

$$
b_{ne,\text{charge}} = -f_{\text{Born}} = \frac{m_n}{2\pi}(4\pi e^2)\frac{\langle r_E^2 \rangle}{6}
$$

$$
= \frac{m_n}{3a_0 m_e}\langle r_E^2 \rangle \simeq (1.158 \times 10^{-2})\,\text{fm}^{-1}\langle r_E^2 \rangle. \tag{10.207}
$$

The Foldy interaction of the neutron with a single electron is

$$
V(q^2) = \int d^3 r\, \pi \frac{ge^2 \hbar^2}{m_p^2 c^2}\delta^3(\mathbf{r} - \mathbf{r}_e)e^{i\mathbf{q}\cdot\mathbf{r}} = \pi \frac{ge^2 \hbar^2}{m_p^2 c^2}, \tag{10.208}
$$

and the scattering length is

$$
b_{ne,\text{Foldy}} = \frac{1}{2}\frac{m_n}{a_0 m_e}g\left(\frac{\hbar}{m_p c}\right)^2 \simeq (1.158 \times 10^{-2})\,\text{fm}^{-1}\left[\frac{3}{2}g\left(\frac{\hbar}{m_p c}\right)^2\right]
$$

$$
= (1.158 \times 10^{-2})\,\text{fm}^{-1} \times (-0.1269)\,\text{fm}^2 = -1.470 \times 10^{-3}\,\text{fm}. \tag{10.209}
$$

Both the electrostatic interaction of the neutron, due to its charge radius, and the Foldy interaction are couplings of the neutron to the electric charge of the atom, including both the electrons and the nucleus. To calculate the scattering length on the full atom, we replace the charge density of a single electron, $-e\delta^3(\mathbf{r} - \mathbf{r}_e)$, with the charge density of the electron cloud and that of the nucleus as well, $eZ[\delta^3(\mathbf{r}) - \rho(\mathbf{r})]$. Thus, the total scattering length of the atom is

$$
b_{ne}(q^2) = -\int d^3 r(b_{ne,\text{charge}} + b_{ne,\text{Foldy}})Z[\delta^3(\mathbf{r}) - \rho(\mathbf{r})]e^{i\mathbf{q}\cdot\mathbf{r}}
$$

$$
= -(b_{ne,\text{charge}} + b_{ne,\text{Foldy}})Z(1 - f(q^2)), \tag{10.210}
$$

where

$$
f(q^2) = \int d^3 r\, e^{i\mathbf{q}\cdot\mathbf{r}}\rho(r) \tag{10.211}
$$

is the Fourier transform of the electron distribution.

Because b_{ne} is a function of q^2, the scattering of the neutron from the atom will not be isotropic. In the total cross section, an average value of $f(q^2)$ will appear and will depend on the incident neutron energy and on the Z of the atom. That factor, $f(E, Z)$, can be reliably computed from atomic models, like Thomas-Fermi or Hartree-Fock.

For scattering at very low energy, $f(Z, E) = 1$ and the b_{ne} terms just cancel between the electrons and the nuclear electric charge, leaving just the nuclear scattering length. This nuclear scattering length was measured very precisely for Pb and Bi by Koester, Nistler, and Waschkowski [Phys. Rev. Lett. **36**, 1021 (1976)]. Their measurement relied on the neutron version of total internal reflection familiar from ordinary optics. The analog of the index of refraction depends on the neutron-nucleus scattering length and on the density of the atoms in the reflector. Measuring the critical angle for total reflection of the neutrons from the surface of lead or bismuth thus determines that scattering length for thermal neutrons. Comparing that result to the measured total cross section at higher (but still low) energies allows the extraction of $b_{ne,\text{charge}} + b_{ne,\text{Foldy}}$. Their result was

$$b_{ne,\text{charge}} + b_{ne,\text{Foldy}} = -(1.378 \pm 0.018) \times 10^{-3} \text{ fm.} \tag{10.212}$$

Comparing this result to the value with that of $b_{ne,\text{Foldy}}$, we see that

$$b_{ne,\text{charge}} = +(0.090 \pm 0.018) \times 10^{-3} \text{ fm,} \tag{10.213}$$

which corresponds to a charge-radius-squared for the neutron

$$\langle r_E^2 \rangle = +(7.7 \pm 1.5) \times 10^{-3} \text{fm}^2. \tag{10.214}$$

10.15 Compton Scattering

10.15.1 Spinless Non-relativistic Case

We consider the scattering of photons by free electrons in Dirac hole theory as a prelude to the covariant formulation of QED. The process of the scattering of light is second order in the electromagnetic interaction, that is, its amplitude is proportional to e^2. We begin with a description of the scattering of a charged, non-relativistic, spinless particle. The "minimal coupling," which ensures gauge invariance, $\mathbf{p} \rightarrow \mathbf{p} - e\mathbf{A}$, gives an interaction Hamiltonian

$$H_{\text{int}} = -\frac{E}{m}\mathbf{A}_{\text{rad}} \cdot \mathbf{p} + \frac{e^2}{2m}\mathbf{A} \cdot \mathbf{A} = H_1 + H_2. \tag{10.215}$$

The scattering amplitude that enters the Golden Rule is

$$M_{\text{fi}} = \sum_j {}' \frac{\langle f|H_1^\dagger|j\rangle \langle j|H_1|i\rangle}{E_i - E_j} + \sum_{j'} {}' \frac{\langle f|H_1|j'\rangle \langle j'|H_1^\dagger|i\rangle}{E_i - E_{j'}} + \langle f|H_2|i\rangle. \tag{10.216}$$

The first term on the right-hand side represents the absorption of the incident photon of momentum $\mathbf{k}$ and subsequent emission of the outgoing photon of momentum $\mathbf{k}'$. In the second term, the outgoing photon is emitted before the absorption of the incoming photon. The third term, the "contact" term, has the absorption and emission occurring simultaneously. See Figure 10.4.

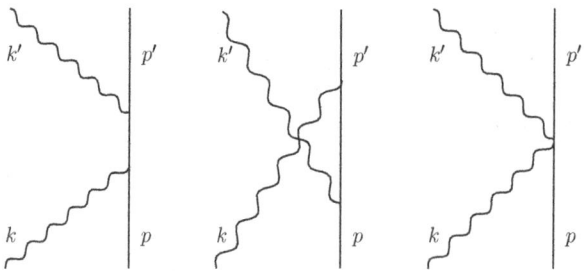

Figure 10.4 Contributions to Compton scattering for a spinless particle. The incoming photon may be absorbed before (left), after (middle), or simultaneous (right) with the emission of the outgoing photon.

We calculate the free-particle matrix elements using the expansion of the electromagnetic potential, Eq. (8.84), with the particles normalized in a box of volume V:

$$\langle j\, |H_1|\, i\rangle = -\frac{e}{m}\sqrt{\frac{4\pi}{V}}\frac{1}{\sqrt{2\omega}}\int d^3x\, \frac{e^{-i\mathbf{p}_j\cdot\mathbf{x}}}{\sqrt{V}}\boldsymbol{\epsilon}_\mathbf{k}\cdot\mathbf{p}\, e^{i\mathbf{k}\cdot\mathbf{x}}\frac{e^{i\mathbf{p}\cdot\mathbf{x}}}{\sqrt{V}}$$

$$= -\frac{e}{V}\sqrt{\frac{4\pi}{V}}\frac{1}{\sqrt{2\omega}}\frac{\boldsymbol{\epsilon}_\mathbf{k}\cdot\mathbf{p}}{m}V\delta_{\mathbf{p}_j-\mathbf{k}-\mathbf{p}}.$$

$$(10.217)$$

Here we have a factor of the volume V times a Kronecker delta because there are discrete values of the momenta in the box and the value of the Kronecker delta is unity if the argument is exactly zero, and vanishes otherwise. Similarly,

$$\left\langle f\, \middle|H_1^\dagger\middle|\, j\right\rangle = -\frac{e}{V}\sqrt{\frac{4\pi}{V}}\frac{1}{\sqrt{2\omega'}}\frac{\boldsymbol{\epsilon}_{\mathbf{k}'}^{*}\cdot\mathbf{p}_j}{m}V\delta_{\mathbf{p}'+\mathbf{k}'-\mathbf{p}_j}. \tag{10.218}$$

Combining these two,

$$\left\langle f\, \middle|H_1^\dagger\middle|\, j\right\rangle\langle j\, |H_1|\, i\rangle = \frac{4\pi e^2}{V}\frac{1}{\sqrt{4\omega\omega'}}\frac{\boldsymbol{\epsilon}_{\mathbf{k}'}^{*}\cdot\mathbf{p}_j}{m}\frac{\boldsymbol{\epsilon}_\mathbf{k}\cdot\mathbf{p}}{m}\delta_{\mathbf{p}_j-\mathbf{k}-\mathbf{p}}\delta_{\mathbf{p}'+\mathbf{k}'-\mathbf{p}_j}. \tag{10.219}$$

Now $\boldsymbol{\epsilon}_\mathbf{k}\cdot\mathbf{k}=0, \boldsymbol{\epsilon}_{\mathbf{k}'}\cdot\mathbf{k}'=0, \mathbf{p}_j=\mathbf{p}'+\mathbf{k}'$, so we can write this as

$$\left\langle f\, \middle|H_1^\dagger\middle|\, j\right\rangle\langle j\, |H_1|\, i\rangle = \frac{4\pi e^2}{V}\frac{1}{\sqrt{4\omega\omega'}}\frac{\boldsymbol{\epsilon}_{\mathbf{k}'}^{*}\cdot\mathbf{p}'}{m}\frac{\boldsymbol{\epsilon}_\mathbf{k}\cdot\mathbf{p}}{m}\delta_{\mathbf{k}'+\mathbf{p}'-\mathbf{k}-\mathbf{p}}\delta_{\mathbf{p}'-\mathbf{k}'-\mathbf{p}_j}. \tag{10.220}$$

The second Kronecker delta fixes $\mathbf{p}_j$ uniquely, so the sum over $\mathbf{p}_j$ is trivial.

In this non-relativistic calculation, we can ignore the rest mass of the electron in the energy denominators, since it cancels in the difference between the initial and intermediate states. Assembling all the pieces,

$$H_{\text{fi}}^{(1)} = \frac{4\pi}{V}\frac{1}{\sqrt{4\omega\omega'}}\frac{e^2}{m^2}\frac{(\boldsymbol{\epsilon}_{\mathbf{k}'}^{*}\cdot\mathbf{p}')(\boldsymbol{\epsilon}_\mathbf{k}\cdot\mathbf{p})}{E_i-\frac{(\mathbf{p}+\mathbf{k})^2}{2m}}\delta_{\mathbf{k}'+\mathbf{p}'-\mathbf{k}-\mathbf{p}}, \tag{10.221}$$

where $E_i=\omega+p^2/(2m)$. If we work in the center-of-mass, $\mathbf{p}+\mathbf{k}=0, p=\omega$, and the denominator is simply $\omega+\omega^2/(2m)$. Note that in the non-relativistic limit, the center-of-mass system is nearly the rest system of the electron.

Similarly, the second term is, in the center-of-mass,

$$H_{\text{fi}}^{(2)} = \frac{4\pi}{V}\frac{1}{\sqrt{4\omega\omega'}}\frac{e^2}{m^2}\left[\frac{(\boldsymbol{\epsilon}_\mathbf{k}\cdot\mathbf{p}')(\boldsymbol{\epsilon}_{\mathbf{k}'}^{*}\cdot\mathbf{p})}{E_i-[2\omega+(\mathbf{p}-\mathbf{k}')^2/(2m)]}\right]\delta_{\mathbf{k}'+\mathbf{p}'-\mathbf{k}-\mathbf{p}}. \tag{10.222}$$

The intermediate state here has both the incident and final photons. We can see that the leading piece of the denominator for $H_{\text{fi}}^{(1)}$ is ω, while for $H_{\text{fi}}^{(2)}$ it is $-\omega$. Thus, in both cases, we find the matrix elements are of order

$$H_{\text{fi}}^{(1,2)} \sim \frac{4\pi}{V} \frac{1}{2\omega} \frac{e^2}{m^2} \frac{p^2}{\omega} \sim \frac{4\pi}{V} \frac{1}{2\omega} \frac{e^2}{m^2} p. \tag{10.223}$$

The matrix element of H_2 needs a little care. If we imagine $\mathbf{A} \cdot \mathbf{A}$ expanded in eigenmodes of the photons, there will be terms of the form

$$\frac{e^2}{2m} \frac{4\pi}{V} \int d^3x \, \frac{e^{-ip' \cdot x}}{\sqrt{V}} \left(\frac{\epsilon_{\mathbf{k}} e^{i\mathbf{k} \cdot \mathbf{x}}}{\sqrt{2\omega}} + \frac{\epsilon_{\mathbf{k}'}^* e^{-i\mathbf{k}' \cdot \mathbf{x}}}{\sqrt{2\omega'}} \right) \cdot \left(\frac{\epsilon_{\mathbf{k}} e^{i\mathbf{k} \cdot \mathbf{x}}}{\sqrt{2\omega}} + \frac{\epsilon_{\mathbf{k}'}^* e^{-i\mathbf{k}' \cdot \mathbf{x}}}{\sqrt{2\omega'}} \right) \frac{e^{ip \cdot x}}{\sqrt{V}}. \tag{10.224}$$

There are thus two contributions, and we have

$$\langle f | H_2 | i \rangle = \frac{e^2}{m} \frac{4\pi}{V^2} \frac{1}{\sqrt{4\omega\omega'}} \epsilon_{\mathbf{k}'}^* \cdot \epsilon_{\mathbf{k}} V \delta_{\mathbf{k}' + \mathbf{p}' - \mathbf{k} - \mathbf{p}}. \tag{10.225}$$

We see now that the contributions from H_1 are smaller by a factor of $\omega/m = p/m$ in the center of mass and can be ignored.

To leading order in ω/m,

$$|H_{\text{fi}}|^2 = \left(\frac{4\pi}{V} \right)^2 \cdot \left(\frac{e^2}{m} \right)^2 \frac{|\epsilon_{\mathbf{k}'}^* \cdot \epsilon_{\mathbf{k}}|^2}{4\omega\omega'} \delta_{\mathbf{k}' + \mathbf{p} - \mathbf{k} - \mathbf{p}}. \tag{10.226}$$

We now take the continuum limit and insert the result into the Golden Rule:

$$d\Gamma = 2\pi \left(\frac{4\pi}{V} \right)^2 \left(\frac{e^2}{m} \right)^2 \frac{|\epsilon_{\mathbf{k}'}^* \cdot \epsilon_{\mathbf{k}}|^2}{4\omega\omega'} \frac{(2\pi)^3}{V} \delta^3(\mathbf{k}' + \mathbf{p}' - \mathbf{k} - \mathbf{p})$$

$$\times \frac{V d^3 k' }{(2\pi)^3} \frac{V d^3 p'}{(2\pi)^3} \delta(E(p') + E(k') - E(p) - E(k))$$

$$= \frac{1}{V} \left(\frac{e^2}{m} \right)^2 \frac{|\epsilon_{\mathbf{k}'}^* \cdot \epsilon_{\mathbf{k}}|^2}{\omega\omega'} k'^2 d\Omega dk' \delta(E(p') + E(k') - E(p) - E(k)). \tag{10.227}$$

We are considering incident photon energies much less than the mass m of the electron, so the laboratory frame and the center of mass are essentially identical. The total energy is $\omega + m = k + m$ ($c = 1$) and $d[E(p') + dE(k')]/dk' \simeq 1$, so

$$d\Gamma = \frac{1}{V} \left(\frac{e^2}{m} \right)^2 |\epsilon_{\mathbf{k}'}^* \cdot \epsilon_{\mathbf{k}}|^2 d\Omega. \tag{10.228}$$

The photon flux is c/V, and with $c = 1$ we have

$$\frac{d\sigma}{d\Omega} = \left(\frac{e^2}{m} \right)^2 |\epsilon_{\mathbf{k}'}^* \cdot \epsilon_{\mathbf{k}}|^2 = r_0^2 |\epsilon_{\mathbf{k}'}^* \cdot \epsilon_{\mathbf{k}}|^2, \tag{10.229}$$

where r_0 is the "classical electron radius" determined by

$$\frac{e^2}{r_0} = mc^2; \quad r_0 = \frac{e^2}{mc^2} = 2.818 \text{ fm}. \tag{10.230}$$

This result was first obtained by J. J. Thomson using classical electrodynamics and is called the Thomson cross section. See Jackson, *Classical Electrodynamics*, 3rd edition, pp. 694–697.

10.15.2 Relativistic Case with Spin

With this low-energy calculation as background, we now proceed to a relativistic description, following Heitler, *The Quantum Theory of Radiation*, pp. 212–219. In the Dirac formalism, the full interaction is simply

$$H_{\text{e-m}} = -e\boldsymbol{\alpha} \cdot \mathbf{A}_{\text{rad}}. \tag{10.231}$$

We are in the radiation or Coulomb gauge. There is no quadratic term. Therefore, the process is entirely second order. We have the same general form as before, but with different matrix elements. Addressing first the contribution when the incident photon is absorbed before the exiting photon is emitted, we have

$$\langle j \,|H_{\text{e-m}}| \, i\rangle = -\frac{e}{V}\sqrt{\frac{4\pi}{V}}\frac{1}{\sqrt{2\omega}}\int d^3x\, w_j^\dagger(\mathbf{p}_j)e^{-i\mathbf{p}_j\cdot\mathbf{x}}e^{i\mathbf{k}\cdot\mathbf{x}}\boldsymbol{\epsilon}_{\mathbf{k}}\cdot\boldsymbol{\alpha}w(\mathbf{p})e^{i\mathbf{p}\cdot\mathbf{x}}$$

$$= -e\sqrt{\frac{4\pi}{V}}\frac{1}{\sqrt{2\omega}}w_j^\dagger(\mathbf{p}_j)\boldsymbol{\epsilon}_{\mathbf{k}}\cdot\boldsymbol{\alpha}w(\mathbf{p})\delta_{\mathbf{p}_j-\mathbf{k}-\mathbf{p}} \tag{10.232}$$

and

$$\left\langle f \left|H_{\text{e-m}}^\dagger\right| j\right\rangle = -e\sqrt{\frac{4\pi}{V}}\frac{1}{\sqrt{2\omega'}}w^\dagger(\mathbf{p}')\boldsymbol{\epsilon}_{\mathbf{k}'}^*\cdot\boldsymbol{\alpha}w_j(\mathbf{p}_j)\delta_{\mathbf{p}'+\mathbf{k}'-\mathbf{p}_j}. \tag{10.233}$$

The contribution to the second-order matrix element is

$$\sum_j{}'e^2\frac{4\pi}{V}\frac{1}{\sqrt{4\omega\omega'}}\frac{w^\dagger(\mathbf{p}')\boldsymbol{\epsilon}_{\mathbf{k}'}^*\cdot\boldsymbol{\alpha}w_j(\mathbf{p}_j)w_j^\dagger(\mathbf{p})\boldsymbol{\epsilon}_{\mathbf{k}}\cdot\boldsymbol{\alpha}w_i(\mathbf{p}_i)}{E_{\text{i}} - E_{\text{int}}}\delta_{\mathbf{p}_j-\mathbf{k}-\mathbf{p}}\delta_{\mathbf{p}'+\mathbf{k}'-\mathbf{p}-\mathbf{k}}. \tag{10.234}$$

As in the previous calculation, we see that $\mathbf{p}_j$ is uniquely determined. The initial energy is $E_{\text{i}} = p_0 + k_0 = p_0 + \omega$, the sum of the initial photon and electron energies. If the intermediate electron state is a positive energy state, then $E_{\text{int}} = \sqrt{(\mathbf{p}+\mathbf{k})^2 + m^2}$. Note that since we are not restricted to low energy, we take the electron's energy to be its full relativistic value, not just its kinetic energy. We postpone consideration of E_{int} when the intermediate state has negative energy.

The second-order matrix element for the other ordering of emission and absorption gives

$$\sum_{j'}{}'e^2\frac{4\pi}{V}\frac{1}{\sqrt{4\omega\omega'}}\frac{w^\dagger(\mathbf{p}')\boldsymbol{\epsilon}_{\mathbf{k}}\cdot\boldsymbol{\alpha}w_{j'}(\mathbf{p}_{j'})w_{j'}^\dagger(\mathbf{p}_{j'})\boldsymbol{\epsilon}_{\mathbf{k}'}^*\cdot\boldsymbol{\alpha}w(\mathbf{p})}{E_{\text{i}} - E_{\text{int}}}\delta_{\mathbf{p}_{j'}+\mathbf{k}'-\mathbf{p}}\delta_{\mathbf{p}'+\mathbf{k}'-\mathbf{p}-\mathbf{k}}. \tag{10.235}$$

If the intermediate electron state j' corresponds to a positive energy state, then $E_{\text{int}} = k_0 + k_0' + \sqrt{(\mathbf{p}-\mathbf{k}')^2 + m^2}$. Again, we postpone consideration of possible negative energy states.

While in the sums over j and j', the intermediate three-momentum is uniquely specified. The sum over the intermediate spin remains. This leaves the question of whether both positive and negative energy states are to be considered. As we shall see, there is an empirical answer. When Klein and Nishina first calculated this process, relativistic Compton scattering, they assumed that the "Dirac sea" was empty and summed over both positive and negative energies. Their result, the Klein-Nishina formula, agrees with experiment.

Without recourse to detailed data, we can settle the question by considering the low-energy limit, the Thomson cross section derived above quantum-mechanically, but known from classical electrodynamics. The two matrix elements calculated in the Dirac case with intermediate positive energy states are essentially the same, as we found above for the two-step processes. The operator α between positive-energy states is roughly $v \simeq p/m$. In the non-relativistic limit, these are negligible relative to the contact term, as we saw above. But here there is no contact term, so we cannot recover the non-relativistic result if we neglect the negative energy states.

10.15.3 Negative Energy State Contribution

The negative energy states correspond to positrons. Thus, the intermediate state somehow has a positron and the original electron present. The time-ordered diagrams are shown in Figure 10.5. For the intermediates corresponding to negative energy states, now properly interpreted as states containing positrons, the energy denominator $(E_i - E_{int})$ will have $E_{int} = E_i + |E_j| + E_f$ or $E_{int} = E_i + |E_{j'}| + E_f$. Thus,

$$E_i - E_{int} = E_i - E_i - |E_j| - E_f = -(E_i - E_j) \tag{10.236}$$

where E_j is negative. This is just the opposite overall sign to what we would have if we blindly summed over positive and negative E_j values, as did Klein and Nishina.

Even the sign gets fixed up! Recall that in doing time-dependent perturbation theory, there was the Dyson time ordering, in which interaction terms at earlier times stood to the right of those at later times. The two signs of E_j correspond in the positron-hole theory description to different time orderings. The operating field that creates the electron (positron) must be commuted with another field to restore the order. Therefore, there is a change of sign. One can think of the results as a consequence of the Pauli principle acting in the "exchange" of the initial electron ($E > 0$) with the intermediate electron (with $E_j < 0$). The net result is that the sum over both signs of the intermediate energy occurs as if the Dirac "sea" were empty.

Note that the "Z" graphs are the ones responsible for the contact $e^2\,\mathbf{A}\cdot\mathbf{A}/m$ term in the non-relativistic interaction. This can be understood because of the virtuality of the intermediate positron and the uncertainty principle. For $E_j < 0$, there is $\Delta p \gtrsim 2mc$. Thus, $\Delta x \lesssim \hbar(2mc)$ for the physical distance between the emission of ω' and the absorption of ω. For $\omega, \omega' \ll m, \lambda \gg \Delta x$. Then the emission and absorption of the two photons appear at essentially the same space-time point, on the scales that are relevant.

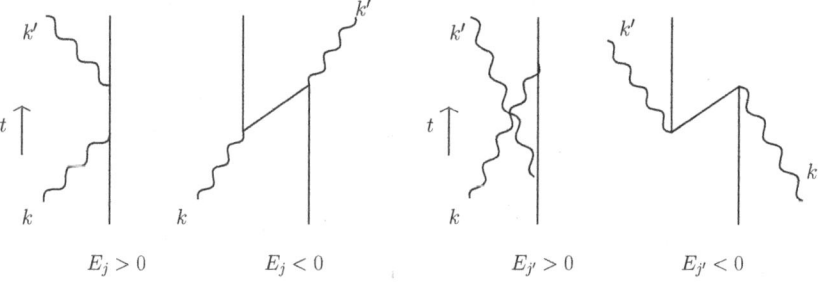

Figure 10.5 Contribution to Compton scattering from two time orderings, for two orderings of the absorption and emission. The top two diagrams show the absorption occuring before emission, while the bottom two show the reverse. Negative energy states on the right are necessary to get the correct answer. Without them, the non-relativistic result is not recovered.

10.15.4 Obtaining Covariant Expression

We return to the calculation but now re-express the terms in a manifestly covariant form. We start with

$$\sum_{j=1}^{4} \frac{w^{\dagger}(\mathbf{p}')\boldsymbol{\epsilon}^{*}_{\mathbf{k}'} \cdot \boldsymbol{\alpha} w_j(\mathbf{p}_j)w^{\dagger}_j(\mathbf{p}_j)\boldsymbol{\epsilon}_{\mathbf{k}} \cdot \boldsymbol{\alpha} w(\mathbf{p})}{E_i - E_j}, \tag{10.237}$$

where we remember that $\mathbf{p}_j = \mathbf{p} + \mathbf{k}$ is uniquely determined, but the sum over j is a sum over the four Dirac spinors with this three-momentum. The awkwardness here is that the denominator depends on whether we have positive- or negative-energy intermediate states. Multiplying numerator and denominator by $E_i + E_j$, the denominator becomes

$$E_i^2 - E_j^2 = (p_0 + k_0)^2 - (\mathbf{p} + \mathbf{k})^2 - m^2 = (p + k)^2 - m^2, \tag{10.238}$$

a conveniently covariant form. Consider then, a factor in the numerator

$$\sum_{j=1}^{4}(E_i + E_j)w_j w^{\dagger}_j = \sum_{j=1}^{4}(p_0 + k_0 + H)w_j w^{\dagger}_j, \tag{10.239}$$

where $H = \boldsymbol{\alpha} \cdot \mathbf{p}_j + \beta m = \boldsymbol{\alpha} \cdot (\mathbf{p} + \mathbf{k}) + \beta m$. Thus, with $\boldsymbol{\gamma} = \beta\boldsymbol{\alpha}$,

$$p_0 + k_0 + H = [(p_0 + k_0)\gamma^0 - \boldsymbol{\gamma} \cdot (\mathbf{p} + \mathbf{k}) + m]\gamma^0 = [m + \not{p} + \not{k}]\gamma^0. \tag{10.240}$$

The sum $\sum_{j=1}^{4} w_j w^{\dagger}_j$ gives the identity matrix, I. Our original expression can be written covariantly:

$$\frac{w^{\dagger}_f(\mathbf{p}_f)\boldsymbol{\epsilon}^{*}_{\mathbf{k}'} \cdot \boldsymbol{\alpha}(m + (\not{p} + \not{k}))\gamma^0 \boldsymbol{\epsilon}_{\mathbf{k}} \cdot \boldsymbol{\alpha} w_i(\mathbf{p}_i)}{(p + k)^2 - m^2}. \tag{10.241}$$

We introduce some conventional notation:

$$w^{\dagger} = \overline{w}\gamma^0 = \frac{1}{\sqrt{2E}}\overline{u}\gamma^0, w = \frac{1}{\sqrt{2E}}u, \tag{10.242}$$

so $\overline{u}u = 2m$. Our expression becomes

$$-\frac{1}{\sqrt{4EE'}}\frac{\overline{u}_f(\mathbf{p}')\not{\epsilon}^{*}_{\mathbf{k}'}(m + \not{p} + \not{k})\not{\epsilon}_{\mathbf{k}}u_i(\mathbf{p})}{m^2 - (p + k)^2}. \tag{10.243}$$

The second ordering of emission and absorption proceeds in a similar, but not identical, fashion. In this instance, the amplitude at the outset is

$$\sum_{j=1}^{4} \frac{w^{\dagger}_f(\mathbf{p}_f)\boldsymbol{\epsilon}_{\mathbf{k}} \cdot \boldsymbol{\alpha} w_{j'}(\mathbf{p}_{j'})w^{\dagger}_{j'}(\mathbf{p}_{j'})\boldsymbol{\epsilon}^{*}_{\mathbf{k}'} \cdot \boldsymbol{\alpha} w_i(\mathbf{p}_i)}{E_i - E_{j'}}. \tag{10.244}$$

Here

$$E_i = p_0 + k_0; \qquad E_{j'} = k_0 + k'_0 + \sqrt{m^2 + (\mathbf{p} - \mathbf{k})^2}. \tag{10.245}$$

So to free the denominator of the square root, we multiply top and bottom by

$$p_0 - k'_0 + \sqrt{m^2 + (\mathbf{p} - \mathbf{k}')^2}. \tag{10.246}$$

We can then write

$$\frac{1}{p_0 - k_0' - \sqrt{m^2 + (\mathbf{p} - \mathbf{k}')^2}} w(\mathbf{p}_{j'}) = \frac{p_0 - k_0' + \sqrt{m^2 + (\mathbf{p} - \mathbf{k})^2}}{(p_0 - k_0')^2 - m^2 - (\mathbf{p} - \mathbf{k}')^2} w(\mathbf{p}_{j'})$$

$$= \frac{p_0 - k' + H}{(p - k')^2 - m^2} w(p_{j'})$$

$$= \frac{(p_0 - k_0') + \boldsymbol{\alpha} \cdot (\mathbf{p} - \mathbf{k}') + \beta m}{(p - k')^2 - m^2} w(p_j)$$

$$= \frac{[(p_0 - k_0')\gamma_0 - \boldsymbol{\gamma} \cdot (\mathbf{p} - \mathbf{k}') + m]\gamma_0}{(p - k')^2 - m^2} w(p_{j'})$$

$$= \frac{[\not{p} - \not{k}' + m]\gamma_0}{(p - k')^2 - m^2} w(p_{j'}). \tag{10.247}$$

Altogether, the diagrams in which the photon emission comes before absorption yield

$$-\frac{1}{\sqrt{4EE'}} \frac{\overline{u}_f(p')\not{\epsilon}_k(m + \not{p} - \not{k}')\not{\epsilon}_{k'}^* u_i(p)}{m^2 - (p - k')^2}. \tag{10.248}$$

If we combine these with the remaining factors from Eq. (10.234), the second-order matrix element is

$$-\frac{1}{V} \frac{4\pi e^2}{\sqrt{(2\omega)(2\omega')(2E)(2E')}} \overline{u}_f(p')$$

$$\left[\frac{\not{\epsilon}_{k'}^*(m + \not{p} + \not{k})\not{\epsilon}_k}{m^2 - (p + k)^2} + \frac{\not{\epsilon}_k(m + \not{p} - \not{k}')\not{\epsilon}_{k'}^*}{m^2 - (p - k')^2} \right] u_i(p)\delta_{\mathbf{p}'+\mathbf{k}'-\mathbf{p}-\mathbf{k}}. \tag{10.249}$$

A reasonable choice is to work in the rest frame of the electron. We can lighten the algebra then by working in a gauge in which the polarization four-vectors are, in fact, three-vectors, and of course $\boldsymbol{\epsilon} \cdot \mathbf{k} = 0, \boldsymbol{\epsilon}' \cdot \mathbf{k}' = 0$. We also then have $\not{\epsilon}\not{p} = -\not{p}\not{\epsilon}, \not{\epsilon}'\not{p} = -\not{p}\not{\epsilon}'$. With this choice,

$$(m + \not{p})\not{\epsilon}u(p) = \not{\epsilon}(m - \not{p})u(p) = 0. \tag{10.250}$$

Writing ϵ for $\epsilon_{\mathbf{k}}$ and ϵ' for $\epsilon_{\mathbf{k}'}$, we have as our matrix element

$$-\frac{1}{V} \frac{4\pi e^2}{\sqrt{(2\omega)(2\omega')(2E)(2E')}} \overline{u}_f(p') \left[\frac{\not{\epsilon}^*\not{k}\not{\epsilon}}{m^2 - (p + k)^2} + \frac{\not{\epsilon}(-\not{k}')\not{\epsilon}^{*'}}{m^2 - (p - k')^2} \right] u_i(p)\delta_{\mathbf{p}'+\mathbf{k}'-\mathbf{p}-\mathbf{k}}$$

$$= -\frac{1}{V} \frac{4\pi e^2}{\sqrt{(2\omega)(2\omega')(2E)(2E')}} \overline{u}_f(p') \left[\frac{\not{\epsilon}^{*'}\not{\epsilon}\not{k}}{2p \cdot k} + \frac{\not{\epsilon}\not{\epsilon}^{*'}\not{k}'}{2p \cdot k'} \right] u_i(p)\delta_{\mathbf{p}'+\mathbf{k}'-\mathbf{p}-\mathbf{k}}. \tag{10.251}$$

This is a typical situation in the evaluation of cross sections in quantum electrodynamics (QED). If we write

$$A = \overline{u}' A u, \tag{10.252}$$

where A is some 4×4 matrix constructed from gamma matrices, we need

$$|\mathcal{A}|^2 = (\overline{u}' A u)^* \overline{u}' A u. \tag{10.253}$$

Now

$$(\overline{u}' A u)^* = u^\dagger A^\dagger \gamma^0 u'. \tag{10.254}$$

Each piece of A is a product of gamma matrices, for example $\not{p}\not{\epsilon}\not{p}$. We use the identity $\gamma^0 \gamma^{\mu\dagger} \gamma^0 = \gamma^\mu$ and find

$$(\overline{u}' \not{p}\not{\epsilon}\not{p} u)^* = \overline{u}(\not{p}\not{\epsilon}\not{p})u'. \tag{10.255}$$

So, for example,

$$|\overline{u}' \not{p}\not{\epsilon}\not{p} u|^2 = \overline{u}' \not{p}\not{\epsilon}\not{p} u \overline{u} \not{p}\not{\epsilon}\not{p} \overline{u}. \tag{10.256}$$

Now $u\overline{u}$ is, in fact, a 4×4 matrix, so we can write this as

$$\mathrm{Tr}\,(u'\overline{u}')\not{p}\not{\epsilon}\not{p}(u\overline{u})\not{p}\not{\epsilon}\not{p}. \tag{10.257}$$

If we sum over the spins of the two Dirac particles, we can replace

$$u\overline{u} = \not{p} + m \tag{10.258}$$

with our normalization $\overline{u}u = 2m$. Such calculations lead to "Diracology," techniques for calculating traces of products of gamma matrices. Setting aside the gamma matrix calculation, writing

$$\mathcal{A} = \overline{u}_{\mathrm{f}}(p') \left[\frac{\not{\epsilon}^{*\prime}\not{\epsilon}\not{k}}{2p \cdot k} + \frac{\not{\epsilon}\not{\epsilon}^{*\prime}\not{k}'}{2p \cdot k'} \right] u_i(p), \tag{10.259}$$

and setting

$$\sum_{spins} |\mathcal{A}|^2 = \sum |\mathcal{A}|^2, \tag{10.260}$$

the Golden Rule tells us

$$d\Gamma = \left(\frac{1}{2} \right) 2\pi \frac{(4\pi e^2)^2}{V^2} \frac{\Sigma |\mathcal{A}|^2}{2k 2k' 2E 2E'} \frac{V d^3 k'}{(2\pi)^3} \frac{V d^3 p'}{(2\pi)^3} \delta_{\mathbf{p}'+\mathbf{k}'-\mathbf{p}-\mathbf{k}} \delta(k' + E' - k - E). \tag{10.261}$$

A few comments are in order. The initial factor of 1/2 is present because we average over the spin alignment of the initial electron. In calculating the density of states, the ρ in the Golden Rule, we have used, the continuum normalization for the final momenta, but have kept the Kronecker delta on the three-momentum because, in squaring the amplitude, this especially simple: the squared of the Kronecker delta is simply the Kronecker delta itself. Finally, rewriting the Kronecker delta in the continuum normalization, we have

$$d\Gamma = \left(\frac{1}{2}\right)2\pi\frac{(4\pi e^2)^2}{V^2}\frac{\Sigma|\mathcal{A}|^2}{2k2k'2E2E'}\frac{Vd^3k'}{(2\pi)^3}\frac{Vd^3p'}{(2\pi)^3}\frac{(2\pi)^3}{V}\delta^4(p'+k'-p-k)$$

$$= \frac{e^4\Sigma|\mathcal{A}|^2}{8kk'EE'V}d^3k'\delta(k'+E'-k-E). \tag{10.262}$$

It is worth generalizing this result by indicating the amplitude by $\mathcal{M}=4\pi e^2\mathcal{A}$ and defining a phase-space factor

$$d\Phi = \frac{d^3k'}{2\omega'(2\pi)^3}\frac{d^3p}{2E'(2\pi)^3}\delta^4(p+k-p'-k') \tag{10.263}$$

to write

$$d\Gamma = \frac{(2\pi)^4}{4\omega EV}|\mathcal{M}|^2 d\Phi, \tag{10.264}$$

where with our choice of the lab frame, $E=m$ and we write ω for k. To get a cross section, we divide by the flux, in our case simply $1/V$ since the incident particle is a photon and we have set $c=1$. Were the incident particle not a photon, the flux would be v/V, with v the incident velocity. The combination $E_1E_2|\mathbf{v}_1-\mathbf{v}_2|$ can be written covariantly as $\sqrt{(p_1\cdot p_2)^2-m_1^2m_2^2}$. Without summing over initial spins, we have the general prescription

$$d\sigma = \frac{(2\pi)^4|\mathcal{M}|^2}{4\sqrt{(p_1\cdot p_2)^2-m_1^2m_2^2}}d\Phi. \tag{10.265}$$

Returning to the case at hand, we integrate over dk' with θ fixed since we seek the cross section as a function of θ.

$$\int dk'\delta(k'+E'-k-E)=(1+\frac{dE'}{dk'})^{-1}. \tag{10.266}$$

From

$$E'^2=(\mathbf{k}-\mathbf{k'})^2+m^2, \tag{10.267}$$

we have

$$1+\frac{dE'}{dk'}=\frac{E'+k'-k\cos\theta}{E'}=\frac{E+k-k\cos\theta}{E'}. \tag{10.268}$$

From the kinematic relation

$$k'=\frac{k}{1+(k/m)(1-\cos\theta)}, \tag{10.269}$$

we obtain, remembering that $E=m$,

$$1+\frac{dE'}{dk'}=\frac{mk}{k'E'} \tag{10.270}$$

and

$$d\Gamma = \frac{e^4\Sigma|\mathcal{A}|^2}{8m^2V}\frac{k'^2}{k^2}d\Omega. \tag{10.271}$$

The flux of incident photons is, with $c=1$, just $1/V$, so

$$\frac{d\sigma}{d\Omega}=\frac{e^4\Sigma|\mathcal{A}|^2}{8m^2}\frac{k'^2}{k^2}. \tag{10.272}$$

Proving that

$$\Sigma |\mathcal{A}|^2 = 2\left(\frac{k}{k'} + \frac{k'}{k}\right) + 8(\epsilon'^* \cdot \epsilon)^2 - 4 \qquad (10.273)$$

is left as Problem 10.10. The result is the Klein-Nishina formula [O. Klein and Y. Nishina, Z. Physik, **52** 853 (1929)],

$$\frac{d\sigma}{d\Omega} = \left(\frac{e^2}{m}\right)^2 \frac{k'^2}{k^2}\left[\frac{1}{4}\left(\frac{k}{k'} + \frac{k'}{k}\right) + (\epsilon'^* \cdot \epsilon)^2 - \frac{1}{2}\right]. \qquad (10.274)$$

10.15.5 Feynman Diagram Approach

The calculation of relativistic Compton scattering using standard perturbation theory is completely legitimate. Nonetheless, it is not the way the calculation is typically done today. The intermediate states in the diagrams above represent real ("on-shell") particles, ones that satisfy $E^2 = p^2 + m^2$. At each vertex in these diagrams, momentum is conserved but energy is not. The energy differences enter the denominators. Not only are such calculations awkward, they do not lend themselves to higher order calculations.

Shin'ichiro Tomonaga, Julian Schwinger, and Richard Feynman pioneered the calculation of quantum electrodynamics to higher order. It is Feynman's technique that has dominated, despite or perhaps because, as Schwinger asserted, "Feynman diagrams brought computation to the masses."

The structure of this covariant form of the matrix element can be understood in terms of vertex factors, propagators, and external lines. One assumes four-momentum conservation at each vertex. The internal lines in Feynman diagrams represent virtual particles, which do not satisfy (in general) $E^2 = m^2 + p^2$. Moreover, a single diagram encompasses both of the processes indicated for a single ordering of the emission and absorption in Compton scattering. In Figure 10.6, we show the diagram for the first term in the Compton amplitude. The propagator is

$$\frac{i}{\not{p} + \not{k} - m + i\epsilon} = i\frac{\not{p} + \not{k} + m}{(p + k)^2 - m^2 + i\epsilon}. \qquad (10.275)$$

Figure 10.6 Feynman diagram for one contribution to Compton scattering.

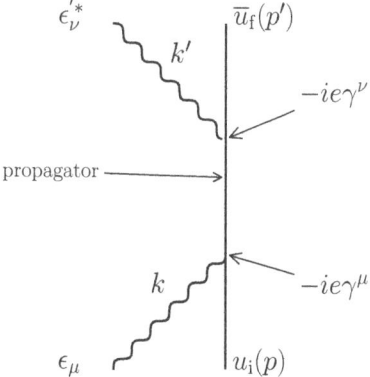

These factors are examples of Feynman rules for calculations in QED and other field theories. The propagator for the photon is

$$-\frac{ig^{\mu\nu}}{q^2 + i\epsilon}. \tag{10.276}$$

The rules are to take $i\mathcal{L}$ and strip-off the spinors, etc. to get the vertex factors. Then what one pieces together by multiplication of all the factors is $i\mathcal{M}$, where $\mathcal{M}$ is the Lorentz invariant amplitude, without the $(2E)^{-1/2}$ factors. Thus for the diagram in Figure 10.6, we obtain the contribution

$$-i\mathcal{M} = \bar{u}_f(p')(-ie\rlap{/}{e}^{'*})\frac{i}{\rlap{/}{p} + \rlap{/}{k} - m + i\epsilon}(-ie\rlap{/}{e})u_i(p). \tag{10.277}$$

The Feynman diagram provides an elegant means of representing the physical process and, at the same time, it gives the prescription for calculating the amplitude.

References and Suggested Reading

For the Foldy-Wouthuysen transformation, we follow the development of Bjorken and Drell, *Relativistic Quantum Mechanics*, pp. 46–52.

For the solution to the Dirac equation for hydrogen, see Bethe and Salpeter, *Quantum Mechanics of One- and Two-Electron Atoms*, Sections 14 and 15.

Many of the fundamental papers in relativistic quantum mechanics are reprinted in *Quantum Electrodynamics*, J. Schwinger (editor), Dover, New York (1958).

For neutron-electron scattering, see V. G. Sears, Physics Reports, **141**, 281, 1986.

Problems

10.1* The Dirac current density is $\mathbf{j} = ec(\psi^\dagger \boldsymbol{\alpha}\psi)$.

    (a) Show that this expression can be transformed into

$$\mathbf{j} = \mathbf{j}_S + c\nabla \times \mathbf{M} + \frac{\partial \mathbf{\Pi}}{\partial t}, \tag{10.278}$$

    where $\mathbf{j}_S$ is a relativistic generalization of the Schrödinger current,

$$\mathbf{j}_S = \frac{e\hbar}{2mi}[\psi^\dagger\beta\nabla\psi - (\nabla\psi)^\dagger\beta\psi] - \frac{e^2}{mc}\mathbf{A}\psi^\dagger\beta\psi, \tag{10.279}$$

    $\mathbf{M}$ is the magnetization, and $\mathbf{\Pi}$ is the electric polarization

$$\mathbf{M} = \frac{e\hbar}{2mc}\psi^\dagger\beta\boldsymbol{\sigma}_D\psi, \qquad \mathbf{\Pi} = \frac{e\hbar}{2imc}\psi^\dagger\beta\boldsymbol{\alpha}\psi. \tag{10.280}$$

    (b) Show that the nonrelativistic (Pauli) approximations to $\mathbf{j}_S$ and $\mathbf{M}$ is equivalent to the forms in part (a) with β set equal to unity and ψ interpreted as a Pauli spinor. (Make an explicit reduction in order to see what is being neglected or discuss "evenness" or "oddness" of the operators.)

(c) Find the nonrelativistic approximation to the polarization $\mathbf{\Pi}$ and show that it is of order v/c times the magnetization $\mathbf{M}$.

10.2 Projection operators for positive and negative energy free-particle states are defined by

$$\Lambda_+(\mathbf{p}) = \sum_{j=1,2} w_j(\mathbf{p})[w_j(\mathbf{p})]^\dagger; \qquad \Lambda_-(\mathbf{p}) = \sum_{j=3,4} w_j(\mathbf{p})[w_j(\mathbf{p})]^\dagger, \qquad (10.281)$$

where the Dirac spinors are normalized as $[w_j(\mathbf{p})]^\dagger w_{j'}(\mathbf{p}) = \delta_{jj'}$.

(a) Show that

$$\Lambda_\pm(\mathbf{p}) = \frac{1}{2}\left[1 \pm \frac{\boldsymbol{\alpha}\cdot\mathbf{p} + \beta m}{E(p)}\right] \qquad (10.282)$$

and also that

$$(\Lambda_\pm)^2 = \Lambda_\pm; \qquad \Lambda_+\Lambda_- = \Lambda_-\Lambda_+ = 0. \qquad (10.283)$$

(b) For invariantly normalized spinors u_j and v_j, defined in Eq. (10.46), with normalization

$$\bar{u}_j(\mathbf{p})u'_j(\mathbf{p}) = \delta_{jj'}; \qquad \bar{v}_j(\mathbf{p})v_{j'}(\mathbf{p}) = -\delta_{jj'}, \qquad (10.284)$$

where $\bar{\psi} = \psi^\dagger\beta$, show that the positive and negative projection operators are

$$\Lambda_+ = \sum_j u_j(\mathbf{p})\bar{u}_j(\mathbf{p})^\dagger = \frac{m + \slashed{p}}{2m}; \qquad \Lambda_- = -\sum_j v_j(\mathbf{p})\bar{v}_j(\mathbf{p})^\dagger = \frac{m - \slashed{p}}{2m}. \qquad (10.285)$$

- Note 1: if the spinors are normalized according to $\bar{u}u = 2m$, the factor $(2m)^{-1}$ in the sums will be absent and will not have the overall normalization of a projection.
- Note 2: the functions w_j in Bjorken and Drell, *Relativistic Quantum Mechanics*, are different from those here for $j = 3, 4$, but their u's and v's are the same. The connection between their w_j's and these is $[w_{3,4}(\mathbf{p})]_{\text{bjd}} = [w_{3,4}(-\mathbf{p})]_{\text{jdj}}$.

10.3 The ground state of a Dirac electron in a fixed point Coulomb potential with charge Ze is a $j = 1/2$ even-parity state with wave function

$$\psi(1/2, m) = Ar^{\gamma-1}e^{-Zr/a_0}\begin{pmatrix} \chi_m \\ iZ\alpha\boldsymbol{\sigma}\cdot\hat{\mathbf{r}}\chi_m/(1+\gamma) \end{pmatrix}, \qquad (10.286)$$

where χ_m is a Pauli spinor with z-component of angular momentum m, $\gamma = \sqrt{1 - \alpha^2 Z^2}$ and A is a normalization constant.

(a) Expand the wave function in terms of free-particle states, $w_j(\mathbf{p})e^{i\mathbf{p}\cdot\mathbf{x}}$. Evaluate the coefficients explicitly only for $\gamma = 1$.

(b) Calculate the probability $P(x)dx$ of finding a negative-energy free electron with magnitude of momentum $x = pa_0/Z$ in the interval dx, regardless of its spin.

(c) Show that the total probability of finding a negative-energy free electron is approximately $8(\alpha Z)^5/(15\pi)$ for $\alpha Z \ll 1$.

(d) What is the physical significance of these negative-energy states? By what order of magnitude would the ground-state energy change if these states were arbitrarily excluded from the wave function?

10.4 (a) Show that with the definitions $\gamma^\mu = (\beta, \beta\alpha)$, the free-particle Dirac equation can be written in the form,

$$(-i\gamma^\mu\partial_\mu + m)\psi = 0. \tag{10.287}$$

This is the Bjorken-Drell form of the Dirac equation. Sakurai's (Pauli's) form, $(\Gamma_\mu\partial_\mu + m)\psi = 0$, follows from $x_\mu = (\mathbf{x}, x_4 = ix_0)$ and $\Gamma_\mu = (-i\beta\alpha, \Gamma_4 = \beta)$.

(b) With the Bjorken-Drell choice of Dirac matrices (obeying $\{\gamma^\mu, \gamma^\nu\} = 2g^{\mu\nu}$) and the Feynman slash notation, $\not{a} = \gamma^0 a^0 - \boldsymbol{\gamma} \cdot \mathbf{a}$, prove the following trace relations:

$$\text{Tr}\,\not{a} = 0; \qquad \text{Tr}\,(\not{a}\not{b}) = 4a \cdot b; \qquad \text{Tr}\,(\not{a}\not{b}\not{c}) = 0$$

$$\text{Tr}\,(\not{a}\not{b}\not{c}\not{d}) = 4[(a \cdot b)(c \cdot d) - (a \cdot c)(b \cdot d) + (a \cdot d)(b \cdot c)]. \tag{10.288}$$

10.5 (a) From the definitions of α and β in terms of γ^μ and the results of Problem 10.4, show that there are analogous traces involving three-vectors:

$$\text{Tr}\,(\boldsymbol{\alpha} \cdot \mathbf{a}\,\boldsymbol{\alpha} \cdot \mathbf{b}) = 4\mathbf{a} \cdot \mathbf{b};$$

$$\text{Tr}\,(\boldsymbol{\alpha} \cdot \mathbf{a}\,\boldsymbol{\alpha} \cdot \mathbf{b}\,\boldsymbol{\alpha} \cdot \mathbf{c}\,\boldsymbol{\alpha} \cdot \mathbf{d}) = 4[(\mathbf{a} \cdot \mathbf{b})(\mathbf{c} \cdot \mathbf{d}) - (\mathbf{a} \cdot \mathbf{c})(\mathbf{b} \cdot \mathbf{d})$$

$$+ (\mathbf{a} \cdot \mathbf{d})(\mathbf{b} \cdot \mathbf{c})]. \tag{10.289}$$

(b) With the definition

$$\gamma^5 = i\gamma^0\gamma^1\gamma^2\gamma^3 = \begin{pmatrix} 0 & I \\ I & 0 \end{pmatrix} = \gamma_5 \tag{10.290}$$

and the Bjorken-Drell matrices, prove

$$\text{Tr}\,\gamma^5\not{a}\not{b}\not{c}\not{d} = 4i\epsilon_{\lambda\mu\nu\sigma}a^\lambda b^\mu c^\nu d^\sigma, \tag{10.291}$$

where $\epsilon_{0123} = +1$. Also prove along the way that that traces of γ^5 with fewer than four gamma matrices vanish.

(c) Prove that

$$\boldsymbol{\alpha} \cdot \mathbf{A}\,\boldsymbol{\alpha} \cdot \mathbf{B}\,\boldsymbol{\alpha} \cdot \mathbf{C} = \mathbf{A} \cdot \mathbf{B}(\boldsymbol{\alpha} \cdot \mathbf{C}) + \mathbf{B} \cdot \mathbf{C}(\boldsymbol{\alpha} \cdot \mathbf{A})$$

$$- \mathbf{A} \cdot \mathbf{C}(\boldsymbol{\alpha} \cdot \mathbf{B}) - i\Gamma^5\mathbf{B} \cdot (\mathbf{C} \times \mathbf{A}), \tag{10.292}$$

where the Pauli representation has

$$\Gamma^5 = i\alpha_1\alpha_2\alpha_3 = -\gamma^5. \tag{10.293}$$

10.6 The Foldy-Wouthuysen transformation produces, in principle, a Dirac wave function free of the ambiguities of negative-energy Fourier components. Operators such as $\mathbf{r}$ and σ_D can be interpreted straightforwardly. If these operators are transformed back to the original Dirac representation according to

$$\mathbf{R} = e^{-iS}\mathbf{r}e^{iS} \qquad \text{and} \qquad \boldsymbol{\Sigma} = e^{-iS}\sigma_D e^{iS}, \qquad \text{etc.} \tag{10.294}$$

they are called average position and spin operators.

(a) For a free Dirac particle and using the expressions

$$H = \beta m + \boldsymbol{\alpha} \cdot \mathbf{p} = \beta m + \mathcal{O}$$

$$S = -\frac{i\beta\mathcal{O}}{2m} = -i\frac{\beta\boldsymbol{\alpha} \cdot \mathbf{p}}{2m}, \tag{10.295}$$

show that, correct to order $1/m$ inclusive, the momentum operator is unchanged, while

$$\mathbf{R} = \mathbf{r} + i\frac{\beta\boldsymbol{\alpha}}{2m}; \qquad \boldsymbol{\Sigma} = \boldsymbol{\sigma}_D - i\frac{\beta\boldsymbol{\alpha} \times \mathbf{p}}{m}. \tag{10.296}$$

(b) Show that

$$\frac{d\mathbf{R}}{dt} = i[H, \mathbf{R}] = \beta\mathbf{p}/m. \tag{10.297}$$

Interpret.

(c) Show that

$$\mathbf{J} = \mathbf{r} \times \mathbf{p} + \frac{1}{2}\boldsymbol{\sigma}_D = \mathbf{R} \times \mathbf{p} + \frac{1}{2}\boldsymbol{\Sigma} \tag{10.298}$$

to order $1/m$, inclusive, where $\mathbf{R} \times \mathbf{p}$ is the average angular momentum. Demonstrate that $\mathbf{R} \times \mathbf{p}$ and $\boldsymbol{\Sigma}$ commute separately with the Hamiltonian to the same order in $1/m$.

Note that, as shown in Section 4.2 or Bjorken and Drell, *Relativistic Quantum Mechanics*, and Problem 3-8 of Sakurai, *Advanced Quantum Mechanics*, the Foldy-Wouthuysen transformation for a free particle can be determined in closed form (even though involving non-local operators). This means that, in principle, this problem can be done exactly, rather than as an expansion in $1/m$. The simplified considerations of the problem are sufficient to illustrate the concepts.

10.7 Consider a Dirac particle of mass m interacting with a Lorentz-scalar, external, static potential $V(\mathbf{r})$. Such a potential appears in the Dirac equation as an addition to the mass term.

(a) Use the Foldy-Wouthuysen reduction procedure to show that the positive-energy interaction Hamiltonian, correct to order m^{-2} inclusive, is equal to

$$H_{\text{int}} = V - \frac{\boldsymbol{\sigma} \cdot (\nabla V \times \mathbf{p})}{4m^2} - \frac{\nabla^2 V}{8m^2} - \frac{V p^2 + p^2 V}{4m^2}. \tag{10.299}$$

Compare the spin-dependent part with the Thomas precession energy (Jackson, *Classical Electrodynamics*, 3rd edition, p. 552, Eq. (11.122)).

(b) For a central potential, $V = V(r)$, transform the spin-dependent interaction into $\boldsymbol{\sigma} \cdot \mathbf{L}$ times a function of r and examine the relative signs and magnitudes of the spin-orbit interactions from a Lorentz-scalar potential and a time-component of a Lorentz-four-vector potential of the same functional form.

10.8* The hyperfine splitting of spectral terms is caused by a coupling between the nuclear magnetic moment and the electron. If the nuclear moment is $\mathbf{M} = \mu_n \mathbf{I}/I = \mu_N g\mathbf{I}$, the corresponding vector potential in the Dirac equation is $\mathbf{A} = (\mathbf{M} \times \mathbf{r})/r^3$.

(a) Use a Foldy-Wouthuysen transformation to show that, to first order in the electronic charge $(-e)$ and the nuclear magnetic moment and correct to order m^{-2}, the hyperfine interaction is

$$H_{int} = (\mu_0 \mu_n / I) \left\{ \frac{2\mathbf{I} \cdot \mathbf{L}}{r^3} + \frac{8\pi}{3} \boldsymbol{\sigma} \cdot \mathbf{I} \delta(\mathbf{r}) + \frac{\boldsymbol{\sigma} \cdot [3\hat{\mathbf{r}}(\hat{\mathbf{r}} \cdot \mathbf{I}) - \mathbf{I}]}{r^3} \right\}, \tag{10.300}$$

where $\mu_0 = e\hbar/(2mc)$ is the Bohr magneton, μ_n is the nuclear magnetic moment, $\boldsymbol{\sigma}/2$ and $\mathbf{L}$ are the electron's spin and orbital angular momenta, and $\mathbf{I}$ is the nuclear angular momentum.

(b) Show that for $L = 0$ states, the energy shift is

$$\Delta E = \frac{8\pi}{3} \mu_0 \mu_n |\Psi(0)|^2 \cdot \{1, -(I+1)/I\} \quad \text{for} \quad F = \{I + 1/2, I - 1/2\}, \tag{10.301}$$

where F is the total angular momentum.

(c) Show that for $p_{1/2}$ states, the energy shift is the same as for $s_{1/2}$ states, except that $\langle r^{-3} \rangle$ replaces $\pi |\Psi(0)|^2$.

The evaluation of the hyperfine interaction for $\ell \neq 0$ is not entirely straightforward. We need the matrix elements of $(\boldsymbol{\sigma} \cdot \hat{\mathbf{r}})(\mathbf{I} \cdot \hat{\mathbf{r}})$ between states of identical ℓ. Following Bethe and Salpeter *Quantum Mechanics of One- and Two-Electron Atoms* [Appendix A, Eq. (A33)], we hypothesize that

$$\mathbf{A} \cdot \mathbf{B} - 3\mathbf{A} \cdot \hat{\mathbf{r}} \mathbf{B} \cdot \hat{\mathbf{r}} = C[L^2 \mathbf{A} \cdot \mathbf{B} - 3\mathbf{A} \cdot \mathbf{L} \mathbf{B} \cdot \mathbf{L}], \tag{10.302}$$

where $\mathbf{A}$ and $\mathbf{B}$ commute with $\mathbf{r}$ and $\mathbf{L}$ and each other. The form on the right side with the L^2 factor guarantees that the righthand side vanishes as the left-hand side does, if we successively take $\mathbf{A} = \mathbf{B} = \hat{\mathbf{x}}, \hat{\mathbf{y}}, \hat{\mathbf{z}}$ and sum. So the problem is to determine the coefficient C. To do this, we need just a single non-zero evaluation. Take $\mathbf{A}$ and $\mathbf{B}$ conveniently and use the Clebsch-Gordan coefficient

$$\langle \ell m 2 0 | \ell m \rangle = \frac{3m^2 - \ell(\ell+1)}{\sqrt{(2\ell-1)\ell(\ell+1)(2\ell+3)}} \tag{10.303}$$

and the Gaunt integral to show that

$$C = -\frac{2}{(2\ell+3)(2\ell-1)}. \tag{10.304}$$

In evaluating the hyperfine interaction, we now encounter $\mathbf{I} \cdot \mathbf{L}$ and $\mathbf{S} \cdot \mathbf{I}$. Use the result of Problem 4.2 to make replacements

$$\mathbf{I} \cdot \mathbf{L} \to \mathbf{I} \cdot \mathbf{J} \frac{\mathbf{L} \cdot \mathbf{J}}{j(j+1)}; \quad \mathbf{I} \cdot \mathbf{S} \to \mathbf{I} \cdot \mathbf{J} \frac{\mathbf{S} \cdot \mathbf{J}}{j(j+1)}. \tag{10.305}$$

Then show that the hyperfine splitting for a single electron in a state defined by j, ℓ, and f with a nucleus of magnetic moment $\mu_n g I$ is

$$E_{hf} = \mu_0 \mu_N g \frac{\ell(\ell+1)}{j(j+1)} (f(f+1) - I(I+1) - j(j+1)) \left\langle \frac{1}{r^3} \right\rangle. \tag{10.306}$$

The results of the problem were first obtained by E. Fermi, Z. Phys. **60**, 320 (1930). For a classical derivation of part (a), see Jackson, *Classical electrodynamics*, 3rd edition, p. 190-1, especially Eqs.(5.73, 5.74).

10.9 Do the Dirac algebra required to prove Eq. (10.273). See Bjorken and Drell, *Relativistic Quantum Mechanics*, pp.127-132 if you want help.

A

Dimensions and Units

The desire to write the Coulomb potential simply as

$$V(r) = \frac{qq'}{r} \tag{A.1}$$

and the need to work with the everyday units of volts and tesla (or, previously, gauss) create a conflict, which may be addressed pragmatically. In circumstances that call for values of electric and magnetic fields, we use SI (Système International) units, e.g. volts/m or tesla ($=10^4$ gauss). In atomic physics, the issue can usually be avoided by incorporating the unit of electric charge into two standard quantities:

The Bohr radius: $a_0 = \hbar^2/(me^2) = 5.29177 \times 10^{-11}$ m.
The Rydberg: $me^4/(2\hbar^2) = 13.606$ eV.
The atomic unit of electric field $e/a_0^2 = 5.142 \times 10^{11}$ V/m.

When we encounter electric fields in atomic physics, as in the Stark effect, the natural unit is simply the electron-volt, but the corresponding natural unit for the Bohr magneton is more complicated. Because the force law in SI units is $\mathbf{F} = q(\mathbf{E} + \mathbf{v} \times \mathbf{B})$ while in Gaussian units it is instead $\mathbf{F} = q(\mathbf{E} + (\mathbf{v}/c) \times \mathbf{B})$, the Bohr magneton is $e\hbar/(2m_e)$ in SI, but $e\hbar/(2m_ec)$ in Gaussian form. When the magnetic field is given in tesla, the Bohr magneton has the value 5.788×10^{-5} eV $\cdot$ T^{-1}. A comprehensive treatment of electric units is given in the Appendix of *Classical Electrodynamics*, 3rd edition.

John David Jackson: A Course in Quantum Mechanics, First Edition. Robert N. Cahn.
© 2024 John Wiley & Sons, Inc. Published 2024 by John Wiley & Sons, Inc.
Companion website: www.wiley.com/go/Jackson/QuantumMechanics

B

Mathematical Tools

B.1 Contour Integration

Contour integration in the complex plane makes it possible to evaluate many definite integrals when the indefinite integrals are unobtainable. We sketch very briefly the essential points of complex variable theory needed to use contour integration.

A function of the complex variable z is said to be analytic at the point z_0 if $f(z)$ can be expressed as a convergent series

$$f(z) = \sum_{n=0}^{\infty} a_n(z - z_0)^n \tag{B.1}$$

for some finite range $|z - z_0| < \delta$.

An integral along a curve C in the complex plane

$$\int_C dz\, f(z) \tag{B.2}$$

can be defined as the limit of a sum of segments Δz times values $f(z)$ in the obvious, but heuristic, way

$$\int_C dz\, f(z) = \lim_{n\to\infty} \sum_{i=1}^{n-1} f(z_i)(z_{i+1} - z_i), \tag{B.3}$$

where the z_i are points along the curve C stretching from z_1 to z_n and the limit is taken so each interval $|z_{i+1} - z_i|$ tends to zero.

For example, if the curve C is a straight line from z_a to z_b,

$$\int_C dz\, z^n = \frac{1}{n+1}[z_b^{n+1} - z_a^{n+1}] \tag{B.4}$$

which follows from setting $z = z_a + t(z_b - z_a)$ and then doing an ordinary integration from $t = 0$ to $t = 1$.

The fundamental theorem (Cauchy), is that if f is analytic throughout a simply connected domain D and C is a closed curve inside D then

John David Jackson: A Course in Quantum Mechanics, First Edition. Robert N. Cahn.
© 2024 John Wiley & Sons, Inc. Published 2024 by John Wiley & Sons, Inc.
Companion website: www.wiley.com/go/Jackson/QuantumMechanics

$$\oint_C dz f(z) = 0. \tag{B.5}$$

This is easy enough to see for, say, a square within the region where the expansion Eq. (B.1) holds. If the vertices of the square (or any quadrilateral) are z_1, z_2, z_3, z_4, each term of the integral will be of the form

$$a_n \oint dz\, z^n = \frac{a_n}{n+1} \left[(z_2^{n+1} - z_1^{n+1}) + (z_3^{n+1} - z_2^{n+1}) \right.$$
$$\left. + (z_4^{n+1} - z_3^{n+1}) + (z_1^{n+1} - z_4^{n+1}) \right] = 0. \tag{B.6}$$

Of course, this integral vanishes by the theorem we are trying to prove, but here we demonstrate it directly. Now for a larger contour, just fill it up with small square contours nestled together. In general, not every square will lie within the region of convergence of the same single z_0, but we can fill the region D with overlapping regions where a convergent expansion exists, always choosing a small enough square to fit within the convergence domain of a single expansion center. Adjacent segments of the squares would cancel each other and only the outer contour, approximated by straight segments, would remain. See Figure B.1. Smaller and smaller squares can be used to nestle the straight-edged shape next to the curve C. This heuristic argument can be made into a real proof.

Residue Theorem

Consider the function $f(z) = 1/z$, which is not analytic at $z = 0$, and a circular contour C given by $|z| = R$. Along this contour, for which the Cauchy theorem does not apply,

$$z = Re^{i\theta}; \quad dz = Rie^{i\theta} d\theta \tag{B.7}$$

so that following the contour C once in the counterclockwise direction

$$\int_C dz f(z) = \int_0^{2\pi} d\theta \frac{Rie^{i\theta}}{(Re^{i\theta})} = \int_0^{2\pi} i d\theta = 2\pi i. \tag{B.8}$$

In fact, for any curve that encircles $z = 0$ once, going counterclockwise will give the same result because that contour can be shrunk or expanded down to this circle without crossing

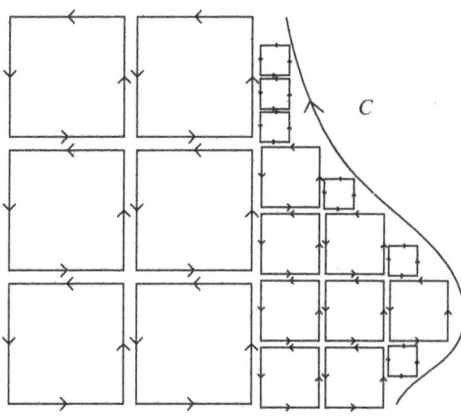

Figure B.1 We approximate the contour C in $\oint_C dz\, f(z)$ by straight line segments at the boundary of a region filled with squares small enough to be within a single region of convergence of the power series for $f(z)$. Then the integral around each square is zero. The contours of adjacent squares cancel except at the boundary, providing a heuristic demonstration that $\oint_C dz\, f(z)$ vanishes. Only a portion of the closed curve C and the filling squares are shown.

any region where the function is not analytic. The difference between the contours can be made of contours for which the integral is zero by Cauchy's theorem.

Consider next

$$\int_C dz\, z^n,$$ (B.9)

where n is an integer other than -1 and again take the same circular path. Now

$$\int_C dz\, z^n = \int_0^{2\pi} id\theta\, Re^{i\theta} R^n e^{in\theta} = \int_0^{2\pi} id\theta\, R^{n+1} e^{i(n+1)\theta}.$$ (B.10)

For $n \neq -1$, this is simply

$$\int_C dz\, z^n = iR^{n+1}(e^{i(n+1)2\pi} - e^0) = 0.$$ (B.11)

Suppose a function can be expanded in some region as

$$f(z) = \sum_{n=-p}^{\infty} a_n(z - z_0)^n.$$ (B.12)

If the most singular term is $(z - z_0)^{-p}$, the function is said to have a pole of order p. The only term that matters for a curve enclosing z_0 (once) is the term $n = -1$. The coefficient a_{-1} is called the residue and we have the relation

$$\int_C dz\, f(z) = 2\pi i \mathrm{Res}\, f(z_0),$$ (B.13)

where the contour C encloses z_0 once, counterclockwise, and no other pole.

If $f(z)$ has a pole of order n, the coefficient a_{-1} can be isolated by evaluating

$$\mathrm{Res}\, f(z)\Big|_{z=z_0} = \lim_{z \to z_0} \frac{1}{(n-1)!} \left(\frac{d}{dz}\right)^{n-1} (z - z_0)^n f(z).$$ (B.14)

Useful Example

As an example of the power of contour integration, consider

$$\int_{-\infty}^{\infty} dx \frac{\sin x}{x}.$$ (B.15)

Since the integrand has the limit 1 at $z = 0$, it is clear there is no pole there. Initially, this integral, viewed in the complex plane, goes along the real axis from $-\infty$ to $+\infty$. Since there are no singularities, we are free to distort the contour, which we do by sliding below the real axis near $z = 0$ so that this point lies above the contour. See Figure B.2. Now write the integral as

$$\int_C dz \frac{e^{iz} - e^{-iz}}{2iz}$$ (B.16)

and evaluate the two separate pieces. For the first piece, we add a semi-circular contour with $|z| = R$ in the upper half-plane to the contour along the real axis from $z = -R$ to $z = R$, avoiding the origin. As R increases, this contour contributes less and less. However, the contour now encircles $z = 0$, where, for this piece, there is a pole. Its residue is simply $1/(2i)$.

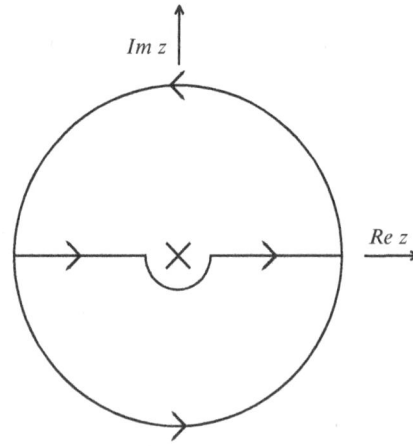

Figure B.2 Contours used to evaluate $\int_{-\infty}^{\infty} dx\,(\sin x)/x$ by treating x as a complex variable, z. The function $\sin z$ is written as the sum of two complex exponentials, divided by $2i$. The contour along the real z is distorted to encircle $z = 0$ from below. For the e^{iz} piece, the contour can be closed in the upper half-plane and if the semi-circular portion is moved off to infinity it makes no contribution. There is a contribution from the pole at $z = 0$. For the e^{-iz} piece, the contour is closed in the lower half-plane. That contour encircles no poles and makes no contribution.

Thus, it contributes to the integral precisely π. Now for the second piece, we must complete the contour with a large semi-circle in the lower half-plane, so that this contribution vanishes as R increases. Now this contour encloses no pole, so it contributes nothing. Thus we find

$$\int_{-\infty}^{\infty} dx \, \frac{\sin x}{x} = \pi. \tag{B.17}$$

Branch Points

If $f(z)$ is not analytic at z, z is said to be a singularity. Not all singularities are poles.

Suppose we want to evaluate

$$\int_{0}^{\infty} dx \, \frac{1}{(x+1)x^{1/2}}. \tag{B.18}$$

Of course this integral is easily determined with the substitution $x = y^2$, but we shall instead use the residue theorem. We consider the complex function

$$f(z) = \frac{1}{(z+1)z^{1/2}}. \tag{B.19}$$

There is a singularity at $z = 0$, which is not a pole, but rather a branch point. This function is not yet completely defined, since we need to give a meaning to $z^{1/2}$. If we write $z = |z|e^{i\theta}$, where θ is defined as shown in Figure B.3, we can chose to define

$$z^{1/2} = |z|e^{i\theta/2}. \tag{B.20}$$

With this definition, the value of $z^{1/2}$ has, different value above and below the branch, cut extending from the origin to $x = +\infty$. Above the branch cut, $z^{1/2} = |z|^{1/2}$ while below the branch $z^{1/2} = -|z|^{1/2}$. Now we evaluate in two different ways

$$\oint dz \, \frac{1}{(z+1)z^{1/2}} \tag{B.21}$$

with the contour shown in Figure B.3, going counterclockwise around the large circular arc. The contribution from the large arc is of order $R^{-1/2}$, with R the radius of the arc. So the only non-zero contribution to the integral is

Figure B.3 Contour used to calculate $\int_0^\infty dx\,(1+x)^{-1}x^{-1/2}$. In the complex plane, we can define $z^{1/2} = |z|e^{i\theta/2}$. There is a branch point at $z = 0$. Just above the branch cut, $z^{1/2} = |z|^{1/2}$. Just below the branch cut $z^{1/2} = -|z|^{1/2}$. The Integral is evaluated by comparing the integral over the closed contour, evaluated directly with its value determined by the residue theorem applied to the pole at $z = -1$.

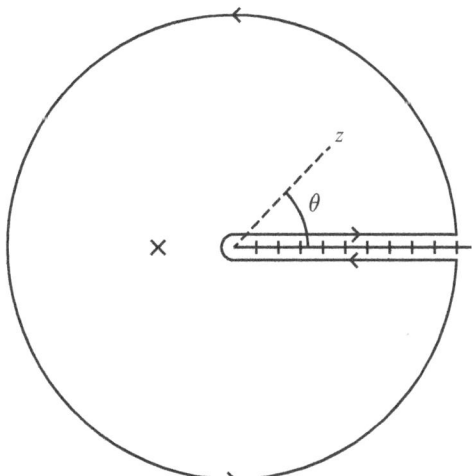

$$2 \int_0^\infty dx\, \frac{1}{(x+1)x^{1/2}}, \tag{B.22}$$

since the function changes sign across the branch point, but so does the direction of integration.

Now we evaluate the same complex integral, noting that there is a first-order pole at $z = -1$. The value of $z^{1/2}$ at the pole is $e^{i\pi/2} = i$. The residue at the pole is $z^{-1/2} = -i$. Thus,

$$\oint dz\, \frac{1}{(z+1)z^{1/2}} = 2\pi i \times (-i) = 2\pi = 2\int_0^\infty dx\, \frac{1}{(x+1)x^{1/2}}. \tag{B.23}$$

Thus,

$$\int_0^\infty dx\, \frac{1}{(x+1)x^{1/2}} = \pi. \tag{B.24}$$

B.2 Green Function for Helmholtz Equation

The essence of Fourier transforms, that if

$$\tilde{f}(k) = \int_{-\infty}^\infty dx\, e^{ikx} f(x), \tag{B.25}$$

then

$$f(x) = \int_{-\infty}^\infty \frac{dk}{2\pi} e^{-ikx} \tilde{f}(k) \tag{B.26}$$

is encapsulated in Dirac's delta function:

$$\int_{-\infty}^\infty dx\, e^{ix(k-k')} = 2\pi \delta(k - k') \tag{B.27}$$

or in three dimensions

$$\int d^3x \, e^{i\mathbf{x}\cdot(\mathbf{k}-\mathbf{k}')} = (2\pi)^3 \delta(\mathbf{k}-\mathbf{k}'). \tag{B.28}$$

To find the Green function for the Helmholtz wave equation

$$(\nabla^2 + k^2)G(\mathbf{x},\mathbf{x}') = -\delta(\mathbf{x}-\mathbf{x}') \tag{B.29}$$

we write $G(\mathbf{x},\mathbf{x}')$ as a Fourier transform

$$G(\mathbf{x},\mathbf{x}') = \int d^3q \, \tilde{G}(q)e^{-i\mathbf{q}\cdot(\mathbf{x}-\mathbf{x}')} \tag{B.30}$$

so that

$$\int d^3q \tilde{G}(\mathbf{q})(-q^2+k^2)e^{-i\mathbf{q}\cdot(\mathbf{x}-\mathbf{x}')} = -\frac{1}{(2\pi)^3}\int d^3q \, e^{-i\mathbf{q}\cdot(\mathbf{x}-\mathbf{x}')}. \tag{B.31}$$

This suggests that

$$\tilde{G}(\mathbf{q}) = \frac{1}{(2\pi)^3}\frac{1}{q^2-k^2}. \tag{B.32}$$

However, because this is singular at $q = \pm k$, it is not yet well defined. Consider

$$\tilde{G}^+(\mathbf{q}) = \frac{1}{(2\pi)^3}\frac{1}{q^2-k^2-i\epsilon} \tag{B.33}$$

in the limit $\epsilon \to 0$. Now doing the angular integral

$$\int d^3q \frac{1}{(2\pi)^3}\frac{1}{q^2-k^2-i\epsilon}e^{-i\mathbf{q}\cdot(\mathbf{x}-\mathbf{x}')} = \frac{4\pi}{(2\pi)^3}\int_0^\infty q^2 dq \frac{\sin qR}{qR}\frac{1}{q^2-k^2-i\epsilon}$$

$$= \frac{1}{(2\pi)^2 R}\int_{-\infty}^\infty dq \frac{q\sin qR}{q^2-k^2-i\epsilon}, \tag{B.34}$$

where $R = |\mathbf{x}-\mathbf{x}'|$. As above, we break $\sin qR$ into two complex exponentials. The term with e^{iqR} is evaluated with a contour along the real z axis and closed far out on the upper half-plane. The sole contribution is from the pole at $q \simeq k + i\epsilon/(2k^2) \equiv k + i\epsilon'$. See Figure B.4.

$$\frac{1}{(2\pi)^2 R}\oint_{UHP} dq \frac{qe^{iqR}}{2i(q^2-k^2-i\epsilon)} = \frac{1}{(2\pi)^2}\oint_{UHP} dq \frac{qe^{iqR}}{2i(q-k-i\epsilon')(q+k)}$$

$$= \frac{1}{(2\pi)^2 R}(2\pi i)\left(\frac{qe^{iqR}}{(2i)(q+k)}\right)_{q=k}$$

$$= \frac{1}{8\pi R}e^{iqR}. \tag{B.35}$$

An equal contribution is given by evaluating the term with e^{-iqR}, with the contour in the lower half-plane encircling the pole at $q = -k - i\epsilon'$. A factor (-1) occurs since the path encircles the pole clockwise. Altogether,

$$G^+(\mathbf{x},\mathbf{x}') = \frac{1}{4\pi R}e^{ikR} \tag{B.36}$$

where $R = |\mathbf{x}-\mathbf{x}'|$.

Figure B.4 Contours used to evaluate the Helmholtz Green function by treating q as a complex variable. The function $\sin qR$ is written as the sum of two complex exponentials, divided by $2i$. There are poles at $q = \pm (k+i\varepsilon')$. For the e^{iqR} piece, the contour can be closed in the upper half-plane, and if the semi-circular portion is moved off to infinity it makes no contribution. There is a contribution from the pole at $q=k+i\varepsilon'$. For the e^{-iqR} piece, the contour is closed in the lower half-plane and encircles the pole at $q=-k-i\varepsilon'$. In both cases, the result is proportional to e^{ikR}, guaranteeing an outgoing wave.

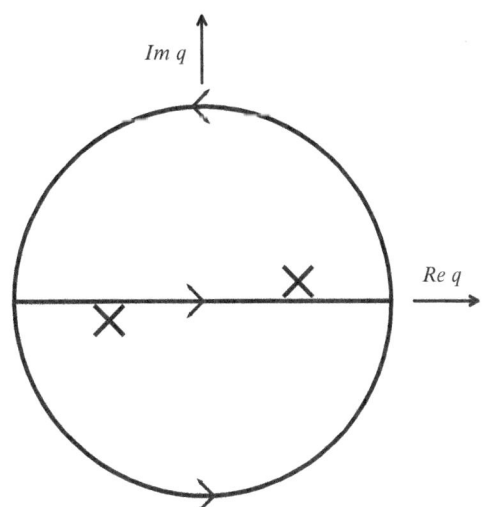

B.3   Wigner 3-j and 6-j Symbols

Reference: Edmonds, *Angular Momentum in Quantum Mechanics*, pp. 45-51, 90-100.

The Wigner 3-j symbol defined by the Clebsch-Gordan coefficient

$$\begin{pmatrix} j_1 & j_2 & j_3 \\ m_1 & m_2 & m_3 \end{pmatrix} = \frac{(-1)^{j_1-j_2-m_3}}{\sqrt{2j_3+1}} \langle j_1 m_1 j_2 m_2 | j_3 - m_3 \rangle \tag{B.37}$$

has these symmetry properties:

Under permutations of the columns, a 3-j symbol picks up a factor (-1) if the sum of the j's is odd and the permutation is odd; otherwise its value is unchanged by the permutation. Similarly, reversing the signs of the m's introduces a factor (-1) if the sum of the j's is odd. The orthogonality relations for the 3-j symbols are

$$(2j_3 + 1) \sum_{m_1 m_2} \begin{pmatrix} j_1 & j_2 & j_3 \\ m_1 & m_2 & m_3 \end{pmatrix} \begin{pmatrix} j_1 & j_2 & j_3' \\ m_1 & m_2 & m_3' \end{pmatrix} = \delta_{j_3 j_3'} \delta_{m_3 m_3'}$$

$$\sum_{j_3 m_3} (2j_3 + 1) \begin{pmatrix} j_1 & j_2 & j_3 \\ m_1 & m_2 & m_3 \end{pmatrix} \begin{pmatrix} j_1 & j_2 & j_3 \\ m_1' & m_2' & m_3 \end{pmatrix} = \delta_{m_1 m_1'} \delta_{m_2 m_2'}. \tag{B.38}$$

Often useful is a combination of four 3-j symbols, whose value depends only on six js:

$$\begin{Bmatrix} j_1 & j_2 & j_3 \\ j_4 & j_5 & j_6 \end{Bmatrix} = \sum_{m_1 \ldots m_6} (-1)^{\sum_i (j_i - m_i)} \begin{pmatrix} j_1 & j_2 & j_3 \\ -m_1 & -m_2 & -m_3 \end{pmatrix} \begin{pmatrix} j_1 & j_5 & j_6 \\ m_1 & -m_5 & m_6 \end{pmatrix}$$

$$\times \begin{pmatrix} j_4 & j_2 & j_6 \\ m_4 & m_2 & -m_6 \end{pmatrix} \begin{pmatrix} j_4 & j_5 & j_3 \\ -m_4 & m_5 & m_3 \end{pmatrix}. \tag{B.39}$$

Note that each m_i occurs once with a positive sign and once with a negative sign. When a 6-j symbol can be inserted in place of a complex combination of 3-j symbols or Clebsch-Gordan coefficients, it makes clearer which variables are actually significant. As an example, consider an expression occurring in the evaluation of the electrostatic interaction between atomic valence electrons:

$$I(\ell_1,\ell_2,\ell_1',\ell_2',L,k) = \frac{1}{2L+1} \sum_{\substack{m_1 m_1' \\ m_2 m_2' M_L}} \langle \ell_1' m_1' \ell_2' m_2' | L M_L \rangle \langle \ell_1 m_1 \ell_2 m_2 | L M_L \rangle$$

$$\times c^k(\ell_1' m_1'; \ell_1 m_1) c^k(\ell_2 m_2; \ell_2' m_2') \tag{B.40}$$

with

$$c^k(\ell_1', m_1', \ell_1, m_1) = (-1)^{m_1'} \sqrt{(2\ell_1'+1)(2\ell_1+1)}$$

$$\begin{pmatrix} \ell_1' & \ell_1 & k \\ 0 & 0 & 0 \end{pmatrix} \begin{pmatrix} \ell_1' & \ell_1 & k \\ -m_1' & m_1 & q \end{pmatrix} \tag{B.41}$$

where implicitly $q = m_1' - m_1$, since otherwise the expression vanishes. We thus find, using liberally the freedom to change factors like $(-1)^{-\ell}$ to $(-1)^{\ell}$ since all the ℓs and ms are integers,

$$I(\ell_1,\ell_2,\ell_1',\ell_2',L,k) = \sqrt{(2\ell_1+1)(2\ell_2+1)(2\ell_1'+1)(2\ell_2'+1)}$$

$$\begin{pmatrix} \ell_1' & \ell_1 & k \\ 0 & 0 & 0 \end{pmatrix} \begin{pmatrix} \ell_2 & \ell_2' & k \\ 0 & 0 & 0 \end{pmatrix}$$

$$\times \sum_{\substack{m_1 m_1' \\ m_2 m_2' M_L q}} (-1)^{\ell_1+\ell_2+\ell_1'+\ell_2'} \begin{pmatrix} \ell_1' & \ell_2' & L \\ m_1' & m_2' & -M_L \end{pmatrix} \begin{pmatrix} \ell_1 & \ell_2 & L \\ m_1 & m_2 & -M_L \end{pmatrix}$$

$$\times (-1)^{m_1'+m_2} \times \begin{pmatrix} \ell_1' & \ell_1 & k \\ -m_1' & m_1 & q \end{pmatrix} \begin{pmatrix} \ell_2 & \ell_2' & k \\ -m_2 & m_2' & q \end{pmatrix}. \tag{B.42}$$

We used the relation $m_1 + m_2 = m_1' + m_2'$ to identify the partners of k in the final two 3-j symbols. Let us identify $\ell_1 = j_1, \ell_2 = j_2, L = j_3, \ell_2' = j_4, \ell_1' = j_5, k = j_6$ and correspondingly $-M_L = m_3, m_2' = m_4, m_1' = m_5, q = m_6$. Then the sum alone becomes

$$\sum_{m_1..m_6} (-1)^{j_1+j_2+j_4+j_5} \begin{pmatrix} j_5 & j_4 & j_3 \\ m_5 & m_4 & m_3 \end{pmatrix} \begin{pmatrix} j_1 & j_2 & j_3 \\ m_1 & m_2 & m_3 \end{pmatrix}$$

$$\times (-1)^{m_5+m_2} \times \begin{pmatrix} j_5 & j_1 & j_6 \\ -m_5 & m_1 & m_6 \end{pmatrix} \begin{pmatrix} j_2 & j_4 & j_6 \\ -m_2 & m_4 & m_6 \end{pmatrix}. \tag{B.43}$$

We see that m_1, m_3, m_4 and q all appear twice with the same sign. This can be remedied by reversing the signs of the ms in the second and fourth 3-j symbols, which introduces a factor $(-1)^{j_1+j_3+j_4+j_6}$, yielding

$$(-1)^{j_1+j_4} \sum_{m_1..m_6} (-1)^{\Sigma_i j_i} \begin{pmatrix} j_5 & j_4 & j_3 \\ m_5 & m_4 & m_3 \end{pmatrix} \begin{pmatrix} j_1 & j_2 & j_3 \\ -m_1 & -m_2 & -m_3 \end{pmatrix}$$

$$\times (-1)^{m_5+m_2} \times \begin{pmatrix} j_5 & j_1 & j_6 \\ -m_5 & m_1 & m_6 \end{pmatrix} \begin{pmatrix} j_2 & j_4 & j_6 \\ m_2 & -m_4 & -m_6 \end{pmatrix}. \tag{B.44}$$

Since $m_1 + m_5 + m_6 = 0 = m_3 + m_4 + m_5$, we can insert a factor $(-1)^{m_1+m_3+m_4+m_6}$. The order of the columns in this expression differs from that in Eq. (B.39) and this can be corrected by odd permutations of all the 3-j symbols, except the one corresponding to $(j_1 j_2 j_3)$. This introduces a factor $(-1)^{j_1+j_2+j_3}$, making the initial coefficient $(-1)^{j_2+j_3+j_4}$, but since $j_2 + j_4 + j_6$ is even according to one of the 3-j symbols, we can instead write the initial factor as $(-1)^{j_3+j_6} = (-1)^{L+k}$. We see that the sum is

$$
(-1)^{L+k} \begin{Bmatrix} \ell_1 & \ell_2 & L \\ \ell'_2 & \ell'_1 & k \end{Bmatrix} \tag{B.45}
$$

and

$$
I(\ell_1, \ell_2, \ell'_1, \ell'_2, L, k) = (-1)^{L+k} \sqrt{(2\ell_1 + 1)(2\ell_2 + 1)(2\ell'_1 + 1)(2\ell'_2 + 1)}
$$

$$
\times \begin{pmatrix} \ell'_1 & \ell_1 & k \\ 0 & 0 & 0 \end{pmatrix} \begin{pmatrix} \ell_2 & \ell'_2 & k \\ 0 & 0 & 0 \end{pmatrix} \begin{Bmatrix} \ell_1 & \ell_2 & L \\ \ell'_2 & \ell'_1 & k \end{Bmatrix}. \tag{B.46}
$$

The 6-j symbol is invariant under any permutation of its column and as well if the upper and lower arguments of two columns are interchanged. If one element is zero, the value can be obtained from

$$
\begin{Bmatrix} j_1 & j_2 & j_3 \\ j_4 & j_5 & 0 \end{Bmatrix} = \frac{\delta_{j_2 j_4} \delta_{j_1 j_5}}{\sqrt{(2j_1 + 1)(2j_2 + 1)}} (-1)^{j_1+j_2+j_3} \tag{B.47}
$$

provided j_1, j_2, j_3 satisfy the triangular inequality and it vanishes otherwise.

C

Selected Solutions

C.1 Chapter 1

1.5 For clarity, we set $\hbar = 1$ here. A general solution to the Schrödinger equation for a free particle can be written as

$$\psi(\mathbf{x}, t) = \int d^3 p \, \phi(\mathbf{p}) e^{i \mathbf{p} \cdot \mathbf{x} - i p^2 t / 2m} \tag{C.1}$$

and by the inverse Fourier transform we must have

$$\phi(\mathbf{p}) = \frac{1}{(2\pi)^3} \int d^3 x' \, \psi(\mathbf{x}', 0) e^{-i \mathbf{p} \cdot \mathbf{x}'}. \tag{C.2}$$

Substituting the latter into the former,

$$\psi(\mathbf{x}, t) = \frac{1}{(2\pi)^3} \int d^3 p \, e^{i \mathbf{p} \cdot \mathbf{x} - i p^2 t / 2m} \int d^3 x' \, \psi(\mathbf{x}', 0) e^{-i \mathbf{p} \cdot \mathbf{x}'}, \tag{C.3}$$

so writing

$$\psi(\mathbf{x}, t) = \int d^3 x' \, \psi(\mathbf{x}', 0) K(\mathbf{x}, \mathbf{x}', t) \tag{C.4}$$

the kernel or Green function is thus

$$K(\mathbf{x}, \mathbf{x}', t) = \frac{1}{(2\pi)^3} \int d^3 p \, e^{i \mathbf{p} \cdot (\mathbf{x} - \mathbf{x}') - i p^2 t / 2m}. \tag{C.5}$$

Note that K depends only on $r = |\mathbf{x} - \mathbf{x}'| \; \tau = t(-t')$. Recalling

$$\int d\Omega e^{i \mathbf{p} \cdot \mathbf{r}} = 4\pi \frac{\sin pr}{pr}, \tag{C.6}$$

John David Jackson: A Course in Quantum Mechanics, First Edition. Robert N. Cahn.
© 2024 John Wiley & Sons, Inc. Published 2024 by John Wiley & Sons, Inc.
Companion website: www.wiley.com/go/Jackson/QuantumMechanics

we have

$$K(r,\tau) = \frac{4\pi}{(2\pi)^3} \int_0^\infty p^2 dp \, \frac{\sin pr}{pr} e^{-ip^2\tau/2m}$$

$$= \frac{1}{2\pi^2 r} \int_0^\infty p dp \, \sin pr \, e^{-ip^2\tau/2m}$$

$$= \frac{1}{2\pi^2 r}(-\frac{d}{dr}) \int_0^\infty dp \, \cos pr \, e^{-ip^2\tau/2m}$$

$$= \frac{1}{4\pi^2 r}(-\frac{d}{dr}) \int_{-\infty}^\infty dp \, e^{ipr} e^{-ip^2\tau/2m}. \tag{C.7}$$

Now the trick is to make this a Gaussian integral. Let

$$\frac{p^2\tau}{2m} = y^2 \tag{C.8}$$

so that

$$\frac{p^2\tau}{2m} - pr = y^2 - r\sqrt{\frac{2m}{\tau}} y = \left(y - r\sqrt{\frac{m}{2\tau}}\right)^2 - r^2\frac{m}{2\tau}. \tag{C.9}$$

Substituting

$$K(r,\tau) = \frac{1}{4\pi^2 r}\sqrt{\frac{2m}{\tau}}\left(-\frac{d}{dr}\right) e^{imr^2/2\tau} \int_{-\infty}^\infty dy \, e^{-iy^2}. \tag{C.10}$$

Now we need

$$\int_{-\infty}^\infty dy \, e^{-iy^2}, \tag{C.11}$$

which we find by inserting a convergence factor, which we remove at the end:

$$\int_{-\infty}^\infty dx e^{-(\epsilon+i)x^2} = \frac{\sqrt{\pi}}{\sqrt{\epsilon+i}} \to e^{-i\pi/4}\sqrt{\pi}. \tag{C.12}$$

Inserting this:

$$K(r,\tau) = \frac{1}{4\pi^2 r}\sqrt{\frac{2m}{\tau}}(\frac{-imr}{\tau})e^{imr^2/2\tau} e^{-i\pi/4}\sqrt{\pi}$$

$$= \left(\frac{m}{2\pi i\tau}\right)^{3/2} \exp\left[-\frac{mr^2}{2i\tau}\right]. \tag{C.13}$$

C.2 Chapter 2

2.6 (a) For $n = 1$ we have

$$[A, B] = [A, B]. \tag{C.14}$$

We assume that A and B commute with $[A, B]$. Now by induction, assume for n

$$[A, B^n] = nB^{n-1}[A, B]. \tag{C.15}$$

Then

$$[A, B^{n+1}] = AB^nB - BB^nA = AB^nB - B^nAB + B^nAB - BB^nA$$
$$= [A, B^n]B + B^n[A, B] = nB^{n-1}[A, B]B + B^n[A, B]$$
$$= (n + 1)B^n[A, B], \tag{C.16}$$

where use the assumption that B commutes with $[A, B]$.

(b) One approach is to consider

$$F(s) = e^{sA}Be^{-sA}. \tag{C.17}$$

Now from the power series expansion of e^{sA}, is it clear that

$$\frac{d}{ds}e^{sA} = Ae^{sA}. \tag{C.18}$$

So, in particular

$$\frac{dF}{ds} = Ae^{sA}Be^{-sA} - e^{sA}Be^{-sA}A = [A, F(s)]. \tag{C.19}$$

Now suppose

$$F(s) = \sum_n s^nG_n. \tag{C.20}$$

But it follows that

$$\frac{dF}{ds} = \left[A, \sum_n s^nG_n\right] = \sum_m ms^{m-1}G_m \tag{C.21}$$

Now equate the coefficient of s^k in the two expressions:

$$[A, G_k] = (k + 1)G_{k+1}. \tag{C.22}$$

Expanding to first order

$$F(s) = 1 + s[A, B] + \cdots \tag{C.23}$$

so $G_1 = [A, B]$ and

$$G_2 = \frac{1}{2}[A, [A, B]]; \ldots G_n = \frac{1}{n!}[A, [A\ldots[A, B]]]. \tag{C.24}$$

(c) Let

$$F(s) = e^{sA}e^{sB}, \tag{C.25}$$

then

$$\frac{dF}{ds} = AF(s) + F(s)B. \tag{C.26}$$

We need to be careful about the order of operators. Now

$$F(s)B = e^{sA}e^{sB}B = e^{sA}Be^{sB}. \tag{C.27}$$

But by part (b)

$$e^{sA}B = (B + [sA, B])e^{sA}$$
$$FB = (B + s[A, B])F, \tag{C.28}$$

so

$$\frac{dF}{ds} = (A + B + s[A, B])F. \tag{C.29}$$

This is a simple first-order differential equation, and the solution that agrees with the value for $s = 0$ is

$$F(s) = e^{sA + sB + \frac{s^2}{2}[A,B]}, \tag{C.30}$$

which gives the required result for $s = 1$.

2.9 (a) First note that the Heisenberg operator $q(t)$ is obtained as

$$q(t) = e^{iHt} q(0) e^{-iHt}, \tag{C.31}$$

so we can write

$$e^{-iHt/\hbar} e^{ikq(0)} |0\rangle = e^{-iHt/\hbar} e^{ikq(0)} e^{iHt/\hbar} e^{-iHt/\hbar} |0\rangle = e^{ikq(-t)} e^{-i\omega t/2} |0\rangle. \tag{C.32}$$

From Problem 2.8, this can be written

$$e^{ikq(0)\cos\omega t - ikp(0)\sin\omega t/(m\omega)} e^{-i\omega t/2} |0\rangle. \tag{C.33}$$

(b) Now use the trick from Problem 2.6 to break the exponential into a product of two exponentials by writing

$$ikq(0)\cos\omega t - ikp(0)\frac{\sin\omega t}{m\omega}$$

$$= \left[ikq(0)\cos\omega t - ikp(0)\frac{\sin\omega t}{m\omega} + \frac{i\hbar k^2}{2m\omega}\sin\omega t \cos\omega t \right]$$

$$- \frac{i\hbar k^2}{2m\omega}\sin\omega t \cos\omega t. \tag{C.34}$$

The term in square brackets has the form $A + B + \frac{1}{2}[A, B]$. Using the results of Problem 2.6, we can write

$$\exp\left(ikq(0)\cos\omega t - ikp(0)\frac{\sin\omega t}{m\omega} \right)$$

$$= \exp(ikq(0)\cos\omega t)\exp\left(-ikp(0)\frac{\sin\omega t}{m\omega} \right)\exp\left(-\frac{i\hbar k^2}{2m\omega}\sin\omega t \cos\omega t \right). \tag{C.35}$$

We want to sandwich this between $\langle t|$ and $|0\rangle$, but we insert a complete set of position states

$$\langle t| = \int dq' \, \langle t|q'\rangle \langle q'|. \tag{C.36}$$

We note that we will have a common phase

$$\exp\left(-i\omega t/2 - i\frac{\hbar k^2 \sin 2\omega t}{4m\omega} + ikq'\cos\omega t \right) \tag{C.37}$$

inside the integral $\int dq'$. In addition, we have the matrix element

$$\left\langle q' \left| \exp\left(-ikp(0)\frac{\sin\omega t}{m\omega} \right) \right| 0 \right\rangle. \tag{C.38}$$

But we recognize that the operator here just translates and the matrix element is found by resolving the ground state at the translated position, $q' - \lambda$, with $\lambda = \hbar k/m\omega$:

$$\left\langle q' - \frac{\hbar k}{m\omega} \sin \omega t \,\middle|\, 0 \right\rangle. \tag{C.39}$$

This is just the ground state wave function evaluated at this time-dependent location. In particular, the square of the wave function gives us

$$|\langle q'|t\rangle|^2 = \sqrt{\frac{m\omega}{\pi\hbar}} \exp\left[-\frac{m\omega}{\hbar}\left(q' - \frac{\hbar k}{m\omega}\sin\omega t\right)^2\right]. \tag{C.40}$$

We see that the wave function sloshes back and forth with a center given by the location expected from a classical description. If we had an excited state rather than the ground state, that wave function would slosh back and forth about the classical location.

C.3 Chapter 3

3.4 We have

$$k(x) = \sqrt{2m(E - V_{\text{eff}})/\hbar^2} = \sqrt{\frac{2m}{\hbar^2}\left(E + \frac{e^2}{r} - \frac{L^2}{2mr^2}\right)}, \tag{C.41}$$

where L^2 is to be defined more precisely below. Let's define the dimensionless $x = r/a_0$, where the Bohr radius is $a_0 = \hbar^2/(me^2)$. Let $E = -(e^2/2a_0)\epsilon$ so ϵ is the binding energy in Rydbergs. Now

$$k^2(x) = \frac{1}{a_0^2}\left[-\epsilon + \frac{2}{x} - \frac{L^2}{x^2}\right]. \tag{C.42}$$

We need

$$I = \int_{x_1}^{x_2} dx \left[-\epsilon + \frac{2}{x} - \frac{L^2}{x^2}\right]^{1/2} = (n + \frac{1}{2})\pi, \tag{C.43}$$

where x_1, x_2 are the classical turning points, that is, where the argument of the square root vanishes. Using $u = 1/r$

$$I = \int_{u_1}^{u_2} \frac{du}{u^2} \sqrt{-\epsilon + 2u - L^2 u^2}. \tag{C.44}$$

The indefinite integral is elementary, but messy, so instead let's use contour integration, which is suggested by the special nature of the endpoints. Let's consider

$$\int_a^b \frac{dz}{z^2}\sqrt{(b - z)(z - a)}. \tag{C.45}$$

There are branch points at $z = a$ and $z = b$. We define the square roots of $b - z$ and $z - a$ so that on the real axis to the right of b both square roots are real. Then, at some z in the complex plane,

$$\sqrt{(b-z)(z-a)} = |b-z|^{1/2}e^{i\phi_b/2}|z-a|^{1/2}e^{i\phi_a/2}. \tag{C.46}$$

See Figure C.1. The branch cut runs along the real axis from a to b and just above the cut the function is i times a real function, while below the cut, the function is $-i$ times the same real function.

Now consider the contour shown in Figure C.2. We evaluate the integral on the contour shown in two ways. First, we calculate it directly from the distinct portions of the contour. Starting at real positive infinity and infinitesimally below the real axis, the integrand is real and positive. However, the contribution along this portion of the contour will be cancelled by the portion of the path that is just above the real axis, but going in the opposite sense. This is true until we reach the branch cut at b. Here the integrand is

$$-i\frac{\sqrt{(b-x)(x-a)}}{x^2}, \tag{C.47}$$

where x is the real part of z. When the contour continues above the cut, the value is simply

$$i\frac{\sqrt{(b-x)(x-a)}}{x^2}. \tag{C.48}$$

Together we obtain a contribution

$$2i\int_a^b dx\,\frac{\sqrt{(b-x)(x-a)}}{x^2}. \tag{C.49}$$

It is often the case in evaluating definite integrals with contour integration that the contribution from the contour at infinity is zero. That is not the case here. It is easy to see that, with our choice of phases defining the branch cut, $\sqrt{(b-z)(z-a)}$ evaluated for very large $|z|$ is simply z, with corrections $\mathcal{O}(1)$. Thus, the contribution to the path integral from the circle at a large distance is

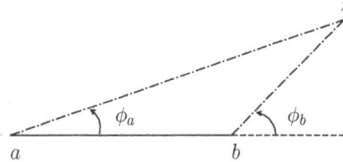

Figure C.1 The definition of the complex function $\sqrt{(b-z)(z-a)}$ obtained by a phase convention making the function real on the real axis for $z > b > a$. As a result, just above the branch cut, which runs along the real axis from a to b, the function is $i|b-z|^{1/2}|z-a|^{1/2}$. Just below the cut it is $-i|b-z|^{1/2}|z-a|^{1/2}$.

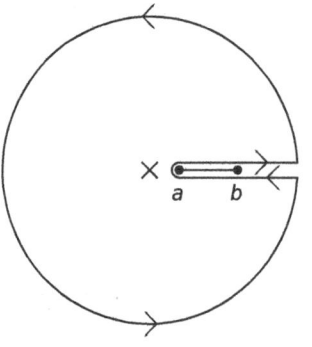

Figure C.2 The contour used to evaluate the integral for Problem 3.4.

$$\oint \frac{dz}{z} = 2\pi i. \tag{C.50}$$

Thus, the total integral along the path is

$$2\pi i + 2i \int_a^b dx \frac{\sqrt{(b-x)(x-a)}}{x^2}. \tag{C.51}$$

On the other hand, we can evaluate the integral using the residue theorem.

$$\oint f(z)dz = \sum_j 2\pi i \operatorname{Res}(z_j). \tag{C.52}$$

To find the residue at a pole z_j, we expand $f(z)$ about z_j and take the coefficient of $1/(z-z_j)$. Our pole is at $z = 0$ and it is a double pole. We expand

$$\frac{\sqrt{(b-z)(z-a)}}{z^2} = -\sqrt{ab}\frac{(1-z/b)^{1/2}(1-z/a)^{1/2}}{z^2}. \tag{C.53}$$

The overall minus sign arises because our function on the real axis below a is real and less than zero. Thus, the residue is

$$-\sqrt{ab}(-\frac{1}{2a} - \frac{1}{2b}). \tag{C.54}$$

Equating the two evaluations of the contour integral, we have

$$2\pi i\sqrt{ab}(\frac{1}{2a} + \frac{1}{2b}) = 2\pi i + 2i \int_a^b dx \frac{\sqrt{(b-x)(x-a)}}{x^2}$$

$$\int_a^b dx \frac{\sqrt{(b-x)(x-a)}}{x^2} = \pi\sqrt{ab}(\frac{1}{2a} + \frac{1}{2b}) - 2\pi = \frac{\pi}{2}\left(\sqrt{\frac{b}{a}} - 2 + \sqrt{\frac{a}{b}}\right)$$

$$= \frac{\pi}{2\sqrt{ab}}(\sqrt{b} - \sqrt{a})^2. \tag{C.55}$$

For our problem

$$b = \frac{1}{L^2}\left(1 + \sqrt{1 - \epsilon L^2}\right)$$

$$a = \frac{1}{L^2}\left(1 - \sqrt{1 - \epsilon L^2}\right)$$

$$ab = \epsilon/L^2$$

$$a + b = 2/L^2 \tag{C.56}$$

and

$$I = \frac{\pi L}{2}\left[\frac{2/L^2 - 2\sqrt{\epsilon/L^2}}{\sqrt{\epsilon/L^2}}\right] = \pi\left[\frac{1}{\sqrt{\epsilon}} - L\right] = \pi(n_r + \frac{1}{2}). \tag{C.57}$$

Then we find

$$\epsilon = -\frac{E}{e^2/(2a_0)} = \frac{1}{(n_r + 1/2 + L)^2}. \tag{C.58}$$

To get the correct expression, we must take $L = \ell + 1/2$, where ℓ is the orbital angular momentum.

The Langer substitution $\ell(\ell + 1) \to (\ell + 1/2)^2$ can be understood as follows:

Separation of the Schrödinger equation in spherical coordinates leads to the one-dimensional equation

$$\frac{d^2u}{dr^2} = \frac{2m}{\hbar^2}\left(E - V(r) - \frac{\ell(\ell+1)\hbar^2}{2mr^2}\right)u = 0, \tag{C.59}$$

where $u(r) = rR(r)$ and $\psi(r, \theta, \phi) = R(r)Y_{\ell m}(\theta, \phi)$. Writing $\ell(\ell + 1) = (\ell + 1/2)^2 - 1/4$ and defining

$$k^2(r) = \frac{2m}{\hbar^2}\left(E - V(r) - \frac{(\ell+1/2)^2\hbar^2}{2mr^2}\right), \tag{C.60}$$

the differential equation becomes

$$\frac{d^2u}{dr^2} + \left(k^2 + \frac{1}{4r^2}\right)u = 0. \tag{C.61}$$

Substituting $u = r^{1/2}w$ gives an equation for w

$$\frac{d^2w}{dr^2} + \frac{1}{r}\frac{dw}{dr} + k^2(r)w = 0. \tag{C.62}$$

If $k^2(r)$ were constant, the solution of this equation for $k^2 > 0$ would be a Bessel function of order zero: $J_0(kr), N_0(kr)$, or $H_0^{1,2}(kr)$, or if $k^{(2)} < 0, I_0(kr)$ or $K_0(kr)$. Now use the WKB ansatz for the function w:

$$w = e^{iS(r)/\hbar}. \tag{C.63}$$

Then

$$w' = \frac{i}{\hbar}S'w; \qquad w'' = \frac{i}{\hbar}S''w - \frac{1}{\hbar^2}(S')^2w \tag{C.64}$$

and we have

$$-(S')^2 + i\hbar S'' + \frac{i\hbar S'}{r} + \hbar^2 k^2(r) = 0. \tag{C.65}$$

If we set $p(r) = \hbar k(r)$, $S = S_0 + \hbar S_1 + \cdots$ and equate terms with common powers of $\hbar$, we find

$$S_0' = \pm p(r), \qquad S_0'' = \pm p'(r)$$

$$S_0 = \pm \int^r p(r')dr'. \tag{C.66}$$

The next term involves

$$-2S_0'S_1' + iS_0'' + \frac{iS_0'}{r} = 0 \tag{C.67}$$

or

$$iS_1' = -\frac{1}{2S_0'}\left(S_0'' + \frac{S_0'}{r}\right) = -\frac{p'}{2p} - \frac{1}{2r} = -\frac{1}{2}\frac{d}{dr}\ln(rp). \tag{C.68}$$

Hence

$$iS_1(r) = \ln\left(\frac{1}{\sqrt{rp(r)}}\right) + \text{const} \tag{C.69}$$

and

$$e^{iS_1(r)} = \frac{A'}{\sqrt{rp(r)}}. \tag{C.70}$$

The WKB-like solution is therefore

$$R(r) = \frac{u(r)}{r} = \frac{w}{\sqrt{r}} = \frac{A}{\sqrt{r^2 k(r)}} e^{\pm i \int^r k(r')dr'} \tag{C.71}$$

for the classically allowed region and

$$R(r) = \frac{u(r)}{r} = \frac{w}{\sqrt{r}} = \frac{A}{\sqrt{r^2 \kappa(r)}} e^{\pm \int^r \kappa(r')dr'} \tag{C.72}$$

in the classically forbidden region.

Now consider the behavior near the origin. Provided $\lim_{r\to 0} r^2 V(r) = 0$, for finite E, we have $\kappa \to (\ell + 1/2)/r$ as $r \to 0$. (For w, the angular momentum barrier dominates for $r \to 0$, even for $\ell = 0$.) Then

$$\xi' = \int_r^{r_1} \kappa(r')dr', \tag{C.73}$$

where r_1 is the beginning of the classically allowed region for w. The dominant portion of ξ' is at small r and

$$\xi' = \int_r^{r_0} \frac{\ell + 1/2}{r'} dr' + \text{const} = \ln\left(\frac{r_0^{\ell+1/2}}{r^{\ell+1/2}}\right) + \text{const} \tag{C.74}$$

and

$$e^{\mp\xi'} \to A'_\pm r^{\pm(\ell+1/2)}. \tag{C.75}$$

Note that

$$\lim_{r\to} r\kappa(r) = \text{constant}. \tag{C.76}$$

Hence

$$R(r) \to \frac{A_\pm}{\sqrt{r}} r^{\pm(\ell+1/2)}. \tag{C.77}$$

This gives $r^{-\ell-1}$ and r^ℓ for the regular and irregular solutions at the origin, in agreement with general rules. This explains the use of $(\ell + 1/2)^2$ in place of $\ell(\ell + 1)$ in the WKB radial quantization.

3.5 (a) Coulomb barrier and alpha decay. To find the transmission coefficient, we must evaluate

$$\Theta = \int_R^{r_1} \kappa(r)dr, \tag{C.78}$$

where

$$\kappa(r) = \sqrt{\frac{2m}{\hbar^2}\left(\frac{zZe^2}{r} - E\right)}. \tag{C.79}$$

The classical turning point is $r_1 = zZe^2/E$. We can thus write

$$\Theta = \sqrt{\frac{2mE}{\hbar^2}}\left(\frac{zZe^2}{E}\right)\int_{R/r_1}^1 dx\sqrt{\frac{1}{x} - 1}, \tag{C.80}$$

where $x = r/r_1$. Putting $x = y^2$ gives

$$\sqrt{\frac{1-x}{x}}\,dx = 2\sqrt{1-y^2}dy, \tag{C.81}$$

so that

$$\Theta = \frac{2zZe^2\sqrt{2m}}{\hbar\sqrt{E}}\int_{\sqrt{R/r_1}}^1 \sqrt{1-y^2}\,dy. \tag{C.82}$$

While we can, of course, do this integral, we are interested in a high barrier so $E \ll zZe^2/R$, and so it makes sense simply to write

$$\int_{\sqrt{R/r_1}}^1 \sqrt{1-y^2}\,dy = \int_0^1 \sqrt{1-y^2}\,dy - \int_0^{\sqrt{R/r_1}} \sqrt{1-y^2}\,dy$$

$$= \frac{\pi}{4} - \sqrt{\frac{R}{r_1}} + \mathcal{O}\left(\left(\frac{R}{r_1}\right)^{3/2}\right). \tag{C.83}$$

Thus, we need to keep only

$$\Theta \simeq \frac{2zZe^2\sqrt{2m}}{\hbar\sqrt{E}}\left(\frac{\pi}{4} - \sqrt{\frac{ER}{zZe^2}}\right). \tag{C.84}$$

Now

$$T = e^{-2\Theta} \qquad \text{for } \Theta \gg 1. \tag{C.85}$$

Thus

$$T = \exp\left(-\frac{A}{\sqrt{E}} + BR^{1/2}\right), \tag{C.86}$$

where

$$A = \frac{\pi zZe^2}{\hbar}\sqrt{2m}, \qquad B = \frac{4}{\hbar}\sqrt{2mzZe^2}. \tag{C.87}$$

(b) The coefficient

$$\left(\frac{A}{Z}\right)^2 = 2\pi^2 \times z^2 \left(\frac{m_\alpha}{m_e}\right) \left(\frac{m_e e^4}{h^2}\right) = 8\pi^2 \times 4 \times 1836 \times 27.21 \text{ eV}$$

$$= 15.67 \text{ MeV}. \tag{C.88}$$

Similarly,

$$ZB^{-2} = \frac{\hbar^2}{32 m_\alpha z e^2} = \frac{a_0}{64} \left(\frac{m_e}{m_\alpha}\right) = 0.1134 \text{ fm} \tag{C.89}$$

$(1 \text{ fm} = 10^{-15}\text{m})$ Thus

$$T = \exp\left[-Z\sqrt{\frac{15.67}{E(\text{MeV})}} + \sqrt{\frac{ZR(\text{fm})}{0.1134}}\right]. \tag{C.90}$$

Nuclear radii are given approximately by $R = r_0 A^{1/3}$, where $r_0 \simeq 1.5$ fm. For alpha emitters with $Z \simeq 80$ to 100, $A \simeq 200 - 250$. Thus $R \simeq 8.8 - 9.4$ fm. We can use an average value of $R = 9.1$ fm in T.

We use an uncertainty principle argument to estimate the frequency of an alpha particle's bouncing back and forth inside the nucleus, ignoring the clustering of alphas that actually occurs. The average speed can be found from demanding that $R\Delta p \simeq \hbar/2$. The velocity of the alpha is then $v \simeq \hbar/(2m_\alpha R)$. The "classical" period is then $\tau_{\text{cl}} \simeq 4R/v$ or

$$\tau_{\text{cl}} = \frac{8m_\alpha c^2}{\hbar c} \cdot \frac{R^2}{c} = \frac{8 \cdot 3727 \text{ MeV}}{197 \text{ MeV} - \text{fm}} \cdot \frac{(9.1)^2 \text{ fm}^2}{3 \times 10^{23} \text{ fm s}^{-1}}$$

$$= 4.2 \times 10^{-20} \text{ s}. \tag{C.91}$$

Another estimation of the period of motion inside is to use $\Delta E \simeq 10$ MeV and say $\Delta E = 2\pi\hbar/\tau_{\text{cl}}$. This gives

$$\tau_{\text{cl}} \simeq \frac{2\pi\hbar c}{\Delta E\, c} = \frac{2\pi \times 197 \text{ MeV fm}}{10 \text{ MeV} \cdot 3 \times 10^{23} \text{ fm/s}} = 4 \times 10^{-22} \text{ s}. \tag{C.92}$$

We thus have a factor of 10^2 difference. Our estimation of $\log_{10} t_{1/2}$ can be expected to have an error of perhaps ± 1.

Our half-life formula from the barrier penetration calculation is

$$t_{1/2} = \frac{\tau_{\text{cl}}}{2} e^{-2\Theta}. \tag{C.93}$$

This follows from the alpha particle striking the nuclear surface every $\tau_{\text{cl}}/2$ and having the chance $e^{-2\Theta}$ of "leaking through" the barrier each time.

$$\log_{10}[t_{1/2}(\text{s})] = -19.7 - 0.4343\left[-Z\sqrt{\frac{15.67}{E(\text{MeV})}} + \sqrt{\frac{ZR(\text{fm})}{0.1134}}\right]$$

$$= -19.7 + \frac{1.719Z}{\sqrt{E(\text{MeV})}} - 1.290\sqrt{ZR(\text{fm})}, \tag{C.94}$$

where the $\tau_{\text{cl}} = 4.2 \times 10^{-20}$ s has been used.

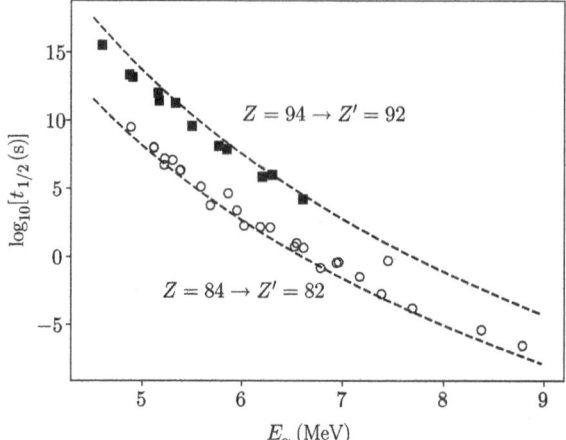

Figure C.3 The decay half-lives and alpha energies for decays of Po and Pu. The dashed lines are the predictions of the WKB model in this problem. The anomalous point for ^{211} Po is the result of the inhibition from a spin change of four units.

With $R = 9.1$ fm we get for Po$\rightarrow$ Pb $+ \alpha$ and Pu$\rightarrow$ U $+ \alpha$

$$Z' = 82: \qquad \log_{10}[t_{1/2}(\text{s})] = -54.9 + \frac{141.0}{\sqrt{E\,\text{MeV}}}$$

$$Z' = 92: \qquad \log_{10}[t_{1/2}(\text{s})] = -57.0 + \frac{158.1}{\sqrt{E\,\text{MeV}}}. \tag{C.95}$$

If a nucleus can decay in more than one way, its lifetime is the reciprocal of the sum of the partial decay rates. Thus $1/\tau = \Gamma_{\text{tot}} = \sum_i \Gamma_i$. Our model is for $\tau_\alpha = 1/\Gamma_\alpha$. The branching ratios are defined by $B_i = \Gamma_i/\Gamma_{\text{tot}}$. Thus $\tau_\alpha = \tau/B_\alpha$. For example, Po^{201} has only 1.6% branching ratio for alpha decay and 98.4% for electron capture. In Figure C.3, the half-lives are corrected for this factor using the given values of the $\log_{10} B_\alpha$.

3.6 (a) The barrier penetration is determined from

$$\Theta = \int_0^{W/eF} dz \sqrt{\frac{2m}{\hbar^2}(W - eFz)} = \sqrt{\frac{2meF}{\hbar^2}} \int_0^{W/eF} dz \sqrt{\frac{W}{eF} - z}$$

$$= \frac{2}{3}\sqrt{\frac{2meF}{\hbar^2}}\left(\frac{W}{eF}\right)^{3/2} = \frac{2}{3}\sqrt{\frac{2m}{\hbar^2}\frac{W^{3/2}}{eF}}. \tag{C.96}$$

Thus the probability of leaking out is

$$P(W) = \exp\left[-\frac{4}{3}\sqrt{\frac{2m}{\hbar^2}\frac{W^{3/2}}{eF}}\right], \tag{C.97}$$

where W is a function of p_z.

(b) Now we know the probability of each electron penetrating the barrier. We need to know the flux of the electrons and, in fact, the flux at each value of p_z. First, let us calculate the relationship between the density of conduction electrons and the Fermi level. In this simple model, we assume that the conduction electrons (the electrons in the conduction band) are free and that their energy is simply $p^2/2m$. Phase space gives the number of states as a function of energy or momentum, $\hbar k$. Here we need a factor of two for the two spin states of the electron:

$$dn = 2\frac{Vd^3k}{(2\pi)^3}. \tag{C.98}$$

Now consider a slab of the sphere in momentum space between p_z and $p_z + dp_z$. This has a volume in k space of $\hbar^{-3}dp_z\pi(p_F^2 - p_z^2)$, and thus the number of electrons in it, per unit physical volume, is

$$\frac{2\pi}{(2\pi\hbar)^3}(p_F^2 - p_z^2)dp_z \tag{C.99}$$

and the velocity of these particles is p_z/m. The number escaping per unit time per unit area is

$$\frac{2\pi}{(2\pi\hbar)^3}(p_F^2 - p_z^2)\frac{p_z}{m}P(W)dp_z = \frac{2\pi}{(2\pi\hbar)^3}(p_F^2 - p_z^2)\frac{dp_z^2}{2m}\exp\left[-\frac{4}{3}\sqrt{\frac{2m}{\hbar^2}}\frac{W^{3/2}}{eF}\right]. \tag{C.100}$$

Let us take as our integration variable $\lambda = (p_F^2 - p_z^2)/(2m)$ and then write

$$W = \phi + \lambda. \tag{C.101}$$

Now the total flow of electrons per unit time and per unit area is

$$\frac{2\pi}{(2\pi\hbar)^3}2m\int_0^{E_F}d\lambda\lambda\exp\left[-\frac{4}{3}\sqrt{\frac{2m}{\hbar^2}}\frac{(\phi + \lambda)^{3/2}}{eF}\right] \tag{C.102}$$

The integral is dominated by small values of λ, so we expand

$$\frac{4\pi m}{(2\pi\hbar)^3}\int_0^{EF}d\lambda\lambda\exp\left[-\frac{4}{3}\sqrt{\frac{2m}{\hbar^2}}\frac{\phi^{3/2} + (3/2)\phi^{1/2}\lambda}{eF}\right] \tag{C.103}$$

and without significant change set the upper limit to infinity

$$\frac{4\pi m}{(2\pi\hbar)^3}\exp\left[-\frac{4}{3}\sqrt{\frac{2m}{\hbar^2}}\frac{\phi^{3/2}}{eF}\right]\int_0^\infty d\lambda\lambda\exp\left[-2\sqrt{\frac{2m}{\hbar^2}}\frac{\phi^{1/2}}{eF}\lambda\right]$$

$$= \frac{4\pi m}{(2\pi\hbar)^3}\exp\left[-\frac{4}{3}\sqrt{\frac{2m}{\hbar^2}}\frac{\phi^{3/2}}{eF}\right]\frac{\hbar^2 e^2 F^2}{8m\phi}$$

$$= \frac{e^2 F^2}{16\pi^2\hbar\phi}\exp\left[-\frac{4}{3}\sqrt{\frac{2m}{\hbar^2}}\frac{\phi^{3/2}}{eF}\right]. \tag{C.104}$$

This is the flux of electrons out of the metal. The current is $(-e)$ times this. For the exciting application to the scanning tunnelling miscroscope, see, for example, J. A. Golovchenko, Science **232**, 48, (4 April 1986); C. F. Quate, Physics Today, **39**, No. 8, 26 (August 1986).

C.4   Chapter 4

4.5 Magnetic resonance with the Heisenberg equations of motion. We have

$$H = -\boldsymbol{\mu} \cdot \mathbf{B} \tag{C.105}$$

where

$$\boldsymbol{\mu} = \gamma \hbar \mathbf{J} = \frac{\hbar \omega_0}{B_0} \mathbf{J} \tag{C.106}$$

and

$$\begin{aligned}
\mathbf{B} &= B_0 \hat{\mathbf{z}} + B_1 (\hat{\mathbf{x}} \cos \omega t - \hat{\mathbf{y}} \sin \omega t) \\
&= B_0 \hat{\mathbf{z}} + \frac{B_1}{2} \left[(\hat{\mathbf{x}} + i\hat{\mathbf{y}}) e^{i\omega t} + (\hat{\mathbf{x}} - i\hat{\mathbf{y}}) e^{-i\omega t} \right].
\end{aligned} \tag{C.107}$$

Thus

$$H = -\hbar \left[\omega_0 J_z + \frac{\Gamma}{2} (J_+ e^{i\omega t} + J_- e^{-i\omega t}) \right], \tag{C.108}$$

where $\Gamma = \omega_0 B_1 / B_0$.

(a) The Heisenberg equations of motion are

$$\frac{d\mathbf{J}}{dt} = \frac{1}{i\hbar} [\mathbf{J}, H]. \tag{C.109}$$

With the aid of the angular momentum commutation relations

$$[J_z, J_\pm] = \pm J_\pm \quad ; \qquad [J_\pm, J_\mp] = \pm 2 J_z, \tag{C.110}$$

we find

$$\begin{aligned}
\frac{dJ_z}{dt} &= i \frac{\Gamma}{2} \left(J_+ e^{i\omega t} - J_- e^{-i\omega t} \right) \\
\frac{dJ_\pm}{dt} &= \mp i \left(\omega_0 J_\pm - \Gamma J_z e^{\mp i\omega t} \right).
\end{aligned} \tag{C.111}$$

The structure of these equations suggest introducing $\tilde{J}_\pm = J_\pm e^{\pm i\omega t}$. With this substitution we have

$$\frac{d\tilde{J}_\pm}{dt} \pm i(\omega_0 - \omega)\tilde{J}_\pm = \pm i \Gamma J_z. \tag{C.112}$$

The equation for J_z becomes

$$\frac{dJ_z}{dt} = i \frac{\Gamma}{2} (\tilde{J}_+ - \tilde{J}_-). \tag{C.113}$$

It is clear that the motion must be periodic, but what is the period? We need to find a homogeneous differential equation for a single amplitude so that we can insert $e^{i\nu t}$ to determine the frequency ν.

Differentiating the equation for J_z we find, writing $\Delta\omega = \omega_0 - \omega$,

$$\frac{d^2 J_z}{dt^2} = -\Gamma^2 J_z + \frac{\Gamma}{2}\Delta\omega(\tilde{J}_+ \mid \tilde{J}_-) \tag{C.114}$$

Differentiating once more,

$$\frac{d^3 J_z}{dt^3} = -\Gamma^2 \frac{dJ_z}{dt} - (\Delta\omega)^2 \left[\frac{i\Gamma}{2}(\tilde{J}_+ - \tilde{J}_-)\right]$$

$$= -\left[\Gamma^2 + \Delta\omega^2\right]\frac{dJ_z}{dt}. \tag{C.115}$$

So we see that the required frequency is given by

$$\nu^2 = \Gamma^2 + \Delta\omega^2. \tag{C.116}$$

So we can write, quite generally,

$$\frac{dJ_z}{dt} = A\cos\nu t + B\sin\nu t \tag{C.117}$$

and

$$J_z(t) = \frac{A}{\nu}\sin\nu t - \frac{B}{\nu}\cos\nu t + C. \tag{C.118}$$

Evidently $J_z(0) = C - B/\nu$, so

$$J_z(t) = J_z(0) + \frac{A}{\nu}\sin\nu t + \frac{B}{\nu}(1 - \cos\nu t). \tag{C.119}$$

To find A and B, we need to consider the differential equations for J_+, J_-, and J_z. Consider first

$$\frac{dJ_z}{dt}(0) = i\frac{\Gamma}{2}(J_+(0) - J_-(0)) = -\Gamma J_y(0) = A. \tag{C.120}$$

Now we want an equation that will introduce $J_x(0) = \frac{1}{2}[J_+(0) + J_-(0)]$.

$$\frac{d^2 J_z}{dt^2} = -\Gamma^2 J_z + \frac{\Gamma}{2}\Delta\omega(\tilde{J}_+ + \tilde{J}_-) = -\nu A\sin\nu t + \nu B\cos\nu t. \tag{C.121}$$

Evaluating at $t = 0$:

$$\nu B = -\Gamma^2 J_z(0) + \Gamma\Delta\omega J_x(0). \tag{C.122}$$

Now inserting our values of A and B,

$$J_z(t) = J_z(0)\left[1 - \frac{\Gamma^2}{\nu^2}(1 - \cos\nu t)\right]$$

$$+ \frac{\Gamma}{\nu}\left[J_x(0)\frac{\Delta\omega}{\nu}(1 - \cos\nu t) - J_y(0)\sin\nu t\right]. \tag{C.123}$$

(b) At $t = 0$, we have only magnetization in the z-direction. The magnetization per atom or molecule is

$$\mu_z(t) = \mu_z(0)\left[1 - \frac{\Gamma^2}{\nu^2}(1 - \cos \nu t)\right]. \tag{C.124}$$

Use $\tau = \Gamma t$ as the unit of time (Γ is the precessional frequency of the system in a magnetic field of strength B_1).

$$\nu t = \sqrt{\Delta\omega^2 + \Gamma^2}t = \tau\sqrt{1 + \Delta\omega^2/\Gamma^2}, \tag{C.125}$$

so

$$\mu_z(t) = \mu_z(0)\left[1 - \frac{2}{1 + \Delta\omega^2/\Gamma^2}\sin^2\left(\frac{\tau}{2}\sqrt{1 + \Delta\omega^2/\Gamma^2}\right)\right]. \tag{C.126}$$

This is shown in Figure C.4. At resonance, the magnetization completely reverses sign. This flipping of all the spins requires energy because of the level splitting in the magnetic field. The absorption of energy by the sample can be detected electrically by having the Q of the resonant circuit providing the oscillating field monitored, or by other suitable means. This is the general principle of the detection of NMR signals.

(c) In many circumstances, the Hamiltonian is time-independent and then the unitary operator that gives time evolution is simple e^{-iHt}. If H is time-dependent, but if it is also true that $[H(t_1), H(t_2)] = 0$, then $U = \exp[\int^t H(t')dt']$. Here, neither is the case. However, in the text the required unitary operator was determined to be

$$U(t) = e^{i\omega t J_z}e^{-i\alpha J_y}e^{i\nu t}e^{i\alpha J_y}, \tag{C.127}$$

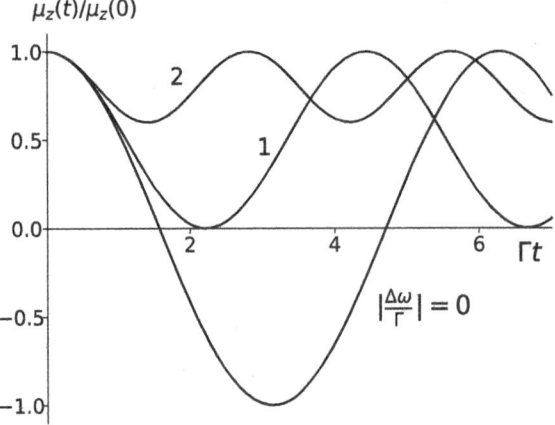

Figure C.4 The magnetization for Problem 4.5 as a function of $\tau = \Gamma t$ for $|\Delta\omega/\Gamma| = 0,1,2..$

where $\alpha = \tan^{-1}(B_1/B_0)$. Consider then

$$
\begin{aligned}
U^\dagger J_z U &= e^{-i\alpha J_y} e^{-i\nu t J_z} e^{i\alpha J_y} e^{-i\omega t J_z} J_z e^{i\omega t J_z} e^{-i\alpha J_y} e^{i\nu t J_z} e^{i\alpha J_y} \\
&= e^{-i\alpha J_y} e^{-i\nu t J_z} [e^{i\alpha J_y} J_z e^{-i\alpha J_y}] e^{i\nu t J_z} e^{i\alpha J_y} \\
&= e^{-i\alpha J_y} e^{-i\nu t J_z} [J_z \cos\alpha - J_x \sin\alpha] e^{i\nu t J_z} e^{i\alpha J_y} \\
&= e^{-i\alpha J_y} J_z \cos\alpha e^{i\alpha J_y} - \sin\alpha e^{-i\alpha J_y} [e^{-i\nu t J_z} J_x e^{i\nu t J_z}] e^{i\alpha J_y} \\
&= e^{-i\alpha J_y} J_z \cos\alpha e^{i\alpha J_y} - \sin\alpha e^{-i\alpha J_y} [J_x \cos\nu t + J_y \sin\nu t] e^{i\alpha J_y} \\
&= \cos\alpha[\cos\alpha J_z + \sin\alpha J_x] - \sin\alpha \sin\nu t J_y \\
&\quad - \sin\alpha \cos\nu t[\cos\alpha J_x - J_z \sin\alpha] \\
&= J_z[\cos^2\alpha + \sin^2\alpha \cos\nu t] + \sin\alpha\cos\alpha(1-\cos\nu t)J_x - \sin\alpha\sin\nu t J_y.
\end{aligned}
$$
$$(\text{C.128})$$

With $\sin\alpha = \Gamma/\nu$ and $\cos\alpha = (\omega_0 - \omega)/\nu$, this can be written as

$$
U^\dagger J_z U = J_z \left[1 - \frac{\Gamma^2}{\nu^2}(1-\cos\nu t)\right] + \frac{(\omega_0 - \omega)\Gamma}{\nu^2} J_x(1-\cos\nu t) - \frac{\Gamma}{\nu} J_y \sin\nu t.
$$
$$(\text{C.129})$$

This is the result found in part (a).

4.11 The neutron and proton bind into a deuteron (with binding energy 2.23 MeV, $J^P = 1^+$) by means of an interaction of the form

$$
V(\mathbf{r}) = V_1(r) + \boldsymbol{\sigma}_1 \cdot \boldsymbol{\sigma}_2 V_2(r) + V_T(r) S_{12}.
$$
$$(\text{C.130})$$

(a) We wish to find the selection rules for matrix elements of S_{12}:

$$
\langle \alpha' L' J' S M' | S_{12} | \alpha L S J M \rangle.
$$
$$(\text{C.131})$$

Because S_{12} is a scalar, it can connect only states with $J' = J$ and $M' = M$. Because S_{12} is even under parity, it can connect only states with the same parity. At the same time, S_{12} is a $\Delta S = 2$ and $\Delta L = 2$ operator. Since S can be only 0 or 1, only $S = S' = 1$ is allowed. Because S_{12} is even under parity, it connects only odd-L to odd-L and even-L to even-L. Moreover, if $L = 0$, then $L' = 2$ and vice-versa and if $L = 1$, then $L' = 1, 3$ and vice versa. Otherwise $L' = L, L \pm 2$.

Thus S_{12} is diagonal in J, M, and S, but off diagonal as well in L.

(b) If $L = 0, S = 1$, then the only values for L', S' are $L' = 2, S' = 1$, which we can indicate by $\langle {}^3D_1 | S_{12} | {}^3S_1 \rangle$. Of course, this is restricted to $J = 1, S = 1$ and is independent of M. There are many ways to proceed, none of them particularly quick. First, note that the $\boldsymbol{\sigma}_1 \cdot \boldsymbol{\sigma}_2$ element cannot contribute to $\langle {}^3D_1 | S_{12} | {}^3S_1 \rangle$ because it is a $\Delta L = 0$ operator. Next consider

$$
(\hat{\mathbf{r}} \cdot (\boldsymbol{\sigma}_1 + \boldsymbol{\sigma}_2))^2 = 4(\hat{\mathbf{r}} \cdot \mathbf{S})^2 = 2 + 2\hat{\mathbf{r}} \cdot \boldsymbol{\sigma}_1 \hat{\mathbf{r}} \cdot \boldsymbol{\sigma}_2,
$$
$$(\text{C.132})$$

so

$$
3\hat{\mathbf{r}} \cdot \boldsymbol{\sigma}_1 \hat{\mathbf{r}} \cdot \boldsymbol{\sigma}_2 = 6(\hat{\mathbf{r}} \cdot \mathbf{S})^2 - 3.
$$
$$(\text{C.133})$$

Once again, we can drop the second piece, which cannot contribute to $\langle {}^3D_1 | S_{12} | {}^3S_1 \rangle$.

Now

$$\hat{\mathbf{r}} \cdot \mathbf{S} = \frac{1}{2}[r_+S_- + r_-S_+] + r_zS_z, \qquad (C.134)$$

where of course r_z is simply $z = \cos\theta$ and $r_\pm = \sin\theta e^{\pm i\phi}$.

Squaring this expression and separating it into pieces with well-defined ΔM_S, we have

$$(\hat{\mathbf{r}} \cdot \mathbf{S})^2 = \frac{1}{4}[r_+^2S_-^2 + r_-^2S_+^2]$$

$$+ \frac{1}{2}r_+r_z(S_-S_z + S_zS_-) + \frac{1}{2}r_-r_z(S_+S_z + S_zS_+)$$

$$+ \frac{1}{4}r_+r_-(S_+S_- + S_-S_+) + r_z^2S_z^2. \qquad (C.135)$$

Now $r_+r_- = 1 - r_z^2$ and once again the "1" cannot contribute. Similarly, $S_+S_- + S_-S_+ = 2(S^2 - S_z^2)$. So effectively

$$(\hat{\mathbf{r}} \cdot \mathbf{S})^2 = \frac{1}{4}[r_+^2S_-^2 + r_-^2S_+^2]$$

$$+ \frac{1}{2}r_+r_z(S_-S_z + S_zS_-) + \frac{1}{2}r_-r_z(S_+S_z + S_zS_+)$$

$$- \frac{1}{2}r_z^2(S^2 - S_z^2) + r_z^2S_z^2, \qquad (C.136)$$

where each line contributes to a distinct value of $|\Delta M_S|$.

Looking up the appropriate Clebsch-Gordan coefficients, we see that

$$\langle{}^3D_1, M = 1| = \sqrt{\frac{3}{5}}Y_{22}^* \langle M_{S'} = -1|$$

$$- \sqrt{\frac{3}{10}}Y_{21}^* \langle M_{S'} = 0| + \sqrt{\frac{1}{10}}Y_{20}^* \langle M_{S'} = 1| \qquad (C.137)$$

and we are free to determine the matrix element for $M = 1$, since the value is independent of M. We have

$$|{}^3S_1, M = 1\rangle = Y_{00}\,|M_{S=1}\rangle. \qquad (C.138)$$

The first term is

$$\frac{1}{4}\sqrt{\frac{3}{5}}\int d\Omega Y_{22}^* r_+^2 Y_{00} \langle M_S = -1|S_-^2|M_S = 1\rangle$$

$$= \frac{1}{4}\frac{1}{4}\sqrt{\frac{3}{5}}\int d\Omega \sqrt{\frac{15}{2\pi}}\sin^2\theta \sin^2\theta \frac{1}{\sqrt{4\pi}}2$$

$$= \frac{3}{4\sqrt{2}}\int_0^1 d\mu(1 - \mu^2)^2 = \frac{3}{4\sqrt{2}}\frac{8}{15} = \frac{\sqrt{2}}{5}. \qquad (C.139)$$

The second is

$$-\frac{1}{2}\sqrt{\frac{3}{10}}\int d\Omega\, Y^*_{21} r_+ r_z Y_{00} \langle M_S = 0 | S_- S_z | S_z S_- | M_S = 1\rangle$$

$$= -\frac{1}{2}\sqrt{\frac{3}{10}}\int d\Omega\left(-\sqrt{\frac{15}{8\pi}}\right)\sin\theta\cos\theta\sin\theta\cos\theta\frac{1}{\sqrt{4\pi}}\sqrt{2}$$

$$= \frac{3}{2\sqrt{2}}\int_0^1 d\mu(1-\mu^2)\mu^2 = \frac{3}{2\sqrt{2}}\frac{2}{15} = \frac{\sqrt{2}}{10}. \quad (C.140)$$

The third is

$$\sqrt{\frac{1}{10}}\int d\Omega\, Y^*_{20} r_z^2 Y_{00}\left\langle M_S = 1 \left| \frac{3}{2}S_z^2 - \frac{1}{2}S^2 \right| M_S = 1\right\rangle$$

$$= \sqrt{\frac{1}{10}}\int d\Omega\sqrt{\frac{5}{4\pi}}(\frac{3}{2}\cos^2\theta - \frac{1}{2})\frac{1}{2}\cos^2\theta\frac{1}{\sqrt{4\pi}}$$

$$= \frac{1}{4\sqrt{2}}\int_0^1 d\mu(3\mu^2-1)\mu^2 = \frac{1}{4\sqrt{2}}\frac{4}{15} = \frac{\sqrt{2}}{30}. \quad (C.141)$$

The sum of the three terms is $\sqrt{2}/3$.
Altogether

$$\langle {}^3D_1, M = 1 | 6(\hat{\mathbf{r}}\cdot\mathbf{S})^2 | {}^3S_1, M = 1\rangle = 6\frac{\sqrt{2}}{3} = 2\sqrt{2}. \quad (C.142)$$

(c) The deuteron is a bound state of a neutron and proton. The interaction Hamiltonian can be written

$$H = V_1(r) + V_2(r)\boldsymbol{\sigma}_1\cdot\boldsymbol{\sigma}_2 + V_T(r)S_{12}. \quad (C.143)$$

The deuteron has $J^P = 1^+$. The even parity is consistent with the lowest energy state being $L = 0$, as expected on general angular momentum grounds. Since p and n have $s = 1/2$ each, the total spin can be $S = 0, 1$.

$$S^2 = \frac{1}{4}(\boldsymbol{\sigma}_1\cdot\boldsymbol{\sigma}_2 + 2\boldsymbol{\sigma}_1\cdot\boldsymbol{\sigma}_2 + \boldsymbol{\sigma}_2\cdot\boldsymbol{\sigma}_2) = \frac{1}{4}(3 + 2\boldsymbol{\sigma}_1\cdot\boldsymbol{\sigma}_2 + 3)$$

$$= \frac{3}{2} + \frac{1}{2}\boldsymbol{\sigma}_1\cdot\boldsymbol{\sigma}_2, \quad (C.144)$$

so

$$\boldsymbol{\sigma}_1\cdot\boldsymbol{\sigma}_2 = 2S^2 - 3. \quad (C.145)$$

Thus for $S = 1$, we find $\boldsymbol{\sigma}_1\cdot\boldsymbol{\sigma}_2 = 1$, while for $S = 0$, $\boldsymbol{\sigma}_1\cdot\boldsymbol{\sigma}_2 = -3$. Since $L = 0$ and $J = 1$ requires $S = 1$, we conclude that V_2 must be attractive, whatever the character of V_1.

The magnetic moment of the deuteron is measured to be $\mu_d = 0.8574\mu_N$, where $\mu_N = e\hbar/(2m_p c)$. The PDG tables give

$$\mu_p = 2.7928473446 \pm 0.0000000008\ \mu_N$$
$$\mu_n = -1.9130427 \pm 0.0000005\ \mu_N. \tag{C.146}$$

So $\mu_p + \mu_n = +0.879801\ \mu_N$ and $\mu_d - (\mu_p + \mu_n = +0.879801) = -0.0224\ \mu_N$. Apparently there is very little contribution from angular motion, consistent with the $L = 0$ picture.

Finally, the deuteron has a quadrupole moment, $Q/e = 2.73 \times 10^{-27}$ cm^2. Since $J = 1$, this is perfectly acceptable, but Q_q is an operator on the spatial coordinates. If $L = 0$ entirely, this rank-two L tensor must vanish. Since the parity is even, the possible values of L are 2, 4,... With $J = 1$ and $S = 1$, only $L = 2$ is allowed. Thus the existence of the quadrupole moment tells us the ground state is a mixture of $^3S_1$ and $^3D_1$.

We can go further to make a rough estimate of the fraction of the deuteron in the d-wave. Let us write the orbital-spin wave functions as $\mathcal{Y}_{LSJM}$. Then deuteron's wave function can be represented by

$$\psi = \frac{u(r)}{r}\mathcal{Y}_{0111M} + \frac{w(r)}{r}\mathcal{Y}_{211M}, \tag{C.147}$$

where $u(r)$ is the radial $\ell = 0$ wave function and $w(r)$ is the $\ell = 2$ wave function. In the deuteron, the d-state probability

$$p_D = \int_0^\infty |w(r)| d^3r \tag{C.148}$$

is small compared to unity. The intent here is to make a quantitative estimate of p_D. The quadrupole moment is defined by evaluating $3z^2 - r^2$ for the $|JJ\rangle$ element. Here we have $J = 1$ and $\mathbf{r}_p = \frac{1}{2}\mathbf{r}$. We write

$$3z^2 - r^2 = r^2\sqrt{\frac{16\pi}{5}}Y_{20}(\theta, \phi). \tag{C.149}$$

Thus

$$Q/e = \frac{1}{4}\int d^3x\, r^2 \times \frac{1}{r^2}\left(u\mathcal{Y}_{0111} + w\mathcal{Y}_{2111}\right)^\dagger \sqrt{\frac{16\pi}{5}}Y_{20}$$
$$\times (u\mathcal{Y}_{0111} + w\mathcal{Y}_{2111}). \tag{C.150}$$

Since w is small, we need only the uw term (the uu term vanishes because we need a $L = 2$ piece). We are considering the $M_S = 1$ state $u\mathcal{Y}_{0111}$, so we need only the $M_S = 1$ from $\mathcal{Y}_{2111}$, i.e. $\sqrt{\frac{1}{10}}Y_{20}^*$.

We thus find

$$Q/e = \frac{1}{4}\int r^2 dr \frac{2uw}{\sqrt{10}}\int d\Omega Y_{20}^2 \sqrt{\frac{16\pi}{5}}\sqrt{\frac{1}{4\pi}} = \frac{1}{5\sqrt{2}}\int_0^\infty r^2 dr\, uw. \qquad (C.151)$$

If we approximate $w = \sqrt{p_D}u$, then

$$Q/e = \frac{1}{5\sqrt{2}}\sqrt{p_D}\langle r^2\rangle_d. \qquad (C.152)$$

A rough estimate of $\langle r^2\rangle$ can be obtained by assuming a "zero-range" wave function. This is equivalent to solving the Schrödinger equation without a potential, writing

$$-\frac{1}{2m}u'' = Eu = -E_b u, \qquad (C.153)$$

where E_b is the binding energy. Then

$$u \propto e^{-\sqrt{2mE_b}r} \propto e^{-\gamma r} \qquad (C.154)$$

and

$$\langle r^2\rangle = \frac{1}{2\gamma^2}. \qquad (C.155)$$

Here we need to use the reduced mass $= m_N/2$ and we find $\gamma^{-1} = 4.3$ fm (remember 197 MeV fm=1). Now

$$p_D = \left[\frac{5\sqrt{2}Q/e}{\langle r^2\rangle}\right]^2 = \left[\frac{10\sqrt{2}\cdot 0.273}{(4.3)^2}\right]^2 = 4.4\times 10^{-2}. \qquad (C.156)$$

This is a crude estimate of the percentage of the d-state in the wave function of the deuteron.

4.12 (a) This is an exercise in integrating by parts. We begin with

$$u'' + Wu = 0, \qquad (C.157)$$

where the prime indicates differentiation with respect to r. Multiply by $r^q du/dr$ and integrate from zero to infinity.

$$\int_0^\infty r^q \frac{du}{dr}\frac{d^2u}{dr^2}dr + \int_0^\infty r^q W(r)u\frac{du}{dr}dr = 0. \qquad (C.158)$$

After multiplying by two, this can be written

$$\int_0^\infty r^q \frac{d}{dr}\left(\frac{du}{dr}\right)^2 dr + \int_0^\infty r^q W(r)\frac{d}{dr}(u^2)dr = 0. \qquad (C.159)$$

Integrate by parts, assuming that u vanishes exponentially at infinity (as is proper for a localized bound state):

$$= \left[r^q \left(\frac{du}{dr} \right)^2 \right]_{r \to 0} - \int_0^\infty \left(\frac{du}{dr} \right)^2 qr^{q-1}dr - \left[r^q W(r)u^2 \right]_{r \to 0}$$

$$- \int_0^\infty u^2 \frac{d}{dr} (r^q W) \, dr = 0. \quad \text{(C.160)}$$

The last integral is the expectation value of $\frac{d}{dr}(r^q W)$ in the state in question. We denote it and other expectation values by $\langle \rangle$.

We deal with the other integral

$$I = \int_0^\infty qr^{q-1} \frac{du}{dr} \frac{du}{dr} dr, \quad \text{(C.161)}$$

integrating by parts:

$$I = -\left[qr^{q-1}u \frac{du}{dr} \right]_{r \to 0} - q \int_0^\infty u \frac{d}{dr} \left(r^{q-1} \frac{du}{dr} \right) dr$$

$$= -\left[qr^{q-1}u \frac{du}{dr} \right]_{r \to 0} - q \int_0^\infty u \left[(q-1)r^{q-2} \frac{du}{dr} + r^{q-1} \frac{d^2u}{dr^2} \right]. \quad \text{(C.162)}$$

We use the Schrödinger equation to write

$$I = -\left[qr^{q-1}u \frac{du}{dr} \right]_{r \to 0} - \frac{q}{2} \int_0^\infty (q-1)r^{q-2} \frac{d}{dr}(u^2)dr + q \langle r^{q-1}W \rangle \quad \text{(C.163)}$$

and integrate by parts once again

$$I = -\left[qr^{q-1}u \frac{du}{dr} \right]_{r \to 0} + \left[\frac{q(q-1)}{2} r^{q-2}u^2 \right]_{r \to 0}$$

$$+ \frac{q(q-1)(q-2)}{2} \langle r^{q-3} \rangle + q \langle r^{q-1}W \rangle. \quad \text{(C.164)}$$

We combine these results to obtain

$$\frac{q(q-1)(q-2)}{2} \langle r^{q-3} \rangle + q \langle r^{q-1}W \rangle + \left\langle \frac{d}{dr}(r^q W) \right\rangle +$$

$$\left[r^q \left(\frac{du}{dr} \right)^2 + r^q W(r)u^2 - qr^{q-1}u \frac{du}{dr} + \frac{q(q-1)}{2} r^{q-2}u^2 \right]_{r \to 0} = 0. \quad \text{(C.165)}$$

We have been careful to keep all the terms to be evaluated at the origin, in case they are not zero. We now turn to that $[..]_{r \to 0}$. We know that if $r^2 V(r) \to 0$ as $r \to 0$, the wave function at the origin goes as $u \propto r^{\ell+1}$.

For $\ell \neq 0$, $W(r) \propto r^{-2}$ because of the centrifugal barrier. Then every term in $[..]_{r \to 0}$ goes as $r^{2\ell+q}$. So if $2\ell + q > 0$, then $[..]_{r \to 0} = 0$. Thus we have for $\ell \neq 0$

$$\frac{q(q-1)(q-2)}{2} \langle r^{q-3} \rangle + 2q \langle r^{q-1}W \rangle + \left\langle r^q \frac{dW}{dr} \right\rangle = 0, \quad \text{(C.166)}$$

provided $r^2 V(r) \to 0$ as $r \to 0$ and $q > -2\ell$.

The usual virial theorem says, briefly, that if we average $\mathbf{x} \cdot \mathbf{F}$ over a particle's path in a conservative force field, where $\mathbf{F} = -\nabla V$, then

$$\int dt\, \mathbf{x} \cdot \mathbf{F} = \int dt\, \mathbf{r} \cdot m\ddot{\mathbf{x}} = -\int dt\, m\dot{\mathbf{x}}^2 = -2\langle KE \rangle \Delta t, \qquad (\text{C.167})$$

where here we average over a long time Δt so that endpoints don't contribute or we average over a cycle of period motion. On the other hand,

$$\int dt\, \mathbf{x} \cdot \mathbf{F} = -\int dt\, \mathbf{r} \cdot \nabla V = -\langle \mathbf{r} \cdot \nabla V \rangle \Delta t. \qquad (\text{C.168})$$

Thus,

$$\langle KE \rangle = \frac{1}{2} \langle \mathbf{r} \cdot \nabla V \rangle. \qquad (\text{C.169})$$

So the analog of $r\frac{dV}{dr}$ here is $q = 1$ and we find

$$2\langle W \rangle + \left\langle r\frac{dW}{dr} \right\rangle = 0. \qquad (\text{C.170})$$

Now W contains $E - V$, the kinetic energy, but also the kinetic energy associated with the angular motion, which we could call $T_\perp$. Thus $\langle W \rangle = \frac{2m}{\hbar^2} \langle T_r \rangle$, where T_r is the kinetic energy associated with radial motion. On the other hand,

$$\frac{dW}{dr} = \frac{2m}{\hbar^2}(-\frac{dV}{dr} + \frac{2\ell(\ell+1)}{2mr^3}), \qquad (\text{C.171})$$

so

$$\begin{aligned}
\left\langle r\frac{dW}{dr} \right\rangle &= \frac{2m}{\hbar^2} \left\langle -r\frac{dV}{dr} + \frac{2\ell(\ell+1)}{2mr^2} \right\rangle \\
&= \frac{2m}{\hbar^2} \left\langle -r\frac{dV}{dr} + 2T_\perp \right\rangle.
\end{aligned} \qquad (\text{C.172})$$

Combining these relations,

$$\begin{aligned}
2\langle W \rangle + \left\langle r\frac{dW}{dr} \right\rangle &= 0 \\
&= \frac{4m}{\hbar^2} \langle T_r \rangle + \frac{2m}{\hbar^2} \left\langle -r\frac{dV}{dr} + 2T_\perp \right\rangle
\end{aligned} \qquad (\text{C.173})$$

and

$$\langle T \rangle = \frac{1}{2} \left\langle r\frac{dV}{dr} \right\rangle, \qquad (\text{C.174})$$

the analog of the classical virial theorem.

(b) The analysis is the same for $\ell = 0$. We still require $q > 0$ if the endpoints are to be ignored and

$$\frac{q(q-1)(q-2)}{2} \langle r^{q-3} \rangle + 2q \langle r^{q-1}W \rangle + \left\langle r^q \frac{dW}{dr} \right\rangle = 0. \qquad (\text{C.175})$$

(c) If $\ell = 0$ and $q = 0$, we no longer can ignore the endpoint contribution at $r = 0$. In particular, in Eq. (C.165), we need to keep $-u'(0)^2/2$. Thus for $\ell = 0, q = 0$

$$u'(0)^2 + \left\langle \frac{dW}{dr} \right\rangle = 0. \tag{C.176}$$

Now the ordinary, three-dimensional wave function for $\ell = 0$ is

$$\Psi(r) = \frac{u(r)}{r} Y_{00} = \frac{1}{\sqrt{4\pi}} \frac{u(r)}{r} \tag{C.177}$$

and

$$\Psi(0) = \frac{1}{\sqrt{4\pi}} u'(0). \tag{C.178}$$

Thus, the virial relation for s-wave is

$$|\Psi(0)|^2 = \frac{1}{4\pi} \frac{2m}{\hbar^2} \left\langle \frac{dV}{dr} \right\rangle = \frac{m}{2\pi\hbar^2} \left\langle \frac{dV}{dr} \right\rangle. \tag{C.179}$$

This general result relating the square of the wave function at the origin to the average value of the radial force finds use in heavy quark spectroscopy, where the decay width for a $Q\bar{Q}$ state with $J^{PC} = 1^{--}$ to decay into charged lepton pairs is given (in the non-relativistic limit of the quarks) by

$$\Gamma_{\ell^+\ell^-} = \frac{16\pi\alpha^2 e_Q^2}{M^2} |\Psi(0)|^2, \tag{C.180}$$

where e_Q is the charge of the quark in units of the proton's charge and M is the mass of the state (factors of $\hbar$ and c are suppressed). For charmonium ($c\bar{c}$) and upsilonium ($b\bar{b}$), $\Gamma_{\ell^+\ell^-}$ is measured to be on the order of 1 keV. With experimental values for M and estimates of $|\Psi(0)|$, one can, for example, deduce that the charge of the b quark is $-1/3$, not $2/3$.

C.5 Chapter 5

5.4 (a) This is an exercise in degenerate perturbation theory, since the $F = 1$ state is degenerate in the absence of the magnetic field. On the basis of the previous problem, we know, using $H = -\mu \cdot \mathbf{B}$,

$$H_1 = \frac{2}{3} g_e g_p \alpha^2 \left(\frac{e^2}{a_0} \right) \frac{m_e}{m_p} \langle \mathbf{s} \cdot \mathbf{I} \rangle + g_e \frac{e\hbar}{2m_e c} S_z B_z - g_p \frac{e\hbar}{2m_p c} I_z B_z. \tag{C.181}$$

The Bohr magneton has the value 5.788×10^{-5} eV-T^{-1}, while the nuclear magneton is smaller by a factor $1836 = m_p/m_e$. So if B is measured in Tesla, the values of a, k_1, and k_2 are

$$a = 5.8844 \times 10^{-6} \text{ eV}$$

$$k_1 = -2(1.00116) \times 5.788 \times 10^{-5} B(\text{T}) \text{ eV} = -1.159 \times 10^{-4} B(\text{T}) \text{ eV}$$

$$k_2 = 2(2.7928) \times (5.788/1836) \times 10^{-5} B(\text{T}) \text{ eV} = 1.761 \times 10^{-7} B(\text{T}) \text{ eV} \quad (\text{C.182})$$

(b) In the absence of H_1, the "ground state" of hydrogen is four-fold degenerate, with $j_1 = 1/2 = s, m_1 = m_s = \pm 1/2$ and $j_2 = I = 1/2, m_2 = m_I = \pm 1/2$. The four sates in the (j_1, m_1, j_2, m_2) basis, can be transformed into the four states of the (j_1, j_2, j, m) basis, where $j = F, m = m_F$. Let us write the various spin states as

$$|1,1\rangle = |F = 1, m_F = 1\rangle = \alpha_e \alpha_p$$

$$|1,-1\rangle = |F = 1, m_F = -1\rangle = \beta_e \beta_p$$

$$|1,0\rangle = |F = 1, m_F = 0\rangle = \frac{1}{\sqrt{2}}[\alpha_e \beta_p + \beta_e \alpha_p]$$

$$|0,0\rangle = |F = 0, m_F = 0\rangle = \frac{1}{\sqrt{2}}[\alpha_e \beta_p - \beta_e \alpha_p]. \tag{C.183}$$

Now evaluate the action of s_z and I_z on these states.

$$s_z |1,1\rangle = \frac{1}{2} |1,1\rangle = I_z |1,1\rangle$$

$$s_z |1,-1\rangle = -\frac{1}{2} |1,-1\rangle = I_z |1,-1\rangle$$

$$s_z |1,0\rangle = \frac{1}{2} |0,0\rangle = -I_z |0,0\rangle$$

$$s_z |0,0\rangle = \frac{1}{2} |1,0\rangle = -I_z |1,0\rangle \tag{C.184}$$

Armed with these results, we can write out the matrix for H_1, with the states ordered as above:

$$H_1 = \begin{pmatrix} \frac{a}{4} - \frac{1}{2}(k_1 + k_2) & 0 & 0 & 0 \\ 0 & \frac{a}{4} + \frac{1}{2}(k_1 + k_2) & 0 & 0 \\ 0 & 0 & \frac{a}{4} & \frac{1}{2}(k_2 - k_1) \\ 0 & 0 & \frac{1}{2}(k_2 - k_1) & -\frac{3a}{4} \end{pmatrix}. \tag{C.185}$$

The $m_F = \pm 1$ states are already diagonal. The energy shifts are

$$\Delta E_{\pm 1} = \frac{a}{4} \mp \frac{1}{2}(k_2 + k_1) = \frac{a}{4} \pm \frac{1}{2}(|k_1| - k_2). \tag{C.186}$$

We need to diagonalize the $m_F = 0$ states. Let

$$x = \frac{k_2 - k_1}{a} = 19.727 \, B(T). \tag{C.187}$$

To find the eigenvalues, solve

$$\begin{vmatrix} \frac{a}{4} - \epsilon & \frac{1}{2}ax \\ \frac{1}{2}ax & -\frac{3a}{4} - \epsilon \end{vmatrix} = 0 \tag{C.188}$$

with the result

$$\epsilon = -\frac{a}{4} \pm \frac{a}{2}\sqrt{1 + x^2}. \tag{C.189}$$

Thus,

$$\Delta E_{0\pm} = a\left[-\frac{1}{4} \pm \frac{1}{2}\sqrt{1 + x^2} \right]$$

$$\Delta E_{\pm 1} = a\left[\frac{1}{4} \pm \frac{1}{2}\left(\frac{|k_1| - k_2}{|k_1| + k_2} \right)x \right] \tag{C.190}$$

and

$$\frac{1}{2}\left(\frac{|k_1| - k_2}{|k_1| + k_2}\right) = 0.49848. \tag{C.191}$$

(c) These results are shown in the Breit-Rabi diagram, Figure C.5.

(d) The eigenvectors of the 4×4 Hamiltonian are found by inserting the energy eigenvalues into the linear equations in the well-known way. For $m_F = \pm 1$, we have the pure states

$$|m_F = \pm 1\rangle = \left|\frac{1}{2}, \frac{1}{2}, 1, m_F = \pm 1\right\rangle \tag{C.192}$$

for all values of the magnetic field. For $m_F = 0$, we have linear combinations of the $F = 1$ and $F = 0$ states:

$$|0\pm\rangle = \alpha_0 |0,0\rangle + \alpha_1 |1,0\rangle, \tag{C.193}$$

where

$$\alpha_0 = \pm \frac{1}{\sqrt{2}} \frac{(\sqrt{1 + x^2} \mp 1)^{1/2}}{(1 + x^2)^{1/4}}; \quad \alpha_1 = \frac{1}{\sqrt{2}} \frac{(\sqrt{1 + x^2} \pm 1)^{1/2}}{(1 + x^2)^{1/4}}. \tag{C.194}$$

At low fields, $x \ll 1$,

$$\alpha_0 \to \pm \frac{1}{\sqrt{2}}(1 \mp 1)^{1/2}, \quad \alpha_1 \to \frac{1}{\sqrt{2}}(1 \pm 1)^{1/2}. \tag{C.195}$$

Then the $|0+\rangle$ state is pure $F = 1$, with energy $\Delta E_1 = a/4$, and the $|0-\rangle$ state is pure $F = 0$, with energy $\Delta E_0 = -3a/4$.

At very high fields, $x \gg 1$, $\alpha_0 \to \pm 1/\sqrt{2}$ and $\alpha_1 \to 1/\sqrt{2}$, and $E_{0\pm} \simeq \pm ax/2$. This means that

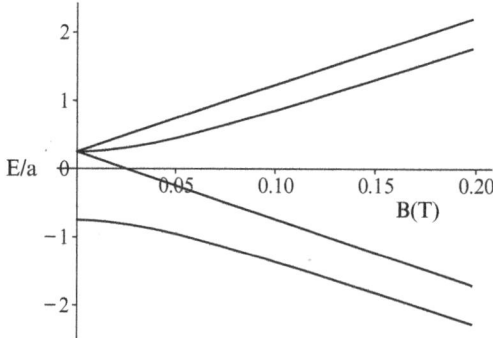

Figure C.5 The Breit-Rabi diagram for 1s hydrogen in the presence of a magnetic field, B, measured in Tesla. At zero field the splitting between $F = 1$ and $F = 0$ is the 21 cm line. The states for $m_F = \pm 1$ undergo no mixing and the splitting is proportional to B. The two $m_F = 0$ states mix. At high fields, the eigenstates are those with well-defined s_z and I_z, as the interaction with the field dominates over the hyperfine interaction.

$$|0+\rangle \simeq |\alpha_e\rangle \, |\beta_p\rangle$$
$$|0-\rangle \simeq |\beta_e\rangle \, |\alpha_p\rangle. \tag{C.196}$$

The higher energy state has the electron's spin parallel to and the proton's spin antiparalllel to B, the lower state has the opposite. This is just what you expect when the $spin-spin$ (hyperfine) interaction can be ignored. Then the magnetic moments line up parallel to the field in the lowest energy state. (Remember that $\mu_e = -(\ )s$, so antiparallel spin s_e means parallel μ_e.)

5.5 See Bethe and Salpeter, *Quantum Mechanics of One- and Two-Electron Atoms*, pp. 205–213.

The spin-orbit splitting of the $n = 2\,^2P_{3/2}$ and $^2P_{1/2}$ states in hydrogen is

$$\Delta = \frac{\alpha^2}{16}\text{Ryd.} \tag{C.197}$$

The Hamiltonian for the magnetic interaction is

$$H' = \mu_B B(L_z + 2s_z). \tag{C.198}$$

This causes mixing between the $j = 3/2$ and $j = 1/2$ states. We need to write explicitly the states $|j = 3/2, m = \pm 1/2\rangle$ and $|j = 1/2, m = \pm 1/2\rangle$ in terms of the $\ell = 1, s = 1/2$ states. This is done conveniently with a table of Clebsch-Gordan coefficients, indicating constituents by $|m_L, m_s\rangle$.

$$|j = 3/2, m = 1/2\rangle = \sqrt{\frac{1}{3}}\,|1, -1/2\rangle + \sqrt{\frac{2}{3}}\,|0, 1/2\rangle$$

$$|j = 1/2, m = 1/2\rangle = \sqrt{\frac{2}{3}}\,|1, -1/2\rangle - \sqrt{\frac{1}{3}}\,|0, 1/2\rangle$$

$$|j = 3/2, m = -1/2\rangle = \sqrt{\frac{1}{3}}\,|-1, 1/2\rangle + \sqrt{\frac{2}{3}}\,|0, -1/2\rangle$$

$$|j = 1/2, m = -1/2\rangle = -\sqrt{\frac{2}{3}}\,|-1, 1/2\rangle + \sqrt{\frac{2}{3}}\,|0, -1/2\rangle \tag{C.199}$$

The magnetic Hamiltonian has both diagonal and off-diagonal pieces. For $m = 1/2$:

$$\langle j = 3/2\,|H'|\, j = 3/2\rangle = \mu_B B[\frac{1}{3}(1-1) + \frac{2}{3}(0+1)] = \frac{2}{3}\mu_B B$$

$$\langle j = 1/2\,|H'|\, j = 1/2\rangle = \mu_B B[\frac{2}{3}(1-1) + \frac{1}{3}(0+1)] = \frac{1}{3}\mu_B B$$

$$\langle j = 1/2\,|H'|\, j = 3/2\rangle = \mu_B B[\frac{\sqrt{2}}{3}(1-1) - \frac{\sqrt{2}}{3}(0+1)] = -\frac{\sqrt{2}}{3}\mu_B B$$

$$= \langle j = 3/2\,|H'|\, j = 1/2\rangle. \tag{C.200}$$

The mixing matrix for $m = 1/2$ is

$$\begin{pmatrix} \dfrac{\Delta}{3} + \dfrac{2}{3}\mu_B B & -\dfrac{\sqrt{2}}{3}\mu_B B \\[2ex] -\dfrac{\sqrt{2}}{3}\mu_B B & -\dfrac{2\Delta}{3} + \dfrac{1}{3}\mu_B B \end{pmatrix}. \tag{C.201}$$

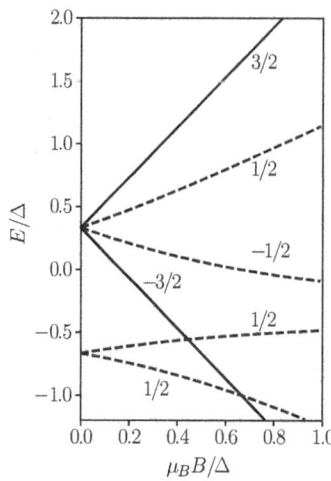

Figure C.6 The splitting of the $n = 2$ p-states in hydrogen due to spin-orbit interaction and an external magnetic field. The upper states are $2p_{3/2}$ and the lower $2p_{1/2}$. Each level is labelled by its value of m. At low field, this is the Zeeman effect. At high field, it is known as the Paschen-Back effect. The spin-orbit splitting Δ is 4.53×10^{-5} eV. The Bohr magneton is $\mu_B = 5.788 \times 10^{-5}$ eV/T, so $\mu_{BB}/\Delta = 1.28B$ (T).

With $a = \mu_B B/\Delta$, the eigenvalue equation is

$$\lambda^2 + \lambda(\frac{1}{3} - a) - \frac{2}{9} - \frac{1}{3}a = 0 \tag{C.202}$$

with the solutions

$$\lambda = \frac{1}{2}\left[a - \frac{1}{3} \pm \sqrt{1 + \frac{2}{3}a + a^2}\right]. \tag{C.203}$$

For $m = -1/2$, we find the same result, with $a \to -a$. The results are shown in Figure C.6.

C.6 Chapter 6

6.2 Neutral argon has closed 1s, 2s, 2p, 3s, and 3p shells. In looking at the excited states, we need only consider the $(3s)^2(3p)^6$ shells and excitations out of them. Excitations from lower shells will have very high energies. In fact, for AI (neutral argon), only the excitations of the $(3p)^6$ shell are relevant. If we remove an electron from the $(3p)^6$ shell and put it in another orbital, we have a two- "electron" problem - a hole in the 3p shell and a $n\ell$ electron, where $n = 4, 5, ...$ and ℓ anything ($\ell \leq n - 1$) and also 3d. Because of penetrating orbits, we expect $(3p)^{-1}4s, (3p)^{-1}4p$ to compete with $(3p)^{-1}3d$. This is borne out by the atomic orbitals of the atoms above argon - ^{19}K has [Ar]4s, ^{20}Ca has [Ar]$(4s)^2$.

(a) For the low-lying excitations, we thus consider the two-particle (inequivalent electrons) configurations: $(3p)^{-1}4s, (3p)^{-1}4p, (3p)^{-1}3d$. We build the Russell-Saunders states as follows:

(i) $(3p)^{-1}4s$ parity$=(-1)^{1+0} = -1$; $S = 0, L = 1$, so $^1P_1^o$; $S = 1, J = 0, 1, 2$, so $^3P_{0,1,2}$. We indicate parity-odd states with a superscript "o." There are thus four distinct states in this multiplet, one singlet P and three triplet P. By Hund's rule (which, it turns out, can be applied to the system of the hole in the p-shell and the promoted s-electron), we expect the states with $S = 1$ (symmetric spin, antisymmetric in space) to lie lowest and within the triplet, the fine structure of a

3p-hole to give inversion, namely the highest J lowest in energy. This group of states lies at an excitation of 93,000 - 95,000 cm^{-1} (about 11 eV or a wavelength of about 100 nm). See the NBS table extract and the figure in part (a) of the next problem.

(ii) $(3p)^{-1}(4p)$, parity $= +1$, $L = 0, 1, 2$, $S = 0, 1$. The states are $^1S_0, ^1P_1, ^1D_2, ^3S_1,$ $^3P_{0,1,2}, ^3D_{1,2,3}$. The total number of states is 2 for $J = 0$, 4 for $J = 1$, 3 for $J = 2$, and 1 for $J = 3$. In the NBS tables, from excitation of 104 k cm^{-1} to 110 k cm^{-1}, there are exactly 10 states with these J values. As can be seen from the figure in the next problem, the $L = 0$ states obey Hund's rule, though that does not necessarily apply for states with holes, as do the $L = 2$ states. The sequence of J values is inverted (both electrons contribute to the fine structure, but the hole has a larger expectation value of $\frac{1}{r}\frac{dV}{dr}$). With the assignments given to the ^3P states, the fine structure appears anomalous. This is evidence of the approximate nature of the Russell-Saunders configuration. As can be seen from the NBS tables under the column labeled "Config.", the $(3p)^{-1}$ hole is described as $p_{3/2}$ or $p_{1/2}$. That $\ell_1^2, s_1^2, j_1, m_1$ state is then combined with the $\ell_2^2, \ell_{2z}, s_2^2, s_{2z}$ state of the excited electron to find the final J, parity. This description is sometimes closer to the truth.

(iii) $(3p)^{-1}(3d)$. Again, odd parity because of the 3p hole. $L = 1, 2, 3$, $S = 0, 1$ so the terms are $^1P_1^o, ^1D_2^o, ^1F_3^o, ^3P_{0,1,2}^o, ^3D_{1,2,3}^o, ^3F_{2,3,4}^o$. A total of 12 states. These are found at excitation between 111 k cm^{-1} and 115.5 k cm^{-1}.

Summary: There are four odd-parity states ($^1P_1, ^3P_J$) arising from promoting one 3p electron to 4s, grouped together and forming the lowest excited multiplet. There are 10 even-parity states from moving a 3p electron to 4p. There are 12 odd-parity states arising from moving the 3p to 3d.

(b) Russell-Saunders assignments for the first states. To make the final Russell-Saunders assignments, we turn to the g-factors.

The g-factor for a term $^{2S+1}L_J$ is

$$g = \frac{3}{2} + \frac{S(S+1) - L(L+1)}{2J(J+1)}. \tag{C.204}$$

For singlet states, $g = 1$. For triplet states

$$g = \frac{3}{2} + \frac{2 - L(L+1)}{2J(J+1)}. \tag{C.205}$$

This yields Table C.1. We have already observed that the four states near 94 k cm^{-1} excitation have the right J values and parities to be the $^3P_{0,1,2}$ and $^1P_1$ states from

Table C.1 The g-factors for for triplet states formed from two electrons.

	J=0	1	2	3	4
$^3S_1$	-	2	-	-	-
^3P_J	-	3/2	3/2	-	-
^3D_J	-	1/2	7/6	4/3	-
^3F_J	-	-	2/3	13/12	5/4

$(3p)^{-1}4s$. The three states with $J \neq 0$ have $g = 1.506$ $(J = 2)$, $g = 1.404$ $(J = 1)$, and $g = 1.102$ $(J = 1)$. These fit well with $^3P_2, ^3P_1$, and $^1P_1$, respectively. This is shown in the figure for the next question.

The next group of 10 states, all even parity, fit into the $(3p)^{-1}4p$ framework. The lowest state at 104102 cm^{-1} has $J = 1$ and $g = 1.985$. This can mean only one thing: $L = 0$. Thus this state must be the $^3S_1$. The next state has $J = 3$ and thus must be the $^3D_3$. The observed g-factor, 1.338, fits nicely with the anticipated 4/3. Similarly, the $J = 2$ state above it is likely $^3D_2$, with a predicted $g = 7/6$, agreeing with the observed 1.112. The next state, with $g = 0.838$, is expected to be the $^3D_1$ from the general size of the fine structure splitting. The expected g is 0.5, a bit off and indicating that the Russell-Saunders description is only approximate. The next state has $J = 2$, $g = 1.305$. It is in the right place to be $^1D_2$, but should have $g = 1$.

Having used up all the D states and the $^3S_1$, we have left $^3P_{0,1,2}$, $^1P_1$, and $^1S_0$ to assign to the remaining five states below 109 k cm^{-1}. They have the correct J values among them. The $J = 2$ state has $g = 1.260$ compared to 3/2 expected for $^3P_2$. One $J = 1$ state has $g = 0.819$, the other $g = 1.380$. We must assign the first to $^1P_1$ (expected $g = 1$) and the second to $^3P_1$ (expected $g = 3/2$). (If these last three g-factors are scaled by 1.16, they would agree very well with expectations.) Only the $J = 0$ states remain unassigned. The one nearest the ^3P_J states is assigned as $^3P_0$, the other as $^1S_0$. The energy level diagram for all 14 states is shown as part of the solution to the next problem.

(c) Singly ionized argon is more complicated. First let us deal with the ground state. The orbital configuration is $(3s)^2(3p)^5$. This corresponds to a 3p hole. Thus the Russell-Saunders ground state is $S = 1/2$, $L = 1$, $J = 1/2, 3/2$, and odd parity. Because the single "particle" is a hole, the fine structure is inverted. The ground state is thus $^2P^o_{3/2}$. The first excited state, at only 1423 cm^{-1}, is $^2P^o_{1/2}$.

With orbital excitations, the states lie above 108,000 cm^{-1} ($\geq$ 13.38 eV above ground state). What are the possibilities?

(i) In the $(3s)^2(3p)^5$ grouping, we can shift a 3s electron to 3p, giving $3s(3p)^6$. The closing of the p-shell will mean that this configuration will be lower than other orbital excitations. Thus we expect $^2S_{1/2}$, even parity, as the first excited state doublet. Empirically, the state at 108.722 k cm^{-1} is $J = 1/2$, even parity, in agreement with expectations.

(ii) Next we must promote an electron from $(3p)^5$ to 3d, 4s, 4p. Empirically, it is the 3d that gives the lowest group of states, as seen in the NBS table. The configuration is $(3s)^2(3p)^4 3d$ or $(3p)^{-2}3d$. We have two equivalent p-electron holes (analogous to carbon) plus the 3d electron. First, we consider the two 3p holes. Just as in carbon, we have $^3P_{2,1,0}, ^1D_2, ^1S_0$ states, with the ^3P_J group lying lowest. These are therefore the states to combine with the 3d electron. The result is $S = 1/2, 3/2$, $L = 3, 2, 1$. The $S = 3/2$ states are spin-symmetric and so will be lower than the $S = 1/2$ states. We therefore expect $^4P_{5/2,3/2,1/2}, ^4D_{7/2,5/2,3/2,1/2}$, and $^4F_{9/2,7/2,5/2,3/2}$ states, 11 in all, all with even parity.

Because of spatial symmetry arguments, the $^4D_{1/2,...7/2}$ states should and do lie lowest (132 -133 k cm^{-1}), forming an inverted set of fine structure lines. Empirically, the ^4P and ^4F states from the 3d excitation lie higher than states caused by a 4s orbital excitation.

(iii) We look at $(3p)^{-2}4s$. Just as above, we couple the $^3P_{0,1,2}$ of the $(3p)^{-2}$ configuration to the 4s. We have $S = 3/2, 1/2$, only $L = 1$. Thus we expect $^4P_{1/2,3/2,5/2}$ and $^2P_{1/2,3/2}$ states. These are even parity states. Hund's rule tells us that $S = 3/2$

Table C.2 Calculated and observed g-factors for singly ionized argon in the configuration $(3p)^{-2}4s$.

J	1/2	3/2	5/3
^4P_J predict	8/3=2.667	26/15=1.733	8/5=1.6
(observed)	(2.650)	(1.722)	(1.598)
^2P_J predict	2/3=0.667	4/3=1.333	-
(observed)	(0.676)	(1.334)	-

Table C.3 Spectral lines observed in neutral argon.

No.	Wavelength(Å)	$(\Delta\bar{\nu})$ cm^{-1}
1	9123	10,961
2	8115	12,322
3	7384	13,543
4	7067	14,149
5	6965	14,357

lies below $S = 1/2$. The data show three states with $J = 5/2, 3/2, 1/2$ at 134.2-135.5 k cm^{-1}and two states with $J = 3/2, 1/2$ at 138.2-139.3 k cm^{-1}. Both are inverted multiples, as expected from the $(3p)^{-2}4s$ configuration.

Finally, the g-factors. The Russell-Saunders formula is

$$g = \frac{3}{2} + \frac{S(S + 1) - L(L + 1)}{2J(J + 1)}. \tag{C.206}$$

Among the low-lying excited states, only the ^4P and ^2P states from $(3p)^{-2}4s$ have observed g-factors. We just compute these.

The agreement is almost too good to be true. See Table C.2. It holds for the higher excited states as well. For example, the $J = 7/2$ state (157,235 cm^{-1}) is designated $^4D^o_{7/2}$. The expected g-factor is 17/12=1.4167; the observed g-factor is 1.427. The $^4P_{5/2}$ state at 155,044 cm^{-1} has an observed g-factor of 1.599 with 1.60 expected. The $^2D_{3/2,5/2}$ at 148621 and 148843 cm^{-1} are expected to have $g = 4/5$ and $g = 6/5$, respectively, as compared to 0.803 and 1.202!

6.3 (a) For neutral argon, the observed lines, in order of decreasing wave length, are given in Table C.3. To find where these come from in the energy level diagram, we note

(i) None can involve the ground state since $\Delta\nu < 93,000$ cm^{-1}.

(ii) Nos. 4 and 5 are close together and so may involve different J states in a fine structure multiplet.

(iii) The transitions are strong and hence probably terminate on the lowest excited states.

This suggests starting with the $^3P^o_2$ state at 93144 cm^{-1} (as given in the NBS table) and adding the above $\Delta\nu$s. The result is

Line No.1	reaches state at	104102
Line No.2	reaches state at	105463
Line No.3	doesn't seem to fit	
Line No.4	reaches state at	107290
Line No.5	reaches state at	107496

Table C.4 Some electric dipole transitions in neutral argon.

Line No.	Initial config. and $^{2S+1}L_J$	Final config.	ΔJ	Parity change	Single orbital transition
1	$(3p)^5 4p\ ^3S_1$	$(3p)^5 4s\ ^3P_2^o$	+1	yes	$4p \rightarrow 4s$
2	$(3p)^5 4p\ ^3D_3$	$(3p)^5 4s\ ^3P_2^o$	−1	yes	$4p \rightarrow 4s$
3	$(3p)^5 4p\ ^3P_2$	$(3p)^5 4s\ ^3P_1^o$	−1	yes	$4p \rightarrow 4s$
4	$(3p)^5 4p\ ^3P_2$	$(3p)^5 4s\ ^3P_2^o$	0	yes	$4p \rightarrow 4s$
5	$(3p)^5 4p\ ^3P_1$	$(3p)^5 4s\ ^3P_2^o$	+1	yes	$4p \rightarrow 4s$

For line No. 3 we try the second excited state ($J = 1$), $^3P_1^o$, at 93751 cm^{-1}. Line No. 3 reaches the state at 107290. The five infrared or red lines are shown on Figure C.7. They all involve a parity change, no spin change, and $\Delta J = 0$ or ± 1. They are, as expected, consistent with being electric dipole radiative transitions.

A summary of these transitions: Note that the first five lines are just some of the strong red and infrared lines. For example, there is an intense line at 7503.9 Å that corresponds to the $^1S_0 \rightarrow ^1P_1^o$ transition among singlet lines.

(b) The observed (by Schuster) lines of AII, in order of decreasing λ are

Lin No.	Wavelength (Å)	$\Delta\nu$ cm^{-1}
1	4880	20,492
2	4806	20,807
3	4765	20,987
4	4736	21,115
5	4727	21,155
6	4348	22,999

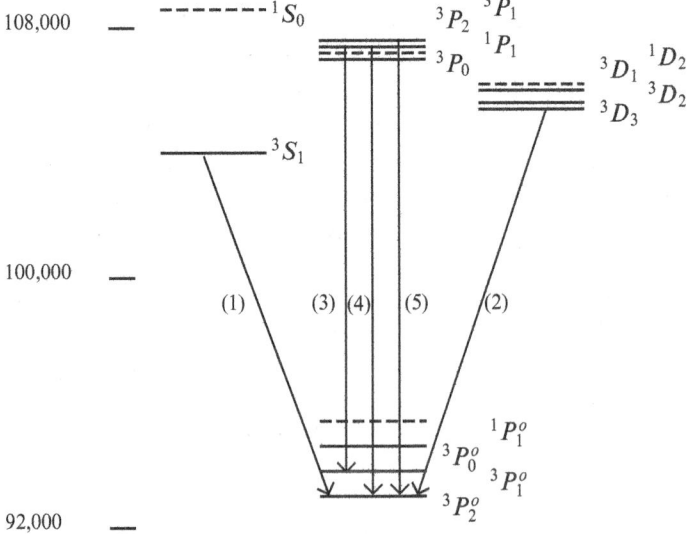

Figure C.7 Five infrared or red lines seen in neutral argon (AI). The energy levels are given in units of cm^{-1} above the ground state.

These transitions are among the excited sates. The spacing between the first and second excited states is 23,606 cm^{-1}. Thus none of these transitions ends on the first excited state. Also, because spin is conserved in E1 transitions, we expect quartet terms to combine with quartet terms, doublets with doublets. The multiplets above the first excited level have the rough locations

second	$^4D_{7/2..1/2}$	132.5 ± 0.2 k cm^{-1}
third	$^4P_{5/2...1/2}$	135.0 ± 0.7 k cm^{-1}
fourth	$^2P_{3/2,1/2}$	138.7 ± 0.5 k cm^{-1}
fifth	$^4F_{9/2...1/2}$	142.8 ± 0.6 k cm^{-1}

These groups of states may well serve as the final states. They all are even parity. The upper states must have odd parity. The lowest of these is $^4P^o_{5/2,3/2,1/2}$ at 155.4 ± 0.3k cm^{-1}. They extend up to 161.1 k cm^{-1}. Since 155-132 =23, it appears that these blue lines fit nicely into this range. Table C.5 shows possible transitions between initial parity-odd states and parity-even final states. We see there are three good matches with energies fitting with discrepancies less than 10 cm^{-1}. These are $^4P_{5/2},^4 P_{3/2},^4 D_{7/2}$ all going to $^4P^o_{5/2}$.

Turning to the doublet states, given the wavelengths, only $^2D^o(4p),^2 P^o(4p)$ to $^2P(4s)$ are possibilities. Examining Table C.6, we find that lines 1, 3, and 5 can be accounted for with $^2D^o_{5/2} \rightarrow^2 P_{3/2},^2D^o_{3/2} \rightarrow^2 P_{3/2}$, and $^2P^o_{3/2} \rightarrow^2 P_{1/2}$

A summary of these transitions:

Figure C.8 shows the transitions in singly-ionized argon for states between 132 k cm^{-1}and 160.5 k cm^{-1}. They correspond to single particle excitations from the 3p-shell to the 4s, 3d, and 4p shells. The odd-parity states at the top are 4p excitations.

Table C.5 Possible transitions among the spin-quartet states in singly-ionized argon. The values corresponding to the observed lines are shown as bold. These correspond to lines 2, 4, and 6 for AII. All values are given in cm^{-1} and correspond to the differences between in initial states listed horizontally and the final states listed vertically.

			$^4P^o(4p)$				$^4D^o(4p)$		
$J\downarrow$	$J \rightarrow$	5/2	3/2	1/2	7/2	5/2	3/2	1/2
	$E\downarrow\rightarrow$	155044	155352	155707	157234	157674	158168	158429
$^4P(3d)$	1/2	147229	7815	8123	8480	10005	10445	10939	11200
	3/2	147504	7540	7848	8205	9730	10170	10664	10925
	5/2	147876	7168	7476	7833	9358	9798	10292	10553
$^4P(4s)$	5/2	134242	**20802**	**21110**	21467	**22992**	23432	23926	24187
	3/2	135086	19958	20266	20623	22148	22588	23082	23343
	1/2	135602	19442	19750	20107	21632	22072	22566	22827
$^4D(3d)$	7/2	132328	22716	23024	23381	24906	25346	25840	26101
	5/2	132482	22562	22870	23227	24752	25192	25686	25947
	3/2	132631	22413	22721	23078	24603	25043	25537	25798
	1/2	132738	22306	22614	22971	24496	24936	25430	25691

Table C.6 Possible transitions among the spin-doublet states in singly-ionized argon. The values corresponding to the observed lines are shown as bold. These correspond to lines 1, 3, and 5 for AII. All values are given in cm^{-1} and correspond to the differences between in initial states listed horizontally and the final states listed vertically.

			$^2P^o(4p)$		$^2D^o(4p)$	
$J\downarrow$	$J\rightarrow$		3/2	1/2	5/2	3/2
		$E\downarrow\rightarrow$	160240	159707	158731	159394
$^2P(4s)$	3/2	138244	21995	21462	**20486**	**21149**
	1/2	139259	**20981**	20448	19471	20135

Table C.7 Electric dipole transitions in singly-ionized argon.

Line No.	Initial config. $^{2S+1}L_J$	Final config.	ΔJ	Parity change	Single orbital transition
1	$(3p)^4 4p\,^2D^o_{5/2}$	$(3p)^4 4s\,^2P_{3/2}$	-1	yes	$4p \rightarrow 4s$
2	$(3p)^4 4p\,^4P^o_{5/2}$	$(3p)^4 4s\,^4P_{5/2}$	0	yes	$4p \rightarrow 4s$
3	$(3p)^4 4p\,^2P^o_{3/2}$	$(3p)^4 4s\,^2P_{1/2}$	-1	yes	$4p \rightarrow 4s$
4	$(3p)^4 4p\,^4P^o_{3/2}$	$(3p)^4 4s\,^4P_{5/2}$	$+1$	yes	$4p \rightarrow 4s$
5	$(3p)^4 4p\,^2D^o_{3/2}$	$(3p)^4 4s\,^2P_{3/2}$	0	yes	$4p \rightarrow 4s$
6	$(3p)^4 4p\,^4D^o_{7/2}$	$(3p)^4 4s\,^4P_{5/2}$	-1	yes	$4p \rightarrow 4s$

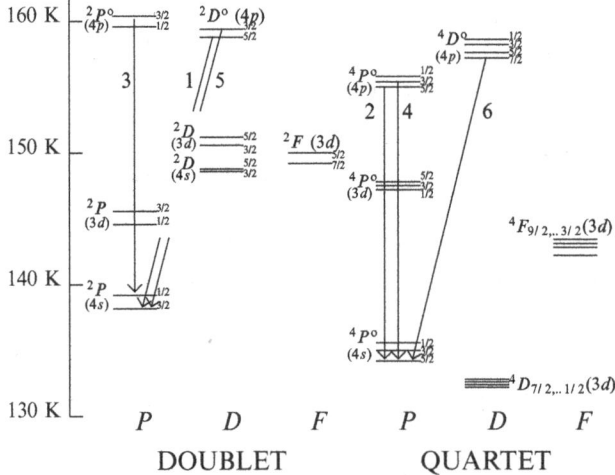

Figure C.8 The E1 transitions in singly-ionized argon.

Transitions are strongest from the 4p to 4s, but there are weaker lines (not shown) 4p to 3d. Since spin is conserved in E1 transitions, there are no transitions between quartet and doublet states.

6.10 (a) In the Hartree-Fock approximation, the inter-electron interaction energy is

$$
E_{ee} = \frac{1}{2}\sum_{ij}\int d^3x_i \int d^3x_j \frac{e^2}{r_{12}}\{|\psi_i(1)|^2|\psi_j(2)|^2 \left\langle m_{s_i}(1)|m_{s_j}(1)\right\rangle \left\langle m_{s_i}(2)|m_{s_j}(2)\right\rangle
$$

$$
- \psi_i^*(1)\psi_j(1)\psi_j^*(2)\psi_i(2)\left\langle m_{s_i}(1)|m_{s_j}(2)\right\rangle \left\langle m_{s_i}(2)|m_{s_j}(1)\right\rangle \}. \qquad \text{(C.207)}
$$

The second integral is the exchange energy E_{ex}. The sums over i and j are over all the different states of the electrons. In the Thomas-Fermi or Fermi gas model of metals, the wave functions are plane waves with all possible values of $\mathbf{k}$ up to $|\mathbf{k}|_{\max} = k_F$, where $k_F = (3\pi^2 n)^{1/3}$. Thus

$$
\sum_i \to \frac{V}{(2\pi)^3}\int_{|\mathbf{k}|<k_F} d^3k \sum_{m_i}
$$

$$
\sum_j \to \frac{V}{(2\pi)^3}\int_{|\mathbf{q}|<k_F} d^3q \sum_{m_j}. \qquad \text{(C.208)}
$$

We then have

$$
E_{ex} = -\frac{e^2 V^2}{(2\pi)^6}\int_{|\mathbf{k}|<k_F} d^3k \int_{|\mathbf{q}|<k_F} d^3q \int d^3x_1 \int d^3x_2
$$

$$
\times \frac{e^{i(\mathbf{k}-\mathbf{q})\cdot\mathbf{x}_1}}{V}\frac{1}{|\mathbf{x}_1 - \mathbf{x}_2|}\frac{e^{-i(\mathbf{k}-\mathbf{q})\cdot\mathbf{x}_2}}{V}. \qquad \text{(C.209)}
$$

Note that $\sum_{m_{s_i}}\sum_{m_{s_j}}\delta_{m_{s_i}m_{s_j}} = 2$, cancelling the 1/2 in E_{ee}. To do the $d^3x_1\, d^3x_2$ integration, change variables to $\mathbf{r} = \mathbf{r}_1 - \mathbf{r}_2$; $\mathbf{R} = (\mathbf{r}_1 + \mathbf{r}_2)/2$. Setting aside the integrals over $\mathbf{k}$ and $\mathbf{q}$, we examine

$$
\int d^3R \int d^3r \frac{e^{i(\mathbf{k}-\mathbf{q})\cdot\mathbf{r}}}{r} = V\int d^3r \frac{e^{i(\mathbf{k}-\mathbf{q})\cdot\mathbf{r}}}{r}. \qquad \text{(C.210)}
$$

Now we have two integrals of the form

$$
\int_{|\mathbf{k}|<k_F} d^3k\, e^{i\mathbf{k}\cdot\mathbf{r}}. \qquad \text{(C.211)}
$$

Since the result must be isotropic, we expand the exponential in spherical harmonics or Legendre polynomials and keep only the isotropic term, $j_0(kr)$, so the integral is

$$\int_{|\mathbf{k}|<k_F} d^3 k e^{i\mathbf{k}\cdot\mathbf{r}} = 4\pi \int_0^{k_F} k^2 dk \frac{\sin kr}{kr} = \frac{4\pi}{r^3} \int_0^{k_F r} x \sin x dx \qquad (C.212)$$

$$= \frac{4\pi}{r^3}[\sin k_F r - k_F r \cos k_F r] = \frac{4\pi k_F^3}{3}\left[\frac{3j_1(k_F r)}{k_F r}\right] \qquad (C.213)$$

$$= \frac{(2\pi)^3}{2} n \left[\frac{3j_1(k_F r)}{k_F r}\right]. \qquad (C.214)$$

Hence,

$$\frac{E_{ex}}{V} = -e^2 \left(\frac{n}{2}\right)^2 \int d^3 r \frac{1}{r} \left[\frac{3j_1(k_F r)}{k_F r}\right]^2. \qquad (C.215)$$

Two convenient integrals derived from Gradshteyn and Rizhik, Eq. 6.574.2 are

$$\int_0^\infty dx\, j_\ell(x)^2 = \frac{\pi}{2}\frac{1}{2\ell+1} \qquad (C.216)$$

$$\int_0^\infty dx \frac{j_\ell(x)^2}{x} = \frac{1}{2}\frac{1}{\ell(\ell+1)}. \qquad (C.217)$$

Using Eq.(C.217), we have

$$\frac{E_{ex}}{V} = -e^2 \frac{9\pi}{4}\frac{n^2}{k_F^2}. \qquad (C.218)$$

With $k_F = (3\pi^2 n)^{1/3}$, the result can be written in various forms:

$$\frac{E_{ex}}{V} = -\frac{3e^2}{4\pi}k_F n = -\frac{e^2 k_F^4}{4\pi^3} = -e^2 \frac{3^{4/3}}{4\pi^{1/3}} n^{4/3}. \qquad (C.219)$$

(b) Now the electrostatic potential energy between two charge distributions $e\rho_1(\mathbf{x})$ and $e\rho_2(\mathbf{x})$ is

$$U = e^2 \int d^3 x_1\, d^3 x_2 \frac{\rho_1(\mathbf{x}_1)\rho_2(\mathbf{x}_2)}{r_{12}}. \qquad (C.220)$$

With $\rho_1 = n/2$, a constant, the energy per unit volume is

$$\frac{U}{V} = \frac{e^2 n}{2} \int d^3 r \frac{\rho_2(\mathbf{r})}{r}, \qquad (C.221)$$

so the comparison with E_{es}/V shows we can identify

$$\rho_{ex} = -\frac{n}{2}\left[\frac{3j_1(k_F r)}{k_F r}\right]^2, \qquad (C.222)$$

where r is the distance away from the electron with which ρ_{ex} is associated. We call

$$g(\mathbf{r}_2 - \mathbf{r}_2) = 1 + \frac{2}{n}\rho_{ex}(\mathbf{r}_1 - \mathbf{r}_2) \qquad (C.223)$$

the pair-correlation function for parallel spins. Since $j_1(x) \to x/3$ as $x \to 0$, $\rho_{ex}(0) = -n/2$. This is the Pauli exclusion principle in action. The exchange density does not cancel the normal density completely because there are two spin states present in $n(\mathbf{x})$.

The volume integral is

$$\int d^3r \rho_{ex} = -\frac{n}{2} \times 4\pi \int_0^\infty dr\, r^2 \left[\frac{3j_1(k_Fr)}{k_Fr}\right]^2. \tag{C.224}$$

In particular, we have

$$\int d^3r \rho_{ex} = -\frac{n}{2} \times 4\pi k_F^{-3} \times 9 \times \frac{\pi}{6} = -1 \tag{C.225}$$

as expected.

See Bethe and Jackiw, *Intermediate Quantum Mechanics*, 3ʳᵈ edition, Chapter 5, for a discussion of how these results can be used to approximate the exchange integral and applied to the Thomas-Fermi model as done first by Dirac. This introduces a contribution in the Thomas-Fermi differential equation $V_{ex} = -e^2(3n(\mathbf{r})/\pi)^{1/3}$.

C.7 Chapter 7

7.9 Attractive square-well potential, s-wave radial Schrödinger equation:

$$\frac{d^2u}{dr^2} + \beta_i^2 u = 0, \tag{C.226}$$

where, with $K_0 = 2mV_0/\hbar^2$,

$$\beta_1 = K = \sqrt{K_0^2 + k^2}; \qquad \beta_2 = k, \tag{C.227}$$

where we use β_1 inside the potential and β_2 outside it.

Define

$$\chi_0 = K_0a; \qquad \chi_0^2 = \frac{2mV_0a^2}{\hbar^2}; \qquad x = ka; \qquad x^2 = \frac{2mEa^2}{\hbar^2} \tag{C.228}$$

and

$$\chi^2 = \chi_0^2 + x^2. \tag{C.229}$$

The solutions to the Schrödinger equation are

$$u_1(r) = A \sin Kr$$
$$u_2(r) = B \sin(kr + \delta). \tag{C.230}$$

Matching logarithmic derivatives at $r = a$ gives

$$K \cot Ka = k \cot(ka + \delta). \tag{C.231}$$

The cross section we seek is

$$\sigma = \frac{4\pi}{k^2} \sin^2 \delta, \tag{C.232}$$

so let's evaluate

$$\sin \delta = \sin(x + \delta)\cos x - \cos(x + \delta)\sin x, \tag{C.233}$$

but

$$\cot(x + \delta) = \frac{\chi \cot \chi}{x}, \tag{C.234}$$

so

$$\frac{\sin \delta}{\sin x \sin(x + \delta)} = \cot x - \cot(x + \delta) = \cot x - (\chi \cot \chi)/x$$

$$\frac{x^2 \sin^2 \delta}{\sin^2 x \sin^2(x + \delta)} = (x \cot x - \chi \cot \chi)^2. \tag{C.235}$$

But

$$\frac{1}{\sin^2(x + \delta)} = 1 + \cot^2(x + \delta) = 1 + \left(\frac{\chi \cot \chi}{x}\right)^2, \tag{C.236}$$

so

$$\frac{\sin^2 \delta}{\sin^2 x} = \frac{(x \cot x - \chi \cot \chi)^2}{x^2 + \chi^2 \cot^2 \chi} \tag{C.237}$$

and

$$\sigma = 4\pi a^2 \frac{\sin^2 x}{x^2} \frac{(x \cot x - \chi \cot \chi)^2}{x^2 + \chi^2 \cot^2 \chi}. \tag{C.238}$$

Now as $E \to 0$, so $x \to 0$;

$$\frac{\sigma}{4\pi a^2} \to \frac{(1 - \chi_0 \cot \chi_0)^2}{\chi_0^2 \cot \chi_0^2} = \left(\frac{\tan \chi_0}{\chi_0} - 1\right)^2 \tag{C.239}$$

This gives the cross section at zero energy as a function of the well depth. See Figure C.9.

The cross section as a function of $x = ka$ is shown in Figure C.10 for $\chi_0^2 = 20$ and $\chi_0^2 = 25$.

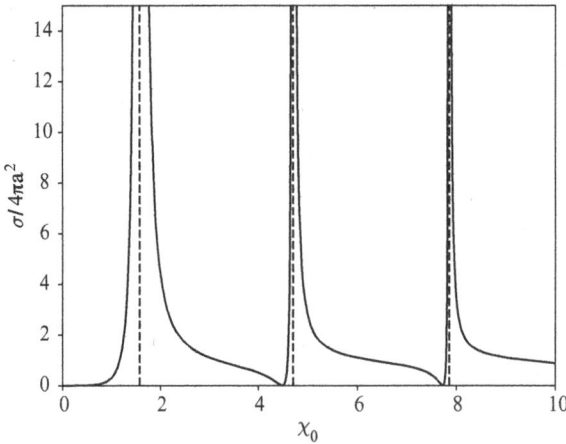

Figure C.9 The Ramsauer-Townsend effect: cross section at very low energy as a function of $\chi_0 = 2maV_0/\hbar^2$, where V_0 is the well depth.

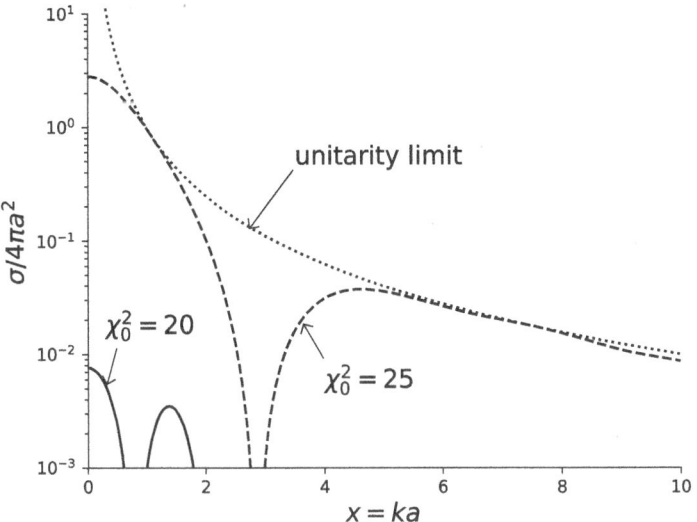

Figure C.10 The s-wave cross section for the square-well problem with $\chi_0^2 = 20, 25$, as well as the unitarity limit.

To isolate δ, we write

$$\tan(x + \delta) = \frac{x}{\chi \cot \chi}.$$ (C.240)

As the depth of the square well is increased from zero, the bound state appears when $\cot \chi_0 = 0$, so this occurs for $\chi_0 = \pi/2$ and the next when $\chi_0 = 3\pi/2$ or $\chi_0^2 \simeq 22.2$. From Levinson's theorem then, we know that for $\chi_0^2 = 25$ the phase shift should start at 2π while for $\chi_0^2 = 20$ it should start at π. This is confirmed by Figure C.11.

In Figure C.11, we also see that for $\chi_0^2 = 20$, the phase shift drifts along near π, giving a very small cross section at the lower energies - this is the Ramsauer effect. A 24% increase in potential strength to $\chi_0^2 = 25$ changes the character of the s-wave cross section completely.

C.8 Chapter 8

8.3 (a) It is sometimes clearer to multiply a vector by a constant vector to obtain a scalar. Thus, we consider

$$4\pi c \mathbf{C} \cdot \mathbf{L} = \int d^3x \, \mathbf{C} \cdot \mathbf{x} \times (\mathbf{E} \times \mathbf{B}) = \int d^3x \, [(\mathbf{C} \cdot \mathbf{E})(\mathbf{x} \cdot \mathbf{B}) - (\mathbf{C} \cdot \mathbf{B})(\mathbf{x} \cdot \mathbf{E})]$$

$$= \int d^3x \, [(\mathbf{C} \cdot \mathbf{E})(\mathbf{x} \cdot (\nabla \times \mathbf{A})) - (\mathbf{C} \cdot (\nabla \times \mathbf{A}))(\mathbf{x} \cdot \mathbf{E})].$$ (C.241)

The rule for integration by parts if the fields fall off at infinity is

$$\int d^3x \, \mathbf{J} \cdot \nabla \times \mathbf{K} = \int d^3x \, \mathbf{K} \cdot \nabla \times \mathbf{J}$$ (C.242)

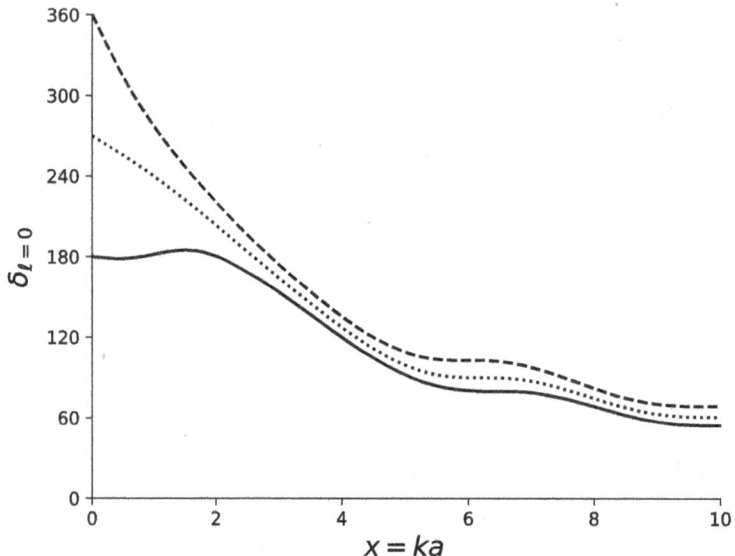

Figure C.11 The s-wave phase shift for the square-well problem with $\chi_0^2 = 20,25$. Solid curve: $\chi_0^2 = 20$; dashed curve: $\chi_0^2 = 25$; dotted curve: $\chi_0^2 = (3\pi/2)^2$.

because the minus sign from integration by parts is offset by the minus sign from reversing the order in the cross product. So the first term gives

$$\int d^3x\, \mathbf{A} \cdot \nabla \times (\mathbf{x}(\mathbf{C} \cdot \mathbf{E})). \tag{C.243}$$

But

$$\nabla \times \mathbf{x} = 0, \tag{C.244}$$

so the first term is simply

$$-\int d^3x\, \mathbf{A} \cdot (\mathbf{x} \times \nabla)(\mathbf{C} \cdot \mathbf{E}). \tag{C.245}$$

The second term is

$$-\int d^3x\, (\mathbf{x} \cdot \mathbf{E})\nabla \cdot (\mathbf{A} \times \mathbf{C}) = \int d^3x\, (\mathbf{A} \times \mathbf{C}) \cdot \nabla(\mathbf{x} \cdot \mathbf{E})$$

$$= \int d^3x\, (\mathbf{A} \times \mathbf{C}) \cdot [\mathbf{E} + x_i \nabla E_i]. \tag{C.246}$$

Together we have

$$\int d^3x\, [\mathbf{C} \cdot (\mathbf{E} \times \mathbf{A}) + (\mathbf{A} \times \mathbf{C})x_i \nabla E_i - \mathbf{A} \cdot (\mathbf{x} \times \nabla)(\mathbf{C} \cdot \mathbf{E})]. \tag{C.247}$$

Note that we can write

$$(\mathbf{A} \times \mathbf{C})x_i \nabla E_i = (\mathbf{A} \times \mathbf{C})_j(x_i \nabla_j E_i - x_j \nabla_i E_i), \tag{C.248}$$

since $\nabla \cdot \mathbf{E} = 0$. If we agree that the ∇ does not act on $\mathbf{A}$,

$$(\mathbf{A} \times \mathbf{C})_j(x_i\nabla_j E_i - x_j\nabla_i E_i) = (\mathbf{x} \times \nabla) \cdot (\mathbf{E} \times (\mathbf{A} \times \mathbf{C}))$$
$$= (\mathbf{x} \times \nabla) \cdot (\mathbf{A}(\mathbf{C} \cdot \mathbf{E}) - \mathbf{C}(\mathbf{A} \cdot \mathbf{E}))$$
$$= \mathbf{A} \cdot (\mathbf{x} \times \nabla)(\mathbf{C} \cdot \mathbf{E}) - \mathbf{C} \cdot (\mathbf{x} \times \nabla)(\mathbf{A} \cdot \mathbf{E}). \text{ (C.249)}$$

Inserting this into Eq. (C.247), we have

$$\int d^3x \, [\mathbf{C} \cdot (\mathbf{E} \times \mathbf{A}) - \mathbf{C} \cdot (\mathbf{x} \times \nabla)(\mathbf{A} \cdot \mathbf{E})]$$
$$= \int d^3x \left[\mathbf{C} \cdot (\mathbf{E} \times \mathbf{A}) - \sum_j A_j \mathbf{C} \cdot (\mathbf{x} \times \nabla)E_j) \right], \tag{C.250}$$

where we recalled that we had agreed that the ∇ did not act on $\mathbf{A}$. However, now we integrate by parts so that ∇ acts on $\mathbf{A}$ instead of $\mathbf{E}$:

$$4\pi c \mathbf{C} \cdot \mathbf{L} = \int d^3x \left[\mathbf{C} \cdot (\mathbf{E} \times \mathbf{A}) + \sum_j E_j \mathbf{C} \cdot (\mathbf{x} \times \nabla)A_j) \right]$$
$$\mathbf{L} = \frac{1}{4\pi c} \int d^3x \left[(\mathbf{E} \times \mathbf{A}) + \sum_j E_j(\mathbf{x} \times \nabla)A_j) \right]. \tag{C.251}$$

The appearance of the second term suggests that it is orbital angular momentum, thus implying that the first term must be due to photon spin.

(b) Using the expansions in Problem 8.2, and recognizing the similarity between $\mathbf{P}$ and $\mathbf{L}$, we find with $c = 1$

$$\mathbf{L}_{spin} = \frac{1}{4\pi} \int d^3x \, \mathbf{A} \times \mathbf{E}$$
$$= \frac{i}{2} \sum_{\mathbf{k}\lambda\lambda'} \Big\{ \epsilon_{\mathbf{k}\lambda} \times \epsilon^*_{\mathbf{k}\lambda'} a_{\mathbf{k}\lambda} a^\dagger_{\mathbf{k}\lambda'} - \epsilon^*_{\mathbf{k}\lambda} \times \epsilon_{\mathbf{k}\lambda'} a^\dagger_{\mathbf{k}\lambda} a_{\mathbf{k}\lambda'}$$
$$+ \epsilon_{\mathbf{k}\lambda} \times \epsilon_{-\mathbf{k}\lambda'} a_{\mathbf{k}\lambda} a_{-\mathbf{k}\lambda'} e^{-2i\omega_k k}$$
$$- \epsilon^*_{\mathbf{k}\lambda} \times \epsilon^*_{-\mathbf{k}\lambda'} a^\dagger_{\mathbf{k}\lambda} a^\dagger_{-\mathbf{k}\lambda'} e^{+2i\omega_k t} \Big\}. \tag{C.252}$$

As in Problem 8.2, the terms with time dependence vanish if we take $(\mathbf{k}, \lambda, \lambda') \rightarrow (-\mathbf{k}, \lambda', \lambda)$. The circular polarization basis has

$$\epsilon_\pm = \frac{1}{\sqrt{2}}(\epsilon_1 \pm i\epsilon_2), \tag{C.253}$$

where $\mathbf{k} \cdot \epsilon_{1,2} = 0$ and $\epsilon_i \cdot \epsilon_j = \delta_{ij}$. As consequences,

$$\epsilon_\pm \times \epsilon^*_\pm = \mp i\hat{\mathbf{k}}; \qquad \epsilon_\pm \times \epsilon^*_\mp = 0. \tag{C.254}$$

In the sum over λ, λ', there is a $\delta_{\lambda,\lambda'}$ and we have, since $\sum_k \hat{\mathbf{k}} = 0$,

$$
\mathbf{L} = \frac{i}{2} \sum_k \hat{\mathbf{k}} \left\{ -ia_{k,+} a_{k,+}^\dagger - ia_{k,+}^\dagger a_{k,+} + ia_{k,-} a_{k,-}^\dagger + ia_{k,-}^\dagger a_{k,-} \right\}
$$

$$
= \sum_k \hat{\mathbf{k}} \left\{ a_{k,+}^\dagger a_{k,+} - a_{k,-}^\dagger a_{k,-} \right\}. \tag{C.255}
$$

Thus, we get plus or minus one unit of angular momentum for helicity plus or minus: the photon has spin one!

C.9 Chapter 9

9.1 There are several ways of doing these radiative transitions. We illustrate the brute-force method on part (a).

(a) The brute-force approach starts with

$$
\Gamma_{\text{fi}} = \frac{4}{3} \frac{e^2 k^3}{\hbar} |\langle f | \mathbf{x} | i \rangle|^2. \tag{C.256}
$$

With initial state $2p_{1/2}$ and final state $1s_{1/2}$, we want to sum over the two final spin values. We can choose either $m = 1/2$ or $m = -1/2$ for the initial state. The result is the same for both. Pick $m = 1/2$. The Clebsch-Gordan series gives

$$
\psi_{2p_{1/2},m=1/2} = R_{2p}(r) \left[\sqrt{\frac{2}{3}} Y_{11}\beta - \frac{1}{\sqrt{3}} Y_{10}\alpha \right]
$$

$$
\psi_{1s_{1/2},m'=\pm 1/2} = R_{1s}(r) Y_{00}[\alpha, \beta], \tag{C.257}
$$

where α is the spin-up state and β is the spin-down state. Now

$$
|\langle f | \mathbf{x} | i \rangle|^2 = |\langle f | z | i \rangle|^2 + |\langle f | x | i \rangle|^2|^2 + |\langle f | y | i \rangle|^2
$$

$$
= |\langle f | z | i \rangle|^2 + |\langle f | \frac{1}{\sqrt{2}} (x + iy) | i \rangle|^2 + |\langle f | -\frac{1}{\sqrt{2}} (x - iy) | i \rangle|^2
$$

$$
= \frac{4\pi}{3} \sum_\mu |\langle f | r Y_{1\mu}^* | i \rangle|^2. \tag{C.258}
$$

Using the explicit wave functions, we have

$$
\sum_{m'} \langle f | \mathbf{x} | i \rangle|^2 = \frac{4\pi}{3} \left| \int_0^\infty dr \, r^3 R_{1s}^* R_{2p} \right|^2 \times \frac{1}{4\pi}
$$

$$
\times \sum_\mu \left\{ \left| \int d\Omega Y_{1\mu}^* (-\frac{1}{\sqrt{3}}) Y_{10} \right|^2 + \left| \int d\Omega Y_{1\mu}^* \sqrt{\frac{2}{3}} Y_{11} \right|^2 \right\}. \tag{C.259}
$$

The two terms in curly brackets come from the β and α terms. The former gives $(1/3)\delta_{\mu,0}$ and the latter $(2/3)\delta_{\mu,1}$, so together we find that the sum over μ of the piece in curly brackets is precisely unity. Thus,

$$| \langle f | \mathbf{x} | i \rangle |^2 = \frac{1}{3} \left| \int_0^\infty dr\, r^3 R_{1s}^* R_{2p} \right|^2 . \tag{C.260}$$

The wave functions, with r measured in units of the Bohr radius a_0, are

$$R_{1s} = 2e^{-r}; \qquad R_{2p} = \frac{1}{2\sqrt{6}} r e^{-r/2}, \tag{C.261}$$

so

$$\int_0^\infty dr\, r^3 R_{1s}^* R_{2p} = \frac{a_0}{\sqrt{6}} \int_0^\infty r^4 e^{-3r/2} dr = \frac{a_0}{\sqrt{6}} \left(\frac{2}{3} \right)^5 4! = \sqrt{\frac{2^{15}}{3^9}} a_0. \tag{C.262}$$

With photon energy as $\hbar\omega_0 = \dfrac{e^2}{2a_0} \left(1 - \dfrac{1}{4} \right) = \dfrac{3}{8} \dfrac{e^2}{a_0}$, we have

$$k a_0 = \frac{3}{8} \alpha; \qquad \left(\alpha = \frac{e^2}{\hbar c} \cong \frac{1}{137} \right). \tag{C.263}$$

The transition rate is

$$\Gamma = \frac{4}{3} \alpha c k^3 \frac{a_0^2}{3} \frac{2^{15}}{3^9} = \left(\frac{2}{3} \right)^8 \left(\frac{\alpha^4 c}{a_0} \right). \tag{C.264}$$

The mean lifetime is

$$\tau = \frac{1}{\Gamma} = \left(\frac{3}{2} \right)^8 \left(\frac{a_0}{\alpha^4 c} \right) = \left(\frac{3}{2} \right)^8 (0.6225 \times 10^{-10} \text{ sec})$$

$$= 1.595 \times 10^{-9} \text{ sec} = \frac{1}{6.268 \times 10^8 \text{ sec}^{-1}}. \tag{C.265}$$

(b) The $2p_{3/2} \rightarrow 1s_{1/2}$ transition can be done most easily by taking $m = 3/2$ for the initial state. Then

$$\psi_{2p_{3/2}} = R_{2p}(r) Y_{11} \alpha \tag{C.266}$$

and the only the radial factor that will enter is

$$-\frac{1}{\sqrt{2}}(x - iy) = \sqrt{\frac{4\pi}{3}} r Y_{11}^* \tag{C.267}$$

and only $m' = 1/2$, hence

$$\sum_{m'} | \langle f | \mathbf{x} | i \rangle |^2 = \frac{4\pi}{3} \left| \int_0^\infty dr\, r^3 R_{1s}^* R_{2p} \right|^2 \times \frac{1}{4\pi} \left| \int d\Omega\, Y_{11}^* Y_{11} \right|^2 . \tag{C.268}$$

This is just the same as for $2p_{1/2} \rightarrow 1s_{1/2}$, ignoring the slight difference in energy due to fine structure.

We can address this problem with the more sophisticated techniques in Section 9.5. If the spatial wave functions depend on L alone, not J, and if the fine structure energy splittings can be ignored, then

$$\sum_{J'M'} \Gamma(n'L'S'J'|nLJSM) = \Gamma(n'L'|nL). \tag{C.269}$$

For parts (a) and (b), there is only one J' value. The answer for (b) is the same as for (a) and is equal to the spinless result. The factor $\ell_>/(2\ell + 1)$ there is equal to $1/3$ here, and so we have

$$\Gamma(2p_j \to 1s_{1/2}) = \frac{4}{3}\alpha ck^3 \times \frac{1}{3} \left| \int_0^\infty dr\, r^3 R_{1s}^* R_{2p} \right|^2. \tag{C.270}$$

This agrees with our brute force calculations.

(c) We here are interested in $2s_{1/2} \to 2p_{1/2}$, i.e. the Lamb shift spontaneous transition ($\nu \cong 1058$ MHz). Using the rule Eq. (9.43), we see that if the $2p_{3/2}$ and $2p_{1/2}$ were degenerate, the two states would share the spinless transition rate in the ratio 2:1, i.e. $\Gamma(2s_{1/2} \to 2p_{1/2}) = (1/3)\Gamma(2s \to 2p)$. Using the spinless result with $\ell_>/(2\ell + 1)$,

$$\Gamma(2s_{1/2} \to 2p_{1/2}) = \frac{4}{3}\alpha ck^3 \times \frac{1}{3} \left| \int_0^\infty dr\, r^3 R_{2p}^* R_{2s} \right|^2. \tag{C.271}$$

The $2s$ wave function with r measured in units of a_0 is

$$R_{2s}(r) = \frac{1}{\sqrt{2}} \left(1 - \frac{r}{2}\right) e^{-r/2} \tag{C.272}$$

and

$$\int_0^\infty dr\, r^3 R_{2p}^* R_{2s} = \frac{a_0}{4\sqrt{3}} \int_0^\infty \left(r^4 - \frac{r^5}{2}\right) e^{-r} dr = -3\sqrt{3}a_0. \tag{C.273}$$

The transition rate is

$$\Gamma = \frac{4}{9} \times 9 \times 3\alpha ck^3 a_0^2 = 12\frac{\alpha^4 c}{a_0} \left(\frac{ka_0}{\alpha}\right)^3. \tag{C.274}$$

Now $(a_0/\alpha^4 c) = 0.6225 \times 10^{-10}$ sec and $a_0 = 0.529 \times 10^{-8}$ cm. Converting the Lamb shift to wave number,

$$k = \frac{2\pi\nu}{c} = 0.2217\ \text{cm}^{-1}; \qquad \left(\frac{ka_0}{\alpha}\right)^3 = 4.156 \times 10^{-21}. \tag{C.275}$$

Finally

$$\Gamma = 12 \times 4.156 \times 10^{-21} \times (0.6225 \times 10^{-10}\text{sec})^{-1} = 8.0 \times 10^{-10}\text{sec}^{-1}. \tag{C.276}$$

More graphically,

$$\tau = 1/\Gamma = 1.25 \times 10^9\ \text{sec} \cong 39.6\ \text{years}. \tag{C.277}$$

(d) Actually, the $2s_{1/2}$ decays by two-photon emission to the $1s_{1/2}$ ground state with a mean lifetime of a few tenths of a second (G. Breit and E. Teller, Astrophys. J, **91**, 115 (1940)). The E1 transition is forbidden by parity. While M1 is allowed by parity, the radial wave functions are orthogonal, so the M1 is forbidden, as well.

A comparison of the line widths in MHz with the splittings is shown in Figure C.12.

For our hydrogen atom, first consider the oscillator strength defined in terms of z and treat $1s \rightarrow 2p$. We mean by the oscillator strength, the expression for f_{ij} summed over all the various m' values in the final state, with a given m value chosen for the initial state. For simplicity, we omit the spin, so there is no fine structure. We have

$$f_{2p,1s} = \frac{2m}{\hbar^2} \times \frac{3}{8}\frac{e^2}{a_0} |\langle 2p_{m=0}| \, z \, |1s\rangle|^2$$

$$= \frac{2m}{\hbar^2} \times \frac{3}{8}\frac{e^2}{a_0} \left| \int_0^\infty R_{2p} r R_{1s} r^2 \, dr \int d\Omega Y_{10} \cos\theta Y_{11} \right|^2. \quad (C.278)$$

The two integrals are, respectively, $\sqrt{2^{15}3^{-9}}a_0$ and $1/\sqrt{3}$. Thus

$$f_{2p,1s} = \frac{2m}{\hbar^2} \times \frac{3}{8}\frac{e^2}{a_0}\frac{2^{15}}{3^9}a_0^2 \times \frac{1}{3} = \frac{2^{13}}{3^9} = 0.4162. \quad (C.279)$$

Table 14 in Bethe and Salpeter, *Quantum Mechanics of One- and Two-Electron Atoms*, p. 265, gives oscillator strengths for hydrogen. Starting with an s state, electric dipole transitions connect only to p states, but these include continuum states in addition to the discrete spectrum. We reproduce a portion of that table in Table C.8.

Figure C.12 Line widths vs. fine structure in $n = 2$ hydrogen. The full-width half-maximum of $2p_{1/2}$ is about one-tenth of the Lamb shift. The hyperfine splittings (not shown) are comparable, e.g. the $F = 0$ to $F = 1$ splitting is 1420 MHz in the $1s$ level and one-eighth of that, 178 MHz, for the $2s$ level.

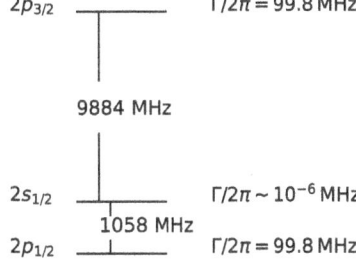

Table C.8 A portion of Table 14 from Bethe and Salpeter, *Quantum Mechanics of One- and Two-Electron Atoms*, p. 265, showing oscillator strengths for hydrogen. Note that the continuum plays a significant role. The sum rule starting with a p state has contributions from both final s-wave and d-wave states.

Initial	1s	2s	2p	2p
Final	np	np	ns	nd
n=1	-	-	−0.139	-
n=2	0.4162	-	-	-
n=3	0.0791	0.4349	0.014	0.696
n=4	0.0290	0.1028	0.0031	0.122
Discrete Spectrum	0.5650	0.6489	−0.119	0.928
Continuum	0.4350	0.3511	0.008	0.183
Total	1.000	1.000	−0.111	1.111

9.3 Multipole expansion of

$$M_{ba} = \langle b| \sum_j e^{-i\mathbf{k}\cdot\mathbf{r}_j} \boldsymbol{\epsilon}^*_{\mathbf{k},\lambda} \cdot \left(\frac{e_j \mathbf{p}_j}{m_j} + i \frac{\mu_j e}{2m_j} \mathbf{k} \times \boldsymbol{\sigma}_j \right) |a\rangle. \tag{C.280}$$

(a) We make use of

$$\frac{\mathbf{p}_j}{m_j} = \dot{\mathbf{x}}_j = \frac{i}{\hbar}[H, \mathbf{x}_j], \tag{C.281}$$

where H is the matter Hamiltonian and the expansion

$$e^{i\mathbf{k}\cdot\mathbf{x}_j} = 1 - i\mathbf{k}\cdot\mathbf{x}_j + \cdots \tag{C.282}$$

We consider first the "1" term. This gives contributions

$$M_{ba} = \langle b| \sum_j \boldsymbol{\epsilon}^*_{\mathbf{k},\lambda} \cdot \left(ie_j[H, \mathbf{x}_j] + i \frac{\mu_j e}{2m_j} \mathbf{k} \times \boldsymbol{\sigma}_j \right) |a\rangle$$

$$= \langle b| \, \boldsymbol{\epsilon}^*_{\mathbf{k},\lambda} \cdot \left(i\omega_{ba} \sum_j e_j \mathbf{x}_j + i\mathbf{k} \times \sum_j \frac{\mu_j e}{2m_j} \boldsymbol{\sigma}_j \right) |a\rangle. \tag{C.283}$$

The first term is the dominant E1 contribution, the second, the spin part of the M1 amplitude:

$$M^{(1)}_{ba}(E1) = \langle b| \, \boldsymbol{\epsilon}^*_{\mathbf{k},\lambda} \cdot i\omega_{ba} \sum_j e_j \mathbf{x}_j \, |a\rangle$$

$$M^{(1)}_{ba}(M1) = \langle b| \, \boldsymbol{\epsilon}^*_{\mathbf{k},\lambda} \cdot i\mathbf{k} \times \sum_j \frac{\mu_j e}{2m_j} \boldsymbol{\sigma}_j \, |a\rangle. \tag{C.284}$$

The next three contributions come from the $-i\mathbf{k}\cdot\mathbf{x}_j$ term.

$$M^{(2)}_{ba} = \langle b| \sum_j -i\mathbf{k}\cdot\mathbf{x}_j \boldsymbol{\epsilon}^*_{\mathbf{k},\lambda} \cdot \left(\frac{e_j \mathbf{p}_j}{m_j} + i \frac{\mu_j e}{2m_j} \mathbf{k} \times \boldsymbol{\sigma}_j \right) |a\rangle \tag{C.285}$$

We have reverted to the $\mathbf{p}_j/m_j$ representation. In the first term, we symmetrize and anti-symmetrize in order to exhibit the first-rank and second-rank tensors:

$$\mathbf{k}\cdot\mathbf{x}\boldsymbol{\epsilon}\cdot\mathbf{p} = \frac{1}{2}(\mathbf{k}\cdot\mathbf{x}\boldsymbol{\epsilon}\cdot\mathbf{p} + \boldsymbol{\epsilon}\cdot\mathbf{x}\mathbf{k}\cdot\mathbf{p}) + \frac{1}{2}(\mathbf{k}\cdot\mathbf{x}\boldsymbol{\epsilon}\cdot\mathbf{p} - \boldsymbol{\epsilon}\cdot\mathbf{x}\mathbf{k}\cdot\mathbf{p})$$

$$= \frac{1}{2}(\mathbf{k}\cdot\mathbf{x}\boldsymbol{\epsilon}\cdot\mathbf{p} + \boldsymbol{\epsilon}\cdot\mathbf{x}\mathbf{k}\cdot\mathbf{p}) + \frac{1}{2}(\mathbf{k}\times\boldsymbol{\epsilon})\cdot(\mathbf{x}\times\mathbf{p}). \tag{C.286}$$

With $\mathbf{x}\times\mathbf{p} = \mathbf{L}$ in the second term and $\mathbf{p} = im[H, \mathbf{x}]$ in the first, we have

$$\mathbf{k}\cdot\mathbf{x}\boldsymbol{\epsilon}\cdot\mathbf{p} = \frac{im}{2}(\mathbf{k}\cdot\mathbf{x}[H, \boldsymbol{\epsilon}\cdot\mathbf{x}] + \boldsymbol{\epsilon}\cdot\mathbf{x}[H, \mathbf{k}\cdot\mathbf{x}]) + \frac{1}{2}(\mathbf{k}\times\boldsymbol{\epsilon})\cdot\mathbf{L}. \tag{C.287}$$

The ordering of the factors in the symmetric (first) term is immaterial because $\mathbf{k}\cdot\boldsymbol{\epsilon} = 0$. This means the first term is actually the commutator of the product

$$\mathbf{k}\cdot\mathbf{x}\boldsymbol{\epsilon}\cdot\mathbf{p} = \frac{im}{2}[H, \mathbf{k}\cdot\mathbf{x}\boldsymbol{\epsilon}\cdot\mathbf{x}] - \frac{1}{2}\boldsymbol{\epsilon}\cdot(\mathbf{k}\times\mathbf{L}). \tag{C.288}$$

Inserting the index j, multiplying by $-ie_j/m_j$, and summing over j gives the first term in Eq. (C.284).

$$[M_{ba}^{(2)}]_{\text{first term}} = \langle b| \, \frac{1}{2} \sum_j \left\{ [H, \mathbf{k} \cdot \mathbf{x}_j \, \boldsymbol{\epsilon}_{\mathbf{k},\lambda}^* \cdot \mathbf{x}_j \,] e_j + \frac{ie_j}{m_j} \boldsymbol{\epsilon}_{\mathbf{k},\lambda}^* \cdot (\mathbf{k} \times \mathbf{L}_j) \right\} |a\rangle$$

$$= \frac{1}{2}\omega_{ba} \langle b| \sum_j e_j \mathbf{k} \cdot \mathbf{x}_j \, \boldsymbol{\epsilon}_{\mathbf{k},\lambda}^* \cdot \mathbf{x}_j \, |a\rangle + \langle b| \, i\sum_j \boldsymbol{\epsilon}_{\mathbf{k},\lambda}^* \cdot \frac{e_j}{2m_j}(\mathbf{k} \times \mathbf{L}_j) \, |a\rangle$$

$$(C.289)$$

The first term gives the electric quadrupole, E2, contribution and the second the orbital piece of the M1.

The second term from Eq.(C.285) is

$$[M_{ba}^{(2)}]_{\text{second term}} = \boldsymbol{\epsilon}_{\mathbf{k},\lambda}^* \cdot \sum_j \frac{\mu_j e}{2m_j} \langle b| \, \mathbf{k} \cdot \mathbf{x}_j (\mathbf{k} \times \boldsymbol{\sigma}_j) \, |a\rangle. \tag{C.290}$$

Once again, we want to break this into symmetric and anti-symmetric parts. It is also suggestive to introduce $\mathbf{b} = \mathbf{k} \times \boldsymbol{\epsilon}$ as a representative of $\mathbf{B}$, as $\boldsymbol{\epsilon}$ represents $\mathbf{E}$. Then

$$\boldsymbol{\epsilon} \cdot (\mathbf{k} \times \boldsymbol{\sigma}) \mathbf{k} \cdot \mathbf{x} = -(\mathbf{k} \times \boldsymbol{\epsilon}) \cdot \boldsymbol{\sigma} \, \mathbf{k} \cdot \mathbf{x}$$

$$= -\frac{1}{2}[\mathbf{b} \cdot \boldsymbol{\sigma} \, \mathbf{k} \cdot \mathbf{x} + \mathbf{b} \cdot \mathbf{x} \, \mathbf{k} \cdot \boldsymbol{\sigma}] - \frac{1}{2}[\mathbf{b} \cdot \boldsymbol{\sigma} \, \mathbf{k} \cdot \mathbf{x} - \mathbf{b} \cdot \mathbf{x} \, \mathbf{k} \cdot \boldsymbol{\sigma}]$$

$$= -\frac{1}{2}[\mathbf{b} \cdot \boldsymbol{\sigma} \, \mathbf{k} \cdot \mathbf{x} + \mathbf{b} \cdot \mathbf{x} \, \mathbf{k} \cdot \boldsymbol{\sigma}] - \frac{1}{2}(\mathbf{k} \times \mathbf{b}) \cdot (\mathbf{x} \times \boldsymbol{\sigma}). \tag{C.291}$$

Now

$$\mathbf{k} \times \mathbf{b} = \mathbf{k} \times (\mathbf{k} \times \boldsymbol{\epsilon}) = -\omega^2 \boldsymbol{\epsilon}, \tag{C.292}$$

so

$$\boldsymbol{\epsilon} \cdot (\mathbf{k} \times \boldsymbol{\sigma}) \mathbf{k} \cdot \mathbf{x} = -\frac{\omega^2}{2}[(\hat{\mathbf{k}} \times \boldsymbol{\epsilon}) \cdot \boldsymbol{\sigma} \, \hat{\mathbf{k}} \cdot \mathbf{x} + (\hat{\mathbf{k}} \times \boldsymbol{\epsilon}) \cdot \mathbf{x} \, \hat{\mathbf{k}} \cdot \boldsymbol{\sigma}]$$

$$+ \frac{1}{2}\omega^2 \boldsymbol{\epsilon} \cdot (\mathbf{x} \times \boldsymbol{\sigma}). \tag{C.293}$$

The second term transforms as a first-rank tensor in the "matter" variables, dotted into the photon. The cross product makes this a polar vector ($\boldsymbol{\sigma}$ is like $\mathbf{x} \times \mathbf{p}$) and thus has odd parity. Therefore, this contributes to E1.

The first term is a second-rank tensor in the matter variables. The presence of $\mathbf{x}$ and $\boldsymbol{\sigma}$ again means odd parity. The combination means this contributes to M2. We ignore this contribution here.

In summary, we have

$$M_{ba}(E1) = i\omega_{ba} \boldsymbol{\epsilon}_{\mathbf{k},\lambda}^* \cdot \sum_j \langle b| \, e_j \mathbf{x}_j - i\frac{\omega_{ba}\mu_j e}{4m_j} \mathbf{x}_j \times \boldsymbol{\sigma}_j \, |a\rangle \tag{C.294}$$

$$M_{ba}(M1) = i\boldsymbol{\epsilon}_{\mathbf{k},\lambda}^* \cdot \mathbf{k} \times \sum_j \langle b| \, \frac{\mu_j e_j}{2m_j}\boldsymbol{\sigma}_j + \frac{e_j}{2m_j}\mathbf{L}_j \, |a\rangle \tag{C.295}$$

$$M_{ba}(E2) = \frac{\omega_{ba}}{2} \sum_j e_j \langle b| \, \boldsymbol{\epsilon}_{\mathbf{k},\lambda}^* \cdot \mathbf{x}_j \mathbf{k} \cdot \mathbf{x}_j \, |a\rangle, \tag{C.296}$$

where $\omega_{ba} = E_b - E_a$ for emission.

(b) The second term in $M_{ba}(E1)$ is called the induced electric dipole (really a spin E1 contribution). Its order of magnitude in atoms, relative to the "static" dipole term, is restoring the appropriate factors of $\hbar$ and c (and recalling that μ_j is dimensionless)

$$\frac{\omega\mu_j}{4m_j} \sim \frac{\mu_e}{4} \cdot \frac{\hbar\omega}{m_ec^2} \lesssim \frac{Z^2\alpha^2}{8} \tag{C.297}$$

since $\hbar\omega = \Delta E \sim \frac{1}{2}Z^2\alpha^2 m_ec^2$.

Unless we are considering an x-ray transition in a heavy element, this induced dipole seems very small. But in principle, it is present. In many-electron atoms, the sum $\sum_j \mathbf{x}_j \times \boldsymbol{\sigma}_j$ will contribute to both $\Delta S = 0$ and $\Delta S \neq 0$ transitions. We can see this by looking at a two-electron case:

$$\mathbf{x}_1 \times \boldsymbol{\sigma}_1 + \mathbf{x}_2 \times \boldsymbol{\sigma}_2 = (\mathbf{x}_1 + \mathbf{x}_2) \times \frac{1}{2}(\boldsymbol{\sigma}_1 + \boldsymbol{\sigma}_2) + (\mathbf{x}_1 - \mathbf{x}_2) \times \frac{1}{2}(\boldsymbol{\sigma}_1 - \boldsymbol{\sigma}_2)$$

$$= (\mathbf{x}_1 + \mathbf{x}_2) \times \mathbf{S} + (\mathbf{x}_1 - \mathbf{x}_2) \times \frac{1}{2}(\boldsymbol{\sigma}_1 - \boldsymbol{\sigma}_2). \tag{C.298}$$

The first term involves $\mathbf{S}$ and so has a selection rule $\Delta S = 0$, but the second term, involving $(\boldsymbol{\sigma}_1 - \boldsymbol{\sigma}_2)$, has matrix elements connecting singlet and triplet states (spin-flip matrix elements).

What about the Hg resonant line at 2537 Å, described in books as a $^3P_1 \to {}^1S_0$ transition? The states involved are $(6s)^2$ and $(6s)(6p)$ configurations, with an effective nuclear charge for the $6p$ electron of $Z_{\text{eff}} \sim 5 - 6$. The estimate is

$$\frac{\hbar\omega}{4mc^2} \simeq \frac{\hbar\omega(\text{eV})}{2} \times 10^{-6} = 2.5 \times 10^{-6}. \tag{C.299}$$

The $^3P_1 \to {}^1S_0$, if a pure Russell-Saunders transition has $\Delta S = 1$, $\Delta J = 1$, and a parity change. It cannot then be a "static" E1, or M1 or E2. In principle, it could be caused by the induced E1 term. But the magnitudes are wrong. In fact, the strength of this transition is about 2-3% of that of the $^1P_1 \to {}^1S_0$ transition at 1850 Å. Our estimate is $((2.5 \times 10^{-6})^2 \simeq 10^{-11}$! So, indeed, the induced spin-flip dipole is present only in principle.

The explanation is that this is a case of intermediate coupling, as explained in an earlier chapter. The $^3P_1$ state actually has an admixture of $^1P_1$, which allows the transition, albeit at a reduced rate.

(c) We can write

$$\mathbf{k} \cdot \mathbf{r}\boldsymbol{\epsilon} \cdot \mathbf{r} = k_a\epsilon_b(r_ar_b - \frac{1}{3}\delta_{ab}), \tag{C.300}$$

since $\mathbf{k} \cdot \boldsymbol{\epsilon}$. In fact, alternatively, we could write

$$(k_a\epsilon_b - \frac{1}{3}\mathbf{k} \cdot \boldsymbol{\epsilon}\,\delta_{ab})r_ar_b = \mathbf{k} \cdot \mathbf{r}\boldsymbol{\epsilon} \cdot \mathbf{r}. \tag{C.301}$$

In contracting two second-rank tensors, one can always add or subtract a zero-rank tensor to one factor.

C.10   Chapter 10

10.1 (a)

$$i\frac{\partial \psi}{\partial t} = \boldsymbol{\alpha} \cdot \frac{1}{i}\nabla\psi - e\boldsymbol{\alpha} \cdot \mathbf{A}\psi + \beta m\psi$$

$$-i\frac{\partial \psi^{\dagger}}{\partial t} = -\frac{1}{i}(\nabla\psi^{\dagger}) \cdot \boldsymbol{\alpha} - e\psi^{\dagger}\boldsymbol{\alpha} \cdot \mathbf{A} + \psi^{\dagger}m\beta, \tag{C.302}$$

where the second line is obtained by taking the Hermitian conjugate of the first. Now multiply the first by $\psi^{\dagger}\boldsymbol{\alpha}\beta$ from the left and the second by $\beta\boldsymbol{\alpha}\psi$ from the right.

$$i\psi^{\dagger}\boldsymbol{\alpha}\beta\frac{\partial \psi}{\partial t} = \psi^{\dagger}\boldsymbol{\alpha}\beta\boldsymbol{\alpha} \cdot \frac{1}{i}\nabla\psi - e\psi^{\dagger}\boldsymbol{\alpha}\beta\boldsymbol{\alpha} \cdot \mathbf{A}\psi + m\psi^{\dagger}\boldsymbol{\alpha}\psi$$

$$-i\frac{\partial \psi^{\dagger}}{\partial t}\beta\boldsymbol{\alpha}\psi = -\frac{1}{i}(\nabla\psi^{\dagger}) \cdot \boldsymbol{\alpha}\beta\boldsymbol{\alpha}\psi - e\psi^{\dagger}\boldsymbol{\alpha} \cdot \mathbf{A}\beta\boldsymbol{\alpha}\psi + m\psi^{\dagger}\boldsymbol{\alpha}\psi \tag{C.303}$$

Now add the two expressions, noting the anti-commutation of β and $\boldsymbol{\alpha}$.

$$i\frac{\partial}{\partial t}(\psi^{\dagger}\boldsymbol{\alpha}\beta\psi) = \psi^{\dagger}\boldsymbol{\alpha}\beta\boldsymbol{\alpha} \cdot \frac{1}{i}\nabla\psi - \frac{1}{i}(\nabla\psi^{\dagger}) \cdot \boldsymbol{\alpha}\beta\boldsymbol{\alpha}\psi$$

$$+ e\psi^{\dagger}\beta\{\boldsymbol{\alpha}, \boldsymbol{\alpha} \cdot \mathbf{A}\}\psi + 2m\psi^{\dagger}\boldsymbol{\alpha}\psi \tag{C.304}$$

Now we use the identity

$$\alpha_i\alpha_j = I\delta_{ij} + i\epsilon_{ijk}\sigma_{Dk} \tag{C.305}$$

to show

$$\psi^{\dagger}\boldsymbol{\alpha}\beta\boldsymbol{\alpha} \cdot \frac{1}{i}\nabla\psi = -\frac{1}{i}[\psi^{\dagger}\beta\nabla\psi - i\psi^{\dagger}\beta\boldsymbol{\sigma}_D \times \nabla\psi] \tag{C.306}$$

and

$$-\frac{1}{i}(\nabla\psi^{\dagger}) \cdot \boldsymbol{\alpha}\beta\boldsymbol{\alpha}\psi = -\frac{1}{i}[-(\nabla\psi^{\dagger})\beta\psi + i(\nabla\psi^{\dagger})\beta \times \boldsymbol{\sigma}_D\psi. \tag{C.307}$$

The sum of these two is

$$i(\psi^{\dagger}\beta\nabla\psi - (\nabla\psi^{\dagger})\beta\psi) - \nabla \times (\psi^{\dagger}\beta\boldsymbol{\sigma}_D\psi). \tag{C.308}$$

Combining terms,

$$i\frac{\partial}{\partial t}(\psi^{\dagger}\boldsymbol{\alpha}\beta\psi) = i(\psi^{\dagger}\beta\nabla\psi - (\nabla\psi^{\dagger})\beta\psi) - \nabla \times (\psi^{\dagger}\beta\boldsymbol{\sigma}_D\psi)$$

$$- 2e\psi^{\dagger}\beta\mathbf{A}\psi + 2m\psi^{\dagger}\boldsymbol{\alpha}\psi. \tag{C.309}$$

Thus

$$\mathbf{j} \equiv e\psi^{\dagger}\boldsymbol{\alpha}\psi = \frac{e}{2mi}(\psi^{\dagger}\beta\nabla\psi - (\nabla\psi^{\dagger})\beta\psi) - \frac{e^2}{m}\mathbf{A}\psi^{\dagger}\beta\psi$$

$$+ \nabla \times \frac{e}{2m}\psi^{\dagger}\beta\boldsymbol{\sigma}_D\psi + \frac{e}{2im}\frac{\partial}{\partial t}(\psi^{\dagger}\beta\boldsymbol{\alpha}\psi) \tag{C.310}$$

as asserted, with

$$\mathbf{j} = \mathbf{j}_S + \nabla \times \mathbf{M} + \frac{\partial \mathbf{\Pi}}{\partial t}$$

$$\mathbf{j}_S = \frac{e}{2mi}[\psi^\dagger \beta \nabla \psi - (\nabla \psi)^\dagger \beta \psi] - \frac{e^2}{m} \mathbf{A} \cdot \psi^\dagger \beta \psi$$

$$\mathbf{M} = \frac{e}{2m} \psi^\dagger \beta \boldsymbol{\sigma} \psi$$

$$\mathbf{\Pi} = \frac{e}{2mi} \psi^\dagger \beta \boldsymbol{\alpha} \psi. \tag{C.311}$$

(b) Since $\beta \boldsymbol{\sigma}_D, \beta \nabla$ and β are "even" operators, the non-relativistic limit will have the same form, but with $\beta \rightarrow I$ and ψ being a Dirac spinor with negligible small components, i.e. a Pauli two-component spinor.

(c) It is immediately apparent that since $\boldsymbol{\alpha} \sim v/c$, the polarization is down by a factor v/c from the magnetic component. Pursuing this further,

$$\mathbf{\Pi} = \frac{e}{2mi} \psi^\dagger \beta \boldsymbol{\alpha} \psi = \frac{e}{2mi}(\phi^\dagger \boldsymbol{\sigma} \chi - \chi^\dagger \boldsymbol{\sigma} \phi). \tag{C.312}$$

But in the non-relativistic limit,

$$\chi = \frac{\boldsymbol{\sigma} \cdot \mathbf{p}}{2m} \phi, \tag{C.313}$$

so

$$\mathbf{\Pi} = \frac{e}{4m^2 i} \left[\phi^\dagger (\boldsymbol{\sigma}\boldsymbol{\sigma} \cdot \mathbf{p} - \boldsymbol{\sigma} \cdot \mathbf{p}\boldsymbol{\sigma} \phi) \right]$$

$$= \frac{e}{2m^2}(\phi^\dagger \mathbf{p} \times \boldsymbol{\sigma} \phi). \tag{C.314}$$

Now if we want to see the contribution this makes to a radiative transition, we include the interaction $\mathbf{A}_{rad} \cdot \partial \mathbf{\Pi}/\partial t$. In the notation of Problem 9.3, $\partial \mathbf{\Pi}/\partial t \cdot \mathbf{A}_{rad}$ will give a matrix element of the form

$$M'_{ba} = \left\langle b \left| \sum_j e^{-i\mathbf{k} \cdot \mathbf{x}_j} \boldsymbol{\epsilon}^*_{k\lambda} \left(\frac{e_j i \hbar \omega_{ba}}{2m_j^2 c^2} \mathbf{p}_j \times \boldsymbol{\sigma}_j \right) \right| a \right\rangle. \tag{C.315}$$

Working in the dipole approximation, the factor $e^{-i\mathbf{k} \cdot \mathbf{x}}$ is simply reduced to unity and the radiative matrix element is

$$M'_{ba} = i\omega_{ba} \left\langle b \left| \sum_j \boldsymbol{\epsilon}^*_{k,\lambda} \cdot \frac{e_j}{2m_j^2 c^2} \mathbf{p}_j \times \boldsymbol{\sigma}_j \right| a \right\rangle, \tag{C.316}$$

where the factor $i\omega_{ba}$ comes from $\partial \mathbf{\Pi}/\partial t$. We can replace

$$\mathbf{p} = \frac{im}{\hbar}[H, \mathbf{r}] \rightarrow im\omega_{ba}\mathbf{r} \tag{C.317}$$

to give us

$$M'_{ba} = -\sum_j \frac{e_j \omega_{ba}^2}{2m_j^2 c^2} \left\langle b \left| \boldsymbol{\epsilon}^*_{k,\lambda} \cdot \mathbf{r}_j \times \boldsymbol{\sigma}_j \right| a \right\rangle. \tag{C.318}$$

This is twice as large as, and the opposite sign to, the induced E1 matrix element of part (a) of Problem 9.3. There is no reason for them to agree. In Problem 9.3, we have a phenomenological Hamiltonian, which we take seriously enough to expand the exponential beyond the first term. On the other hand, here we have included only the "1" from the expansion of $e^{-i\mathbf{k}\cdot\mathbf{x}}$.

10.8 Hyperfine interaction in non-relativistic limit via Foldy-Wouthuysen reduction. Nucleus has spin $\mathbf{I}$ and magnetic moment $\mathbf{M} = \mu_n \mathbf{I}/I$. The vector potential is

$$\mathbf{A} = \frac{\mathbf{M} \times \mathbf{r}}{r^3} = -\mathbf{M} \times \nabla\left(\frac{1}{r}\right). \tag{C.319}$$

(a) Now we have

$$H = \beta m + \mathcal{O}; \qquad \mathcal{O} = \boldsymbol{\alpha} \cdot \left[\mathbf{p} - e\mathbf{M} \times \nabla\left(\frac{1}{r}\right)\right]. \tag{C.320}$$

We can ignore the nuclear Coulomb field here. To the extent that we keep only terms linear in $\mathbf{A}$ and Φ, we can just add the reduced Hamiltonians, computed separately. Note that we have put $q_{\text{electron}} = -e$.

We just need to evaluate

$$\frac{\mathcal{O}^2}{2m} - \frac{p^2}{2m} \tag{C.321}$$

and omit the $e^2 A^2$ term. Thus,

$$\begin{aligned}
\beta H_{\text{int}} &= -\frac{e}{2m}\left\{\boldsymbol{\alpha}\cdot\mathbf{p}\,\boldsymbol{\alpha}\cdot\mathbf{M}\times\nabla\left(\frac{1}{r}\right) + \boldsymbol{\alpha}\cdot\mathbf{M}\times\nabla\left(\frac{1}{r}\right)\boldsymbol{\alpha}\cdot\mathbf{p}\right\} \\
&= -\frac{e}{2m}\left\{-i\nabla\cdot\mathbf{M}\times\nabla\left(\frac{1}{r}\right) + 2[\mathbf{M}\times\nabla\left(\frac{1}{r}\right)]\cdot\mathbf{p} \right. \\
&\qquad \left. + \boldsymbol{\sigma}_D\cdot\left[\nabla\times\left(\mathbf{M}\times\nabla\left(\frac{1}{r}\right)\right)\right]\right\}.
\end{aligned} \tag{C.322}$$

The first term is proportional to $\mathbf{M}\cdot\nabla\times\nabla(1/r)$ and vanishes as the curl of a gradient. The second term can be written

$$-2\left(\mathbf{M}\times\frac{\mathbf{r}}{r^3}\right)\cdot\mathbf{p} = -2\frac{\mathbf{M}}{r^3}\cdot(\mathbf{r}\times\mathbf{p}) = -\frac{2}{r^3}\mathbf{M}\cdot\mathbf{L}. \tag{C.323}$$

The third term needs care in handling because of its singular nature. The vector potential assumes a point nucleus. Actually, we must consider a limit as the size of the nucleus goes to zero. As a model, take a uniform sphere of magnetization such that the total moment is $\mathbf{M}$. With a as the radius, we find the replacement (Jackson, *Classical Electrodynamics*, 3rd edition, p. 200, Eq.(5.111))

$$-\nabla\left(\frac{1}{r}\right) \rightarrow \frac{\hat{\mathbf{r}}}{a}\left(\frac{r_<}{r_>^3}\right). \tag{C.324}$$

Now for $r > a$, outside the nucleus

$$\begin{aligned}
\nabla\times(\mathbf{M}\times\nabla\frac{1}{r}) &= \mathbf{M}\nabla^2\left(\frac{1}{r}\right) + (\mathbf{M}\cdot\nabla)\frac{\hat{\mathbf{r}}}{r^2} \\
&= -\frac{(3\mathbf{M}\cdot\hat{\mathbf{r}})\hat{\mathbf{r}} - \mathbf{M}}{r^3}.
\end{aligned} \tag{C.325}$$

For $r < a$, inside the nucleus

$$\nabla \times \left(\mathbf{M} \times \left(\frac{-\mathbf{r}}{a^3} \right) \right) = -\frac{\mathbf{M}}{a^3}(\nabla \cdot \mathbf{r}) + \frac{(\mathbf{M} \cdot \nabla)}{a^3}\mathbf{r} = -\frac{2\mathbf{M}}{a^3}. \tag{C.326}$$

Integrating over the volume of the sphere,

$$\int_{r<a} \left(-2\frac{\mathbf{M}}{a^3} \right) d^3x = -\frac{8\pi}{3}\mathbf{M}. \tag{C.327}$$

Thus, in the limit $a \to 0$, we can write

$$\nabla \times \mathbf{A} = \frac{8\pi}{3}\delta(\mathbf{r})\mathbf{M} + \frac{3(\mathbf{M} \cdot \hat{\mathbf{r}})\hat{\mathbf{r}} - \mathbf{M}}{r^3}\Theta(r - 0^+). \tag{C.328}$$

(c.f. Jackson, *Classical Electrodynamics*, 3rd edition, p. 188, Eq. (5.64), where this result is derived for an arbitrary distribution of magnetic moment density.)

The interaction Hamiltonian is therefore

$$\beta H_{\text{int}} = \frac{e}{2m}\left\{ \frac{2}{r^3}\mathbf{M} \cdot \mathbf{L} + \frac{8\pi}{c}\mathbf{M} \cdot \sigma_D\delta(\mathbf{r}) + \frac{1}{r^3}\sigma_D \cdot [3(\mathbf{M} \cdot \hat{\mathbf{r}})\hat{\mathbf{r}} - \mathbf{M}] \right\}. \tag{C.329}$$

With $\mathbf{M} = \mu_n\mathbf{I}/I$ and $\mu_0 = e/(2m)$, we get the given expression, taking the non-relativistic limit, $\beta \to 1$, $\sigma_D \to \sigma$.

(b) The standard dipole field $(3(\mathbf{M} \cdot \hat{\mathbf{r}}) - \mathbf{M})/r^3$ has zero angular average. Thus it does not enter for $\ell = 0$ states. Note that the r^{-3} behavior is cut off inside the nucleus; there is no question of $0 \times \infty =$ finite. Thus restricted to $\ell = 0$,

$$(\Delta H)_{\ell=0} = \frac{8\pi}{3}\mu_0\mu_n\frac{\sigma \cdot \mathbf{I}}{I}\delta(\mathbf{r}). \tag{C.330}$$

To compute $\sigma \cdot \mathbf{I}$, consider

$$\mathbf{F} = \mathbf{I} + \frac{1}{2}\sigma$$

$$F_{op}^2 = I_{op}^2 + \frac{3}{4} + \sigma \cdot \mathbf{I}$$

$$F^2 = \left(I \pm \frac{1}{2} \right)\left(I + 1 \pm \frac{1}{2} \right) = I(I + 1) \pm \left(I + \frac{1}{2} \right) + \frac{1}{4} \tag{C.331}$$

Thus, the eigenvalues are

$$\sigma \cdot \mathbf{I} = I \qquad \left(F = I + \frac{1}{2} \right)$$

$$\sigma \cdot \mathbf{I} = -(I + 1) \qquad \left(F = I - \frac{1}{2} \right). \tag{C.332}$$

Thus, the shift due to the hyperfine interaction for $\ell = 0$ is

$$\Delta E = \frac{8\pi}{3}\mu_0\mu_n|\Psi(0)|^2 \times \{1, -(I + 1)/I\} \quad \text{for } \left\{ F = I + \frac{1}{2}, F = I - \frac{1}{2} \right\} \tag{C.333}$$

(c) Take $\mathbf{A} = \mathbf{B} = \hat{\mathbf{z}}$ and evaluate

$$\int d\Omega Y^*_{\ell m}(\Omega)(1 - 3\cos^2\theta)Y_{\ell m}(\Omega) = -2\sqrt{\frac{4\pi}{5}} \int d\Omega Y^*_{\ell m}(\Omega)Y_{20}(\Omega)Y_{\ell m}(\Omega) \tag{C.334}$$

using the Gaunt integral

$$\int d\Omega \, Y_{\ell_1 m_1}(\theta,\phi)Y_{\ell_2 m_2}(\theta,\phi)Y^*_{\ell_3,m_3}(\theta,\phi)$$
$$= \sqrt{\frac{(2\ell_1+1)(2\ell_2+1)}{4\pi(2\ell_3+1)}} \langle \ell_3 0|\ell_1 0 \ell_2 0\rangle \langle \ell_1 m_1 \ell_2 m_2|\ell_3 m_3\rangle. \tag{C.335}$$

We thus find

$$\int d\Omega Y^*_{\ell m}(\Omega)(1 - 3\cos^2\theta)Y_{\ell m}(\Omega) = \frac{2(3m^2 - \ell(\ell+1))}{(2\ell-1)(2\ell+3)}. \tag{C.336}$$

On the other hand,

$$\int d\Omega Y^*_{\ell m}(\Omega)(L^2 - 3L_z^2)Y_{\ell m}(\Omega) = \ell(\ell+1) - 3m^2. \tag{C.337}$$

Thus our factor C is given by

$$C = -\frac{2}{(2\ell+3)(2\ell-1)}. \tag{C.338}$$

Now in the non-relativistic approximation, our Hamiltonian is, for $\ell \neq 0$,

$$H_{\text{int}} = \mu_0 \frac{2}{r^3}\frac{\mu_n}{I}\left\{\mathbf{I}\cdot\mathbf{L} - \frac{2[3(\mathbf{I}\cdot\mathbf{L})(\mathbf{L}\cdot\mathbf{S}) - L^2(\mathbf{I}\cdot\mathbf{S})]}{(2\ell+3)(2\ell-1)}\right\}. \tag{C.339}$$

Now we make the replacements that project $\mathbf{I}$ onto $\mathbf{J}$.

$$H_{\text{int}} = \mu_0 \frac{2}{r^3}\mu_N g\frac{\mathbf{I}\cdot\mathbf{J}}{J^2}\left\{\mathbf{J}\cdot\mathbf{L} - \frac{2[3(\mathbf{J}\cdot\mathbf{L})(\mathbf{L}\cdot\mathbf{S}) - L^2(\mathbf{J}\cdot\mathbf{S})]}{(2\ell+3)(2\ell-1)}\right\} \tag{C.340}$$

To address the troublesome part, let us explicitly write $j = \ell \pm 1/2$, so

$$J^2 = \ell(\ell+1) + \frac{1}{4} \pm \frac{1}{2}(2\ell+1) \tag{C.341}$$

and

$$2\mathbf{J}\cdot\mathbf{L} = J^2 + L^2 - S^2 = 2\ell(\ell+1) - \frac{1}{2} \pm \frac{1}{2}(2\ell+1)$$
$$2\mathbf{J}\cdot\mathbf{S} = J^2 + S^2 - L^2 = 1 \pm \frac{1}{2}(2\ell+1)$$
$$2\mathbf{L}\cdot\mathbf{S} = J^2 - L^2 - S^2 = -\frac{1}{2} \pm \frac{1}{2}(2\ell+1). \tag{C.342}$$

To simplify the notation, let $\alpha = \pm\frac{1}{2}(2\ell + 1)$. Then

$$
3(\mathbf{J} \cdot \mathbf{L})(\mathbf{L} \cdot \mathbf{S}) - \mathbf{L}^2(\mathbf{J} \cdot \mathbf{S}) = \frac{1}{4}\left[3(2\ell(\ell + 1) - \frac{1}{2} + \alpha)(-\frac{1}{2} + \alpha)\right.
$$

$$
\left. -2\ell(\ell + 1)(1 + \alpha)\right]
$$

$$
= \frac{1}{4}\left[-5\ell(\ell + 1) + \frac{3}{4} + 3\alpha^2 + 3\alpha(2\ell(\ell + 1) - 1)\right.
$$

$$
\left. -2\alpha\ell(\ell + 1)\right]
$$

$$
= \frac{1}{4}\left[-2\ell(\ell + 1) + \frac{3}{2} + \alpha[4\ell(\ell + 1) - 3]\right]
$$

$$
= \frac{1}{8}\left[(2\ell + 3)(2\ell - 1)(-1 + 2\alpha)\right]. \qquad \text{(C.343)}
$$

Combining,

$$
\mathbf{J} \cdot \mathbf{L} - \frac{2[3(\mathbf{J} \cdot \mathbf{L})(\mathbf{L} \cdot \mathbf{S}) - \mathbf{L}^2(\mathbf{J} \cdot \mathbf{S})]}{(2\ell + 3)(2\ell - 1)}
$$

$$
= \ell(\ell + 1) - \frac{1}{4} + \frac{1}{2}\alpha - \frac{1}{4}(-1 + 2\alpha) = \ell(\ell + 1). \quad \text{(C.344)}
$$

Inserting this, and noting $2\mathbf{I} \cdot \mathbf{J} = f(f + 1) - I(I + 1) - j(j + 1)$, we have

$$
H_{\text{hf}} = \mu_0 \mu_N g \frac{\ell(\ell + 1)}{j(j + 1)}(f(f + 1) - I(I + 1) - j(j + 1))\left\langle\frac{1}{r^3}\right\rangle, \qquad \text{(C.345)}
$$

in agreement with Bethe and Salpeter, *Quantum Mechanics of One- and Two-Electron Atoms*, Eq. (22.9).

Now consider $p_{1/2}$, i.e. $\ell = 1, j = 1/2, f = I \pm 1/2$. Then

$$
H_{\text{hf}} = \mu_0 \mu_N g \frac{2}{3/4}((I \pm 1/2)(I + 1 \pm 1/2)) - I(I + 1) - 3/4)\left\langle\frac{1}{r^3}\right\rangle
$$

$$
= \mu_0 \mu_N g \frac{8}{3}((I \pm 1/2)(I + 1 \pm 1/2)) - I(I + 1) - 3/4)\left\langle\frac{1}{r^3}\right\rangle
$$

$$
= \mu_0 \mu_N g \frac{8}{3}(\pm(I + 1/2) - 1/2)\left\langle\frac{1}{r^3}\right\rangle. \qquad \text{(C.346)}
$$

So noting that $g\mu_N I = \mu_n$, we see that the hyperfine splittings for $s_{1/2}$ and $p_{1/2}$ are the same except for the replacement $\pi\psi(0)|^2 \to \langle r^{-3}\rangle$.

Bibliography

Abramowitz, M. and Stegun, I., *Handbook of Mathematical Functions with Formulas, Graphs, and Mathematical Tables*, U. S. Government Printing Office (1964).

Bateman, H. and Erdélyi, A., *Higher Transcendental Functions*, 3 volumes, McGraw Hill, New York (1953–5).

Baym, G., *Lectures on Quantum Mechanics*, CRC Press (1969).

Bell, J. S., *Speakable and Unspeakable in Quantum Mechanics*, Cambridge Press, 2nd edition, (2004).

Bethe, H. A., *Intermediate Quantum Mechanics*, W. A. Benjamin, New York, (1964).

Bethe, H. A. and Jackiw, R., *Intermediate Quantum Mechanics*, 3rd edition, Addison-Wesley, New York, (1997).

Bethe, H. A. and Salpeter, E. E., *Quantum Mechanics of One- and Two-Electron Atoms*, Springer-Verlag, Berlin (1957), paperback, Plenum, New York (1977).

Bjorken, J. D. and Drell, S. D., *Relativistic Quantum Mechanics*, McGraw-Hill, New York, (1964).

Born, M., *Atomic Physics*, 7th edition, Hafner, New York, (1962).

Brink, D. M. and Satchler, G. R., *Angular Momentum*, 3rd edition, Oxford Press (1993).

Cohen-Tannoudji, C., Diu, B., and Laloë, F., *Quantum Mechanics*, Volume I, 2nd edition, Wiley (2020).

Condon, E. U. and Shortley, G. H., *Theory of Atomic Spectra*, Oxford University Press, Oxford (1953).

Courant, R. and Hilbert, D., *Methods of Mathematical Physics*, Interscience, New York, (1953).

Dennery, P. and Kryzwicki, A., *Mathematics for Physicists*, Dover, Mineola (1996).

Dirac, P. A. M., *Principles of Quantum Mechanics*, 4th edition, revised, Oxford University Press, Oxford (1967).

Edmonds, A. R., *Angular Momentum in Quantum Mechanics*, Princeton University Press (1957).

Fano, U. and Racah, G., *Irreducible Tensorial Sets*, Academic Press, New York (1959).

Frauenfelder, H., and Henley, E. M., *Subatomic Physics*, Prentice-Hall, 2nd edition, (1991).

Fröman, N. and Fröman, P. O., *JWKB Approximation: Contributions to the Theory*, North-Holland, Amsterdam (1965).

Gasiorowicz, S., *Quantum Physics*, 3rd edition, Wiley, (2003).

Gottfried, K., *Quantum Mechanics: Fundamentals*, Vol.I, W. A. Benjamin, New York, (1966).

Gottfried, K. and Yan, T.-M., *Quantum Mechanics: Fundamentals*, Springer, New York (2003).

Gradshteyn, I. S. and Ryzhik, I. M., *Tables of Integrals, Series, and Products*, edited by D. Zwillinger, Academic Press, New York (2014).

Heitler, W., *The Quantum Theory of Radiation*, Oxford University Press, Oxford (1954).

Jeffreys, H. and Jeffreys, B. S., *Methods of Mathematical Physics*, Cambridge, 3rd edition, (1956).

Landau, L. D. and Lifshitz E. M., *Quantum Mechanics*, 2nd edition, Translated by J B. Sykes and J. S. Bell, Pergamon Press, Oxford (1965).

Magnus, W. and Oberhettinger, F., *Formulas and Theorems for the Functions of Mathematical Physics*, Chelsea, New York (1949).

Margenau, H. and Murphy, G. M., *The Mathematics of Physics and Chemistry*, Vol.1, 2nd edition, Van Nostrand, Princeton (1964).

Merzbacher, E., *Quantum Mechanics*, 2nd edition, Wiley, New York, (1970).

Messiah, A., *Quantum Mechanics*, North-Holland, Amsterdam, (1961).

Morse, P. M. and Feshbach, H., *Methods of Theoretical Physics*, 2 volumes, McGraw-Hill, New York (1953).

Mott, N. F. and Massey, H. S. W., *The Theory of Atomic Collisions*, 2nd edition, Oxford University Press, Oxford (1952).

Rose, M. E., *Elementary Theory of Angular Momentum*, Wiley, New York (1957).

Sakurai, J. J. and Napolitano, J., *Modern Quantum Mechanics*, 3rd edition, Cambridge University Press (2021).

Sakurai, J. J., *Advanced Quantum Mechanics*, Addison-Wesley, Reading (1967).

Schiff, L. I., *Quantum Mechanics*, 3rd edition, McGraw-Hill, New York (1968).

Sommerfeld, A., *Partial Differential Equations in Physics*, Academic Press, New York (1949).

von Neumann, J., *Mathematical Foundations of Quantum Mechanics*, trans. R. T. Beyer, Princeton, (1955).

Watson, G. N., *A Treatise on the Theory of Bessel Functions*, Cambridge University Press (1966).

Weinberg, S., *Lectures on Quantum Mechanics*, 2nd edition, Cambridge University Press, Cambridge, (2015).

Whittaker, E.T. and Watson, G. N., *A Course of Modern Analysis*, Cambridge University Press, 4th edition (1965).

Wigner, E. P., *Group Theory and its Application to the Quantum Mechanics of Atomic Spectra*, Academic Press, (1959).

Index

John David Jackson: A Course in Quantum Mechanics, First Edition. Robert N. Cahn.
© 2024 John Wiley & Sons, Inc. Published 2024 by John Wiley & Sons, Inc.
Companion website: www.wiley.com/go/Jackson/QuantumMechanics